The Mentalities of Gorillas and Orangutans

Comparative Perspectives

Research on the mental abilities of chimpanzees and bonobos has been widely celebrated and used in reconstructions of human evolution. In contrast, scant attention has been paid to the abilities of gorillas and orangutans. This volume aims to complete the picture of hominoid cognition by bringing together the work on gorillas and orangutans and setting it in comparative perspective. The introductory chapters set the evolutionary context for comparing cognition in gorillas and orangutans to that of chimpanzees, bonobos, and humans. The remaining chapters focus primarily on the kinds and levels of intelligence displayed by orangutans and gorillas compared to other great apes, including performances in the classic domains of tool use and tool-making, imitation, self-awareness, social communication, and symbol use. The final chapter suggests that many of the abilities commonly attributed exclusively to chimpanzees and bonobos were already present in the common ancestor of all the great apes. All those wanting more information on the mental abilities of these neglected, but important primates will find this book a treasure trove.

SUE TAYLOR PARKER is Professor of Anthropology at Sonoma State University, California. She has published extensively and has co-edited three other books, *"Language" and Intelligence in Monkeys and Apes* (1990) with K. Gibson, *Self-Awareness in Animals and Humans* (1994) with R. Mitchell and L. Boccia and *Reaching Into Thought* (1996) with A. Russon and K. Bard. She has also co-authored *Origins of Intelligence: The Evolution of Cognitive Development in Monkeys, Apes, and Humans* (1997) with M. McKinney.

ROBERT W. MITCHELL is Associate Professor in the Department of Psychology at Eastern Kentucky University. He co-edited *Deception: Perspectives on Deceit in Humans and Nonhumans* (1986) with Nicholas Thompson, *Anthropomorphism, Anecdotes and Animals* (1993) with Nicholas Thompson and H. Lyn Miles, and *Self-Awareness in Animals and Humans* with Sue Taylor Parker (1994).

H. LYN MILES is Professor of Anthropology at the University of Tennessee at Chattanooga, and Adjunct Professor of Psychology at the Georgia Institute of Technology. She co-edited *Anthropomorphism, Anecdotes and Animals* (1993) with Robert Mitchell, and is Director of Project Chantek, investigating the cognitive and communicative development of an organgutan learning to use sign language.

The Mentalities of Gorillas and Orangutans

Comparative Perspectives

Edited by

SUE TAYLOR PARKER

ROBERT W. MITCHELL

H. LYN MILES

PUBLISHED BY THE PRESS SYNDICATE OF THE UNIVERSITY OF CAMBRIDGE
The Pitt Building, Trumpington Street, Cambridge, United Kingdom

CAMBRIDGE UNIVERSITY PRESS
The Edinburgh Building, Cambridge CB2 2RU, UK www.cup.cam.ac.uk
40 West 20th Street, New York, NY 10011–4211, USA www.cup.org
10 Stamford Road, Oakleigh, Melbourne 3166, Australia
Ruiz de Alarcón 13, 28014 Madrid, Spain

First published 1999

Printed in the United Kingdom at the University Press, Cambridge

Typeset in Ehrhardt 9.5/12pt [VN]

A catalogue record for this book is available from the British Library

ISBN 0 521 58027 7 hardback

Contents

Contributors

DAVID R. BEGUN
Department of Anthropology, University of Toronto, Toronto, Ontario, M5S 1A1, Canada

JOHN D. BONVILLIAN
department of Psychology, University of Virginia, Charlottesville, VA 22903, USA

SARAH T. BOYSEN
The Ape Cognition Project, The Ohio State University, OH 43210–1222, USA

RICHARD W. BYRNE
Scottish Primate Research Group, St. Andrews University, St. Andrews, Fife, KY16 9JU, Scotland

JOSEPH CALL
Department of Psychology, University of Liverpool, Liverpool, L69 3BX, England

SIÂN EVANS
Dumond Conservancy for Tropical Forests and Primates, 14805 SW 2nd Street, Miami, FL 33170, USA

ELIZABETH A. FOX
Department of Biological Anthropology and Anatomy, Duke University, Durham, NC, USA

BIRUTÉ M. F. GALDIKAS
Department of Archeology, Simon Fraser University, Vancouver, British Columbia, V5A 1S6, Canada

JUAN CARLOS GÓMEZ
Scottish Primate Research Group
St. Andrews University, St. Andrews, Fife, KY16 9JU, Scotland

JAY GOULD
Department of University of West Florida, Pensacola, FL 32514, USA

PETER HALLIDAY
University of Kent, Canterbury, UK

YOLANDA MULDONADO HALLIDAY
Howlett's Park Zoo, Canterbury, England

MARY KERR
Gorilla World, San Francisco Zoo, San Francisco, CA 94116, USA

VALERIE A. KUHLMEIER
The Ape Cognition Project, The Ohio State University, OH 43210–1222, USA

HAL MARKOWITZ
Department of Psychology, San Francisco State University, San Francisco, CA 94132, USA

H. LYN MILES
Department of Sociology and Anthropology, University of Tennessee at Chattanooga, Chattanooga, TN 37403, USA

ROBERT MITCHELL
Department of Psychology, Eastern Kentucky University, Richmond, KY 40475–3108, USA

SUE TAYLOR PARKER
Department of Anthropology, Sonoma State University Rohnert Park, CA 94928, USA

FRANCINE G. P. PATTERSON
The Gorilla Foundation, Woodside, CA 94062, USA

ANNE E. RUSSON
Department of Psychology, Glendon College, York University, Toronto, Ontario, M4N 3M6, Canada

DENA SARAUW
Department of Psychology, Lehman College, CUNY, Bronx, NY 10468, USA

CAREL P. VAN SCHAIK
Department of Biological Anthropology, Box 90383, Duke University, Durham NC 27708–0383, USA

KATERINA SEMENDEFERI
Department of Anthropology, University of California, San Diego, La Jolla, CA 92093–0532, USA

GARY L. SHAPIRO
Orangutan Foundation International, Los Angeles, CA 90049, USA

ARNOLD F. SITOMPUL
Department of Biology, University of Indonesia, Depok 16424, Jawa Barat, Indonesia

KARYL B. SWARTZ
Department of Psychology, Lehman College, CUNY, Bronx, NY 10468, USA

JOANNE E. TANNER
School of Psychology, St. Andrews University, St. Andrews, Fife, KY16 9JU, Scotland

ANDREW WHITEN
Scottish Primate Research Group, St. Andrews University, St. Andrews, Fife, KY16 9JU, Scotland

Preface

The title *The mentalities of gorillas and orangutans: Comparative perspectives* was inspired by Köhler's famous book *The mentality of apes* (1927), in which apes were implicitly equated with chimpanzees. This book focuses on two other great apes that were less known in Köhler's day. It is the fourth in a related series of edited volumes on cognition of great apes. The first volume, *"Language" and intelligence in monkeys and apes* (Parker & Gibson, 1990), emphasized the importance of using models from developmental psychology to compare the cognitive and symbolic abilities of monkeys, apes, and humans. The second volume, *Self-awareness in animals and humans* (Parker, Mitchell, & Boccia, 1994), used developmental frameworks to compare manifestations of self-awareness in monkeys, apes, and humans. It was aimed at broadening the scope of research in this subject to extend beyond the classical mark test for mirror self-recognition. The third volume, *Reaching into thought* (Russon, Bard, and Parker, 1996), focused on cognitive abilities of great apes using models from developmental psychology, but extended the scope to include studies in wild populations.

This most recent volume continues the tradition of using models from developmental psychology to compare the cognitive abilities of great apes, and that of including studies from both captive and wild populations. It also expands the preview to include studies of taxonomy and phylogeny (Begun, this volume) and the brain (Semendeferi, this volume). It differs from the other volumes in focusing primarily on gorillas and orangutans. The aim of this volume is to redress the chimpocentric imbalance in attention to chimpanzees and bonobos at the expense of the other great apes. This bias follows in part from the greater availability of chimpanzees in captivity and the greater number of field studies of chimpanzees in the wild. It also arises from the greater appeal of studying our closest relatives, and, finally, from our fascination with another species that hunts, uses tools, and engages in territorial patrols and warfare. Conversely, the relative neglect of gorillas and orangutans follows from their rarity in captivity which has led to studies of single subjects rather than the multiple subjects favored in experimental research. Also, of course, the apparent lack of tool use, hunting, and warfare in these apes has made them less intriguing to those focused on human evolution.

The current volume includes recent discoveries revealing that orangutans use a variety of tools in the wild (Fox, Sitompul, & Van Schaik this volume), and that gorillas regularly use a variety of tools in captivity (Boysen, Kuhlmeier, Halliday, & Halliday and Parker et al., this volume). It reveals that gorillas and orangutans show the same rate of mirror self-recognition as chimpanzees (Swartz, Sarauw, & Evans, this volume), and the same kinds of deception (Mitchell, this volume). It also reveals that – like that of chimpanzees – early sign-language development in gorillas parallels that of human infants (Bonvillian & Patterson, and Miles, this volume). Other studies in this volume focus on the abilities of orangutans to imitate gestures and tool use (Call, and Russon, this volume).

The overall conclusion of this volume is that understanding of human evolution requires systematic comparison of ourselves with all the great apes not just with bonobos and chimpanzees (as well as with more distantly related primates). Only through systematic comparison of these ingroup and outgroup species can we reconstruct the emergence of bigger brains, longer childhoods, and longer lives, let alone such behaviors as intelligent tool use, mirror self recognition, imitation of novel behaviors, and intentional deception (Parker & Mitchell, this volume). In sum, we hope this work will help overcome the chimpocentric approach to comparative studies (Beck, 1982).

REFERENCES

Beck, B. (1982) Chimpocentrism: Bias in cognitive ethology. *Journal of Human Evolution*, **11**, 3–17.

Köhler, W. (1927). *The mentality of apes*, London: Kegan Paul, Trench & Co. Ltd.

Parker, S. T. & Gibson, K. R. (eds.) (1990) "*Language*" and intelligence in monkeys and apes. Cambridge University Press.

Parker, S. T., Mitchell, R. W., & Boccia, M. L. (eds.) (1994) *Self-awareness in animals and humans*. Cambridge University Press.

Russon, A. E. Bard, K., & Parker, S. T. (eds.) *Reaching into thought: the minds of the great apes*. Cambridge University Press.

Acknowledgments

The editors of this volume would like to thank some of the many people who helped contribute to its completion. First, we thank our respective Universities and Departments, the Department of Anthropology at Sonoma State University, the Department of Psychology at Eastern Kentucky University, and the Department of Sociology and Anthropology at the University of Tennessee at Chattanooga. We also thank our students who have helped shape our understanding of primate behavior. We thank the many colleagues who have given their support and feedback, our mates and partners who have given us free reign to work on this project. We acknowledge and thank the zoological gardens that have encouraged research on their rare gorilla and orangutan exhibits, especially the San Francisco Zoo. Finally, we thank an anonymous reviewer for helpful suggestions about including a broader scope of work in the volume, and our editor at Cambridge, Dr. Tracey Sanderson, for her steady guidance.

PART I

Comparative evolutionary and developmental perspectives on gorillas and orangutans

1

Hominid family values: morphological and molecular data on relations among the great apes and humans.

DAVID R. BEGUN

"What we've got here is, failure to communicate." (Cool Hand Luke. Warner Bros., 1967)

INTRODUCTION

Recent fossil discoveries, methodological advances, and ongoing analyses of hominoid comparative molecular biology and morphology have led to dramatic changes in our current understanding of relations among the great apes and humans. The late nineteenth-century view of these relations held that humans and the African apes were most closely related, and that the Asian great ape, the orangutan, was a more distant relative (Darwin, 1871; Huxley, 1959). For most of the twentieth century however, the great apes have been placed not only in their own family (the Pongidae), but also in their own separate evolving lineage, to the exclusion of humans. In the past thirty years, a few molecular biologists have been questioning the evolutionary reality of the pongid lineage, suggesting a return to the Darwin–Huxley view, placing African apes closer to humans, but based on the results of comparative molecular biology. During that time, the techniques of molecular systematics have been greatly improved and refined, to the point where researchers are today actually comparing nucleotide sequences, the fundamental, specific arrangement of DNA molecules, as opposed to overall DNA similarity (hybridization) or the products of DNA metabolism (proteins).

This work in molecular biology has been joined by recent developments in method and theory in palaeontology and comparative anatomy. Foremost among these is the widespread adoption of cladistic methodology, but other developments include more comprehensive analysis of hominoid comparative anatomy, and a more complete hominoid fossil record. One of the outcomes of this revolution in hominoid systematics is the perception that a great divide exists between "traditional" and "modern" hominoid classifications, and that the former has or inevitably will be replaced by the latter. We could switch the words "traditional" with "morphological," and "modern" with "molecular," reflecting the widely held view, even among some morphologists, that, in systematics, molecules have priority over morphology (e.g., Gould, 1985). The fact is, however, that morphology-based hominoid systematics has progressed enormously during the time in which molecular systematics has been coming into its own. Since the 1980s, new discoveries and the standardization of approaches to phylogeny reconstruction have made morphological systematics quite distinct from earlier research. Unfortunately the "phylogenetic baggage" of a century or more of work in hominoid systematics, often of very high quality, has made some researchers doubtful of their own results (e.g., Groves, 1986; Pilbeam, Rose, Barry, & Shah 1990; Pilbeam, 1996). In contrast, the impressiveness of recent advances in molecular techniques, which currently allow the relatively quick sequencing of nuclear genes and complete mitochondrial genomes in multiple individuals, may have produced some excess of confidence in the potential of molecular approaches to yield the "right" answer (e.g., Gould, 1985; Easteal, Collet, & Betty, 1995). What is needed in this area of research is more communication between the practitioners of both approaches. There is more disagreement within molecular systematics and within morphological systematics than between the two (see below). Compared to the conclusions of hominoid systematics through the 1970s, the conclusions of morphologists and molecular systematists today are much closer than they are divergent. So, rather than focusing on the so-called morphology vs. molecules debate (Patterson, 1987; Cartmill & Yoder, 1994; Shoshani, Groves, Simons, & Gunnel, 1996), this chapter highlights the significant amount of consensus between these fields that has been emerging in the past five years.

In an attempt to move away from the morphology–molecule dichotomy, this chapter is organized along research questions rather than methodology. The main question is, what are the evolutionary relations among living great apes and humans? We can resolve this big question into a series of smaller issues given the results of research over the past century and a quarter. These are, from the phylogenetic perspective: What is hominid, nowadays? What is an orangutan, anyway? How are African apes and humans related to each other? The last question is the most complex, and has generated the most controversy, particularly with regard to the issue of how humans fit in. Before coming to these questions, however, one approach common to both molecular and morphological systematics, cladistics, is outlined, as it has had perhaps the most profound influence on this field of any development in modern systematics.

BACKGROUND TO CLADISTIC APPROACHES

A cladistic approach to systematics is simply one in which the only information considered relevant to the classification of organisms is commonality of descent. Organisms share characteristics of external appearance, genetics, morphology, and behavior for a number of reasons, only one of which is directly attributable to the pattern of their evolutionary relationships (see below). Because a classification of organisms must be derived from phylogenetic conclusions deduced from observations of traits organisms share, the method and theory of classification focuses primarily on distinguishing phylogenetically informative characters, or *shared derived characters* (*synapomorphies*) from others (see below). Systematics is not palaeobiology or neontology. It is simply classification, which helps to organize and standardize the database of organisms known to humans, and allows or facilitates communication among colleagues. It does not seek to characterize organisms in any way beyond their evolutionary relations. The more interesting qualities of organisms are fleshed out in studies of ecology, behavior, evolution, and adaptation.

This view of the role of systematics is not universal (Hull, 1979), but it is becoming more common. Given this limited goal, cladistics can be viewed as a well-defined protocol for determining evolutionary relationships. Although there have been many refinements, the primary reference to cladistic methodology is Hennig (1966). While many of the ideas proposed by Hennig were already part of mainstream systematics (e.g., Simpson, 1961), Hennig standardized the approach to systematics that is nearly universally accepted today, providing a well-defined, if slightly cumbersome, vocabulary (Table 1.1), and a straightforward methodology.

One important limitation of the cladistic approach is the fact that it is incapable of providing evidence for ancestor-descendent relationships. In fact, it seeks only to find what are referred to as sister group relations. Sister taxa are those which share a common ancestor not shared by any other organism, that is, taxa that are more closely related to one another than they are to anything else. Establishing closeness of relationship is a matter of identifying character states[1] shared only by sister taxa, which are most easily explained as having been passed on to these taxa from a common ancestor. Thus, in the process of identifying sister taxa, a set of ancestral character states, or ancestral morphotype, is assembled, but a specific ancestor cannot be identified. At best, one could say that a known taxon does not differ from a reconstructed ancestral morphotype and is a candidate for the ancestry of another taxon. *Ardipithecus ramidus* (White, Suwa, & Asfaw, 1994), for example, appears to conform very closely to the reconstructed ancestral morphotype for the human lineage, but cannot technically be identified as an ancestor using this approach (Figure 1.1). An hypothesis of ancestor-descendant relationship is one that includes a processual component, some statement that involves time, selection, adaptation, environmental change or any combination of these (Figure 1.1). Sister relationships are much simpler hypotheses, involving only pattern recognition. A cladistic approach to systematics thus produces a cladogram, or dendrogram of relations, which can be viewed as the skeleton upon which an hypothesis of phylogeny (including ancestors and descendants) is fleshed out.

In a nutshell, sister relations are determined by identifying shared character states derived from the last common ancestor of the set of taxa under analysis. In order to carry out this type of analysis, taxa must be compared to other taxa known to be outside their evolutionary lineage, that is, an outgroup. This is a necessary initial assumption, but it is usually fairly straightforward. Outgroups are usually taxa most closely related

[1] Character states are specific configurations of a character. For example, in hominoids the character "os centrale" has the character states "separate" and "fused to the scaphoid."

Table 1.1 *Cladistic terminology used in this chapter*

Autapomorphy	Uniquely derived characters, having evolved since the divergence of a taxon from its **sister taxon**.
Characters	Units of analysis in phylogeny reconstruction.
Character States	The condition of a character in a taxon (large, small, pronounced, weak, etc.).
Clade	An evolutionary lineage, and the unit of an evolutionary classification.
Homology	Shared characters and/or character states having been inherited from a common ancestor.
Homoplasy	Shared characters and/or character states having been inherited independently from a common ancestor (parallelism) or from different ancestors (convergence).
Monophyletic taxon	A taxon having evolved from a common ancestor. A monophyletic taxon normally includes all descendants of that common ancestor, and is identified on the basis of its **synapomorphies**.
Outgroup	A relative of the group of organisms under analysis, usually the **sister taxon**, used to determine the **polarity** of **character states**.
Paraphyletic taxon	A taxon that excludes some of the descendants of a common ancestor. Paraphyletic taxa are usually identified on the basis of **symplesiomorphies**, and not considered valid evolutionary taxa by cladists.
Polarity	The evolutionary significance of a character state, either apomorphic (derived) because it is not present in the outgroup, or plesiomorphic (primitive) as indicated by its presence in the outgroup.
Polyphyletic taxon	A taxon with multiple ancestors, usually based on homoplasies. Polyphyletic taxa are infrequently identified today.
Sister taxon	The closest relative of a taxon, usually at the same hierarchical level (species, genera, families, etc.).
Symplesiomorphy	Shared primitive character state also found in the **outgroup** and presumed to have been inherited from a common ancestor with the outgroup and not from the last common ancestor of the taxa under analysis.
Synapomorphy	Shared derived characters, not found in the **outgroup** and presumed to have been inherited from the last common ancestor of the taxa under analysis.

Source: Hennig, 1966.

(a)

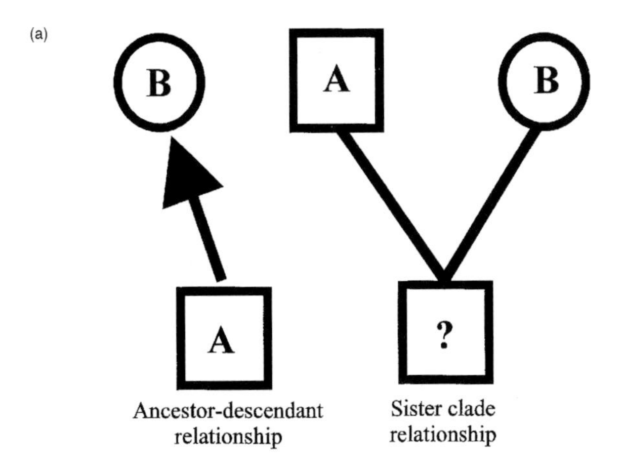

(b)

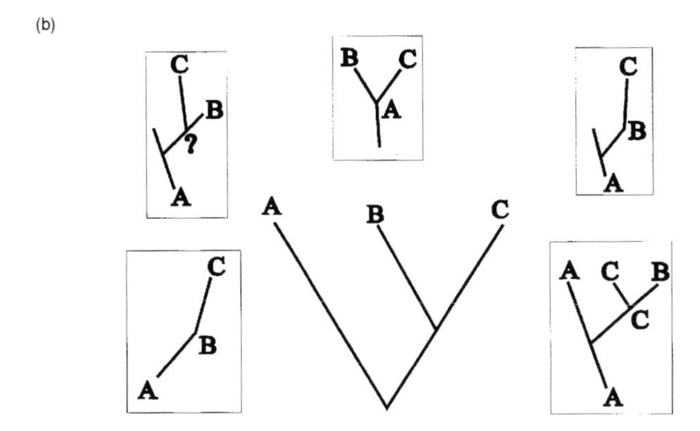

Figure 1.1 Phylogenies and cladograms.
(a) A phylogeny (left) depicts ancestor–descendant relationships. An ancestor–descendant hypothesis requires support from data directly related to the interpretation of evolutionary relationships (morphology, molecular biology, behavior, etc.) as well as information on the process of evolutionary change and the passage of time, represented by the arrow. A cladogram (right) simply represents a hierarchy of evolutionary relationships, without regard to the specifics of the process responsible for the observed diversity. Cladistic hypotheses are thus simpler but less comprehensive than phylogentic hypotheses, and should always precede the latter.
(b) A cladogram surrounded by five alternative phylogenies. Each phylogenetic hypothesis is consistent with the cladogram. Clockwise from the left: a unilinear hypothesis; the ancestors persists following two divergences, with the common ancestor of B and C unknown; B and C diverge for A; A and the common ancestor of B and C are briefly contemporary, and then B evolves into C; A, B, and C persist to the present following a divergence of the common ancestor of B and C from A and then B from C. Many other phylogenies are also consistent with this single cladogram. Deciding among them requires information on adaptation, selection, paleoenvironment, relative dating, and other paleontological variables.

to the set of taxa under analysis, but not included among these taxa. For catarrhines, the outgroup is widely viewed to be platyrrhines, for the Hominoidea, there is little dispute in recognizing the Cercopithecoidea as the outgroup, and for hominids it is widely viewed as hylobatids (see below). In order to identify character states that were present in the last common ancestor of living catarrhines, the outgroup, in this case platyrrhines, is examined for these traits. If catarrhine character states are found in platyrrhines, the most obvious explanation is that they come from a more ancient ancestor shared by both platyrrhines and catarrhines. For example, many catarrhines have tails and fore and hindlimbs of roughly equal length, moderate encephalization compared to prosimians, large canines, well-developed snouts, and divergent halluces. All of these traits are also found in most platyrrhines, and so it can be deduced that these were present in the common ancestor of platyrrhines and catarrhines and do not therefore serve to distinguish catarrhines and platyrrhines. These are *shared primitive characters*, or *symplesiomorphies*. Many or all catarrhines also have two premolars, an ossified ectotympanic tube, some reduction in the tail, and no humeral entepicondylar foramen. These character states are not found or are not common in platyrrhines, suggesting that they appeared in the ancestor of the catarrhines after it diverged from the ancestor of the platyrrhines. These would be shared derived characters of the catarrhines. However, because platyrrhines have also evolved since the appearance of their last common ancestor, it is possible that character states not found in platyrrhines have been lost or modified from their ancestral condition. Thus, it is advisable to look to the next available outgroup to confirm the conclusion that these character states are indeed derived. As it happens, many fossil and living prosimians also resemble platyrrhines in lacking most of these character states, although one of them, tail reduction, is ambiguous. In this case, it seems that catarrhines retain a simpler tail from a more ancient ancestor, while many platyrrhines have evolved more elaborate tails.

The method outlined above, using an outgroup to establish whether character states typical of the ingroup are primitive or derived (i.e., the polarity of character states), is called the outgroup criterion, and it is by far the most common. This is due to its relative simplicity, though a few assumptions are required. The most important assumption concerns *homology*. It is assumed that character states that are identified as the same in different taxa are homologous, that is, that they are the same character in an evolutionary sense, having evolved once in a common ancestor and having been retained little modified in descendants. This is a necessary initial assumption, and it is also fairly safe, from a methodological standpoint, for two reasons. The assumption of homology is falsifiable. The pattern of derived character states found in ingroup taxa need not conform perfectly to the pattern of relationships deduced from it. A minority of characters will always be inconsistent with the pattern of relationships suggested by most of the characters. These inconsistent characters are considered *a posteriori* to have arisen independently, by convergence or parallelism, and are thus identified as *homoplasies*. The assumption of homology, which was necessary to deduce a pattern of relationships, is falsified for these characters. The other strength of the assumption of homology is parsimony. It is more parsimonious to assume that a character state arose once than to assume that it arose independently more than once. Parsimony, however, is not a description of the evolutionary process. In fact, as just noted, cladistic analyses always reveal homoplasies, which demonstrates the degree to which evolution is not parsimonious, that is, the frequency with which similar character states arise by multiple pathways. Parsimony is a point of logic only. It simply refers to the longstanding preference in science for simpler explanations over more complex ones. All else being equal, a simpler explanation is to be preferred, so long as it does not violate well-established principles. Cladistics has been criticized for representing evolution as parsimonious (most recently in Marks, 1994), but this misunderstanding could not be further from the truth. Cladistic analysis, ironically, has revealed more parallelism and convergence among organisms than had previously been thought possible, reinforcing the notion of the enormous complexity of the evolutionary process.[2]

The final goal of a cladistic analysis is the hierarchical representation of *monophyletic* clades, the cladogram. Strictly monophyletic clades are lineages which include all the descendants of a single ancestor. Another way of describing monophyletic clades is that they contain taxa that are all more closely related to each other than to any taxon in another clade. In contrast to monophyletic clades, *paraphyletic* clades exclude some descendants, or they group together taxa some of which are in fact more closely related to outsiders than to others in the clade (Figure 1.2). In the hominoids, for example, Pongidae is paraphyletic when used in the traditional sense because it excludes humans even though humans are more closely related to some pongids (African apes) than these pongids are to other pongids (orangutans). Pongidae would be monophyletic if it included humans, but humans are designated as hominids, which has taxonomic priority[3] over pongids, so that the correct nomen for this monophyletic clade is Hominidae (see below). Recognition of this paraphyly has lead many systematists to reject the nomen Pongidae. In contrast, paraphyly in other hominoids has lead to an increase in nomina. The traditional use of the nomen *Australopithecus* (including all non-*Homo* fossil humans) is also paraphyletic, because it excludes *Homo* despite the fact that *Homo* is more closely related to some *Australopithecus* than some species of this genus are to others in the same genus. The trend here has been to recognize other distinct genera (*Ardipithecus*, *Paranthropus*) thereby removing them from the *Australopithecus* clade.

[2] Other methods of identifying derived character states have been proposed. These include palaeontological and ontogenetic criteria, which rely on the order of appearance of character states in the fossil record or in individual growth and development to polarize character states. Because of the additional assumptions these criteria require, concerning the process of ontogeny and the degree to which it accurately reflects evolutionary relations, and the relative completeness and reliability of the fossil record, these approaches are much less commonly used than the outgroup criterion. In this chapter, nearly all the described research proposing various evolutionary relations relies on the outgroup criterion.

[3] Taxonomic priority is a simple but essential rule of taxonomic nomenclature. It requires that taxa considered to be the same be designated by the first used nomen. In this case, since great apes and humans belong in the same family (see below), the first used family nomen Hominidae has precedence. This rule prevents the inevitable confusion that would arise from using different "preferred" nomina to describe the same taxon.

(a)

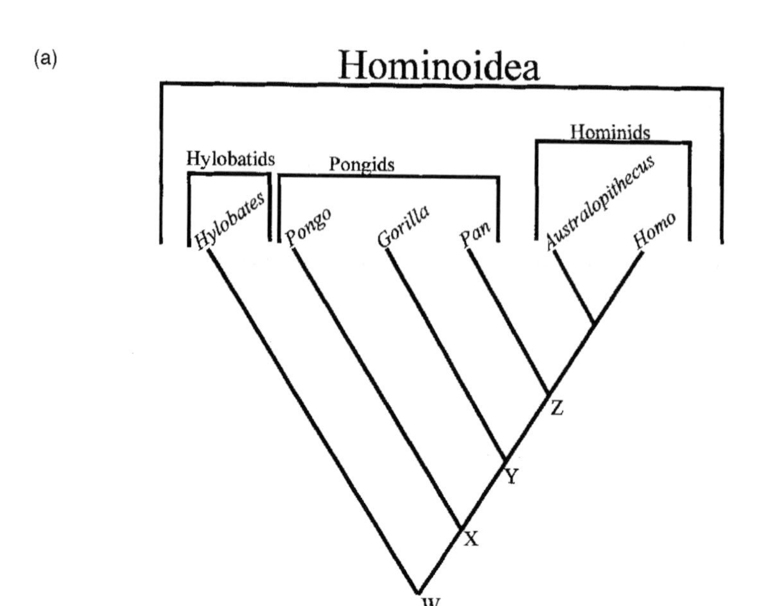
Hominoidea
Hylobatids
Pongids
Hominids
Hylobates
Pongo
Gorilla
Pan
Australopithecus
Homo
Z
Y
X
W

(b)

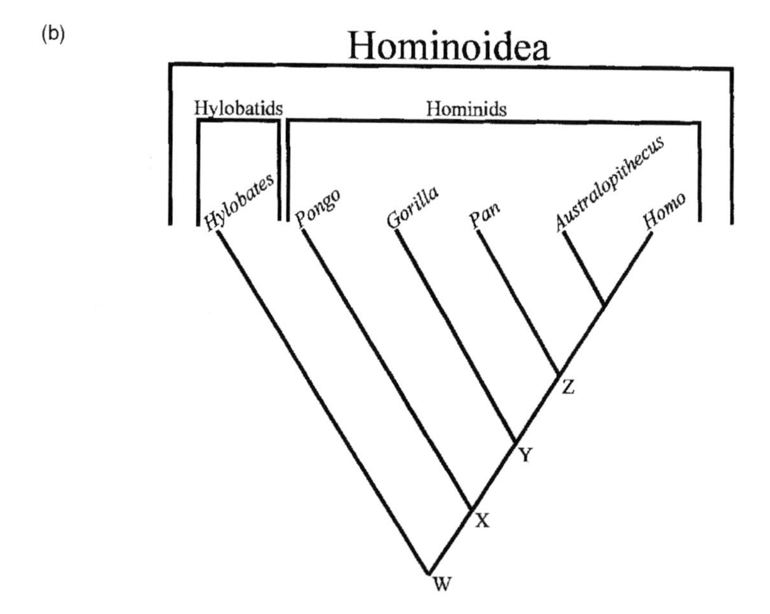
Hominoidea
Hylobatids
Hominids
Hylobates
Pongo
Gorilla
Pan
Australopithecus
Homo
Z
Y
X
W

Though cladistics was developed and refined by neontological morphologists, it is now widely used by both paleontologists and molecular systematists. The methodology is the same in all areas of research. Character analysis uses the outgroup criterion to polarize character states. Parsimony analysis determines the branching sequence requiring the fewest number of independent origins of character states while still being consistent with the largest number of characters. The remainder of this chapter surveys the recent literature on hominoid systematics, and reviews some of the implications for understanding the evolutionary history of some of the more basic attributes of living great apes and humans.

WHAT IS A HOMINID, NOWADAYS?

A consensus has been achieved among systematists today that the family Hominidae includes more than the genus *Homo*, which was the case when the term was coined. But which additional taxa are allowed to enter the hominid club is the subject of some debate. Although it was not immediately clear to any researcher that *Australopithecus* should be classified as a hominid (Dart, 1925; Woodward, 1925; Keith, 1931), the hominid status of *Australopithecus* soon became widely accepted (Le Gros Clark, 1934). Simpson (1945), in his classic work on mammalian systematics, separated humans and their fossil ancestors, the Hominidae, from all apes, which he placed in the Pongidae. Today, the overwhelming majority of researchers working on hominoid systematics would include some or all of the great apes among the hominids. Widening the definition of hominids was first proposed convincingly based on immunological distances (Goodman, 1963). Goodman (1963) suggested that the African apes should be grouped with humans in the Hominidae, reflecting the closer evolutionary relationship among African apes and humans than between them and either orangutans or hylobatids. As Goodman (1963) noted, this view mirrors that of Darwin (1871), who concluded that it was likely that humans are more closely related to African apes than to

Figure 1.2 (*Opposite*) Monophyletic and paraphyletic taxa.
Pongid is a paraphyletic taxon because it includes the African apes and excludes humans, even though African apes and humans are more closely related to each other than either group is to orangutans. Another way of looking at this is that pongid is paraphyletic because it does not include all the descendants of the common ancestor of included taxa. In this case, Y and Z must be pongids because they are ancestors of the African apes and they postdate the common ancestor of the pongids (X), yet the pongids as defined here do not include all the descendants of Y and Z (the hominids are excluded). Separating hominids and pongids also fails to recognize the close evolutionary relationship among these taxa compared to hylobatids.
A simple monophyletic solution to the paraphyly of the pongids. As defined here, the hominids unites all the great apes and humans in a single family, and also serves to communicate their close evolutionary ties. All hominids have the same evolutionary relationship to hylobatids, and all the descendants of earlier hominids are included in the taxon. Note as well that the stem hominoid (W) is neither a hylobatid nor a hominid, because if it were one of these, that taxon would become paraphyletic. This illustrates one of the methodological limitations of cladistic analysis in recognizing ancestor–descendant relationships.

Asian apes. Goodman and colleagues (Goodman et al., 1989; Bailey, et al., 1992; Goodman et al., 1994) have recently gone one step further in proposing that the hominids should also include all living hominoids, including the hylobatids (gibbons and siamangs). In a reversal of the more usual order in which revisions are proposed in hominoid systematics, the conclusions of Goodman and colleagues concerning hylobatids corresponds to the previously proposed classification of hominoids provided by Szalay and Delson (1979), based almost exclusively on morphological data. Other researchers who use family-level taxonomic categories (most avoid the taxonomic issues and concentrate on phylogeny) continue to separate hylobatids from hominids (e.g. Groves, 1993; Wilson & Reeder, 1993; Hayashi et al., 1995).

The molecular work proposing a redefinition of the Hominidae was eventually followed by morphological support (Delson & Andrews, 1975; Szalay & Delson, 1979). Though it is fair to say that the acceptance of this revision was slow among palaeontologists and palaeoanthropologists specifically interested in hominoid systematics (e.g. Simpson, 1963; Ciochon, 1983; Fleagle, 1988; Martin, 1990), it was relatively quickly accepted by many morphologists and molecular systematists (e.g., Andrews & Cronin, 1982; De Bonis, 1983). Simpson (1963), Ciochon (1983), Fleagle (1988), and Martin, (1990) all in fact agree with the phylogenetic hypothesis proposed by Goodman (1963), but prefer to retain Hominidae for humans and their ancestors, to communicate the uniqueness of humans, and to maintain a certain amount of stability in hominoid nomenclature. While there is something to be said for stability, stability at the expense of an accurate representation of evolutionary relationships has long been rejected in systematics (e.g., Darwin, 1859; Simpson, 1961; Hennig, 1966; Hull, 1979; Brooks & McClennan, 1991). An overemphasis on taxonomic stability results in more confusion than that which arises from the periodic changing of nomina to accommodate strongly supported phylogenetic revisions.

More recently, the Hominidae has been expanded to include the orangutan and its fossil relative *Sivapithecus* (Andrews, 1985). This is currently very widely accepted, especially among researchers working on great ape systematics and relationships among great apes and humans (Tattersall, Delson & Van Couvering, 1988; Groves, 1989; Andrews, 1992; Begun, 1992a, 1994a; Perrin-Pecontal, Gouy, Nignon & Trabuchet, 1992). Most recently, other diagnostically great ape taxa from the Miocene (*Dryopithecus*, *Ouranopithecus*, *Lufengpithecus*, and *Oreopithecus*) have also been added to the list of known hominids (Begun, 1994a; Begun, Ward, & Rose, 1997).

The one area in which a strong reluctance to include great apes among the hominids persists is Plio-Pleistocene hominid research. Researchers working on Pliocene hominid fossils most commonly refer to these specimens as "early hominids" (Leakey, Feibel, McDougall, & Walker, 1995; White et al., 1994; Bromage & Schrenk, 1995), "basal hominids" (White et al., 1994), or even "the oldest hominids" (Wood, 1994). Two recent papers however point to a possible change in this dichotomy between researchers on hominoid systematics and those who work primarily on fossil humans. Aiello and Wood (1994) use the term hominine (an informal subfamily designation) for fossil and living humans, suggesting that the great apes are to be included with humans in the

Hominidae, though they do not use this latter term. Lieberman, Wood, and Pilbeam, (1996) in their analysis of phylogenetic relations among fossil humans, appear conscious of this discrepancy as well, and are consistent in their qualification "Pliocene hominids," suggesting that earlier or other hominids may exist. My sense from these published indicators and from conversations with Plio-Pleistocene fossil hominid researchers is that the more narrow, traditional definition of the Hominidae will disappear in the near future. Slowly, the core of paleoanthropology is also coming to accept the idea that the hominids include at least some if not all of the living great apes and their fossil relatives. Most paleoanthropologists and nearly all molecular systematists however, now agree on key aspects of the pattern of evolution among great apes and humans.

WHAT IS AN ORANGUTAN, ANYWAY?

Darwin (1871) suggested that orangutans represent a separate lineage within the great apes and humans, based on comparative anatomy and biogeography. Later Nuttal (1904) reached the same conclusion based on biochemistry. Despite this, for most of the first two-thirds of the twentieth century, orangutans were grouped with African apes, and sometimes even with hylobatids, in the Pongidae (Schultz, 1963; Simpson, 1945; Pilbeam, 1969). It is nearly universally accepted now that this is not the case. As noted above, Goodman (1963) and later Sarich and Wilson (1967) showed on the basis of immunology that African apes are more closely related to humans than they are to orangutans. While there have been a few studies suggesting either that the pongids really do form a monophyletic group (Kluge, 1983) or that orangutans are actually more closely related to humans than are African apes (Schwartz, 1984), these proposals have not withstood further testing, and have been rejected even by some former advocates (e.g., Schwartz, 1990). The vast majority of molecular and morphological research on this topic has repeatedly confirmed the conclusion linking humans to African apes to the exclusion of orangutans (summarized in Andrews & Cronin, 1982; Groves, 1986; Andrews, 1987; Goodman et al., 1994; Ruvolo, 1994; Easteal et al., 1995).

Orangutans are quite distinct from other hominids as measured either by their morphology or their molecules. This suggests that the amount of time that has elapsed since the divergence of the orangutan lineage from that of other hominids has been relatively great. The fossil record provides a relatively clear indication of a minimum of 13 ma (mega-annum, or, million years) of separate evolutionary history for the orangutan (Pilbeam, 1982; Andrews and Cronin, 1982; Kappelman, et al., 1991). This is the approximate age of the oldest member of the orangutan lineage or clade, *Sivapithecus*. Recently some doubt has been cast on whether or not *Sivapithecus* is indeed more closely related to orangutans than to any other hominid (Röhrer-Ertl, 1988; Pilbeam et al., 1990; Pilbeam, 1996), but most analyses of fossil data continue to support this link (Alpagut et al., 1997; Ward, 1997; Begun et al., 1997). 13 ma is a minimum age for the orangutan clade, which could be considerably older. Since 13 ma is widely used as a calibrating date for molecular clocks[4] (Horai et al., 1992, 1995; Bailey et al., 1992;

Adachi & Hasegawa, 1995; Easteal et al., 1995; Kim & Takenaka, 1996), the divergence times suggested by these analyses must be regarded as minimum estimates. This point is particularly relevant to interpretations of the divergence times of African apes and humans, some of which are much younger than the fossil evidence suggests (Easteal et al., 1995; see below).

A minimum of 13 million years of separate evolution in the orangutan clade may help to explain the large amount of genetic diversity in living orangutans. Morphologically, Bornean and Sumatran orangutans (*Pongo pygmaeus pygmaeus* and *Pongo pygmaeus abelii*) are very similar to one another but not identical (reviewed in Groves, 1986). It is not too difficult to distinguish among the two subspecies in cranial metrics (given relatively large samples) nor is it difficult to identify living individuals by subspecies (Groves, 1986). Though they more closely resemble one another than do bonobos and chimpanzees (*Pan paniscus* and *Pan troglodytes*), the genetic differences between Sumatran and Bornean orangutans are equal to or greater than those that distinguish the two species of *Pan*, according to most accounts (Lucotte and Smith, 1982; Ferris, Brown, Davidson, & Wilson, 1981; Bruce & Ayala, 1979; Groves, 1986; Courtenay, Groves, & Andrews, 1988; Caccone & Powell, 1989; Janczewski, Goldman, & O'Brien, 1990; Ruvolo, 1994). This has prompted Groves (1993) to wonder about the possible advisability of recognizing two separate species of orangutan. Noting that intra-subspecific differences may be as great as those between the subspecies, Groves (1993) rejects this option for the moment. If nothing else, the discrepancy between genetic and morphological differences between orangutan subspecies indicates some de-coupling of the rates of change in these systems, a phenomenon noted long ago in other hominid taxa (e.g., King & Wilson, 1975). In the final analysis, it is very clear from nearly all lines of evidence that orangutans have a long period of evolutionary history independent of other hominids, a point to be considered when assessing the morphological, genetic, and behavioral differences among living great apes, or the lack thereof.

HOW ARE AFRICAN APES AND HUMANS RELATED TO ONE ANOTHER? PART 1: WHERE DO HUMANS FIT IN?

Molecular approaches

While there is nearly universal agreement on the monophyly of the African apes and humans, there remains some disagreement on relations within this clade. This situation is changing, however. Up to about 1994 there were a significant number of systematists who suggested that the gorilla–chimpanzee–human trichotomy could not be or has not yet been resolved (Groves, 1986, 1989; Marks, Schmid, & Sarich, 1988; Holmquist, Miyamoto, & Goodman, 1988; Maeda, Wu, Bliska, & Reneke, 1988; Groves and Patterson, 1991; Ellis et al., 1990; Kawamura et al., 1991; Marks, 1992; Ruano, Rogers, Ferguson-Smith, & Kidd, 1992; Rogers, 1993; Retief et al., 1993). Among those systematists who felt the trichotomy could be resolved, most preferred a chimpanzee–

[4] Clocks that use rate and amount of genetic change between lineages to determine divergence times (see Easteal, et al., 1995 for detailed discussion).

human clade (Brown, Prager, Wang, & Wilson, 1982; Goodman, Braunitzer, Stangl, & Schrank, 1983; Sibley & Alquist, 1984, 1988; Sibley Constock, & Alquist, 1990; Slightom, Chang, Koop, & Goodman, 1985; Miyamoto, Slightom, & Goodman, 1987; Hasegawa, Kishino, & Yano, 1987; Savatier et al., 1987; Ueda et al., 1989; Caccone & Powell, 1989; Gonzalez, et al., 1990; Galili & Swanson, 1991; Ruvolo et al., 1991; Bailey et al., 1992; Begun, 1992a; Horai et al., 1992; Kawaguchi, O'hVigin, & Klein, 1992; Perrin-Pecontal et al., 1992). The smallest number of systematists were those who preferred a chimp–gorilla clade (Ferris, et al., 1981; Templeton, 1983; Hixon & Brown, 1986; Andrews & Martin, 1987; Smouse & Li, 1987; Djian & Green, 1989).[5] However, it is also the case that many of the systematists who expressed a preference in resolving this dichotomy felt that the issue was not fully resolved, and that more evidence would be needed to increase confidence in their conclusions. In essence, then, most molecular systematists up to about 1994 had published that the issue was still unresolved, although more were willing to go out on a limb and favor a chimpanzee–human clade than were willing to favor a chimpanzee–gorilla clade. Although 2 morphological analyses found some evidence for a chimpanzee–human clade (Groves, 1986; Begun, 1992a), few morphologists, including Groves himself (Groves, 1986, 1989; Groves & Patterson, 1991) had much confidence in this conclusion. Most followed Andrews and Martin (1987) and Andrews (1992) in uniting the African apes to the exclusion of humans, based primarily on the shared occurrence of certain details of enamel structure and characters related to knuckle-walking. One exception was Pilbeam (cited in Caccone and Powell, 1989), who suggested that these features may have been secondarily lost in humans, but present in our ancestors. This idea was developed in more detail independently in Groves and Patterson (1991) and Begun (1992a) (see below). Since that time, the molecular systematists have come closer to a consensus (indeed, a consensus is claimed by many already!), and more morphologists are inclined to agree as well. That is, an increasing number of researchers are coming to the conclusion that a chimpanzee–human clade is the most likely (see below).

Two major changes in molecular systematics seem to be responsible for the more consistent results that are beginning to emerge on the question of African ape–human relations. The first is the more widespread use of new techniques (principally PCR [polymerase chain reaction], for DNA amplification) that makes it much easier to sequence nuclear genes, mitochondrial genes, and even entire mitochondrial genomes.[6] This led to a plethora of new sequence data. The second is the now more-or-less universal application of cladistic methods to determine phylogenetic relations from genetic data. Previously the strongest support for a chimpanzee–human clade came from the DNA hybridization data (Sibley & Alquist, 1984, 1987), though these have been

[5] These citations exclude chromosome studies, which have produced widely divergent results, and which have been criticized recently (Borowik, 1995).

[6] Molecular systematists can now determine the sequence of nucleotides that make up the genetic code of DNA in the nucleus, inherited from both parents, as well as the DNA in mitochondria, an organelle within the cell, inherited only from the mother (sperm, though they have mitochondria, do not pass them on during fertilization of the egg). This difference in inheritance accounts in part for the differing rates at which mitochondrial and nuclear DNA evolve.

criticized for methodological shortcomings (Marks et al., 1988; but see Sibley et al., 1990). Caccone and Powell (1989), however, reached the same conclusion from the DNA hybridization data in an analysis considered more reliable (Ruvolo, 1994). Unfortunately, DNA hybridization data are not amenable to cladistic analysis, since the results produce a measure of overall difference rather than discovering discrete "character" differences (Andrews, 1987). When Andrews (1987) attempted to carry out a cladistic analysis of the few molecular data sets that could be analyzed in this way at that time, his results were mixed, and overall he found studies favoring a chimpanzee–human clade unconvincing. In the intervening years, all sequence data (proteins, nuclear DNA, ribosomal RNA, mitochondrial DNA) have been analyzed cladistically.

A number of recent sequence data analyses are quite unambiguous in supporting a chimpanzee–human clade. Goodman et al. (1994) reviews and reanalyzes previous work, primarily that of Bailey et al. (1992), concluding that the chimpanzee–human link is very strong based on the β-globin gene cluster. Their reanalysis of other sequence data, including 10 different nuclear genomic regions (in addition to the β-globin gene cluster), 1 RNA sequence, and 2 mitochondrial DNA (mtDNA) sequences, offers strong support for the chimpanzee–human clade, with many more parallelisms being required to support a chimpanzee–gorilla clade (Goodman et al., 1994, Table 1.1). These authors recognize significant amounts of homoplasy in these data, and interpret it as an indication that the separation of gorillas from the common ancestor of chimpanzees and humans occurred close in time to the separation of chimpanzees and humans.

Ruvolo and colleagues (Ruvolo, 1994, 1995; Ruvolo et al., 1994) also offer strong and convincing evidence for a chimpanzee–human clade. This work, based primarily on mitochondrial cytochrome oxidase subunit (COII) sequences, takes more account of variability with the inclusion of between 4 to 6 African ape individuals in the sample. All the hominoid individuals cluster by taxon, and chimpanzees also cluster with humans to the exclusion of gorillas (Ruvolo, 1994; Ruvolo et al., 1994). This study also includes 4 bonobos, which cluster with chimpanzees, supporting the monophyly of *Pan*. Ruvolo (1994) also considers other molecular studies that fail to support a chimpanzee–human clade, or that support one but only weakly. She points out, as have others (Goodman et al., 1994; Easteal et al., 1995), that in many instances the genes under study have changed too slowly to allow for a resolution of the African ape–human clade. There is, in fact, substantial evidence to suggest significant amounts of variability in the rate at which different parts of the nuclear and mitochondrial genomes change through time (Chang and Slightom, 1984; Goodman, Koop, Czelusnaik, & Weiss, 1984; Vawter & Brown, 1986; Li & Tanimura, 1987; Hasegawa et al., 1987; Hayasaka, Gojobori, & Horai, 1987; Miyamoto et al., 1987; Holmquist et al., 1988; Maeda et al., 1988; Djian & Green, 1989; Perrin-Pecontal et al., 1992; Horai et al., 1992, 1995; Retief et al., 1993; Ellsworth, Hewett-Emmet, & Li, 1993; Deeb et al., 1994; Livak, Rogers, & Lichter, 1995; Adachi & Hasegawa, 1995; Ruvolo, 1995; Easteal et al., 1995; Mohammad-Ali, Eladari, & Galibert, 1995; Queralt et al., 1995; Nachman, Brown, Stoneking, & Aquadro, 1996; Arnason, Xiuteng, & Gullberg, 1996; Kim & Takenaka, 1996). Mohammad-Ali et al. (1995) also reject the results of an analysis that failed to support a chimpanzee–human clade based on rate variability. In this case, it was the work of Oetting, Stine, Townsend, and King, (1993), who found support for a human–gorilla clade based on their analysis of tyrosinase-related genes. Mohammad-Ali et al. (1995) suggest that the relatively short nucleotide sequence analyzed by Oetting, et al. (1993) is known to have evolved fast, making the identification of homologous changes at given *loci* very difficult.

A number of other recent studies also support a chimpanzee–human clade from nucleotide sequence data (Perrin-Pecontal et al., 1992; Ueda et al., 1989; Gonzalez et al., 1990; Galili & Swanson, 1991; Horai et al., 1992, 1995; Deeb et al., 1994; Adachi & Hasegawa, 1995; Van der Kuyl, Kuiken, Dekker, & Goudsmit, 1995; Dorit, Akashi, & Gilbert, 1995; Meneveri et al., 1995; Mohammad-Ali et al., 1995; Kim & Takenaka, 1996). In addition to the previously cited work on DNA hybridization in support of this grouping (Caccone & Powell, 1989; Sibley et al., 1990), a cladistic analysis of discrete chromosomal data treating these essentially as morphological character states also supports this conclusion (Borowik, 1995).

Only a few recent molecular studies support a chimpanzee–gorilla clade. As noted above, a larger number fail to resolve the trichotomy, or weakly support either a chimpanzee–gorilla or chimpanzee–human clade, but this is probably related to the rate at which the *loci* evolve, the fact of a relatively close evolutionary relationship among the African apes and humans, and a number of other methodological and/or theoretical constraints (Hasegawa et al., 1987; Ruvolo, 1994, 1995). Although the work of Djian and Green (Djian & Green, 1989; Green & Djian, 1995), Livak et al. (1995), Queralt, et al. (1995), and Ellis, et al. (1990) (in part) has been used to argue in favor of a chimpanzee–gorilla clade, in 3 of the 4 studies, the authors make a distinction between gene trees and phylogenetic trees,[7] and are unwilling to strongly support a particular phylogeny. Livak et al. (1995), for example, conclude that the chimpanzee dopamine D_4 receptor gene sequence is closer to that of gorillas than to humans. However, they suggest that because this is incongruent with other sequence data (see above), the divergences among the three great apes must have occurred very close in time (see also Rogers [1993]). This suggestion is not inconsistent with other conclusions that nevertheless support a chimpanzee–human clade (Goodman et al., 1994). Ellis et al. (1990) found X-chromosome pseudoautosomal boundary sequences more similar between chimpanzees and gorillas, while finding Y-chromosome pseudoautosomal boundary sequences to be more similar between chimpanzees and humans. Although this result is ambiguous, these authors also point out that the total number of position synapomorphies in all sequences shared by humans and chimpanzees is 4, whereas only 2 are shared by chimpanzees and gorillas. While not very strong, overall this data would tend to support a chimpanzee–human clade as well. Queralt et al. (1995), who also found a

[7] Gene trees, like character trees, provide evidence for relationships based on individual *loci* or characters. Phylogenetic trees should combine data from more than one *locus* or from many characters. They depict relationships supported by the largest number of *loci* or characters. Those *loci* or characters that link taxa in a pattern of relationship different from the phylogenetic hypothesis are deduced to be homoplasies, having evolved independently in the separate lineages that share them.

closer link between chimpanzees and gorillas than between chimpanzees and humans, nevertheless fail to find support for a chimpanzee–gorilla clade, again citing the need for many more genes. The same research group previously reported an analysis of the same primate sequences (Retief et al., 1993) in which a human–gorilla link was supported. The main difference between the two studies was the inclusion of nonprimate data in the 1995 paper. It should also be noted that these same analyses grouped marmosets with green monkeys (Retief et al., 1993) and gibbons with African apes and humans to the exclusion of orangutans (Queralt et al., 1995), thus contributing to the uncertainty regarding the exact phylogenetic significance of these results.

The only analysis so far that claims an unambiguous linkage between chimpanzees and gorillas is the involucrin gene sequence data of Djian and Green (1989). These authors (Djian & Green, 1989; Green & Djian, 1995) claim that the involucrin gene data is particularly well suited to resolving relationships among very closely related hominoids because of its rapid rate of change.[8] Ruvolo (1994) disagrees with this conclusion. She argues that the segment of the involucrin gene sequence that links chimpanzees to gorillas, the "modern" repeat region, is susceptible to a number of molecular evolutionary processes that may cause it to generate an unreliable phylogeny. There has been no resolution to this disagreement (Green & Djian, 1995; Ruvolo, 1995). As noted earlier, because of the likelihood that individual gene trees will not always correspond to phylogenetic trees due to homoplasy, the greater the number of genes analyzed, the more likely a consensus will emerge (Ruvolo, 1994, 1995; Stewart, 1993; Queralt et al., 1995; Retief et al., 1993; Maeda et al., 1988; Pamilo & Nei, 1988). Even if the gene tree for involucrin does link gorilla and chimpanzee sequences, it is still in the significant minority compared to those that link humans to chimpanzees.

A few researchers are particularly concerned with the issue of intraspecific variability in sequence data (polymorphism), and in the related issue of homoplasy (Rogers, 1993; Rogers & Comuzzie, 1995; Marks, 1992, 1994). Marks (1994), for example, points out that a great deal of homoplasy exists in the nuclear DNA sequence data if a chimpanzee–human clade is accepted, but much less with a chimpanzee–gorilla–human trichotomy. However, more fully resolved phylogenetic trees (trees with as many dichotomous branchings as possible) are usually more homoplasious than are less fully resolved trees. By definition, collapsing separate branches into a single origin will decrease contradictory data, because much of the data is, in fact, ignored. For example, collapsing the orangutan branch into the African ape and human branch, making orangutans, African apes, and humans unresolved, decreases the amount of homoplasy by the number of homoplasies in the orangutan clade, and there are many (Begun & Guleç, 1998). This does not mean that an African ape–orangutan–human trichotomy is a better explana-

[8] The claim that gene sequences evolve too rapidly, too slowly, or at just the right rate to be helpful in analyzing particular relationships (e.g., Ruvolo, 1994; Mohammad-Ali, et al., 1995; Djian & Green, 1989) may seem overly convenient and is difficult to prove, but it mirrors similar claims about anatomical regions and their phylogenetic significance (e.g., Begun, 1995). It is clear that, depending on the rate of change and the time since divergence, some sequences will be more useful than others for resolving particular relationships, in the same way that some isotopes are more useful than others in calculating absolute ages, depending on decay rates and geologic time.

tion. Because more data support a specific resolution, it just means that there is a lot of homoplasy. In other words, trichotomies result in less homoplasy than dichotomies for methodological reasons; a priori assumptions about the amount of homoplasy resulting from the evolutionary process are neither necessary nor desirable.

Marks (1992, 1994) shares with Rogers (1993) and Rogers and Comuzzie (1995) a more serious concern about polymorphism and the possibility that, by chance alone, different alleles could find themselves distributed among descendant populations in a pattern that differs from the actual evolutionary relations among those descendant populations (see Rogers, 1993 and Marks, 1992 for a comprehensive discussion of this problem). Concern about polymorphism has been widely expressed in the molecular literature (Ruvolo, 1994, 1995; Pamilo & Nei, 1988; Wu, 1991). This is one of the main reasons for the widespread recognition that a gene tree may be different from the species tree (see above), and it holds for morphological data as well, in the distinction between a single-character tree and a phylogeny based on many characters (Begun, 1994b). Rogers (1993) and Marks (1992) both make the claim that the chimpanzee–human–gorilla trichotomy is unresolvable or difficult to resolve currently because of the confounding influence of polymorphism. Rogers (1993) and Rogers and Comuzzie (1995) recognize that the evidence of a chimpanzee–human clade is stronger, i.e., supported by a larger number of genetic *loci*, but suggest that the number of contradictory *loci* indicate a very short time between the gorilla divergence and the chimpanzee–human divergence, a conclusion essentially in agreement with Goodman et al. (1994). In the final analysis, more nucleotide positions support a chimpanzee–human clade than any other single clade, as noted by Goodman, et al. (1994), and for this reason this clade is to be preferred, regardless of the amount of homoplasy implied.

Finally, one last objection to the results of molecular systematics commonly expressed by anthropologists is that the number of individuals sampled by molecular systematists is very small, leading to the possibility that the gene sequence discovered may not be representative of the species as a whole (Marks, 1994; Ruvolo, 1995). However, as Ruvolo (1994) clearly shows, based on the theoretical work of Takahata (1989), increasing the number of individuals will not increase the reliability of a gene tree, because the polymorphism that exists today does not reflect ancestral polymorphism. Modern alleles are, in fact, more likely to have evolved from a single ancestral allele following coalescence of ancestral polymorphisms. This point is also made strongly by Pamilo and Nei (1988) and Wu (1991). So, although ancestral polymorphism can confound a phylogenetic analysis when the divergence times are short, larger sample sizes of modern hominoids will not help resolve this difficulty.[9]

Though genes clearly do not evolve at the same rates, and a number also probably do

[9] Because descendant populations tend to represent only a small percentage of the original genetic diversity of the ancestral species, most ancestral polymorphism will not be represented in a descendant population. Ancestral polymorphisms may only be represented by a single allele in a descendant population. However, even if more than one allele is passed on to a new species, since allelic frequencies fluctuate between 0% and 100%, ancestral polymorphic *loci* will eventually coalesce to a single allele, from which subsequent variation in the descendant species will evolve. The longer the time since separation between species, the less likely they will be to share ancestral polymorphisms.

not evolve at a constant or predictable rate (Zuckerkandl & Pauling, 1962; Sarich & Wilson, 1967; Miyamoto et al., 1987; Ellsworth et al., 1993; Ruvolo, 1995; Nachman et al., 1996; Kim & Takenaka, 1996), studies have shown that some gene sequences and some DNA hybridization data can be used to estimate times of divergence based on amounts of difference among the taxa analyzed (Sibley et al., 1990; Hasegawa et al., 1987; Hayasaka et al., 1988; Caccone & Powell, 1989; Ruvolo, 1994, 1995; Adachi & Hasegawa, 1995; Easteal et al., 1995; Horai et al., 1992, 1995).

There are basically two schools of thought on this issue with regard to African ape–human relations. One holds that the divergence between gorillas and the ancestor of chimpanzees and humans was very close to that between chimpanzees and humans, while the other holds that it was further away. As previously noted, Rogers (1993) and Rogers and Comuzzie (1995) suggest that the degree of incongruity in sequence data results suggests that the divergence times were very close together. In fact, though both Rogers (1994) and Marks (1994) concede that humans and chimpanzees may share a most recent common ancestry, both believe that the divergence times among the three are so close together that molecular data may never be adequate to resolve the issue. For both authors, if the time between the divergence of the gorilla and that between chimpanzees and humans is less than about 1 ma, then the molecular data may prove inconclusive. It is ironic that the paleontological data may, in fact, be more useful in this instance. Morphological change in lineages of many mammals are quite distinguishable within time ranges well under 1 ma. In fact, in rapidly evolving lineages such as Pleistocene rodents, differentiation is detectable within 0.1 ma. Even within hominoids, there is obvious evolutionary change within 0.5 ma intervals among the australopithecines and within the genus *Homo*.

Goodman et al. (1994) also believe that divergence times among the African apes and humans were close, based on the overall degree of similarity of the three genera and the amount of homoplasy in the data. Easteal et al. (1995) also place the divergences of gorillas, humans, and chimpanzees very close together, though they support the chimpanzee–human clade. These authors are so impressed with the overall genetic similarity of the African apes and humans that they advocate placing them all, along with all fossil humans, in the genus *Homo*. In contrast to these studies, Ruvolo (1994, 1995) provides evidence for a substantial period of separation of the chimpanzee–human and gorilla clades, based both on results using mtDNA and her interpretation of the DNA hybridization results of Caccone and Powell (1989). The very recent work of Kim and Takenaka (1996), finding about twice the amount of difference between gorillas and the chimpanzee group as between chimpanzees and humans in the Y-chromosomal TSPY gene sequence, also suggests a substantial separation between the two events. To these conflicting interpretations can be added the conclusions of researchers advocating a molecular-clock approach.

The earliest molecular-clock estimates produced a divergence-time estimate between chimpanzees and humans of roughly 5 ma (Sarich & Wilson, 1967), based on protein immunodiffusion. There has been a great deal of discussion of these results in the molecular and paleontological literature, much of which is beyond the scope of this

Table 1.2 *Recent hominid divergence times estimates*

Research	Orangutan	Gorilla	Chimp–Human	Chimp–Bonobo
Sequence data				
Nuclear DNA				
Hasegawa et al. (1987)	9–15	4.1–7.8	3.1–7.0	
Easteal et al. (1995)	7.3–9.8	*c.* 5	3.2–4.5	
Holmes et al. (1988)	14.5[a]	16.2–9.2	5.9–8.9	1.9–4.4
mtDNA				
Adachi and Hasegawa (1995)	13–16[a]	7–9[b]	4–5	2–3[b]
Horai et al. (1995)	13[a]	6.3–6.9	4.7–5.1	2.1–2.5
Hybridization data				
Sibley and Alquist (1987)	13–17[a]	7.7–11	5.5–7.7	2.4–3.4
Caccone and Powell (1989)12–16[a]	6–8			
Ruvolo (1995)		7.4–8.9	5–6	
Paleontology[c]	13–16		> 4.4	

Dates in millions of years. See text for discussion.
[a] Age given by authors using paleontological data.
[b] Applying the rough correction used by the authors on their chimpanzee–human divergence estimate.
[c] Data from Kappelman et al. (1991), Cande and Kent (1995) and WoldeGabriel et al. (1994).

chapter. More recently, as noted above, a large number of studies have supported the notion of a molecular clock on empirical and theoretical grounds for DNA hybridization and nucelotide sequence data. The most recent of these are reviewed here.

Sibley and Alquist (1987) estimated times of divergence within the African apes and humans using DNA hybridization data. They estimated the time of divergence between gorillas and the chimpanzee–human ancestor to be between 7.7 and 11 ma, and that between humans and chimpanzees to be between 5.5 and 7.7 ma (Table 1.2). This was based on the assumption that the relationship between time and overall average DNA divergence is linear within hominoids, and a calibration range using fossil evidence to date the first appearance of the orangutan lineage (13–17 ma) and the first appearance of the cercopithecoid lineage (25 ma). Caccone and Powell (1989), in an analysis generally accepted as the most sophisticated DNA hybridization work on hominoid relations (Ruvolo, 1995), obtain essentially the same results based on the same assumptions.

Other divergence dates provided recently are based on DNA sequences. Hasegawa et al. (1987) used data from the η globin pseudogene (a gene that does not code for a functional polypeptide) to provide 95% confidence intervals for the divergence dates, yielding about 9–15 ma for the orangutan divergence, 4.1–7.8 for that of the gorilla, and 3.1–7.0 for that of humans and chimpanzees. This was based on an estimate of 38 ma for the divergence of platyrrhines and catarrhines, based on the fossil record. These authors also note that their estimates of great ape–human divergence times may be too recent, given evidence they find of a rate of slowdown in the Hominoidea. Nevertheless, in another study, Hasegawa et al. (1990) found an even later date of 3.2–4.6 ma for the divergence between chimpanzees and humans, and 4.3–5.9 ma for that of the gorilla, based on a short sequence of mtDNA. In their most recent reanalysis, Adachi and Hasegawa (1995) obtain dates of 3.0–4.2 ma for divergence between chimpanzees and humans, 5.1–6.5 for gorilla, and 1.4–2.2 for the two species of *Pan*, assuming a divergence time of 13–16 ma for orangutans. However, as earlier, they note that these dates may be too recent given evidence of rate variation. In the final analysis, they conclude that a reasonable estimate for the human–chimpanzee divergence would be 4–5 ma.

Horai et al. (1992) also used mtDNA sequence data to obtain a divergence estimate of 4.2–5.2 ma for chimpanzee–humans, 7.0–8.4 ma for gorillas, and 2.0–3.0 for bonobos and chimpanzees, again assuming a divergence time of 13 ma for orangutans, from the fossil record. Although their analysis of evolutionary relations among hominoids was based on a long sequence of mtDNA, they chose to estimate divergence times based on a subset of these data containing relatively conservative sequences. This was to avoid including too many different genes, possibly evolving at different rates, and also to avoid including portions of the sequence that have evolved more rapidly and may have undergone changes that are not detectable. Their results overall are quite compatible with those of Hasegawa and colleagues (see above), though these authors have suggested that the results of Horai et al. (1992) give divergence times that may be too old, for a number of methodological reasons (Adachi & Hasegawa, 1995). In a more recent refinement of their work, based on a larger portion of the mtDNA sequence, Horai et al. (1995) estimated the divergence times at 6.3–6.9 for gorilla, 4.7–5.1 for human–chimpanzee, and 2.1–2.5 for chimpanzee–bonobo, again calibrating based on an orangutan divergence of 13 ma. The gorilla divergence is slightly closer to that of humans and chimpanzees compared to their previous analysis, but otherwise their results are similar to other work (Table 1.2).

In the most recent and most intriguing work, Easteal et al. (1995) cite a range of dates from 2.5–6.0 ma for the divergence of humans and chimpanzees, based on a review of both hybridization and sequence data (gorillas are not included in their analysis). The ages they obtain are compatible with those of previous researchers when they use the same estimates of base substitution rates, but they also make the case that these rates should be higher than generally accepted, at least within the Hominoidea (Easteal et al., 1995). In fact, the range of rates they prefer gives an estimate of the date of divergence of chimpanzees and humans between 3.2 and 4.5 ma. However, their preference for these rates is based on their conclusion that DNA evolution proceeds at the same rate in all mammals, and that these ranges result in divergence estimates that are the least incompatible with paleontological data for other mammalian divergences. These results contrast with the work of both Sibley and Alquist (1987) on DNA in general, and Goodman et al. (1983) on protein coding sequences, which demonstrate significantly different rates in different mammalian lineages. Other work supporting the idea of rate variability in molecular evolution are cited above. Despite their unorthodox approach, Easteal et al. (1995) go so far as to question the sister relationships of *Sivapithecus* and *Pongo* and of *Australopithecus* and *Homo*, saying, in essence, that these fossils are too old, regardless of their anatomy, to be uniquely related to these modern taxa (see below). This claim ignores a massive corpus of molecular, comparative anatomical and paleontological data which is, ironically, so thoroughly reviewed by Easteal et al. (1995).

As can be seen from Table 1.2, either a short or a long separation between the divergence of the gorilla and divergence of chimpanzees and humans can be argued from these results. The data are simply not conclusive, partly because not all authors include all taxa in their analysis. It appears that, more than the order of branching events, the main area of disagreement among molecular evolutionary biologists studying hominoid evolution is the age of these events and the amount of time separating each one. This has a direct bearing on our understanding of the amount of time African apes and humans shared a common ancestor and the amount of time during which a unique ancestor was shared by humans and chimpanzees, which may in turn have implications for the ways in which behavioral similarities and differences among these taxa are interpreted. Once again, more communication between molecular biologists and paleontologists will help to resolve this issue. The paleontological record does not yet provide evidence on the divergence of gorilla or the chimpanzee-bonobo split, because there are no known fossil relatives of the living African apes (Andrews, 1992; Begun, 1992a, 1994a,b). However, the well-established geologic age of the earliest known humans at close to 4.5 ma years ago makes it clear that chimpanzees and humans must have diverged at least 0.5 to 1.0 million years earlier (WoldeGabriel et al., 1994).

Morphological approaches

Like molecular biologists, comparative anatomists and paleontologists have extensively studied relations among the African apes and humans. Though there has been a great deal of discussion on this issue by paleontologists, most feel that the issue is unresolved. Three recent groups of studies are the most comprehensive so far dealing with African ape and human relations from the anatomical and/or paleontological data. Andrews and colleagues have been the strongest proponents of the African ape clade hypothesis, though most recently there seems to have been a retreat from this position. Andrews (1987) lists synapomorphies of various branching points of the great ape and humans, including synapomorphies of the African ape clade, which excludes humans. Here, and in Andrews and Martin (1987), two character complexes are stressed above all others. These are those characters related to knuckle-walking and those related to enamel structure. On the basis of these important characters, Andrews (1987) and Andrews and Martin (1987) support the notion of an African ape clade that excludes humans. More recently, Andrews (1992) and Andrews and Pilbeam (1996) have published phylogenies in which uncertainty is expressed as to the resolution of the African ape and human clade based on the paleontological evidence.

The characters described in these publications have been analyzed by Groves (1986, 1989, 1993) and Groves and Patterson (1991), using the cladistic methods outlined above, resulting in the most comprehensive analysis of evolutionary relations among African apes and humans based on the comparative anatomy of modern taxa. Begun (1992a, 1994a, 1995) and Begun et al. (1997) present evidence from an even larger list of character states from both fossil and recent hominoids.

In Groves (1986), an extensive compilation of anatomical features relevant to the systematics of the great apes and humans is presented. This list, dominated by soft-tissue characters, is based exclusively on living great ape taxa. Groves (1986) lists 25 derived character states shared between humans and chimpanzees, as compared to only 7 for chimpanzees and gorillas, and 12 for humans and gorillas. Despite the dominance of character states linking humans and chimpanzees, Groves (1986, 1989) prefers a trichotomous branching pattern for the origins of the African apes and humans. He cites a number of models of evolutionary pattern consistent with this hypothesis, which he states also explains the large number of shared derived character states shared by humans and gorillas, and chimpanzees and gorillas. In an expanded and somewhat modified data set, Groves and Patterson (1991) prefer two dichotomies, but suggest that chimpanzee–human and chimpanzee–gorilla are equally probable. Finally, Groves (1993), while not heartily endorsing the view that humans and chimpanzees form a clade, seems to favor this alternative, while still regarding it as controversial.

The main difference between Groves (1986) and Groves and Patterson (1991) is the inclusion in the later paper of character states related to knuckle-walking and enamel histology, cited by Andrews and Martin (1987) and Andrews (1987) as linking chimpanzees to gorillas. However, as noted by Groves (1986) and Groves and Patterson (1991), there are problems with these character states. The main difficulty is in their number. There are 10 knuckle-walking and 4 enamel character states said to link chimpanzees and gorillas, though in fact each group of character states varies as a single trait. The knuckle-walking characters are not only very strongly interdependent, they vary within the African apes in degree of development, in a manner that is predictable and explicable based on body size and biomechanics (Inouye, 1991, 1992). Additionally, there are characters of the forelimb and hand shared by African apes and humans that suggest that humans, too, have evolved from a knuckle-walker, making this character complex primitive for the African apes and humans (Lewis, 1973; Begun, 1992a, 1994a). Regarding the enamel character states, in addition to the fact that these co-vary, more recent research on variability in great ape enamel suggests that the patterns described in Martin (1985) and Andrews and Martin (1987) are not so clear-cut (Beynon, Dean, & Reid, 1991). In addition, the conclusion that thin enamel is a shared derived character of the African apes (Andrews & Martin, 1987) is weakened by the recent discovery of a fossil human with relatively thin enamel, *Ardipithecus* (White et al., 1994), suggesting that thin enamel may also be primitive for the great apes and humans. In the final analysis, the two character complexes that are said to phylogenetically link African apes are questionable in their number and their polarity, and have to be

Table 1.3 *Synapomorphies of the chimpanzee–human clade. These characters are also known in one or more fossil great ape*[a]

Spatulate upper central incisors	Reduced caliber patent incisive canals
Short nasal premaxilla	Narrow lateral orbital pillar @ frontozygomatic
Incisive foramen distal to P^3 suture	Flared P^4 and upper molar crowns
Reduced M^3	Lower position of inion
Short molars lacking cingula	Large lesser palatine foramen
Elongated naso-alveolar clivus	

Modified from Begun, 1994a; Begun and Güleç, 1998.
[a] Twenty-five other characters of the chimpanzee-human clade are given by Groves (1986, Table 4a, p. 191). These include characters related to development, internal organ size, morphology and position, reproductive biology, and a few additional skeletal characters for which there is no hominid fossil record.

considered at best ambiguous, and probably currently unreliable, in the absence of more information from the fossil record.

Without the dental and knuckle-walking characters, the morphological evidence of modern hominids as clearly favors a chimpanzee–human clade as does the molecular evidence. This conclusion is supported by more recent work incorporating fossil taxa. Begun (1992a) listed 12 derived character states shared by humans and chimpanzees that are also represented in the fossil great ape *Dryopithecus*, which was used as the outgroup for polarizing characters of the great ape and human clade. The main difference between this work and previous approaches was the inclusion of fossil taxa to polarize character states, instead of having to rely on highly specialized modern hominids.

This approach revealed evidence to suggest that many character states formerly thought to be primitive retentions of *Australopithecus* and the African apes are, in fact, character states linking chimpanzees to *Australopithecus*, and were never present in the common ancestor shared with gorillas (Begun, 1992a). In an expanded and more detailed analysis, (Begun, 1994a), 10 synapomorphies of the chimpanzee–human clade were retained, and only 2 were identified as potentially linking chimpanzees to gorillas (Table 1.3). More recent work including more characters and more fossil taxa continue to strongly support a chimpanzee–human clade (Begun and Kordos, 1997; Begun et al., 1997). Begun et al. (1997) list 260 characters, more than three times as many as in Begun (1994a), and more than twice the number listed in Groves and Patterson (1991). The chimpanzee–human clade is one of the most strongly supported clades in the most parsimonious cladogram of Begun et al. (1997). A substantial increase in homoplasy is required to link chimpanzees with gorillas to the exclusion of humans (Begun et al., 1997). Pilbeam (1996) has also recently concluded that the morphological and molecular data support this conclusion, based on his review of the molecular research and most of the morphological studies cited here.

HOW ARE AFRICAN APES AND HUMANS RELATED TO ONE ANOTHER? PART 2: ON GORILLAS AND ON THE MONOPHYLY OF *PAN*.

How many species of *Pan*?

Another issue within the systematics of the African apes and humans is the question of the relations between the species of the genus *Pan*. Recent molecular work indicates that the amount of genetic difference between bonobos and chimpanzees is very small, less than or equal to the average difference between the two subspecies of living orangutans (Groves, 1986, 1993; Ferris et al., 1981; Janczewski et al., Ruvolo, 1994), and, in at least one gene cluster, only about twice that within modern humans (Bailey et al., 1992). Molecular and chromosome data also unambiguously link bonobos to chimpanzees (Djian & Green, 1989; Caccone & Powell, 1989; Horai et al., 1992; Bailey et al., 1992; Ruvolo, 1994; Deeb et al., 1994; Morin, et al., 1994; Adachi & Hasegawa, 1995; Van der Kuyl et al., 1995; Borowik, 1995), as does the morphological data (Groves, 1986). Bonobos differ from chimpanzees in details of craniodental anatomy and proportions (Coolidge, 1933; Fenart & Deblock, 1973; Johanson, 1974; Cramer, 1977; Kinzey, 1984; Shea, 1984, 1985; Shea & Coolidge, 1988; Morbeck & Zihlman, 1989; Shea, Leigh, & Groves, 1993), limb proportions (Zihlman & Cramer, 1978; Zihlman, 1984; Morbeck & Zihlman, 1989), soft-tissue anatomy (Zihlman, 1984), and behavior (Wrangham, McGrew, De Waal, & Heltne, 1994 and chapters therein; Susman, 1984 and chapters therein), but not overall body size, despite their other name (pygmy chimpanzee) (Jungers & Susman, 1984; Morbeck and Zihlman, 1989). There seems little doubt that chimpanzees and bonobos are most closely related to one another, and that the living subspecies of chimpanzees are more closely related to one another than any is on its own to bonobos, thus justifying a species-level distinction between the two within the monophyletic clade *Pan*.

A number of studies have recently assessed variability in chimpanzees, particularly across subspecies boundaries. Morbeck and Zihlman (1989) have shown that, although variable, chimpanzees of the three generally recognized subspecies (*Pan troglodytes troglodytes*, *Pan troglodytes schweinfurthii*, and *Pan troglodytes verus*) can be distinguished from bonobos by a number of skeletal criteria. Shea et al. (1993) have also shown that variability in chimpanzee crania, though considerable, nevertheless permits subspecies-level identification based on cranial characteristics, and also shows a greater difference between *Pan paniscus* and *Pan troglodytes* than within either species. A number of genetic studies show substantial amounts of polymorphism within chimpanzees (Ferris, et al., 1981; Janczewski et al., 1990; Caccone & Powell, 1989; Kocher & Wilson, 1991; Ruano et al., 1992; Ruvolo, 1994; Morin et al., 1994; Arnason et al., 1996; and Nachman, et al., 1996), which has very recently called into question the taxonomic status of at least one subspecies of chimpanzee (Wrangham et al., 1994; Morin et al., 1994). Morin et al. (1994) found particularly high mtDNA differences between *Pan troglodytes verus* and the other two subspecies, equal to about two thirds of the difference between bonobos and chimpanzees. In their view, if confirmed by similar differences at other *loci*, *Pan troglodytes verus* should be elevated to the rank of a species. These results have been confirmed recently from the analysis of other portions of the *Pan troglodytes* mtDNA sequence (Arnason, Xiufeng, & Gullberg, 1996; Nachman et al., 1996), which strengthens the cases for a species-level distinction. Ruvolo (1994), however, did not find a link between *Pan troglodytes troglodytes* and *Pan troglodytes schweinfurthii* to the exclusion of *Pan troglodytes verus* using data from the mtDNA COII gene. Morin, et al. (1994) suggest that acceptance of a species-level difference should await confirmation from at least three unlinked nuclear *loci*, following the suggestion of Wu (1991) concerning the number of *loci* necessary to infer species phylogeny given the presence of ancient polymorphism. If *Pan troglodytes troglodytes* and *Pan troglodytes schweinfurthii* are, in fact, more closely related to each other than either is to *Pan troglodytes verus*, then either a species-level distinction for *Pan troglodytes verus* or the elimination of the subspecies-level distinction between *Pan troglodytes troglodytes* and *Pan troglodytes schweinfurthii* would be equally appropriate. Given the close genetic relations not only among all African apes and humans, but also within the genus *Pan*, especially compared to other primates (Ferris et al., 1981; Janczewski et al., 1990; Caccone & Powell, 1989), it is probably more advisable to recognize only two subspecies, *Pan troglodytes troglodytes* (which has priority over *Pan troglodytes schweinfurthii*) and *Pan troglodytes verus*. This also has the effect of communicating the closer evolutionary relations among the different types of chimpanzees compared to bonobos than would a taxonomy that included three species of the genus *Pan*.

How many species of *Gorilla*?

Many of the same studies cited above concerning variability in chimpanzees have also examined variability in gorillas. The most comprehensive analysis of anatomical variability and taxonomy in the genus *Gorilla* is still that of Groves (1970a, b). Groves recognizes a large number of skeletal and soft-tissue characters that distinguish among the three generally recognized subspecies of gorilla, *Gorilla gorilla gorilla*, *Gorilla gorilla beringei*, and *Gorilla gorilla graueri*. Although not so well studied to date, molecular data tend to confirm this taxonomy. Some anatomical and molecular evidence suggests a closer evolutionary relationship between the two eastern subspecies, *Gorilla gorilla beringei* and *Gorilla gorilla graueri* (Schultz, 1934; Vogel, 1961; Garner & Ryder, 1992; Ruvolo, 1994). Ruvolo (1994) found a greater degree of difference between *Gorilla gorilla gorilla* and the other two subspecies than exists between the two species of *Pan* in the COII mtDNA gene sequence. If this difference is confirmed at additional *loci*, and it is deemed advisable to recognize this pattern taxonomically, then the appropriate nomina would be *Gorilla gorilla* for the western lowland or coast gorilla, *Gorilla beringei beringei* for the mountain gorilla, and *Gorilla beringei graueri* for the eastern lowland gorilla. This modification, however, must await further analysis.

CONCLUSIONS ON THE TAXONOMY AND ANCESTRY OF THE GREAT APES.

A great deal of effort on the part of both molecular systematists and paleoanthropologists has been expended in an attempt to understand evolutionary relations among the great apes and humans. It is probably fair to say that this has been a difficult and sometimes contentious area of analysis, but we are finally approaching a consensus on many important issues. Given the studies cited above, a taxonomy of the living Hominoidea that many systematists and anthropologists could support would be as follows:

Hominoidea
- Hylobatidae
  - *Hylobates* species
- Hominidae
  - Ponginae
    - *Pongo pygmaeus*
  - Homininae
    - Hominini
      - *Pan troglodytes*
      - *Pan paniscus*
      - *Homo sapiens*
    - Gorillini
      - *Gorilla gorilla*

Although the inclusion of fossil taxa would require additional higher-level taxonomic categories, such as for example Hominina for the subtribe that includes humans and their fossil relatives, it may be advisable to limit these taxa. After all, although it is imperative that a classification accurately reflect phylogenetic relations as they are currently understood, there is no rule that requires every branch to be represented in a taxonomy by a nomen. In other words, we do not have to name every node in a cladogram. It may be wise, in order to avoid a proliferation of nomina, to recognize a limitation to the amount of information, even if exclusively phylogenetic, that can be included in a taxonomy.

The ancestral hominid morphotype

What do the evolutionary relations represented by this classification tell us about the origins of basic adaptations within the great apes and humans? It is likely that the ancestral hominid, as defined here, that is, the common ancestor of the great apes and humans, was a large arboreal quadruped with strongly developed forelimb suspensory adaptations and a predominantly frugivorous diet, based on the habits of most of its living members, and based on fossil evidence of early hominids (e.g., Begun, 1994a; Begun, et al., 1997). All known fossil and living great apes are larger than hylobatids, and it is very likely that the ancestral hominid was as well. Most fossil and living great apes are either known to be frugivorous (that is, lacking a specialized diet of insects, leaves, or meat) or have been reconstructed as such based on their anatomy and dental microwear (Kay, 1981; Teaford & Walker, 1984; Andrews & Martin, 1991; Begun, 1994a; Ungar & Kay, 1995; Kay & Ungar, 1997). All living apes have strongly developed forelimb suspensory adaptations, and many of the essential characters of this complex were present in fossil great apes as well (Straus, 1963; Rose, 1983, 1988, 1994, 1997; Sarmiento, 1987; Begun, 1992b, 1993, 1994a; Begun & Kordos, 1997; Begun, et al., 1997; Harrison & Rook, 1997; though see Benefit & McCrossin, 1995; McCrossin & Benefit, 1997 for a contrasting view). No fossil great ape was so specialized an arboreal quadruped as the quadrumanus orangutan, nor were any as terrestrial as living African apes. The ancestral hominid was therefore probably more generalized, but with a much stronger suspensory component than in most living generalized arboreal quadrupeds, such as Old World monkeys, which generally walk on branches rather than swinging below them. The ancestral hominid may have been more like large, more arboreal, and more suspensory living monkeys, such as *Nasalis*, though with a more highly developed suspensory forelimb, possibly related to their larger body size (larger arboreal quadrupeds are more stable below branches than atop them [Grand, 1978]). At any rate, the specializations of each lineage of living hominid appeared after their divergence from the common ancestor.

This hominid was probably also relatively encephalized and probably also showed some degree of maturational delay, as is typical of living forms, and as indicated in a few fossil forms as well. *Sivapithecus* shows evidence of maturational delay in dental eruption data (Kelley, 1997), while new evidence of brain size in *Dryopithecus* suggests a degree of encephalization (brain size relative to body size) typical of living great apes (Kordos & Begun, 1998). Since encephalization and rates of development are strongly correlated in primates (Smith, 1989, 1991; Martin, 1990), both these fossil taxa are likely to have shared both characters. In these fossil great apes, other aspects of life history related to development and brain size, such as gestation period, litter size, lifespan, age at sexual maturity, age at first reproduction, are also likely to have been most like that typical for living great apes. This is therefore probably also true of the ancestral hominid. Other important characters, such as sexual dimorphism and even the most basic aspects of social organization are not clear, given the diversity of these characters in living and fossil forms, and the unusual nature of both in the nearest outgroup, the gibbons and siamangs. As Parker (1996) notes, these social characteristics are probably more flexible and more likely to evolve rapidly than are the more conservative characters of life history. Social characteristics are thus more likely to characterize individual species, or even subspecies, rather than groups of related forms, because of the unique ecological circumstances of each population. Because of this, it is not possible to speculate with any degree of confidence about mating systems, group size, group composition, ranging behavior, or other detailed aspects of social organization in the common ancestor of the great apes and humans.

The ancestral hominine morphotype

As far as the African apes and humans are concerned, a bit more detail is possible in reconstructing the ancestral pattern of morphology and behavior. In a number of ways there was probably little change from the ancestral hominid morphotype. The ancestral hominine was probably also a forelimb dominated suspensory arboreal quadruped, but with a more significant terrestrial component to its locomotory repertoire. The African apes and humans share terrestriality as an important aspect of their positional behavior compared to other hominoids. The form of terrestriality characteristic of the ancestral hominine was most likely a type of knuckle-walking more or less indistinguishable from that characteristic of living African apes (Begun, 1994a). Pilbeam (1996) has recently reached the same conclusion based on his synthesis of molecular and morphological data. There are important differences among the African apes in their knuckle-walking techniques and frequencies (Doran & Hunt, 1994; Tuttle, 1967, 1969; Inouye, 1991, 1992), but most of this variation is related to body size. Gorillas are probably autapomorphic (uniquely derived) in their body sizes, judging from the body sizes of living *Pan*, *Homo*, and fossil humans, so that, in its details, the ancestral pattern of knuckle-walking may have more closely resembled that seen in modern chimpanzees. On the other hand, gorillas and humans are more terrestrial than chimpanzees and bonobos. All other things being equal, it could be argued that the ancestral pattern of positional behavior was more terrestrial, with chimpanzees and bonobos becoming secondarily more arboreal. However, it does seem more likely that gorillas have become more specialized in a number of ways, and that their similarities to humans are convergences (see below).

The ancestral hominini morphotype

It seems likely that the degree to which chimpanzees are distinguished from gorillas and orangutans in their use of tools, their dietary diversity, their habitat diversity, the complexity of their social relations, their hunting, and their levels of violence, are all traceable to a common ancestry with humans (Begun, 1994a). Parker (1996) notes that there are a number of cognitive similarities among great apes and humans that do not characterize other primates, and suggests that these were also present in the ancestor of the hominids (as defined here). Specifically, Parker (1996) considers intelligent tool use to have been part of the behavioral repertoire of the ancestral hominid, having been lost in orangutans and gorillas. Though both these taxa have been observed to use tools in the wild, this behavior is much rarer than in chimpanzees (Parker, 1996). However, gorillas and orangutans are proficient and intelligent tool users in captivity, as are ex-captive orangutans in the wild (Parker, 1996). This suggests to Parker (1996) that tool use is atavistic in these species. In other words, it is a capacity inherited from the common ancestor of the hominids but no longer commonly expressed in the wild in orangutans and gorillas.

The large brain and delayed development of the hominid ancestor (see above) certainly suggest similarities in cognition, learning, teaching, and intelligence to living hominids. However, I consider it significant that only chimpanzees and humans regularly and frequently make and use tools on their own. In my view, there is in essence a shared derived character of the intelligence of chimpanzees and humans that expresses itself in their spontaneous and naturalistic intelligent tool use. Orangutans, gorillas, and other primates lack this character state. It is well known that all great apes can be taught to express themselves using modes of communication that are much more complex than those that occur in the wild, and more like human language. Yet we would not conclude from the presence of these abilities in captive chimpanzees, gorillas, and orangutans that the common ancestor of hominids had language skills more like those of humans than living great apes. Intelligence is an incredibly complicated phenomenon about which we know almost nothing. The expression of various abilities, each somehow related to intelligence, under experimental or otherwise unusual circumstances, probably reflects the overwhelming complexity and sophistication of hominid intelligence rather than the presence of each of these abilities in the ancestor of the hominids.

In many ways, the common ancestor of chimpanzees and humans was very much like modern chimpanzees (Begun, 1994a). Chimpanzees are intermediate between other nonhuman hominids and humans in their dietary and habitat diversity and in their behavioral and social complexity. It is not difficult to imagine a relatively minor ecological circumstance that would send one population of chimpanzee-like hominini down the path leading to humans. A slightly greater dependance on terrestrial resources or even a slight increase in the ability to cross open country between patches of forest could lead to selection for bipedalism and megadontia, eventually transforming a chimpanzee-like ancestor into an early human.

While it is interesting and relevant to speculate on the origins of humans, an equally intriguing issue concerns the evolutionary history of chimpanzees as it compares to that of humans. Why have chimpanzees changed so little and humans so much? Why has there been so much more change genetically in the genus *Pan*, as exemplified by its chromosomal and nucleotide diversity, and yet so little change in morphology or behavior? It may be that living chimpanzees continue to inhabit ecological settings like that of their ancestors and ours. While some dietary diversification has occurred, leading to the divergence of chimpanzees and bonobos, the necessary changes have been much less drastic than those for which we have direct fossil evidence in humans. This comparative lack of speciation may contribute to the genetic diversity of chimpanzees. Without selection favoring small populations of chimpanzees with particular genetic characteristics, chimpanzee populations are able to accumulate genetic diversity by mutation. There is no fossil record of chimpanzees and bonobos that could help to answer this question. More information on ecological differences between chimpanzees and bonobos, genetic diversity in the genus *Pan*, and on the Plio-Pleistocene ecology of equatorial Africa should continue to provide important clues until fossil evidence to test these ideas is found.

The ancestral pongine and gorillini morphotypes

Fossil evidence from the sister clade of modern orangutans, *Sivapithecus*, suggests that orangutans may have more recently become more strongly or exclusively arboreal (Pilbeam et al., (1990). The specialized postcranium of orangutans may be a comparatively recent development that may reflect their much more restricted geographic distribution and possible ecological specialization compared to their Pleistocene relatives. Unfortunately, no postcranial remains of Pleistocene orangutans are yet known to test this idea. On the other hand, in many details the facial structure of orangutans has remained little changed, apart from the dentition, which has undergone significant modification since *Sivapithecus* (Brown & Ward, 1988). Overall, though, there is no strong evidence for a very dramatic shift in feeding patterns of dietary strategies in the orangutan lineage (Begun & Guleç, 1998).

Gorillas may be intermediate between chimpanzees and humans in the amount of evolutionary change their lineage has experienced since their divergence. All three genera of living hominines are primarily frugivorous, though gorillas are more dependent on folivorous resources and mountain gorillas are considered to be folivores (Schaller, 1963; Fossey, 1983; Tutin & Fernandez, 1993). This increase in folivory is probably a more recent development in these gorillas. Gorillas are also much larger than most fossil and living hominoids, and more terrestrial than all hominids except humans. Larger body size, increased folivory, and increased terrestriality are all interrelated in mammals in general and in gorillas in particular (Grand, 1978; Kay, 1984; Kay & Covert, 1984; Temerin, Wheatley, & Rodman, 1984). Gorillas can feed effectively on more fibrous food of higher volume relative to nutritional value because their large body sizes can accommodate very large digestive tracts. Because they are so large, gorillas must be more terrestrial than most other hominoids. Because this complex of interrelated characters is unique to gorillas, it is likely that they evolved together after gorillas separated from the chimpanzee–human clade. These changes allow gorillas to exploit resources insufficiently rich in nutrients to be used by chimpanzees or humans without expanding the dietary regimen.

There is currently no fossil evidence of the gorilla lineage, though two fossil forms have been associated with gorillas in the past. Dean and Delson (1992) suggested that *Ouranopithecus*, a fossil hominid from the late Miocene of Greece, may be related to gorillas, but this view has been challenged (Begun, 1995). Pilbeam (1986) and Andrews (1992) have suggested that *Samburupithecus*, a late Miocene hominoid from Samburu Hills, Kenya, may be related to gorillas, though Pilbeam (1996) has more recently concluded that it shares no synapomorphies with gorillas; a view I share. Gorillas probably evolved from a somewhat smaller, probably more arboreal, and probably more frugivorous ancestor than most modern gorillas. Though there is diversity in gorilla diet, the dentition of gorillas does suggest overall a more folivorous diet, a more recent specialization of this lineage as well. The degree of sexual dimorphism, which is quite high in gorillas, may be primitive for the great apes, given that orangutans and probably fossil humans are also quite sexually dimorphic. Reduction in this trait may have occurred independently in *Pan* and *Homo*. Gorillas may have taken the opposite path from that taken by humans. Rather than focusing on a more open country adaptive strategy, gorillas have evolved adaptations that allow them to exploit forest resources in a more focused or specialized manner than in the ancestral hominine.

The future

With the crystallizing consensus on relations within the Hominoidea, it appears that barriers to communication within and between disciplines are breaking down. Researchers can now turn their attention to testing ideas about the evolutionary histories of each hominid lineage, building on a framework provided by the cladistic analysis of character states and the resulting reconstruction of ancestral morphotypes. This process of hypothesis testing and refinement is already beginning, using new fossil discoveries, newly described nucleotide sequences, continued analysis of hominid comparative anatomy, and, hopefully, more detailed data on the behavior, ecology, and cognition of living hominids. More information on modern orangutans and gorillas is particularly important to complement the larger corpus of data that exists for chimpanzees. As we begin to know more about the evolutionary history of orangutans, gorillas, chimpanzees, bonobos, and humans, we will be in a better position to understand to what extent the details of the lives of each can be explained from their evolutionary history, and to what extent more immediate effects (disease, environmental degradation, interspecific competition, population genetics) are involved.

REFERENCES

Adachi, J. & Hasegawa, M. (1995). Improved dating of the human/chimpanzee separation in the mitochondrial DNA tree: heterogeneity among amino acid sites. *Journal of Molecular Evolution,* **40**, 622–628.

Aiello, L. C. & Wood, B. A. (1994). Cranial predictors of hominine body mass. *American Journal of Physical Anthropology,* **95**, 409–426.

Alpagut, B., Andrews, P., Fortelius, M., Kappelman, K., Temizsoy, I., & Lindsay, W. (1996). A new specimen of *Ankarapithecus meteai* from the Sinap formation of centralAnatolia. *Nature,* **382**, 349–351.

Andrews, P. (1985). Family group systematics and evolution among catarrhine primates. In E. Delson (ed.), *Ancestors: The hard evidence* (pp. 14–22). New York: Alan R. Liss.

Andrews, P. (1987). Aspects of hominoid phylogeny. In C. Patterson (ed.), *Molecules and morphology in evolution: Conflict or compromise* (pp. 23–53). Cambridge University Press.

Andrews, P. (1992). Evolution and environment in Miocene hominoids. *Nature,* **360**, 641–646.

Andrews, P. & Cronin, J. (1982). The relationships of *Sivapithecus* and *Ramapithecus* and the evolution of the orang-utan. *Nature,* **297**, 541–546.

Andrews, P. & Martin, L. (1987). Cladistic relationships of extant and fossil hominoids. *Journal of Human Evolution,* **16**, 101–118.

Andrews P. & Martin L. (1991). Hominoid dietary evolution. *Philosophical Transactions of the Royal Society London*, **B 334**, 199–209.

Andrews, P. & Pilbeam, D. (1996). The nature of the evidence. *Nature,* **379**, 123–124.

Arnason, U., Xiufeng, X., & Gullberg, A. (1996). Comparison between the complete mitochondrial DNA sequence of *Homo* and the common chimpanzee based on nonchimeric sequences. *Journal of Molecular Evolution,* **42**, 145–152.

Bailey, W. J., Hayasaka, K., Skinner, C. G., Kehoe, S., Sieu, L. C., Slightom, J. L., & Goodman, M. (1992). Reexamination of the African hominoid trichotomy with the additional sequences from the primate beta-globin gene cluster. *Molecular Phylogenetics and Evolution,* **1**, 97–135.

Begun, D. R. (1992a). Miocene fossil hominids and the chimp-human clade. *Science,* **247**, 1929–1933.

Begun D. R. (1992b). Phyletic diversity and locomotion in primitive European hominids. *American Journal of Physical Anthropology,* **87**, 311–340.

Begun D. R. (1993). New catarrhine phalanges fron Rudabánya (Northeastern Hungary) and the problem of parallelism and convergence in hominoid postcranial morphology. *Journal of Human Evolution,* **24**, 373–402.

Begun, D. R. (1994a). Relations among the great apes and humans: New interpretations based on the fossil great ape *Dryopithecus*. *Yearbook of Physical Anthropology*, **37**, 11–63.

Begun, D. R. (1994b). The significance of *Otavipithecus namibiensis* to interpretations of hominoid evolution. *Journal of Human Evolution*, **27**, 385–394.

Begun, D. R. (1995). Late Miocene European orang-utans, gorillas, humans, or none of the above. *Journal of Human Evolution*, **29**, 169–180.

Begun, D. R. & Güleç, E. (1998). Restoration of the type and palate of *Ankarapithecus meteai*: taxonomic, phylogenetic, and functional implications. *American Journal of Physical Anthropology*, **105**, 279–314.

Begun D. R. & Kordos L. (1997). Phyletic affinities and functional convergence in *Dryopithecus* and other Miocene living hominids. In D. R. Begun, C. V. Ward, & M. D. Rose (eds.), *Function, phylogeny and fossils: Miocene hominoid origins and adaptations* (pp. 294–316). New York: Plenum.

Begun D. R., Ward C. V., & Rose M. D. (1997). Events in hominoid evolution. In D. R. Begun, C. V. Ward, & M. D. Rose (eds.), *Function, phylogeny and fossils: Miocene hominoid evolution and adaptations* (pp. 389–415). New York: Plenum Publishing Co.

Benefit B. R. & McCrossin M. L. (1995). Miocene hominoids and hominid origins. *Annual Review of Anthropology*, **24**, 237–256.

Beynon, A. D., Dean, M. C., & Reid, D. J. (1991). On thick and thin enamel in hominoids. *American Journal of Physical Anthropology*, **86**, 295–309.

Borowik, O. A. (1995). Coding chromosomal data for phylogenetic analysis: Phylogenetic resolution of the *Pan–Homo–Gorilla* trichotomy. *Systematic Biology*, **44**, 563–570.

Bromage, T. G. & Schrenk, F. (1995). Biogeographic and climatic basis for a narrative of early hominid evolution. *Journal of Human Evolution*, **28**, 109–114.

Brooks, D. R. & McClennan, D. A. (1991). *Phylogeny, ecology, and behaviour*. University of Chicago Press.

Brown, B. & Ward, S. (1988). Basicranial and facial topography in *Pongo* and *Sivapithecus*. In *Orang-utan Biology*, ed. J. H. Schwartz, (pp. 247–260). New York: Oxford University Press.

Brown, W. M., Prager, E. M., Wang, A., & Wilson, A. C. (1982). Mitochondrial DNA sequences of primates: tempo and mode of evolution. *Journal of Molecular Evolution*, **18**, 225–239.

Bruce, E. J. & Ayala, F. J. (1979). Phylogenetic relationships between man and the apes: electrophoretic evidence. *Evolution*, **33**, 1040–1056.

Caccone, A. & Powell, J. R. (1989). DNA Divergence among hominoids. *Evolution*, **43**, 925–942.

Cande, S. C. & Kent, D. V. (1995). Revised calibration of the geomagnetic polarity time scale for the late Cretaceous and Cenozoic. *Journal of Geophysical Research*, **100**, 6093–6095.

Cartmill, M. & Yoder, A. D. (1994). Molecules and morphology in primate systematics: an introduction. *American Journal of Physical Anthropology*, **94**, 1.

Chang, L.-Y. E. & Slightom, J. (1984). Isolation and nucleotide sequence analysis of the beta-type globin pseudogene from human, gorilla and chimpanzee. *Journal of Molecular Biology*, **180**, 767–784.

Ciochon, R. L. (1983). Hominoid cladistics and the ancestry of modern apes and humans. A summary statement. In R. L. Ciochon & R. S. Corruccini (eds.), *New interpretations of ape and human ancestry* (pp. 783–843). New York: Plenum Press.

Coolidge, H. J. (1933). Pan paniscus: Pygmy chimpanzee from south of the Congo. *American Journal of Physical Anthropology*, **18**, 1–57.

Courtney, J., Groves, C., & Andrews, P. (1988). Inter- and intra-island variation? An assessment of the differences between Bornean and Sumatran Orang-utans. In J. H. Schwartz (ed.), *Orang-utan biology* (pp. 19–29). New York: Oxford University Press.

Cramer, D. L. (1977). Craniofacial morphology of *Pan paniscus*. In F. S. Salzay (ed.), *Contributions to primatology* (pp. 1–64). New York: Karger.

Dart, R. A. (1925). *Australopithecus africanus*: the man-ape of South Africa. *Nature,* **115**, 195–199.

Darwin, C. (1859). *On the origin of species by means of natural selection or the preservation of favored races in the struggle for life*. London: Murray.

Darwin, C. (1871). *The descent of man*. London: Murray.

De Bonis, L. (1983). Phyletic relationships of Miocene hominoids and higher primate classification. In R. L. Ciochon & R. S. Corruccini (eds.), *New Interpretation of Ape and Human Ancestry* (pp. 625–649). New York: Plenum Press.

Dean, D. & Delson, E. (1992). Second gorilla or third chimp? *Nature,* **359**, 676–677.

Deeb, S. S., Jorgensen, A. L., Battisti L., Iwasaki L., & Motulsky A. G. (1994). Sequence divergence of the red and green visual pigments in great apes and humans. *Proceedings of the National Academy of Sciences,* **91**, 7262–7266.

Delson, E. & Andrews, P. (1975). Evolution and interrelationships of the catarrhine primates. In F. S. Szalay & W. P. Luckett (eds.), *Phylogeny of the primates* (pp. 405–445), New York: Plenum Publishing Corporation.

Djian, P. & Green, H. (1989). Vectorial expansion of the involucrin gene and the relatedness of the hominoids. *Proceedings of the National Academy of Sciences,* **86**, 8447–8451.

Doran, D. M. & Hunt, K. D. (1994). Comparative locomotor behavior of chimpanzees and bonobos: species and habitat differences. In R. W. Wrangham, W. C. McGrew, F. B. M. de Waal & P. G. Heltne (eds.), *Chimpanzee cultures* (pp. 93–107). Cambridge, MA: Harvard University Press.

Dorit, R. L., Akashi, H., & Gilbert, W. (1995). Absence of polymorphism at the ZFY locus on the human Y chromosome. *Science,* **268**, 1183–1185.

Easteal, S., Collet, C., & Betty, D. (1995). *The mammalian molecular clock.* Austin: R. G. Landes Company.

Ellis, N., Yen, P., Neiwanger, K., Shapiro, L. J., & Goodfellow, P. N. (1990). Evolution of the pseudoautosomal boundary in old world monkeys and great apes. *Cell,* **63**, 977–986.

Ellsworth, D. L., Hewett-Emmett, D., & Li, W. H. (1993). Insulin-like growth factor II intron sequences support the hominoid rate-slowdown hypothesis. *Molecular Phylogenetics and Evolution,* **2**, 315–321.

Fenart, R. & Deblock, R. (1973). *Pan paniscus-Pan troglodytes* Craniometric. etude Comparative et ontogénetique selon les méthodes classiques et vestibulaire. *Annales du Musee Royal de L'Afrique Centrale,* **Serie IN-8, Sciences Zoologiques**.

Ferris, S. D., Brown, W. M., Davidson, W. S., & Wilson, A. C. (1981). Evolutionary tree for apes and humans based on cleavage maps of mitochondrial DNA. *Proceedings of the National Academy of Sciences,* **78**, 6319–6323.

Fleagle, J. G. (1988). *Primate adaptation and evolution*. New York: Academic Press.

Fossey, D. (1983). *Gorillas in the mist*. Boston: Houghton Mifflin.

Galili, U. & Swanson, K. (1991). Gene sequences suggest inactivation of a-1,3–galactosyltransferase in catarrhines after the divergence of apes from monkeys. *Proceedings of the National Academy of Sciences,* **88**, 7401–7404.

Garner, K. J. & Ryder, O. A. (1992). Some applications of PCR to studies in wildlife genetics. *Symposium of the Zoological Society of London,* **64**, 167–181.

Gonzalez, I. L., Sylvester, J. E., Smith, T. F., Stambolian, D., & Schmickel, R. D. (1990). Ribosomal RNA gene sequences and hominoid phylogeny. *Molecular Biology and Evolution,* **7**, 203–219.

Goodman, M. (1963). Man's place in the phylogeny of the primates as reflected in serum proteins. In S. L. Washburn (ed.), *Classification and human evolution* (pp. 204–234). Chicago: Aldine Press.

Goodman, M., Braunitzer, G., Stangl, Z., & Schrank, B. (1983). Evidence on human origin from haemoglobins of African apes. *Nature,* **303**, 546–548.

Goodman, M., Koop, B. F., Czelusniak, J., & Weiss, M. L. (1984). The η-globin gene. Its long evolutionary history in the β-globin gene family of mammals. *Journal of Molecular Biology,* **180**, 803–823.

Goodman, M., Bailey, W. J., Hayasaka, K., Stanhope, M. J., Slightom, J., & Czelusniak, J. (1994). Molecular evidence on primate phylogeny from DNA sequences. *American Journal of Physical Anthropology,* **94**, 3–24.

Goodman, M., Koop, B. F., Czelusniak, J., Fitch, D. H. A., Tagel, D. A., & Slightom, J. L. (1989). Molecular phylogeny of the family of apes and humans. *Genome,* **31**, 316–335.

Gould, S. J. (1985). A clock of evolution. *Natural History,* **4**, 12–25.

Grand T. I. (1978). Adaptations of tissue and limb segments to facilitate moving and feeding in arboreal folivores. In G. G. Montgomery (ed.), *The ecology of arboreal folovores* (pp. 231–241). Washington, DC: Smithsonian Institution Press.

Green, H. & Djian, P. (1995). The involucrin gene and hominoid relationships. *American Journal of Physical Anthropology,* **98**, 213–216.

Groves, C. P. (1970a). Population systematics of the gorilla. *Journal of the Zoological Society (London),* **161**, 287–300.

Groves, C. P. (1970b). *Gorillas.* London: Arthur Baker.

Groves, C. P. (1986). Systematics of the great apes. In D. W. Swindler & J. Erwin (eds.), *Comparative primate biology* (pp. 187–217). New York: Alan R. Liss.

Groves, C. P. (1989). *A theory of primate and human evolution.* Oxford: Clarendon Press.

Groves, C. P. (1993). Speciation in living hominoid primates. In W. H. Kimbel & L. B. Martin (eds.), *Species, species concepts, and primate evolution* (pp. 109–121). New York: Plenum Press.

Groves, C. P. & Patterson, J. D. (1991). Testing hominoid phylogeny with the PHYLIP programs. *Journal of Human Evolution,* **20**, 167–183.

Harrison T. & Rook L. (1997). Enigmatic anthropoid or misunderstood ape: The phylogenetic status of *Oreopithecus bambolii* reconsidered. In D. R. Begun, C. V. Ward & M. D. Rose (eds.), *Function, phylogeny and fossils: miocene hominoid origins and adaptations* (pp. 327–362). New York: Plenum Press.

Hasegawa, M., Kishino, H., & Yano, T.-a. (1987). Man's place in Hominoidea as inferred from molecular clocks of DNA. *Journal of Molecular Evolution,* **26**, 132–147.

Hasegawa, M., Kishino, H., Hayasaka, K., & Horai, S. (1990). Mitochondrial DNA evolution in primates: transition rate has been extremely low in lemur. *Journal of Molecular Evolution,* **31**, 113–121.

Hayasaka, K., Gojobori, T., & Horai, S. (1988). Molecular phylogeny and evolution of primate mitochondrial DNA. *Molecular Biology and Evolution,* **5**, 626–644.

Hayashi, S., Hayasaka, K., Takenaka, O., & Horai, S. (1995). Molecular phylogeny of gibbons inferred from mitochondrial DNA sequences: Preliminary report. *Journal of Molecular Evolution,* **41**, 359–365.

Hennig, W. (1966). *Phlogenetic systematics.* Chicago: University of Illinois Press.

Hixon, J. E. & Brown, W. M. (1986). A comparison of the small ribosomal RNA genes from the mitochondrial DNA of great apes and humans: sequence, structure, evolution, and phylogenetic implications. *Molecular Biology and Evolution,* **3**, 1–18.

Holmquist, R., Miyamoto, M. M., & Goodman, M. (1988). Higher primate phylogeny. Why can't we decide? *Molecular Biology and Evolution,* **5**, 201–216.

Horai, S., Hayasaka, K., Kondo, R., Tsugane, K., & Takahata, N. (1995). Recent African origin of modern humans revealed by complete sequences of hominoid mitochondrial DNAs. *Proceedings of the National Academy of Sciences,* **92**, 532–536.

Horai, S., Hayasaka, K., Kondo, R., Inoue, T., Ishida, T., Hayashi, S., & Takahata, N. (1992). Man's place in Hominoidea revealed by mitochondrial DNA geneology. *Journal of Molecular Evolution,* **34**, 32–43.

Hull, D. L. (1979). The limits of cladism. *Systematic Zoology,* **28**, 416–440.

Huxley, T. H. (1959). *Man's place in nature*. Ann Arbor: University of Michigan Press.

Inouye, S. E. (1991). Ontogeny and allometry in African ape fingers. In A. Ehara, T. Kimura, O. Takenaka & M. Iwamoto (eds.), *Primatology yoday* (pp. 537–538). New York: Elsevier Science Publishers.

Inouye, S. E. (1992). Ontogeny and allometry of African ape manual rays. *Journal of Human Evolution,* **26**, 459–485.

Janczewski, D. N., Goldman, D., & O'Brien, S. J. (1990). Molecular divergence of Orang Utan (*Pongon pygmaeus*) subspecies based on isozyme and two-dimensional gel electrophoresis. *Journal of Heredity,* **81**, 375–387.

Johanson, D. C. (1974). Some metric aspects of the permanent and deciduous dentition of the pygmy chimpanzee. *American Journal of Physical Anthropology,* **41**, 39–48.

Jungers, W. L. & Susman, R. L. (1984). Body size and skeletal allometry in African apes. In ed. R. L. Susman (ed.), *The pygmy chimpanzee: Evolutionary biology and behavior* (pp. 131–177). New York: Plenum Press.

Kappelman, J., Kelley, J., Pilbeam, D., Sheikh, K. A., Ward, S., Anwar, M., Barry, J. C., Brown, B., Hake, P., Johnson, N. M., Raza, S. M., & Shah, S. M. I. (1991). The earliest occurrence of *Sivapithecus* from the middle Miocene Chinji Formation of Pakistan. *Jouranl of Human Evolution,* **21**, 61–73.

Kawaguchi, H., O'hUigin, C., & Klein, J. (1992). Evolutionary origin of mutations in the primate cytochrome P450c21 gene. *American Journal of Human Genetics,* **50**, 766–780.

Kawamura, S., Tanabe, H., Watanabe, Y., Kurosaki, K., Saitou, N., & Ueda, S. (1991). Evolutionary rate of immunoglobin alpha noncoding region is greater in hominoids than in old world monkeys. *Molecular Biology and Evolution,* **8**, 743–752.

Kay R. F. (1981). The nut-crackers – a theory of the adaptaitions of the *Ramapithecinae*. *American Journal of Physical Anthropology,* **55**, 141–151.

Kay R. F. (1984). On the use of anatomical features to infer foraging behaviour in extinct primates. In P. S. Rodman & J. G. H. Cant (eds.), *Adaptations for foraging in primates* (pp. 21–53). New York: Columbia University Press.

Kay R. F. & Covert H. H. (1984). Anatomy and behavior of extinct primates. In D. J. Chivers, B. A. Wood & A. Bilsborough (eds.), *Food aquisition and processing in primates* (pp. 467–508). Cambridge University Press.

Kay R. F. & Ungar P. S. (1997). Dental evidence for diet in some Miocene catarrhines with comments on the effects of phylogeny on the interpretation of adaptation. In D. R. Begun, C. V. Ward, & M. D. Rose (eds.), *Function, phylogeny and fossils: Miocene hominoid evolution and adaptations* (pp. 131–151). New York: Plenum Press.

Keith, A. (1931). *New discoveries relating to the antiquity of man*. London: Williams and Norgate.

Kelley, J. (1997). Paleobiological significance of life history in miocene hominoids. In D. R. Begun, C. V. Ward, & M. D. Rose (eds.), *Function, phylogeny and fossils: Miocene hominoid evolution and adaptations* (pp. 173–208). New York: Plenum Press.

Kim, H.-S. & Takenaka, O. (1996). A comparison of TSPY genes from Y-chromosomeal DNA of the great apes and humans: sequence, evolution and phylogeny. *American Journal of Physical Anthropology,* **100**, 301–309.

King, M.-C. & Wilson, A. C. (1975). Evolution at two levels in humans and chimpanzees. *Science,* **188**, 107–113.

Kinzey, W. (1984). The dentition of the pygmy chimpanzee (*Pan paniscus*). In R. L. Susman (ed.), *The pygmy chimpanzee. Evolutionary biology and behaviour* (pp. 68–87). New York: Plenum Press.

Kluge, A. G. (1983). Cladistics and the classifiacation of the great apes. In R. L. Ciochon & R. S. Corruccini (eds.), *New interpretations of ape and human ancestry* (pp. 151–177). New York: Academic Press.

Kocher, T. D. & Wilson, A. C. (1991). Sequence evolution of mitochondrial DNA in humans and chimpanzees: control region and a protein coding region. In S. Osawa & T. Honjo (eds.), *Evolution of life: fossils, molecules and culture* (pp. 391–413). Tokyo: Springer-Verlag.

Kordos, L. and Begun, D. (1998) Encaphalization and endocranial morphology in *Dryopithecus brancoi*: Implications for brain evolution in early hominids. *American Journal of Physical Anthropology supplement*, **26**, 141–142.

Le Gros Clark, W. E. (1934). *Early forerunners of man*. Baltimore: William and Wood.

Leakey, M. G., Feibel, C. S., McDougall, I., & Walker, A. (1995). New four-million-year-old hominid species from Kanapoi and Allia Bay, Kenya. *Nature*, **376**, 565–570.

Lewis, O. J. (1973). The hominid *os capitatum*, with special reference to the fossil bones from Sterkfontein and Olduvai Gorge. *Journal of Human Evolution*, **2**, 1–13.

Li, W.-H. & Tanimura, M. (1987). The molecular clock runs more slowly in man than in apes and monkeys. *Nature*, **326**, 93–96.

Lieberman, D. E., Wood, B. A., & Pilbeam, D. R. (1996). Homoplasy and early *Homo*: an analysis of the evolutionary relationships of *H. habilis senso stricto* and *H. rudolfensis*. *Journal of Human Evolution*, **30**, 97–120.

Livak, K. J., Rogers, J., & Lichter, J. B. (1995). Variability of dopamine D4 receptor (DRD4) gene sequence within and among nonhuman primates species. *Proceedings of the National Academy of Sciences*, **92**, 427–431.

Lucotte, G. & Smith, D. G. (1982). Distance électrophoretiques entre l'homme, le chimpanzé (*Pan troglodytes*) et le gorille (*Gorilla gorilla*) basées sur la mobilité de enzymes érythrocytaires. *Human Genetics*, **60**, 16–18.

Maeda, N., Wu, C.-I., Bliska, J., & Reneke, J. (1988). Molecular evolution of intergenic DNA in higher primates: pattern of DNA changes, molecular clock, and evolution of repetitive sequences. *Molecular Biology and Evolution*, **5**, 1–20.

Marks, J. (1992). Genetic relationships among the apes and humans. *Current Opinion in Genetics and Development*, **2**, 883–889.

Marks, J. (1994). Blood will tell (won't it?): a century of molecular discourse in anthropological systematics. *American Journal of Physical Anthropology*, **94**, 59–79.

Marks, J., Schmid, C. W., & Sarich, V. M. (1988). DNA hybridization as a guide to phylogeny: relations of the Hominoidea. *Journal of Human Evolution*, **17**, 769–786.

Martin, L. (1985). Significance of enamel thickness in hominoid evolution. *Nature*, **314**, 260–263.

Martin, R. D. (1990). *Primate origins and evolution*. Princeton University Press.

McCrossin M. L. & Benefit B. R. (1997). On the relationships and adaptations of *Kenyapithecus*, a large-bodied hominoid from the middle Miocene of eastern Africa. In D. R. Begun, C. V. Ward, & M. D. Rose (eds.), *Function, phylogeny and fossils: Miocene hominoid origins and adaptations* (pp. 241–267). New York: Plenum Press.

Meneveri, R., Agrest, A., Rocchi, M., Marozzi, A., & Ginelli, E. (1995). Analysis of GC-repetitive nucleotide sequences in great apes. *Journal of Molecular Evolution*, **40**, 405–412.

Miyamoto, M., Slightom, J. L., & Goodman, M. (1987). Phylogenetic relationships of humans and African apes as ascertained from DNA sequences. (7.1 kilobase pairs) of the psi eta-globin region. *Science*, **238**, 369–373.

Mohammad-Ali, K., Eladari, M.-E., & Galibert, F. (1995). Gorilla and orangutan c-myc nucleotide sequences: inference on hominoid phylogeny. *Journal of Molecular Evolution,* **41**, 262–276.

Morbeck, M. E. & Zihlman, A. L. (1989). Body size and proportions in chimpanzees, with special reference to *Pan troglodytes schweinfurthii* from Gombe National Park, Tanzania. *Primates,* **30**, 369–382.

Morin, P. A., Moore, J. J., Chakraborty, R., Jin, L., Goodall, J., & Woodruff, D. S. (1994). Kin selection, social structure, gene flow, and the evolution of chimpanzees. *Science,* **265**, 1193–1202.

Nachman, M. W., Brown, W. M., Stoneking, M., & Aquadro, C. F. (1996). Nonneutral mitochondrial DNA variation in humans and chimpnazees. *Genetics,* **142**, 953–963.

Nutall, G. (1904). *Blood, immunity, and blood relationships.* London: Cambridge University Press.

Oetting, W. S., Stine, O. C., Townsend, D., & King, R. A. (1993). Evolution of the Tryosinae related gene (TYRL) in primates. *Pigmental Res.,* **6**, 171–177.

Pamilo, P. & Nei, N. (1988). Relationships between gene trees and species trees. *Molecular Biology and Evolution,* **5**, 568–583.

Parker S. T. (1996). Using cladistic analysis of comparative data to reconstruct the evolution of cognitive development in hominids. In E. P. Martins (ed.), *Phylogenies and the comparative method in animal behavior* (pp. 361–398). New York: Oxford University Press.

Patterson, C. (ed.) (1987). *Molecules and morphology in evolution: Conflict or compromise?* Cambridge University Press.

Perrin-Pecontal, P., Gouy, M., Nignon, V.-M., & Trabuchet, G. (1992). Evolution of the primate β-globin gene region: Nucleotide sequence of the δ-β-globin intergenic region of *Gorilla* and phylogenetic relationships between African apes and man. *Journal of Molecular Evolution,* **34**, 17–30.

Pilbeam, D. R. (1969). Teriary Pongidae of east Africa: Evolutionary relationships and taxonomy. *Bulletin of the Peabody Museum of Natural History,* **31**, 1–185.

Pilbeam, D. R. (1982). New hominoid skull material from the Miocene of Pakistan. *Nature,* **295**, 232–234.

Pilbeam, D. R. (1986). Hominoid evolution and hominoid origins. *American Anthropologist,* **88**, 295–312.

Pilbeam, D. (1996). Genetic and morphological records of the Hominoidea and hominid origins: a synthesis. *Molecular Phylogenetics and Evolution,* **5**, 155–168.

Pilbeam, D. R., Rose M. D., Barry J. C., & Shah S. M. I. (1990). New *Sivapithecus* humeri from Pakistan and the relationship of *Sivapithecus* and *Pongo. Nature,* **384**, 237–239.

Queralt, R., Adroer, R., Oliva, R., Winkfein, R. J., & Dixon, G. H. (1995). Evolution of protoamine P1 genes in mammals. *Journal of Molecular Evolution,* **40**, 601–607.

Retief, J. D., Winkfein, R. J., Dixon, G. H., Adroer, R., Querlt, R., & Oliva, R. (1993). Evolution of protoamine P1 genes in primates. *Journal of Molecular Evolution,* **37**, 426–434.

Rogers, J. (1993). The phylogenetic relationships among *Homo*, *Pan*, and *Gorilla*: a population genetics perspective. *Journal of Human Evolution,* **25**, 201–215.

Rogers, J. (1994). Levels of genealogical hierarchy and the problem of hominoid phylogeny. *American Journal of Physical Anthropology,* **94**, 81–88.

Rogers, J. & Comuzzie, A. G. (1995). When is ancient polymorphism a potential problem for molecular phylogenetics? *American Journal of Physical Anthropology,* **98**, 216–217.

Röhrer-Ertl, O. (1988). Research history, nomenclature, and taxonomy of the orang-utan. In J. H. Schwartz (ed.), *Orang-utan biology* (pp. 7–18). New York: Oxford University Press.

Rose, M. D. (1983). Miocene hominoid postcranial morphology: monkey-like, ape-like, neither, or both? In R. L. Ciochon & R. S. Corruccini (eds.), *New interpretations of ape and human ancestry* (pp. 405–417). New York: Plenum Press.

Rose M. D. (1988). Another look at the anthropoid elbow. *Journal of Human Evolution,* **17**, 193–224.

Rose M. D. (1994). Quadrupedalism in some Miocene catarrhines. *Journal of Human Evolution,* **26**, 387–411.

Rose M. D. (1997). Functional and phylogenetic features of the forelimb in Miocene hominoids. In D. R. Begun, C. V. Ward, & M. D. Rose (eds.), *Function, phylogeny and fossils: Miocene hominoid evolution and adaptations* (pp. 79–100). New York: Plenum Press.

Ruano, G., Rogers, J., Ferguson-Smith, A., & Kidd, K. K. (1992). DNA sequence polymorphisms within hominoid species exceeds the number of phylogenetically informative characters for a HOX2 locus. *Molecular Biology and Evolution,* **9**, 575–586.

Ruvolo, M. (1994). Molecular evolutionary processes and conflicting gene trees: the hominoid case. *American Journal of Physical Anthropology,* **94**, 89–113.

Ruvolo, M. (1995). Seeing the forest and the trees: replies to Marks; Rogers and Comuzzie; Green and Djian. *American Journal of Physical Anthropology,* **98**, 218–232.

Ruvolo, M., Disotell, T. R., Allard, M. W., Brown, W. M., & Honeycutt, R. L. (1991). Resolution of the African hominoid trichotomy by use of a mitochondrial gene sequence. *Proceedings of the National Academy of Sciences. USA,* **88**, 1570–1574.

Ruvolo, M., Pan, D., Zehr, S., Goldberg, T., Disotell, T. R., & von Dornum, M. (1994). Gene trees and hominoid phylogeny. *Proceedings of the National academy of Sciences,* **91**, 8900–8904.

Sarich, V. M. & Wilson A. C. (1967). Immunological time-scale for hominoid evolution. *Science,* **158**, 1200–1203.

Sarmiento S. (1987). The phylogenetic position of *Oreopithecus* and its significance in the origin of the Hominoidea. *American Museum Novitates,* **2881**, 1–44.

Savatier, P., Trabuchet, G., Cheblune, Y., Faure, C., & Verdier, G. (1987). Nucleotide sequence of the beta-globin genes in *Gorilla* and macaque: The origin of nucleotide polymorphisms in humans. *Journal of Molecular Evolution,* **24**, 309–318.

Schaller, G. B. (1963). *The mountain gorilla: Ecology and behavior.* University of Chicago Press.

Schultz, A. C. (1934). Some distinguishing characters of the mountain gorilla. *Journal of Mammalogy,* **15**, 51–61.

Schultz, A. H. (1963). Age changes, sex differences, and variability as factors in the classification of primates. In S. L. Washburn (ed.), *Classification and human evolution* (pp. 85–115). Chicago: Aldine.

Schwartz, J. H. (1984). Hominoid evolution: A review and a reassessment. *Current Anthropology,* **25**, 655–672.

Schwartz, J. H. (1990). *Lufengpithecus* and its poteneial relationship to an orang-utan clade. *Journal of Human Evolution,* **19**, 591–605.

Shea, B. T. (1984). An allometric perspective on the morphological and evolutionary relationships between pygmy (*Pan paniscus*) and common (*Pan troglodytes*) chimpanzees. In R. L. Susman (ed.), *The pygmy chimpanzee: Evolutionary biology and behavior*, (pp. 89–130). New York: Plenum Press.

Shea, B. T. (1985). Bivariate and multivariate growth allometry: statistical and biological considerations. *Journal of Zoology (London),* **206**, 367–390.

Shea, B. T. & Coolidge, N. J. J. (1988). Craniometric differentiation and systematics in the genus *Pan. Journal of Human Evolution,* **17**, 661–685.

Shea, B. T., Leigh, S. R., & Groves, C. P. (1993). Multivariate craniometric variation in chimpanzees. Implications for species identification in paleoanthropology. In W. H. Kimbel, & L. B. Martin (eds.), *Species, species concepts, and primate evolution* (pp. 265–296). New York: Plenum Press.

Shoshani, J., Groves, C. P., Simons, E. L., & Gunnel, G. F. (1996). Primate phylogeny: Morphological vs. molecular results. *Molecular phylogenetics and Evolution,* **5**, 101–153.

Sibley, C. G. & Alquist, J. E. (1984). The phylogeny of the hominoid primates as indicated by DNA-DNA hybridization. *Journal of Molecular Evolution,* **20**, 2–15.

Sibley, C. G. & Alquist, J. E. (1987). DNA hybridization evidence of hominoid phylogeny: Results from an expanded data set. *Journal of Molecular Evolution*, **26**, 99–121.

Sibley, C. G., Constock, J. A., & Alquist, J. E. (1990). DNA hybridization evidence of hominoid phylogeny: a reanalysis of the data. *Journal of Molecular Evolution*, **26**, 202–236.

Simpson, G. G. (1945). The principles of classification and a classification of mammals. *Bulletin of the American Museum of Natural History*, **85**, 1–350.

Simpson, G. G. (1961). *Principles of animal classification*. New York: Columbia University Press.

Simpson, G. G. (1963). The meaning of taxonomic statements. In S. L. Washburn (ed.), *Classification and human evolution* (pp. 1–35). Chicago: Aldine.

Slightom, J. L., Chang, L.-Y. E., Koop, B. F., & Goodman, M. (1985). Chimpanzee fetal $^{G}\gamma$ and $^{A}\gamma$ globin gene nucleotide sequences provide further evidence of gene conversions in hominine evolution. *Molecular Biology and Evolution*, **2**, 370–389.

Smith B. H. (1989). Dental development as a measure of life history in primates. *Evolution*, **43**, 683–688.

Smith B. H. (1991). Dental development and the evolution of life history in the Hominidae. *American Journal of Physical Anthropology*, **86**, 157–174.

Smouse, P. E. & Li, W-H. (1987). Likelihood analysis of mitochondrial restriction-cleavage patterns for the chimpanzee–human–gorilla trichotomy. *Evolution*, **41**, 1162–1176.

Stewart, C. B. (1993). The powers and pitfalls of parsimony. *Nature*, **361**, 603–607.

Straus W. L. (1963). The classification of *Oreopithecus*. In S. L. Washburn (ed.), *Classification and human evolution*, (pp. 146–177). Chicago: Aldine.

Susman, R. L. (ed.) (1984). *The pygmy chimpanzee: Evolutionary biology and behavior*. New York: Plenum Press, pp. 131–177.

Szalay, F. & Delson, E. (1979). *Evolutionary history of the primates*. New York: Academic Press.

Takahata, N. (1989). Gene geneology in three related populations: consistency probability between gene and population trees. *Genetics*, **122**, 957–966.

Tattersall, I., Delson, E., & Van Couvering, J. (1988). *Encyclopedia of Human Evolution and Prehistory*. New York: Garland.

Teaford M. F. & Walker A. C. (1984). Quantitative differences in the dental microwear between primates with different diets and a comment on the presumed diet of *Sivapithecus*. *American Journal of Physical Anthropology*, **64**, 191–200.

Temerin B., Wheatley P., & Rodman P. S. (1984). Body size and foraging in primates. In P. S. Rodman & J. G. H. Cant (eds.), *Adaptations for foraging in primates* (pp. 217–248). New York: Columbia University Press.

Tutin, C. E. G. & Fernandez, M. (1993). Composition of the diet of chimpanzees and comparisons with that of sympatric lowland gorillas in the Lope Reserve, Gabon. *American Journal of Primatology*, **30**, 195–211.

Tuttle, R. H. (1967). Knuckle-walking and the evolution of hominoid hands. *American Journal of Physical Anthropology*, **26**, 171–206.

Tuttle, R. H. (1969). Knuckle-walking and the problem of human origins. *Science*, **166**, 953–961.

Ueda, S., Wantanabe, Y., Saitou, N., Omoto, K., Hayashida, H., Miyata, T., Hisajima, H., & Honjo, T. (1989). Nucleotide sequences of immunoglobin-epsilon pseudogene in man and apes and their phylogenetic relationships. *Journal of Molecular Biology*, **205**, 85–90.

Ungar P. S. & Kay R. F. (1995). The dietary adaptations of European Miocene catarrhines. *Proceedings of the National Academy of Science USA*, **92**, 5479–5481.

Van der Kuyl, A., Kuiken, C. L., Dekker, J. T., & Goudsmit, J. (1995). Phylogeny of African monkeys based upon mitochondrial 12S rRNA sequences. *Journal of Molecular Evolution*, **40**, 173–180.

Vawter, L. & Brown, W. M. (1986). Nuclear and mitochondrial DNA comparisons reveal extreme rate vatriation in the molecular clock. *Science,* **234**, 194–196.

Vogel, C. (1961). Zur systematischen untergliederung der gatung *Gorilla* anhand von untersuchungen der mandibel. *Zeitschrift für Saugetierkunde,* **26**, 65–128.

Ward, S. (1997). The taxonomy and phylogenetic relationships of *Sivapithecus* revisited. In D. R. Begun, C. V. Ward, & M. D. Rose (eds.), *Function, phylogeny and fossils: Miocene hominid origins and adaptations* (pp. 269–290). New York: Plenum Press.

White, T., Suwa, G., & Asfaw, B. (1994). *Australopithecus ramidus:* A new species of early hominid from Aramis, Ethiopia. *Nature,* **371**, 306–312.

Wilson, D. E. & Reeder, D. M. (eds.) (1993). *Mammal species of the world. A taxonomic and geographic reference.* Washington: Smithsonian Institution Press.

WoldeGabriel, G., White, T. D., Suwa, G., Renne, P., de Heinzelin, J., Hart, W. K., & Heiken, G. (1994). Ecological and temporal placement of early Pliocene hominids at Aramis, Ethiopia. *Nature,* **371**, 330–333.

Wood, B. A. (1994). The oldest hominid yet. *Nature,* **371**, 280–281.

Woodward, A. S. (1925). The fossil anthropoid ape from Taungs. *Nature,* **155**, 235–236.

Wrangham, R. W., McGrew, W. C., de Waal, F. B. M., & Heltne, P. G. (ed.) (1994). *Chimpanzee cultures.* Cambridge, MA: Harvard University Press.

Wu, C.-I. (1991). Inferences of species phylogeny in relation to segregation of ancient polymorphisms. *Genetics,* **127**, 429–435.

Zihlman, A. L. (1984). Body build and tissue composition in *Pan pansicus* and *Pan troglodytes*, with comparisons to other hominoids. In R. L. Susman (ed.), *The pygmy chimpanzee: Evolutionary biology and behavior* (pp. 179–200). New York: Plenum Press.

Zihlman, A. L. & Cramer, D. L. (1978). Sexual differences between pygmy (*Pan paniscus*) and common chimpanzees (*Pan troglodytes*). *Folia Primatologica,* **29**, 86–94.

Zuckerkandl, E. & Pauling, L. (1962). Molecular disease, evolution, and genetic heterogeneity. In M. Kasha & B. Pullman (eds.), *Horizons in biochemistry* (pp. 189–225). New York: Academic Press.

2

The life history and development of great apes in comparative perspective

SUE T. PARKER

Life-history strategy theory is the comparative study of the shape of the life cycle and the ontogeny of form and function as they relate to reproductive success in individuals and populations. According to this theory, life cycles have been shaped by selection in such a manner that life-time resources are allocated among the functions of growth, maintenance, defense, and reproduction in a manner that maximizes life-time reproductive potential (Pereira, 1993). Such design occurs within developmental and phylogenetic constraints, of course (Smith et al., 1985).

Life-history theory expands the focus of evolutionary biologists beyond morphology and behavior to include developmental and behavioral processes through the life cycle as parts of an integrated whole. Basic data for life-history studies include population data on the following features: body size, gestation length, age at weaning, birth interval, number of offspring per litter, offspring per year, neonatal body weight, adult body weight, age at onset of reproduction, lifespan, and mortality at various stages in the life cycle (Promislow & Harvey, 1990; Stearns, 1992).

Some biologists have characterized two contrasting life-history strategies. K strategists are large, long-lived animals that mature slowly and produce few offspring in each reproductive effort, but invest heavily in each offspring. They generally are adapted to ecological niches that are fairly stable over their lifespans. (Most anthropolid primates are *K* strategists.) In contrast, *r* strategists are small, short-lived animals that mature rapidly and produce large numbers of offspring in one or a few reproductive efforts, investing little energy in their offspring. They generally are adapted to unstable ecological niches that are subject to booms and busts, and are able to respond to resource booms with rapid population growth (MacArthur & Wilson, 1967). This typology gives a good intuitive guide for contrasting life histories in related species, but is limited in its explanatory value (Stearns, 1992). See Figure 2.1 for a graphic depiction of contrasting life histories of gorillas and mouse lemurs.

Body size is significant to life history because differences in body size are associated with differences in the ratio between body volume and body surface area. These variables affect an organism's ability to retain and/or dissipate body heat and hence the amount of calories it needs relative to each unit of body weight. Because body weight

Figure 2.1 Graphic comparison and contrast of the life histories of gorillas and mouse lemurs (Fleagle, 1988).

(volume) increases more rapidly than body surface area, smaller animals have a greater surface area relative to their volume than larger animals.

This means that smaller animals have a higher metabolic rate and need more calories per unit of body weight than larger animals to maintain a constant body temperature. This is the reason that smaller species in a related group of sister species of primates tend to eat higher-energy foods (e.g., insects, meat, green shoots, nectar, or honey) whereas larger species within the same group or adaptive array tend to eat lower-energy foods. Likewise, larger body size in animals tends to correlate with larger brains, longer gestation periods, and longer lifespans (e.g., Schmidt-Nielsen, 1984). Brain size, however, is of equal, if not greater, importance in life history. Brain size in mammals seems to be the pacemaker in life history. This is the reason that closely related species with similar brain indices (e.g., log of brain weight: log of body weight) tend to have similar life-history patterns despite differences in body size (Martin, 1983).

Body size and brain size also correlate with developmental status at birth. Portmann (Portmann, 1990) describes two contrasting patterns of neonatal development. Altricial species of mammals are born after a brief gestation period in an undeveloped state (immoblile, hairless with eyes and ears closed) in large litters. Altricial neonates receive brief but intense parental investment. They are usually born in nests or burrows which protect them from predators. In contrast, precocial species are born after a more extended gestation period in a well-developed state (often ready to move on their own) and are usually singletons. Despite their head start, however, they have larger brains and develop more slowly than altricial species. Most anthropoid primates are precocial. Portmann (1990) classifies human neonates as secondarily altricial, arguing that they are born many months early as compared to other anthropoid primates owing to their large brains relative to birth canal size.

ADAPTIVE ARRAYS OF PRIMATES

Since Linnaeus' time, biologists have known that humans are a species of mammals in the order primates which also includes apes, monkeys, and prosimians. Since Darwin's time, they have known that the primates share certain characteristics they inherited from their common ancestor who diverged from other mammals more than 55 million years ago at the base of the Tertiary Period.

Notable among these characteristics is the climbing-by-grasping behavioral complex which is made possible by the combination of opposable inner digits on the hands and/or feet and elongated digits with tactile pads stabilized by fingernails. Also common to most primates is the stereoscopic vision complex which involves forward orientation of the eyes and crossing of the optic nerve into both sides of the brain for binocular vision. Monkeys and apes and some diurnal lemurs have color vision.

More recently evolved branches of primates share other derived characteristics: all the apes, for example, share the brachiating anatomy of the shoulder, trunk, and arms which allows them to engage in arm-over-arm suspensory locomotion under branches (Clark, 1959). Each branch of the primate family tree shares certain uniquely derived characteristics that it inherited from its common ancestor. Species and higher categories of primates are classified into taxonomic groups on the basis of their common ancestry. See Table 2.1 for a summary taxonomy of primate genera (Richard, 1985).

Since the 1960s, the phylogeny of many branches of the primate family tree have been elucidated by comparative molecular studies. In addition to confirming many traditional taxonomic relationships, these studies have provided the first dates for the divergence of the great apes and hominids. According to early "molecular clocks" based on comparative immunological studies of albumin molecules, the sequence of evolution of the great apes was, first, orangutans at about 8 to 10 million years ago, and then gorillas, chimpanzees, and humans at about 5 million years ago (Goodman, 1963; Sarich, 1968; Sarich & Cronin, 1976). Recently, several DNA studies have suggested that chimpanzees and bonobos are more closely related to humans than gorillas are (Ruvolo, 1994) (see Begun, this volume) though this is still a matter of dispute (Marks, 1995).

The next branch out on the primate family tree contains the gibbons and siamangs, the so-called lesser apes, who diverged from a common ancestor with the great apes approximately 18 million years ago. Next farther out is the branch of the Old World monkeys, which includes macaques and baboons in one group and leaf-eating monkeys in the other. This group branched off its common ancestor with the apes at about 22 million years ago.

Table 2.1 *Taxonomy of primate genera*

Suborder	Infraorder	Superfamily	Family	Subfamily	Genus	Common name
				Alouattinae	*Alouatta*	Howler monkey
				Aotinae	*Aotus*	Owl or night monkey
				atelinae	*Ateles*	Spider monkey
					Brachyteles	Woolly spider monkey
					Lagothrix	Woolly monkey
	Platyrrhini	Ceboidea	Cebidae	Callicebinae	*Callicebus*	Titi monkey
				Cebinae	*Cebus*	Capuchin monkey
				Pitheciinae	*Cacajao*	Uakari
					Chiropotes	Bearded saki
					Pithecia	Saki
				Saimiriinae	*Saimiri*	Squirrel monkey
Haplorhini				Cercopithecinae	*Allenopithecus*	Swamp monkey
					Cercocebus	Mangabey
					Cercopithecus	Guenon
					Erythrocebus	Patas
					Macaca	Macaque
					Miopithecus	Talapoin

	Cercopithecoidea	Cercopithecidae		*Papio*	Savanna baboon
				Theropithecus	Gelada baboon
				Colobus	Colobus monkey
				Nasalis	Proboscis monkey
Catarrhini			Colobinae	*Presbytis*	Langur
				Pygathrix	Douc langur
				Rhinopithecus	Golden monkey
		Hominidae	Gorillinaer	*Gorilla*	Gorilla
				Pan	Chimpanzee
	Hominoidea*		Homininae	*Homo*	Human
		Hylobatidae		*Hylobates*	Gibbon
		Pongidae		*Pongo*	Orangutan

Source: Richard (1985), p. 5.

Farther out still is the branch of the New World monkeys from Central and South America, which includes squirrel monkeys and cebus monkeys. This group had a common ancestor with the Old World monkeys and apes approximately 35 million years ago. Finally, various prosimians, lorises, lemurs, and others, diverged from the common ancestor of all the other primates about 55 million years ago shortly after diverging from a common ancestor with other mammals. See Figure 2.2 for a family tree or phylogeny of monkeys and apes. (We have selected a phylogeny that reconstructs a common ancestor of chimpanzees and humans [Friday, 1992], though this issue is irrelevant to our evolutionary reconstruction.)

Closely related species that have radiated from a common ancestor form adaptive arrays. Figure 2.2 reveals a series of adaptive arrays arising from a series of common ancestors that existed at increasingly earlier times: the common ancestor of hominid species, of African apes, of all great apes, of all apes, of all Old World monkeys, of all monkeys, etc. Adaptive arrays of species in the same genus (in monkeys) or family (in apes) usually display similar life-history patterns, but typically differ in diet, body size, and social organization. Examples of adaptive arrays in living primates include cebus, marmosets, guenons, macaques, gibbons, and African apes (Fleagle, 1988). Comparing the life-history patterns within the ingroup (great apes) with those of the nearest outgroups (lesser apes and Old World monkeys) provides insights into their evolutionary history (Brooks & McLennan, 1991). Comparing the life histories of the ingroup (great apes) with those of distantly related species with similar life histories (cebus monkeys) provides insights into possible adaptive significance of the patterns (Coddington, 1988).

The following sections present a summary of comparative data on the life histories and developmental patterns of the great apes and other closely related adaptive arrays of primates: lesser apes, macaques, and cebus monkeys. Life-history data are critical for understanding the functional significance of cognitive development because they reveal the contexts in which individuals develop. The life histories of great apes can best be appreciated in the context of the life histories of their close relatives. Unfortunately, because the duration of most field or laboratory studies comprises such a small percentage of the lifespan of the subjects, accurate life-history data for long-lived species such as monkeys and apes are difficult to attain.

Furthermore, since various investigators may define terms slightly differently and use slightly different data-collection methodologies, data are not always strictly comparable. Despite these difficulties, which are reflected in the gaps in life-history tables and in discrepancies between tables of various investigators, consistent patterns do emerge. Harvey, Martin, & Clutton-Brock (1987) have published the most comprehensive survey of primate life-history features done so far, but many gaps remain. See Figure 2.3 for a graphic depiction of selected primate life histories (Smith, 1992).

Regarding definitions, it is important to note that lifespan refers to the maximum lifespan (Cutler, 1975). This figure is important because it expresses the genetic potential of the species, whereas average lifespan reflects mortality throughout the life cycle. Age at sexual maturity refers to age of first reproduction rather than puberty. For obvious reasons this is more difficult to determine for males than for females. Gestation

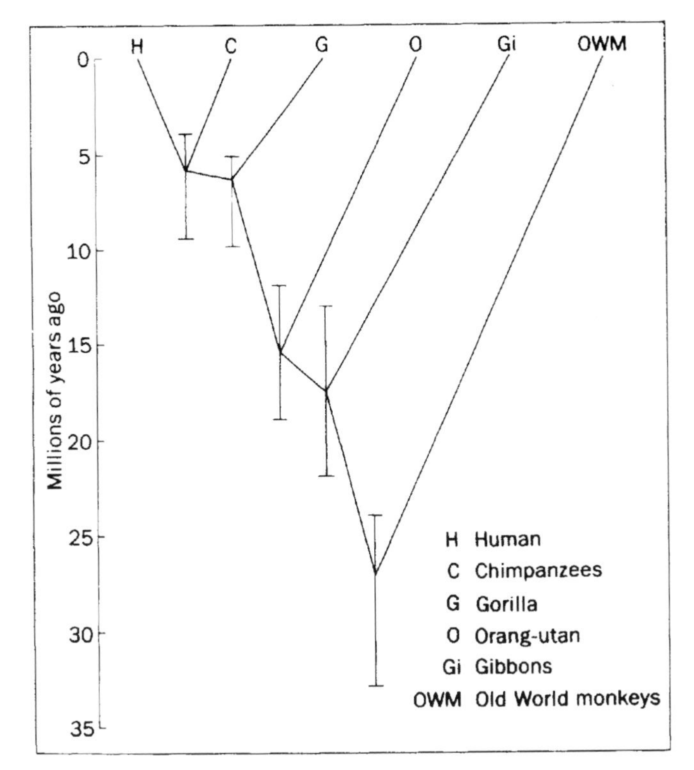

An evolutionary tree for primates based on globin and mitochondrial gene sequences. The pattern represents a consensus of recent studies, but the relationships between humans, chimpanzees and gorillas remain questionable. The bars indicate ranges of estimated times for the branch points.

Figure 2.2 Phylogeny of primates based on globin and mitchondrial genes (Friday, 1992).

periods can be fairly accurately determined from the last estrus in most species. Birth intervals are difficult to determine owing to various forms of sampling error and to variations occurring as a result of infant mortality (Galdikas & Wood, 1990).

Life-history data reveal interesting patterns of similarities and differences within and across primate species. With few exceptions, primate females bear one infant at a time. (This pattern correlates with the two pectoral teats which are characteristic of primates.) Gestation periods show little variation within species or even among species within the same genera, whereas weaning and birth intervals show considerable variation within genera and even within species. Like gestation period, adult brain weight shows relatively little variation within genera and even families.

The evolutionary conservatism of gestation periods and brain weight contrasts with the variation in adult female body weight, which often varies considerably among sister species within the same genus or family by as much as a factor of three. Even greater intrageneric and interfamilial variation occurs in adult male body weight. In many

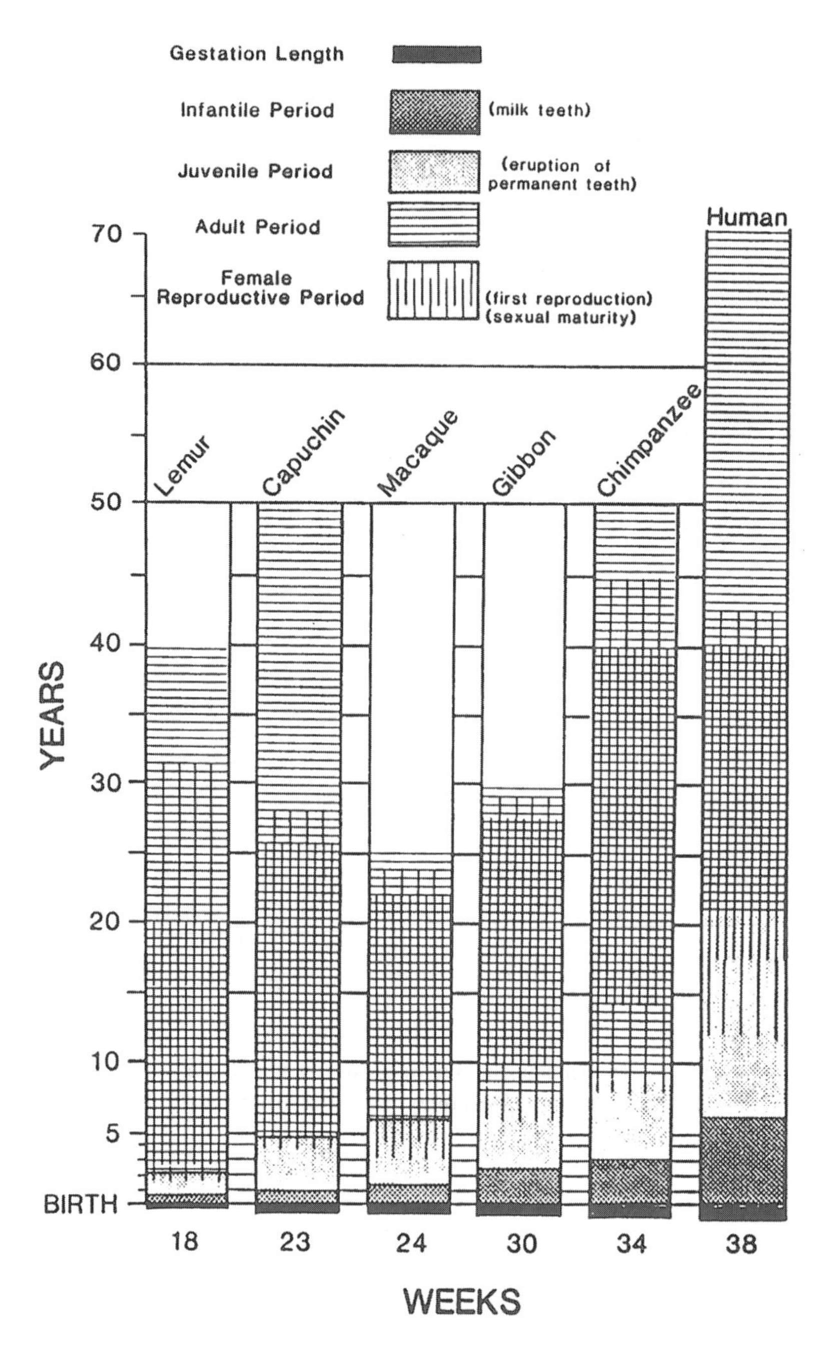

Figure 2.3 Graphic comparison and contrast of life stages in selected primate species (Smith, 1992).

species, especially mid-sized to large species (greater than 5 kg) males are as much as twice as large as females. Males of these sexually dimorphic species may achieve their greater body size through one of two forms of bimaturism, earlier female maturation or prolonged male growth, males maturing 1 to 3 years later than females. This pattern is especially pronounced in baboons and in the great apes.

The most striking differences within the primate order occur between the great apes and monkeys in the following life-history features: body size, gestation length, age at sexual maturity, lifespan, and brain size. Whereas monkeys are similar to other medium-sized mammals, the great apes, like a few other large mammals, most closely approximate the human pattern in all these features.

The Great Apes

The number of living species of great apes has recently been revised upward from 4 to 6: 1 or possibly 2 species of orangutans (*Pongo pygmaeus*), 2 species of gorillas (*Gorilla gorilla* and *Gorilla beringei*) (Morell, 1994; Ruvolo, 1994), and 3 species of chimpanzees, *Pan troglodytes*, *Pan verus* or common chimpanzees, and *Pan paniscus*, the bonobos or pygmy chimpanzees (Morin et al., 1994). Orangutans today are confined to the islands of Borneo and Sumatra. All the other great apes live in Africa. Chimpanzees have the widest distribution of the three, living in a variety of habitats in both eastern and western Africa, whereas the bonobos have the more restricted distribution, living only in the Congo Basin forests. Gorillas are spottily distributed across the center of Africa from the mountains of Ruwanda to the lowlands of eastern and western Africa, different subspecies living in different forest habitats.

Although the precise figures given in different sources vary, it is clear that great ape species differ substantially in their adult body sizes: gorillas are the largest (females 67–98 kg; males 130–218 kg); orangutan the next largest (females 33–45 kg; males 34–91 kg); common chimpanzees next (females 26–50 kg; males 32–70 kg); and bonobos smallest (females 27–38 kg; males 37–61 kg) (from Tuttle, 1986, who gives the largest range of variation for each species). The variation in body size within genera may reflect variation across subspecies or even species which were previously unrecognized, and/or variation in definitions of adult status by different investigators. Such variation in body size within an adaptive array, i.e., a group of sister species who have diverged from a common ancestor, is often seen in birds and mammals (Fleagle, 1988).

Within an adaptive array, a related group of sister species, smaller body size is associated with relatively high-energy omnivorous diets, whereas larger body size is associated with relatively low-grade herbivorous diets. This pattern is apparent in the African apes: chimpanzees being the most omnivorous, eating primarily fruits and shoots and animal food; gorillas being the most herbivorous; and bonobos, being intermediate. This kind of adaptive radiation in body size may reduce competition among species living in the same habitat (Stanley, 1973). A similar adaptive radiation in body size and diet can be seen in the living lesser apes (Raemaekers, 1984) and among fossil hominids known as Australopithecines (Pilbeam & Gould, 1974).

Many of the differences in body shape and other features among species in an adaptive array like the African apes are consequences of biomechanical scaling of features relative to body size (Shea, 1984): Larger species in an adaptive array are not scaled-up versions of their smaller sister species. Some body features do not scale exactly in proportion to increased body size. Exactly proportional scaling up of organs or life plans is known as isometry, whereas nonproportional scaling up of organs or life plans is known as allometry.

Allometric scaling among sister species can either be less than would be expected from isometric scaling (i.e., negative allometry) or greater than expected (i.e., positive allometry). Certain organs (for example, blood vessels, lungs, gut, molar teeth, cross-section of long bones) which function in relation to their surface area must scale up more than expected in order to compensate for the increase of body volume versus surface area; whereas other organs (for example, brains, special senses, endocrine glands) which function in other ways scale up less than would be expected because they do not need to keep abreast with increased volume (Schmidt-Nielsen, 1984). The surface area of molar teeth in gorillas, for example, is larger than would be expected from proportionate scaling because greater surface area is necessary to process the volume of food required by the greater volume of the animal, whereas the brain size of gorillas is smaller than would be expected from proportionate scaling to increased body size (Pilbeam & Gould, 1974).

Allometric differences in features also occur within species as a consequence of size differences between adults of different sexes and as a consequence of size differences between individuals at different stages in the life cycle. The effect of allometry must be factored out of sex, age, and species differences before other evolutionary phenomena can be identified. See Figure 2.4 for a graphic comparison of interspecific, intraspecific (e.g., sexual dimorphic), and growth allometry.

Despite their differences in body size and body shape, all of the great apes display the brachiation complex and engage in arm-over-arm suspensory locomotion when they are small enough to be supported by branches. This character must have arisen in the common ancestor of all the apes. Only the African apes, gorillas, bonobos, and chimpanzees, walk on their knuckles when they are on the ground, which suggests that knuckle-walking arose in the common ancestor of the African apes. Gorillas are primarily terrestrial, whereas chimpanzees and bonobos spend considerable time in trees as well as on the ground. Female orangutans are primarily arboreal, whereas the large adult males are significantly terrestrial; their mode of terrestrial locomotion is known as fist-walking. All of the great apes make sleeping nests by bending and interweaving small leafy branches into a central web either in trees or on the ground. This sleeping adaptation correlates with the loss of sleeping pads on the rump (ischeal callosities) which are characteristic of Old World monkeys and lesser apes.

As indicated above, sister species within an adaptive array often diverge in their dietary patterns as they diverge in body size. Dietary adaptations, in turn, have profound effects on the ranging patterns and hence the social organization of species (Wrangham, 1986): The size of foraging groups is constrained by the dispersion and density of foods. The size of annual home ranges is also constrained by this factor.

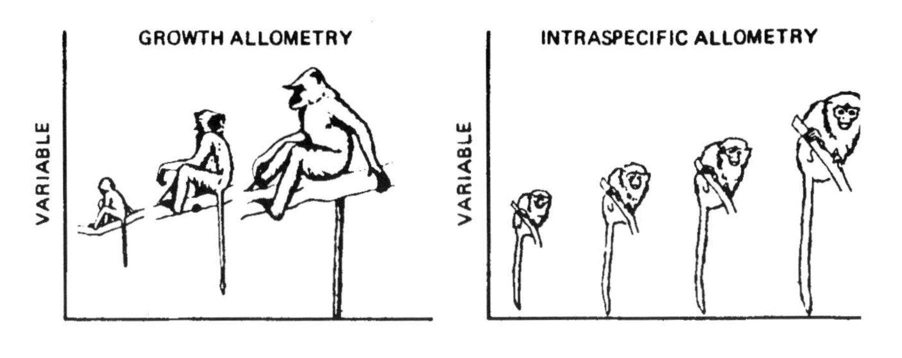

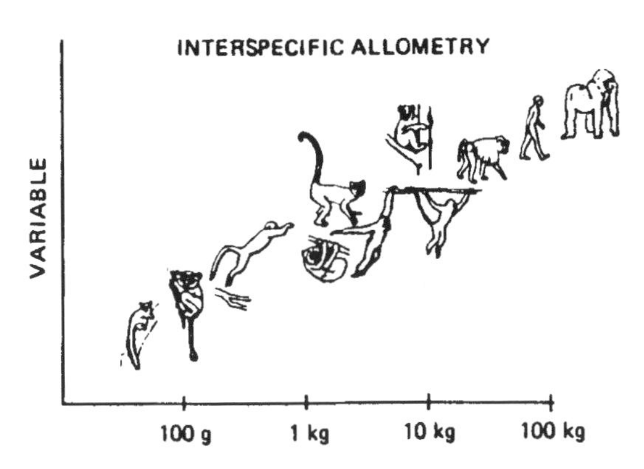

Figure 2.4 Graphic depiction of allometry in primates (Fleagle, 1988).

Gorillas feed on abundant foods that are fairly evenly dispersed in time and space. They therefore have smaller home ranges (6–34 km^2) than common chimpanzees who feed on dispersed foods that are highly clumped in space and time. Chimpanzee home ranges are large (6–278 km^2). Gorillas live in small stable groups. Chimpanzees live in social groups that fluctuate in size and composition with fluctuations in food resources; females and their offspring are the most constant subunit. The dispersion of orangutans foods is so great that the large males are forced into a semi-solitary life with only transitory social and sexual encounters. Smaller female orangutans travel with their infants for several years in home ranges as small as 1 km^2. Chimpanzee and bonobo males remain in their natal groups (are philopatric) whereas females disperse. Both sexes disperse in gorillas and orangutans (Pusey & Packer, 1987).

The reproductive strategies of the males of the various great apes also relate to the dispersion and density of their food sources, their home-range size, and the dispersion of females (Wrangham, 1986). Of the four species, only gorillas have home ranges small enough and group foraging patterns compact enough to allow males to guard females from competing males. Adult silverback gorilla males guard their harems from extra group males who sometime raid them (Fossey, 1983) Owing to their flexible fission–

fusion social organization, chimpanzee males are unable to keep harems. When females are in estrus, male chimpanzees try to entice and/or bully them into going off alone, but when this tactic fails they must compete directly with other males for copulations with promiscuous females (Goodall, 1986). Most conceptions occur during enforced consortships (Tutin & MacGinnis, 1981).

Orangutan males have at least two mating strategies depending upon their age and body size and the availability of territories: large males may defend large territories overlapping those of two or more females or they may range over huge areas. Territorial resident males advertise their presence to competitors with their loud calls and fight with intruding large males. The smaller males engage in "sneaker strategies" and both large and small males coerce females into sex (MacKinnon, 1974). Female choice plays an important role in orangutan reproduction; most conceptions occur during consortships which are possible only when a female is proceptive (Galdikas, 1981). Bonobos, like common chimpanzees, have a promiscuous sexual life, but the relations between males and females are more relaxed and egalitarian (Thompson-Handler, Malenky, & Badrian, 1984).

Those ape species in which males defend harems and engage in polygynous mating, i.e., gorillas and orangutans, show a high degree of sexual dimorphism in body, canine size, and bodily displays. Those apes , i.e., chimpanzees and bonobos, in which males do not keep harems, engage in promiscuous sex. They display a low degree of sexual dimorphism in body size, but show large testis size reflecting sperm competition (Short, 1980). The amount of parental investment by male great apes also varies with mating strategies and with the related factor of degree of confidence in their paternity. Gorilla males invest heavily in their offspring, chimpanzee males less heavily, and orangutans virtually not at all.

Female mating strategies among the great apes vary in relation to group composition and male strategies. Female gorillas, who usually have only one potential partner at a time, show very minimal genital swelling during estrus and are receptive only a few (1–3) days a month when they are not pregnant or lactating. Females generally initiate sex (Harcourt, Stewart, & Fossey, 1981). Female chimpanzees, who usually have many more potential partners and rivals than gorilla females, show major genital swelling and are receptive for many days each month when they are not pregnant or lactating, though they may copulate after they are impregnated. Bonobo females are even more extreme in this regard, having even longer receptive periods and copulating even while they are pregnant and lactating. Homosexual behavior is common among bonobo females, apparently operating as a social bonding mechanism among unrelated females (Thompson-Handler et al., 1984). Relations between bonobo groups are also more relaxed than those among chimpanzee groups (De Waal & Lanting, 1997). Orangutan females apparently show no estrus and may be constantly receptive (MacKinnon, 1974). Females of all four species display prolonged parental investment in their offspring (Harcourt et al., 1981) and, consequently, long interbirth intervals (see below). Females of all three group-living species also move between groups, whereas males more often remain in their natal groups.

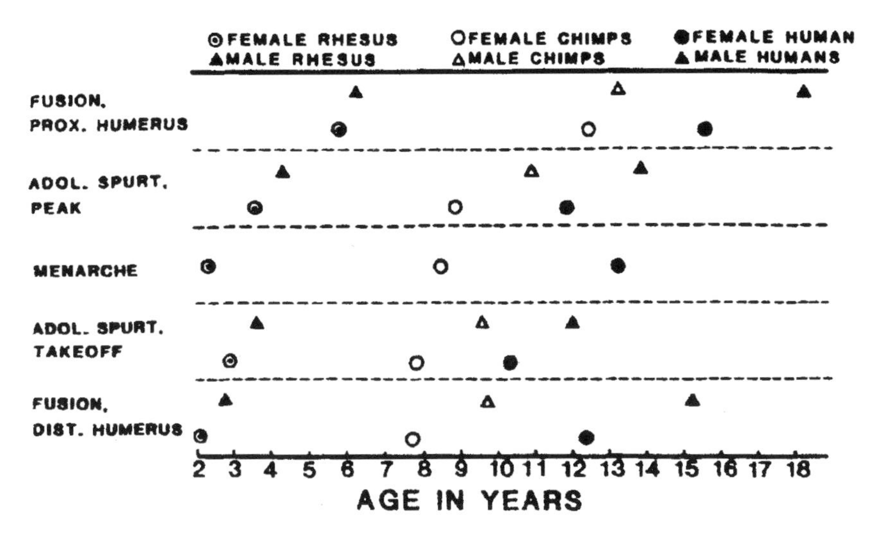

Figure 2.5 Graphic comparison of development in selected primate species (Watts, 1985).

As this brief summary suggests, certain adaptations such as locomotor complexes, brain size, gestation length, and age at maturity are fairly conservative evolutionarily within primate genera and families, birth intervals are flexible adaptations, whereas other adaptations such as diet and body size, reproductive strategies, and social behaviors underlying social organization tend to evolve more rapidly and be less similar among sister species, with the notable exceptions of marmosets and lesser apes.

Development in Great Apes and Humans

Field workers have described the following stages in the life cycle of three of the great apes species: infancy, juvenility, adolescence, subadulthood, adulthood. Infancy ends with weaning and the capacity to survive without a caretaker (Pereira, 1993). In females, juvenility ends and adolescence begins with a growth spurt just before menarche (it generally involves a period of adolescent sterility for females) (Watts, 1985). Adulthood begins with onset of adult reproductive roles, in females this occurs with the birth of the first offspring. Owing to bimaturism, males have a longer developmental period in all the great apes (except perhaps bonobos in which both sexes mature at similar rates), extending up to 20 years in the case of orangutans. See Figure 2.5 from Watts (1985) comparing development in macaques, chimpanzees, and humans.

Development can be assessed by a variety of physical measures including bone age, dental age, bone ossification, weight for height and/or age, and sitting height (Bogin, 1988; Eveleth & Tanner, 1990; Tanner, 1990). Brain development is another measure particularly pertinent to comparative developmental studies of cognition. These various measures do not necessarily correlate with one another. Recent evidence suggests that dental growth is the most stable marker of development. This is so because it is highly heritable and more strongly resistant to environmental effects than bone growth (Smith, 1989). Such other measures as age at reproductive maturity are more variable.

Gestation periods

All the great apes have similar gestation periods: orangutans, 275 days; gorillas, 270 days; and chimpanzees 253 days; humans 280 days. The gestation period in humans is shorter than it should be in comparison to other similar mammals. If humans were able to give birth to a larger brain neonate, our gestation period would be closer to 21 months (Portmann, 1990).

Birth intervals and age at weaning

Weaning occurs when locomotorically competent infants are capable of feeding themselves. Interbirth intervals encompass lactation, estrus periods, and gestation. Recent comparative data on interbirth intervals in great apes are as follows: about 4.5 years (54 months) for bonobos (De Waal & Lanting, 1997); 5.5 years (67 ± 1 months) for gorillas; 7.7 years (93 ± 1 months) for common chimpanzees; 7.7 years (93 ± 2 months) for orangutans as compared to 3.5 years (43 ± 1 months) for a human hunter-gatherer group, the Gainj (Galdikas & Wood, 1990). The intervals in wild chimpanzees and orangutans is longer than initially believed. Birth intervals apparently vary in response to the age and nutritional status of the mother (Goodall, 1986; Galdikas, 1995). Birth intervals and age at first reproduction are apparently the more flexible features of life history..

Gorilla infants can develop rapidly because they enjoy more accessible, high-protein diets as compared to chimpanzee infants. They also expend less energy foraging owing to their shorter day ranges and more terrestrial life style (Doran, 1997). The relatively short interval for gorillas may also reflect the relatively modest apprenticeship involved in feeding on plant foods (Bryne & Bryne, 1991) as opposed to the longer apprenticeships in common chimpanzees involved in tool-aided feeding on social insects and hard-shelled nuts. Finally, more rapid infant development and shorter birth interval may have been favored by the higher incidence of male take over and infanticide among gorillas as compared to other great apes.

Brain development

Data on brain development in great apes is sparse. Portmann (1990) cites (unreferenced) data from Schultz indicating that newborn great apes have achieved the following proportions of their adult brain weights by birth: humans, 23%; chimpanzees 45%; gorillas, 59%, and orangutans, 40%. (If we divide the gorilla neonatal brain weight of 227 (Martin, 1983) by average adult brain weight of 443 for females and 535 for males (Schultz, 1972), we get 51% and 59% respectively) (Also see Semendeferis chapter for other possible differences in gorilla brains.) Portmann (1990) argues that, contrary to popular understanding, the degree of brain development at birth is greater in precocial as compared to altricial mammals. Portmann (1990) notes that 80–90% of

Table 2.2 *Dental eruption ages (in years) for selected primates*

Species:	di1	dp4	M1	M2	M3
cebus	0	0.34	1.15	2.17	3. +
macaques	0.05	0.34	1.37	3.15	5.4
gibbons	0.05	0.42	1.75	?	7.5
chimpanzees	0.25	0.75	3.19	6.45	10.71
gorillas	0.17	0.75	3.5	6.58	10.58
orangutans	0.42	0.81	3.5	5.0	10.0
humans	0.79	2.2	6.15	11.5	20.4

di1 = first deciduous incisor; dp4 = last deciduous premolar; M1 = first molar; M2 = second molar; M3 = third molar.
Source: Smith, Crummett, & Grandt, 1994.

brain growth in hominoids has been achieved by late childhood, and 95–98% by late adolescence.

Dental development

Data on dental development, in contrast, is rich. Two major dental markers are age of eruption of the first molar tooth which corresponds to age at weaning, and completion of dentition which corresponds with completed brain growth (Smith, 1992). Eruption of molar 1 also correlates with gestation length, interbirth interval, female age at first breeding, and male age at sexual maturity (Smith, 1992). See Table 2.2 for dental eruptions times (Smith, Crummett, & Brandt, 1994).

It is clear from inspection of this table that dental maturation occurs almost twice as soon in great apes as in humans and about twice as soon in monkeys as in great apes. It is also apparent that dental maturation in gorillas occurs earlier than in chimpanzees and orangutans.

Body size

As indicated above, great ape body sizes range from 218 to 26 kg: gorillas (females 67–98 kg; males 130–218 kg); orangutans (females 33–45 kg; males 34–91 kg); chimpanzees (females 26–50 kg; males 32–70 kg); and bonobos (females 27–38 kg; males 37–61 kg) (from Tuttle, 1986). Changes in body weight during development of humans, chimpanzees, gorillas, and orangutans are compared in Figure 2.6 (from Portmann, 1990). The faster rate of development in gorillas and orangutans as compared to chimpanzees is consistent with their greater adult size, and in gorillas, with their earlier maturation.

Portmann notes that humans are born with a greater percent of adult body size than great apes, and display a different pattern of growth. Whereas orangutans and gorillas

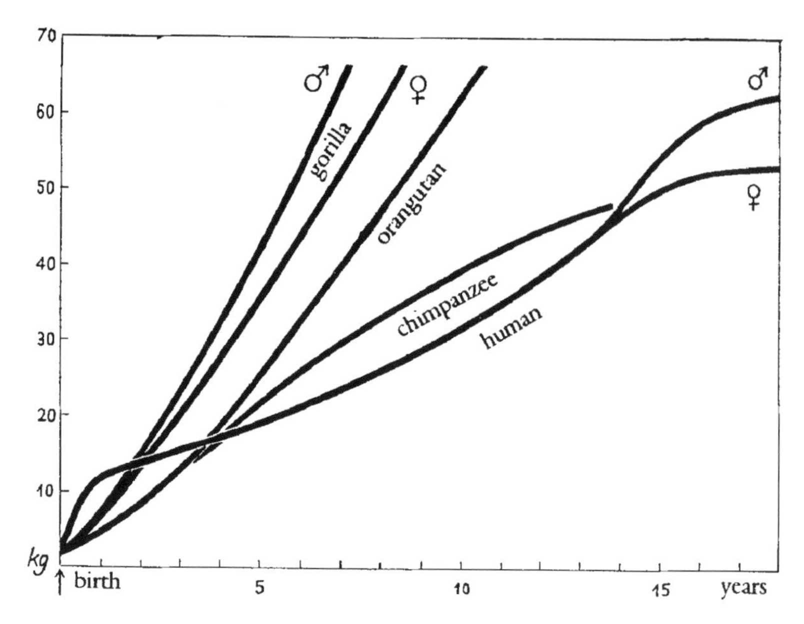

Figure 2.6 Comparison of developmental rates in selected primate species (Portmann, 1990).

develop at a more or less constant rate after birth, humans grow more rapidly (at a fetal rate) immediately after birth, then drop to a slower rate, and accelerate again at puberty.

Motor development

Motor development of infants is a life-history feature prerequisite to the transition from infancy to juvenility. In contrast to monkeys, great ape infants are born with weak clinging abilities, and human infants without effective clinging abilities. Ape mothers offer their infants some support during the first weeks of life, whereas human mothers offer complete support. Ape mothers carry and/or are clung to by their infants continuously for the first two or three months except for brief periods that their mothers park them near by. By the end of infancy, young apes are usually able to forage and sleep independently. This occurs by the time their mothers bear another infant.

Comparative data on human and captive chimpanzee locomotor development can be found in Reisen and Kinder's (1952) large-scale study of common chimpanzee motor development. Reisen (1952) tested 13 nursery-reared chimpanzees on 202 items identified by Gesell and Thompson in their study of human infants. Reisen describes a complex pattern of differences and similarities between the two species by body region. Generally, these items develop in a different sequence (though some necessarily precede others developmentally). The primary difference is that static postures develop before dynamic postures in human infants, whereas the two develop simultaneously in chimpanzees.

Reisen describes marked differences between chimpanzee and human infants in hand and foot development, an early and obvious difference being the strong foot grasp in the neonatal chimpanzee. He notes that chimpanzee infants show higher frequencies of flexion of the hand (and also of the arms and legs) than extension or hands-open positions which predominate in human infants. This correlates with the clinging functions of the hands. A subsequent study of captive chimpanzees and bonobos using the Gesell and Thompson scales found the same rate of locomotor development in these two species (Brakke & Savage-Rumbaugh, 1991).

Comparative data on locomotor development in wild bonobos, chimpanzees, and mountain gorillas reveal developmental changes in locomotion across three subperiods of infancy: class I (0–5 months); class II (6–23 months); and class III (2–5 years) in three species (Doran, 1992; 1997). Six forms of locomotor behavior in infants were described: quadrupedalism (palmigrade and knuckle-walking); guadrumanous climbing and scrambling; suspensory behavior; bipedalism (supported and unsupported), leaping and diving, and somersaulting.

Class I infant chimpanzees and gorillas show similar trends: in both species, infants begin to crawl at about 3.5 to 4 months, palmigrade quadrupedalism in both begins at about 4.5 to 5.5 months (gorillas beginning slightly later). Forelimb-dominated behaviors are the norm in this age class. Chimpanzees in this age class are more independent than gorillas, climbing on lianas, whereas gorillas climb on their mothers.

Class II infants (6–23 months of age) begin to differ significantly in their locomotor behaviors. In gorillas of this age, quadrupedalism is already the predominant form of locomotion, accounting for more than 50% of locomotion, whereas in chimpanzees of this age it is an infrequent form of locomotion, accounting for less than 15% of their locomotion. Moreover, gorilla infants are already using the knuckle-walking form of quadupedalism about 50% of the time by 10 months, whereas chimpanzee infants only begin to use this form at that rate at about 30 months. This pattern reflects the fact that gorilla infants are moving on the ground most of the time, whereas chimpanzee infants are rarely moving on the ground.

Class III infants (2–5 years) continue to show these differences. Gorilla infants continue to engage in more quadrupedal behaviors on the ground. Chimpanzee infants continue to engage in more climbing and suspensory behaviors above ground though they have begun to knuckle-walk more frequently in their quadrupedal locomotion. By 3 years of age, gorilla infants travel largely on their own; by 4 years they are using an essentially adult form of locomotion. In contrast, chimpanzees do not move independently of their mothers until they are about 5 years old. Various locomotor landmarks occur earlier in captive than in wild gorillas (e.g., Maple & Hoff, 1982) perhaps owing to better nutrition (Doran, 1997). It might also reflect population or species differences (Morin, et al., 1994) given that all the captive studies have been of lowland rather than mountain gorillas. See Figure 2.7 comparing frequencies of various forms of locomotion in chimpanzee and gorilla infants of various ages.

Gorilla and chimpanzee infants show similar locomotor behaviors at similar body sizes which they achieve at different ages. Owing to the more rapid growth of gorillas, 1–year-old gorillas are similar in size to 2-year-old chimpanzees (and similar to juvenile bonobos). At this body size of roughly 9 kg, both species engage predominantly in

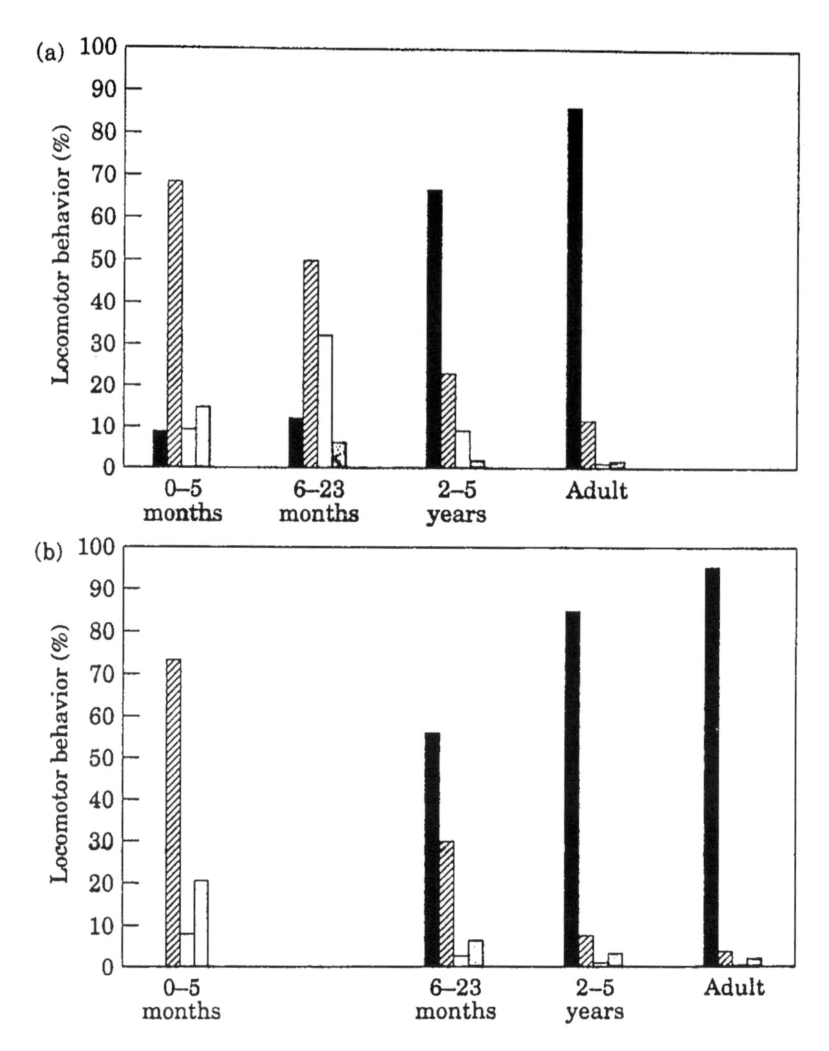

Figure 2.7 Species differences in locomotive behavior. (a) Chimpanzee, (b) gorilla. ■, quad; ▨, climb; □, susp.; ▩, bipedal.

quadrupedal knuckle-walking locomotion. In addition to differences in body size, differences in body proportions – broader scapulae and shorter fingers – render gorillas less suited to suspensory locomotion than chimpanzees (Doran, 1997).

Doran (1992) was unable to classify bonobo development into the foregoing categories owing to lack of precise age data. Instead, she classified bonobos into infants, juveniles, adolescents, and adults. She concluded that both bonobos and chimpanzees show a developmental trend toward decreased suspensory behavior and increased quadrupedalism. They also show significant differences in the frequencies of various locomotor behaviors. Bonobos of all ages engage more frequently in quadrupedalism and less frequently in quadrumanous climbing and scrambling than do chimpanzees. Bonobo juveniles and adolescents engage in more suspensory behaviors than do chimpanzees.

Although adult bonobos and chimpanzees are roughly the same size, bonobos develop body size more slowly than chimpanzees do. By three years of age, bonobos have achieved approximately 2.3 times their birth weight, whereas chimpanzees have achieved about 4.5 times theirs (Kuroda, 1989).

Comparable data on locomotor development in orangutan infants are not available. Given that females are strictly arboreal, it seems likely that orangutan infants engage primarily in climbing and suspensory locomotion. For the same reason, it seems likely that they develop their locomotor behaviors later than other great apes. Galdikas (1995) reports that 4-year-old wild orangutans are still nursing and nesting with their mothers, and being occasionally carried. Mothers definitively reject and wean their offspring at about 7 years of age. The slower development of orangutans may also reflect their adaptation for processing toxic foods, an adaptation associated with slower development in other species (Horn, 1978).

Over all, differences in locomotor development seem to occur after the first six months of life, and seem to be related primarily to differences in body size and substrate use; the more terrestrial species developing adult forms of locomotion earlier than the more arboreal species.

Summary of great ape development

Over all, great apes mature at an earlier age than humans by several measures: dentition, body weight, reproduction, and motor behavior. Within great apes, dental development is quite similar. Available data suggest that locomotor development and weaning occur earlier in bonobos and gorillas than in chimpanzees and orangutans. Sexual maturity also occurs earliest in female gorillas and latest in female orangutans.

Over all, these data suggest that among great apes, gorillas and bonobos are more precocial than chimpanzees and orangutans These data also suggest that orangutans are the least precocial of all the great apes. The fact that orangutans are the most distantly related to both humans and the African apes suggests that these species differences do not reflect phylogeny directly. It seems likely the faster development and shorter interbirth intervals are adaptive responses to lessened pressures for parental investment among species that enjoy reliable food resources (Doran, 1997).

The significance of these comparative life-history data are highlighted through comparison with the life histories of other closely related taxa, specifically, the lesser apes, the Old World monkeys, and the New World monkeys. Only by comparing the life histories of our ingroup with those of more distantly related outgroups can we see which are shared derived characters unique to the great apes.

The Lesser Apes

The lesser apes, members of the family Hylobatidae, include at least 8 species of lesser apes: the gibbons (genus *Hylobates*) and the siamangs (classified as genus *Symphalanges* by some and *Hylobates* by others). Gibbon species live throughout Southeast Asia and range from China to Assam and across Indonesia and the Mentawai Islands. Siamangs live in Malaysia and Sumatra (Preuschoft, Chivers, Brockelman, & Creel, 1984). The family of lesser apes branched off their common ancestor with great apes approximately 22 million years ago and differentiated into gibbons and siamangs approximately 6 million years ago (Cronin, Sarich, & Ryder, 1984).

Lesser apes are small highly specialized brachiators who are strictly arboreal. Although they have lost their tails as part of their brachiation complex, lesser apes have the sitting pads or ischeal callosities that are also typical of Old World monkeys. Unlike great apes, they do not build nests but sleep sitting up on these calluses.

Like the great apes, lesser apes comprise an adaptive array, falling into three groups by body weight: about 5 kg, 7 kg, and 10 kg (Raemaekers, 1984). Unlike the great apes, lesser apes show little sex difference in body weight or canine size: Gibbon females average between 5 and 7 kg, whereas gibbon males average between 6 and 7 kg. Siamangs are slightly larger, both males and females averaging 11 kg. Their monomorphism in body size correlates with their monogamous mating systems and joint defense of feeding territories. Although most species are lacking in sexual dimorphism, some species are dichromatic in coat color, whereas other species show two different color phases without regard to sex, and others show no variation in color.

Lesser apes live in a variety of forest types, gibbons feeding primarily on fruits, whereas siamangs, as suggested by their greater body weight, feed more on leaves. Siamang and lar gibbons, for example, differ in the proportion of fruit and young leaves and shoots in their diets in accord with differences in body weight and metabolic rate. Consequently, they also differ in biomass densities, the biomass of siamangs being twice that of lar gibbons (Raemaekers, 1984). All lesser apes live in small feeding territories which they jointly defend against other groups. Defense is accompanied by loud calling and elaborate swinging displays by the males. Both sexes disperse at puberty.

The life history of lesser apes is intermediate between that of the great apes and Old World monkeys in gestation length, brain size, and lifespan. Gestation in gibbons is about 210 days; in siamangs about 230 days. adult brain size is 107–130 grams; maximum lifespan about 30 years. As among great apes, age at weaning and age at sexual maturity show more variation across species. Females have their first offspring between 7 and 9 years of age. They wean their offspring when they are 1–2 years old. Like most primates, lesser apes bear one offspring at a time, but, unlike most primates, both gibbon and siamang males provide considerable parental investment, carrying and playing with their offspring. Their high investment correlates with their monogamy and consequent high confidence of their paternity. Because the monogamous bond is maintained by adult intolerance for members of the same sex, adults drive their own offspring out of their territory when they become sexually mature. When new pairs form, they generally remain together for the lifetime of the first to be deceased.

The lesser apes have been very little studied in regard to their development since Carpenter's (1940) study. Reynolds (1967) gives the span of dental development as

2.5–8 years. In a recent comparative study of primate dentition, Smith et al. (1994) give 1.75 years for M1 eruption and 7.5 years for M3 eruption.

The Macaque Monkeys

The Old World (catarrhine) monkeys live in Africa, Asia, and the islands off southeast Asia. They branched off a common ancestor with the apes about 35 million years ago. They include two families, the leaf-eating monkeys (Colibidae) and the cheek-pouch monkeys (Cercopithecidae). The Cercopithecidae include macaques, baboons (*Papio*), guenons (*Cercopithecus*), and mangabeys (*Cercocebus*). These two adaptive radiations branched off a common ancestor approximately 12 million years ago, several million years after that common ancestral species branched off its common ancestor with the apes.

The macaques are a large genus containing as few as 9 or as many as 19 species, depending on the taxonomist (Lindburg, 1980). Their geographic range extends from Morocco in North Africa across India and the Himalayas into southeast Asia and Japan and China. This enormous range reflects the variety of habitats exploited by members of this genus, from sea-shore to high mountains, from forests to mangrove swamps to scrublands to cities. In tropical rainforests, they are often sympatric with gibbons and with other Old World monkeys and with gibbons. They are primarily frugivorous with some omnivorous tendencies (Lindburg, 1980).

Like great apes, macaques form an adaptive array of body sizes, ranging from the small rhesus macaques of India (females about 3 kg; males 6 kg) to the large Barbary macaque of the Atlas Mountains of Morocco and Japanese macaques of Japan (females 10 kg; males 12 kg). Some of the larger ground-living forms have very short tails, for example, stumptailed macaques of China, pigtail macaques of southeast Asia, the Japanese macaques, and the Barbary macaques. Macaques are larger than most guenons (genus *Cercopithecus*) and smaller than most baboons (*Papio*).

Most macaques live in fairly large multimale, multifemale social groups and engage in promiscuous mating, males and females forming consortships during the mating season. Females are philopatric, usually remaining in their natal troops whereas males tend to immigrate to other groups at puberty (Pusey & Packer, 1987). In some macaque species, female lineages form ranked matrilines.

Like other Old World monkeys, macaques have shorter gestation periods, earlier maturity, and shorter lifespans than either the lesser or the great apes. Females in most species have their first offspring when they are 4 years old and live to be less than 30 years old. Their gestation periods range from 162 to 177 days. Their bone and dental growth occurs earlier than that of the great apes and cebus monkeys. The first molar erupts early in the second year of life and the third molar by the sixth year (Smith et al., 1994; Watts, 1990).

Likewise, the motor development of macaques occurs earlier than that of the great apes: stumptail macaques begin to walk in an unsteady quadrupedal baby walk as early as 2 weeks of age and are confidently running by 4 or 5 weeks. Simple palmar grasping develops by 3 weeks and pincer grasping by 4 weeks of age (Parker, 1973). See Table 2.3

Table 2.3 *Developmental milestones in five genera of nonhuman primates*

Parameter	*Cebus*	*Saimiri*	*Macaca*	*Papio*	*Pan*
Weight of brain	49 [5]	63 [5]	61 [37] 48–69 [4]	30–41 [4]	36–46 [4]
First off mother	9 weeks[a]	3 weeks[a] [38]	2 weeks [23]	3 days [34] 2 weeks [33]	16–24 weeks [21]
≥ 50% of daytime off carrier	19 weeks[a]	12 weeks[a]	10 weeks [23, 39]	17 weeks [30]	1 year [35]
Voluntary grasping appears	5 weeks [16]	3 weeks[a]	3 weeks [2, 24, 31, 32]	2 weeks [34]	16–24 weeks [21]
Precision grip appears	13 weeks	absent	4–8 weeks [2, 32]	4 weeks [36]	10 months[b]

Brain weight was measured at birth and is given as a percentage of the adult female's brain weight. Sources are listed in brackets.
[a] Fragaszy, Scollay, and Baer, unpublished data.
[b] Bard, personal communication.
Source: Fragaszy (1990).

(from Fragaszy, 1990) for a comparison of several monkey species with common chimpanzees in some motor milestones.

Because of their precocious sexual development, which occurs as early as 2 years of age, rhesus macaques may not be a good model for Old World monkeys for phylogenetic reconstructions (Watts, 1990). I have chosen to focus on macaques because, like common chimpanzees, they have been widely studied in the wild and especially in the laboratory. Since Harlow established his primate research laboratory at the University of Wisconsin in the thirties, rhesus monkeys have been one of the most common laboratory primates. Consequently, more is known about growth and cognitive development in this group than in any other group of Old World monkeys excepting perhaps baboons who have been well studied in the wild. For this reason I use them as an outgroup for evolutionary reconstructions.

The Cebus Monkeys

In this section, I focus on cebus monkeys because they have been well studied in the wild and in captivity and because they show some interesting convergences with great apes in life history and cognitive abilities. These features, which are shared by great apes and cebus but not by lesser apes and Old World monkeys, render cebus useful for understanding the adaptive significance of great ape life history and cognition (Parker & Gibson, 1977).

The New World (platyrrhine) monkeys live in Central and South America. They are comprised of two families, the marmosets (Callithrixidae) and all the other neotropical monkeys (Cebidae). The Cebidae include all the larger species of New World monkeys including spider (*Ateles*) and howler (*Alouatta*) monkeys as well as the tiny squirrel (*Siamiri*) monkeys and the cebus. Some of the Cebidae have prehensile tails, i.e., tails that will support their full body weight. The New World monkeys branched off a common ancestor with the Old World monkeys about 35 million years ago, after which they underwent an adaptive radiation. It seems likely that their common ancestor came across the Atlantic Ocean on floating vegetation at that time when South American and Africa were still closer together.

The *Cebus* are a genus of Cebidae comprised of four species who live in Central America and the Amazon forest. They are all small animals (females 2–3 kg; males 3–4 kg) with little to slight sexual dimorphism in body size. They are larger than marmosets and squirrel monkeys, but smaller than spider and howler monkeys. At one time, they were widely used as organ-grinders' monkeys. They inhabit a variety of forest types from tropical rainforest to dry deciduous forests, feeding primarily in the trees, but also foraging on the ground. They live in medium-sized multimale multifemale groups characterized by male dispersal (Strier, 1996). Generally they are sympatric with other monkeys; in some locations two cebus species are sympatric (e.g., Terborgh, 1983). Cebus are primarily frugivorous, but also eat a wide variety of plant and animal foods, sympatric species specializing on different foods during the nonfruiting season (Terborgh, 1983).

Cebus life histories are unusual for monkeys of their small body size. They have gestation periods, maturation, and lifespans more similar to those of apes than those of monkeys. Females are reproductively mature at 6 years; males mature at 12 years, and animals live as long as 47 years. Infants are weaned at about 18 months (Robinson, 1988). Their bone maturation is late compared to that of squirrel monkeys and macaques (Watts, 1990). Likewise, their motor development is retarded compared to that of macaques. They begin to crawl at 2 months and to walk competently at about 3 months (Fragaszy, 1990). Their dental development, however, is early compared to that of macaque monkeys, M1 erupting at just after the first year, and M3 just after the third year (Smith et al., 1994). This life-history pattern correlates with proto tool use in extractive foraging on embedded foods (Parker & Gibson, 1977).

SUMMARY AND CONCLUSIONS

The foregoing summary of anthropoid life histories illustrates the well-known phenomenon that life-history and developmental patterns in primates roughly parallel degrees of genetic relatedness. Comparative life-history data reveal a stepwise increase in the most conservative features from macaques to lesser apes to great apes to humans: gestation periods range from about 168 for macaques to 215 for gibbons to 240 for great apes to 270 for humans; age at first reproduction of females range from 3 years for macaques to 8 years for gibbons to 10 years for great apes to 18 years for humans; maximum lifespans range from 30 years for macaques and gibbons to 40 years for great apes to 95 years for humans. Brain sizes range from about 95 grams for macaques to about 115 for gibbons to about 450 for great apes and 1250 for humans. Cebus are similar to the apes in lifespan and age at reproductive maturity, but more like macaques in gestation period and brain size.

Comparison data on primate life histories reveal close parallels in great ape and human development as compared to monkey development. They also reveal close parallels in the life histories of the four great ape species, and patterns of difference which seem to correlate with ecological niches rather than degrees of phylogenetic divergence. As previously indicated, the longer lifespan and slower maturation of the great apes as compared to monkeys is closely associated with their greater intellectual abilities.

REFERENCES

Bogin, B. (1988). *Patterns of human growth*. Cambridge University Press.

Brakke, K. E. & Savage-Rumbaugh, E. S. (1991). Early postural behavior in Pan: Influences on development. *Infant Behavior and Development*, **14**, 265–288.

Brooks, D. & McLennan, D. (1991). *Phylogeny, ecology, and behavior*. University of Chicago Press.

Byrne, R. W. & Byrne, J. M. E. (1991). Hand preferences in the skilled gathering tasks of mountain gorillas (*Gorilla g. beringei*). *Cortex*, **27**, 521–546.

Clark, W. W. L. G. (1959). *The antecedents of man*. New York: Harper and Row.

Coddington, J. A. (1988). Cladistic tests of adaptational hypotheses. *Cladistics*, **4**, 3–22.

Cronin, J. E., Sarich, V. M., & Ryder, O. (1984). Molecular perspectives on the evolution of the lesser apes. In H. Preuschoft, D. J. Chivers, W. Y. Brockelman, & N. Creel (eds.), *The lesser apes* (pp. 468–485). Edinburgh University Press.

Cutler, R. G. (1975). Evolutionof longevity in primates. *Journal of Human Evolution*, **5**, 169–201.

De Waal, F. & Lanting, F. (1997). *Bonobos: The forgotton ape*. Berkeley: University of California Press.

Doran, D. M. (1992). The ontogeny of chimpanzee and pygmy chimpanzee locomotor behavior: A case study of paedomorphism and its behavioral correlates. *Journal of Human Evolution*, **23**, 139–157.

Doran, D. M. (1997). Ontogeny of locomotion in mountain gorillas and chimpanzees. *Journal of Human Evolution*, **32**, 323–344.

Eveleth, P. B. & Tanner, J. M. (1990). *Worldwide variation in human growth*. (2nd edn). Cambridge University Press.

Fleagle, J. G. (1988). *Primate adaptation and evolution*. New York: Academic Press.

Friday, A. E. (1992). Human evolution: the evidence from DNA sequencing. In S. Jones, R. Martin, & D. Pilbeam (eds.), *The Cambridge encyclopedia of human evolution* (pp. 316–321). Cambridge University Press.

Fossey, D. (1983). *Gorillas in the mist*. New York: Houghton-Mifflin.

Fragaszy, D. M. (1990). Early behavioral development in capuchins (Cebus). In D. M. Fragaszy, J. Robinson, & E. Visalberghi (eds.), *Adaptation and adaptability of capuchin monkeys* (vol. 54, pp. 119–128). Basle: Karger.

Galdikas, B. (1981). Orangutan reproduction in the wild. In C. E. Graham (ed.), *Reproductive biology of the great apes* (pp. 281–299,). New York: Academic Press.

Galdikas, B. (1995). *reflections of Eden: My years with the orangutans of Borneo*. New York: Little Brown.

Galdikas, B. & Wood, J. W. (1990). Birth spacing patterns in humans and apes. *American Journal of Physical Anthropology*, **83**, 185–191.

Goodall, J. (1986). *Chimpanzees of the Gombe*. Cambridge, MA: Harvard University Press.

Goodman, M. (1963). Serological analysis of the systematics of recent hominoids. *Human Biology*, **35**, 377–436.

Harcourt, A., Stewart, K. J., & Fossey, D. (1981). Gorilla reproduction in the wild. In C. E. Graham (ed.), *Reproductive biology of the great apes* (pp. 265–279). New York: Academic Press.

Harvey, P., Martin, R. D., & Clutton-Brock, T. (1987). Life histories in comparative perspective. In B. Smuts, D. Cheney, R. Seyfarth, R. Wrangham, & T. Struhsaker (eds.), *Primate societies* (pp. 181–196). University of Chicago Press.

Horn, H. S. (1978). Optimal tactics of reproduction and life history. In J. R. Krebs & N. B. Davies (eds.), *Behavioral ecology, an evolutionary approach* (first edn, pp. 411–429). New York: Oxford University Press.

Kuroda, S. (1989). Developmental retardation and behavioral characteristics of pygmy chimpanzees. In P. G. Heltne & L. A. Marquardt (Eds.), *Understanding chimpanzees*, (pp. 184–193). Cambridge, MA: Harvard University Press.

Lindburg, D. G. (ed.). (1980). *The macaques: Studies in ecology, behavior and evolution*. New York: Van Nostrand Reinhold.

MacArthur, R. H. & Wilson, E. O. (1967). *Theory of island biogeography*. Princeton University Press.

MacKinnon, J. R. (1974). The ecology and behaviour of wild orangutans (*Pongo pygmaeus*). *Animal Behaviour*, **22**, 3–74.

Maple, T. & Hoff, M. (1982). *Gorilla behavior*. New York: Van Nostrand Reinhold.

Marks, J. (1995). Learning to live with a trichotomy. *American Journal of Physical Anthropology*, **90**, 237–246.

Martin, R. D. (1983). *Human brain evolution in an ecological context*. New York: American Museum of Natural History.

Morell, V. (1994). Will primate genetics split one gorilla into two? *Science*, **265**, 1661.

Morin, P. A., Moore, J. J., Chakraborty, R., Jin, L., Goodall, J., & Woodruff, D. S. (1994). Kin selection, social structure, gene flow, and the evolution of chimpanzees. *Science*, **265**, 1193–1201.

Parker, S. T. (1973). Piaget's sensorimotor series in an infant macaque: The organization of unstereotyped behavior in the evolution of intelligence. Unpublished Ph.D. thesis, University of California, Berkeley, CA.

Parker, S. T. & Gibson, K. R. (1977). Object manipulation, tool use, and sensorimotor intelligence as feeding adaptations in cebus monkeys and great apes. *Journal of Human Evolution*, **6**, 623–641.

Pereira, M. (1993). Juvenility in animals. In M. E. Pereira & L. A. Fairbanks (eds.), *Juvenile primates: Life history, development and behavior* (pp. 17–27). New York: Oxford University Press.

Pilbeam, D. & Gould, S. J. (1974). Size and scaling in human evolution. *Science*, **186**, 892–901.

Portmann, A. (1990). *A zoologist looks at humankind* (Judith Schaefer, trans.). New York: Columbia University Press.

Preuschoft, H., Chivers, D. J., Brockelman, W. Y., & Creel, N. (eds.). (1984). *The lesser apes: Evolutionary and behavioural biology*. Edinburgh University Press.

Promislow, D. E. & Harvey, P. H. (1990). Living fast and dying young: A comparative analysis of life-history variation among mammals. *Journal of the Zoological Society, London*, **220**, 417–437.

Pusey, A. E. & Packer, C. (1987). Dispersal and philopatry. In B. Smuts, D. Cheney, R. Seyfarth, R. Wrangham, & T. Struhsaker (eds.), *Primate societies* (pp. 250–265). University of Chicago Press.

Raemaekers, J. (1984). Large vs. small gibbons: Relative role of bioenergetics and competition in their ecological segregation in sympatry. In H. Preuscroft, D. J. Chivers, & N. Creel (eds.), *The lesser apes* (pp. 209–218). Edinburgh University Press.

Reisen, A. H. & Kinder, E. F. (1952). *the postural development of infant chimpanzees*. New Haven, CT: Yale University Press.

Reynolds, V. (1967). *The apes*. New York: Harper.

Richard, A. (1985). *Primates in nature*. New York: W. H. Freeman.

Robinson, J. C. (1988). Demography and group structure in wedge-capped capuchin monkeys (*Cebus olivaceus*). *Behaviour*, **104**, 202–234.

Ruvolo, M. (1994). Molecular evolutionary processes and conflicting gene trees: The hominoid case. *American Journal of Physical Anthropology*, **94**, 89–113.

Sarich, V. (1968). Immunological time scale for hominoid evolution. *Science*, **158**, 1200.

Sarich, V. & Cronin, J. E. (1976). Molecular systematics of the primates. In M. Goodman & R. E. Tashian (eds.), *Molecular Anthropology* (pp. 141–170). New York: Plenum Press.

Schmidt-Nielsen, K. (1984). *Scaling: Why is animal size so important?* Cambridge University Press.

Schultz, A. (1972). *The life of primates*. New York: Universe Books.

Shea, B. T. (1984). An allometric perspective on the morphological and evolutionary relationships between pygmy (*Pan paniscus*) and common (*Pan troglodytes*) chimpanzees. In R. Susman (ed.), *The pygmy chimpanzee* (pp. 89–129). New York: Plenum.

Short, R. V. (1980). The origins of human sexuality. In C. R. Austin & R. V. Short (eds.), *Human sexuality* (vol. 8, pp. 1–33). Cambridge University Press.

Smith, B. H. (1989). Dental development as a measure of life history in primates. *Evolution*, **43**(3), 683–688.

Smith, B. H. (1992). Life history and the evolution of human maturation. *Evolutionary Anthropology*, **1**, 134–142.

Smith, B. H., Crummett, T. L., & Brandt, K. L. (1994). Ages of eruption of primate teeth: A compendium for aging individuals and comparing life histories. *Yearbook of Physical Anthropology*, **37**, 177–231.

Smith, J. M., Burian, R., Kaufman, S., Alberch, P., Campbell, J., Go, Lande, R., Godwin, B., Lande, R., Raup, D., & Wolpert, L. (1985). Developmental constraints and evolution. *Quarterly Review of Biology*, **60**(3), 265–285.

Stanley, S. M. (1973). An explanation for Cope's rule. *Evolution*, **27**, 1–26.

Stearns, S. C. (1992). *The evolutiòn of life histories*. Oxford University Press.

Strier, K. B. (1996). Male reproductive strategies in New World monkeys. *Human Nature*, **7**(2), 105–123.

Tanner, J. M. (1990). *Foetus into man*. (revised and enlarged edn). Cambridge, MA: Harvard University Press.

Terborgh, J. (1983). *Five New World primates*. Princeton University Press.

Thompson-Handler, N., Malenky, R. K., & Badrian, N. (1984). Sexual behavior of *Pan paniscus* under natural conditions in the Lomako Forest, Equateur, Zaire. In R. L. Susman (ed.), *The pygmy chimpanzee* (pp. 347–368). New York: Plenum Press.

Tutin, C. & MacGinnis, P. R. (1981). Chimpanzee reproduction in the wild. In C. E. Graham (ed.), *Reproductive biology of the great apes* (pp. 239–263). New York: Academic Press.

Tuttle, R. (1986). *Apes of the world: Their sociatization, behavior, mentality, and ecology*. Park Ridge, NJ: Hayes.

Watts, E. (1985). Adolescent growth and development of monkeys, apes, and humans. In E. Watts (ed.), *Nonhuman primate models for human growth and development* (pp. 41–65). New York: Alan R. Liss.

Watts, E. (1990). A comparative study of neonatal skeletal development in cebus and other primates. In D. M. Fragaszy, J. G. Robinson, & E. Visalberghi (eds.), *Adaptation and adaptability of capuchin monkeys* (vol. 54, pp. 217–224). Basle: Karger.

Wrangham, R. (1986). Evolution of social structure. In B. Smuts, D. Cheney, R. Seyfarth, R. Wrangham, & T. T. Struhsaker (eds.), *Primate societies* (pp. 382–305). University of Chicago Press.

3

The frontal lobes of the great apes with a focus on the gorilla and the orangutan

KATERINA SEMENDEFERI

INTRODUCTION

Major developments have taken place in the field of neuroscience during the last two decades. We have seen an explosion in imaging techniques used in the study of the structure and function of the living human brain, a variety of improved staining and immunohistological techniques used in post-mortem material, different tracing methods, and stereological and quantitative tools among others. Most of these innovative techniques, however, have not yet been incorporated in the comparative study of the primate brain nor in questions concerned with the evolution of the human brain.

There is a large body of knowledge about the brains of several monkey species and in particular about that of the macaque, as well as steadily growing information about the human brain. Nevertheless, very little is known about the brains of the apes and in particular the brains of gorillas and orangutans. The statement made by Tuttle (1986), more than ten years ago is still very much correct:

> are there specific areas of ape central nervous systems (CNS) that underpin these capabilities and are they homologous with areas in the human CNS which are thought to make speech and other symbolically mediated behaviors possible? We are greatly hampered in the exploration of these questions by the fact that, to some extent, the "ape condition" must be interpolated from experiments on monkeys and clinical observations on humans. (p. 175)

What is the underlying neural circuitry responsible for the complex behavior of the hominoids? What can be said about the evolution of the hominoid brain and cognition? What are the anatomical features in the complex organ of the mind, the brain, that can be identified as species-specific and which of these features are shared with our closest living relatives, the orangutan, the gorilla, the chimpanzee, and the bonobo? Such questions are still waiting to be answered. Earlier quests for answers to similar questions focused mostly on relating overall brain size to body size across several species. Today, we know that, although overall brain size is important, any attempt to better understand neural adaptations demands more complex approaches to the study of specific components of the brain.

Some parts of the brain have been more closely associated with higher cognitive functions than others. Creative thinking, planning of future actions, decision-making, artistic expression, aspects of emotional behavior, as well as working memory, language, and motor control, are functions attributed mostly to the frontal lobes. While many of these behaviors constitute a major part of our notion of being human, some of them are also thought to be present in all or at least some of the great apes. The frontal lobes could be indeed a starting-point for questions relating to species-specific cognitive abilities and attributes. What is known regarding the size, sulcal patterns and organization of the brains and in particular the frontal lobes of the great apes? How large is the frontal lobe and its main sectors (dorsal, mesial, orbital) in the gorilla and the orangutan?

Although earlier in this century, deficits relating to the frontal lobe used to be lumped together as the "frontal lobe syndrome," it is now better understood that the frontal lobe is a composite of many distinct neural circuits and underlying cytoarchitectonic areas. A plethora of experimental studies has been performed on many of these areas in several monkey species during the last decades, while, more recently, noninvasive techniques have been used to study the human brain. Some of these latter techniques, together with novel approaches to the study of post-mortem brain specimens of the apes, provide insights to the understanding of these, our closest relatives.

Our present knowledge of the structure of the brain of the apes constitutes only a fragment of the understanding of the complexity that characterizes the organ of the mind. Here we will review older available information on the frontal lobes of the more frequently forgotten relatives, the gorillas and the orangutans. We will also discuss contemporary approaches to the study of the brain of the apes that include the application of some of the most recent developments in the neurosciences.

FRONTAL LOBE AND COGNITIVE PROCESSES

The frontal lobe of all primates, with the exception of the prosimians, has a similar overall external layout. It lies anterior to the central sulcus and can be divided into two major components: a posterior sector containing the motor cortices and an anterior sector containing the prefrontal association cortices (Figure 3.1). The motor cortices include several separable components: the primary motor cortices and the premotor cortices, which are again subdivided into lateral and mesial (or supplementary motor areas). The prefrontal cortex lies rostral to the motor-related cortices and extends to the frontal pole, the most anterior sector of the frontal lobe. The great apes and humans have large frontal lobes (Semendeferi, Damasio, Frank, & Van Hoesen, 1997a) characterized by major primary and numerous secondary sulci and gyri (Figure 3.2).

The frontal lobe is comprised of motor and premotor cortices as well as limbic and higher-order association cortices, the latter usually designated as prefrontal cortices. In humans, the frontal lobe initiates movement and participates in problem-solving, decision-making, planning, generation of willful responses, regulation of social behaviors, emotion, affect and autonomic function, as well as regulation of language and thought process (Damasio, 1985).

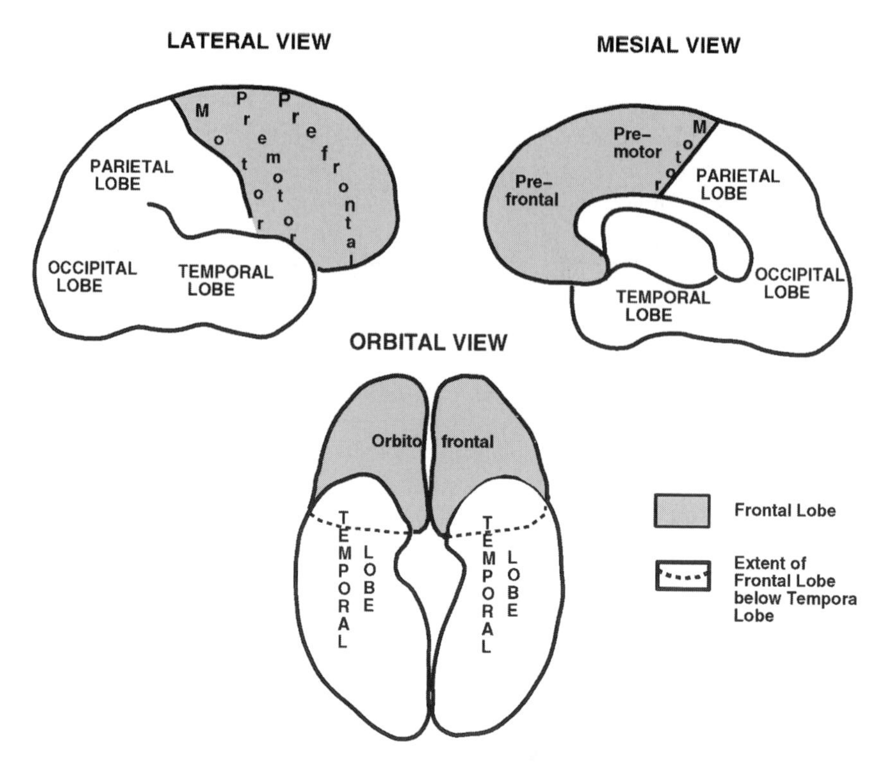

Figure 3.1 Diagram of the lobes of the hominoid brain. Frontal lobe includes motor, premotor, and prefrontal cortices.

The motor cortices (Brodmann's area 4 or M1; Figure 3.3) are somatotopically organized for contralateral body segments, with the phonatory apparatus and hand areas accorded the largest share. The premotor region (Brodmann's cytoarchitectonic area 6) is part of a network used during motor programming; it receives projections from sensory cortices and subcortical structures (basal ganglia) and serves as an integrating region for incoming information related to movement, before final peripheral excitation through corticospinal tract neurons takes effect in the primary motor region. The mesial aspect of area 6 was recently shown to be anatomically discontinuous with the rest of the premotor cortex (Zilles et al., 1995) and constitutes a functionally separate region as well. It contains a contralateral body map, but the representation is horizontal (head anterior) instead of the vertical distribution seen in M1. It is usually designated as supplementary motor area or M2. Also on the mesial side of the frontal lobe lies the cingulate cortex, whose anterior sector has been involved in motivation among other limbic functions. Immediately anterior to area 6 we find area 8, which on the lateral and mesial surfaces of the hemisphere is known as the frontal eye field, because it is involved in voluntary eye movements.

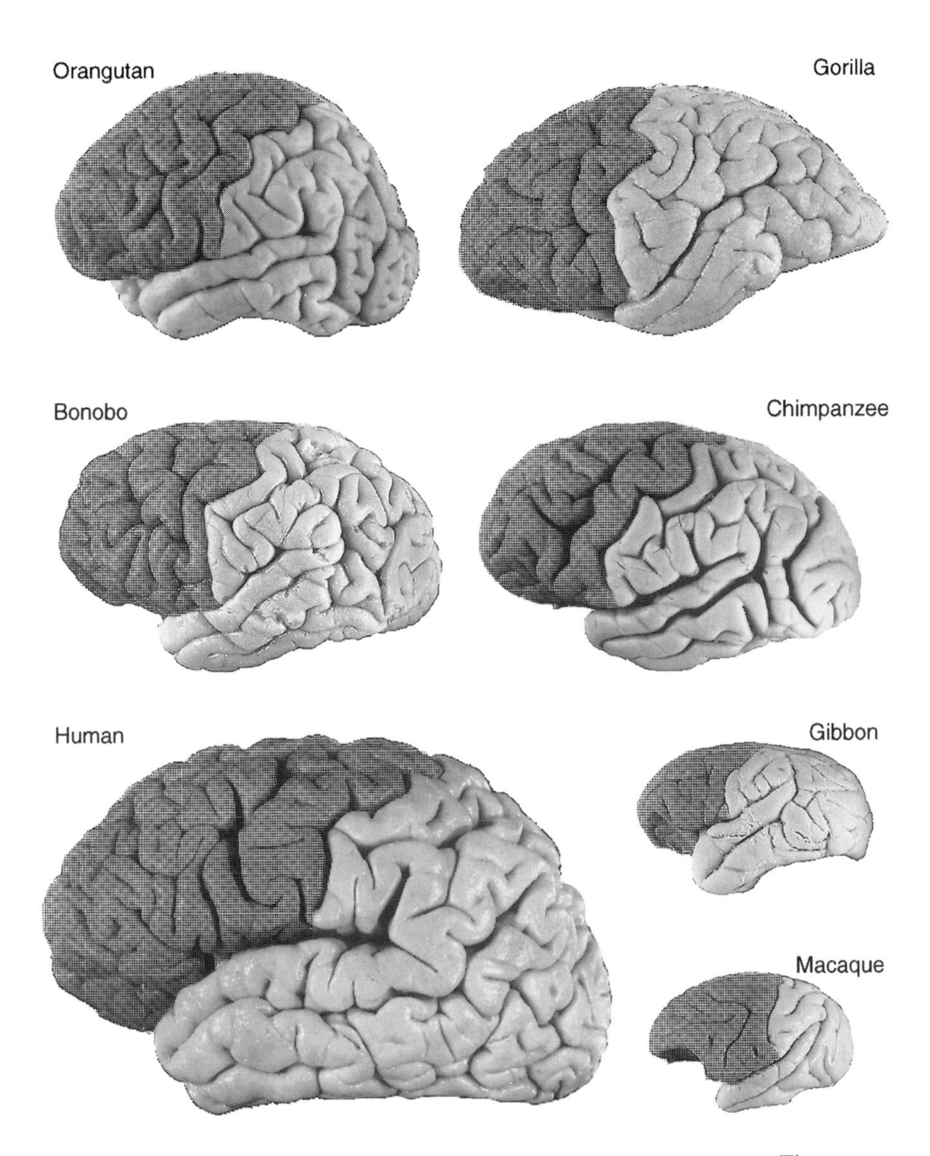

Figure 3.2 Photographs of the lateral view of the brain of all hominoids and the macaque. The frontal lobes are highlighted. The relative scale is only an approximation.

Areas 44 and 45 on the left or dominant hemisphere are usually designated as Broca's area and have been known for more than a century to be important in speech and language. If damaged, aphasia is caused, while damage to the equivalent areas on the nondominant hemisphere may alter the prosodic qualities of speech. Furthermore, it is now well established that language production and comprehension involves multiple brain regions. Some are directly involved with the transient reconstruction and explicit

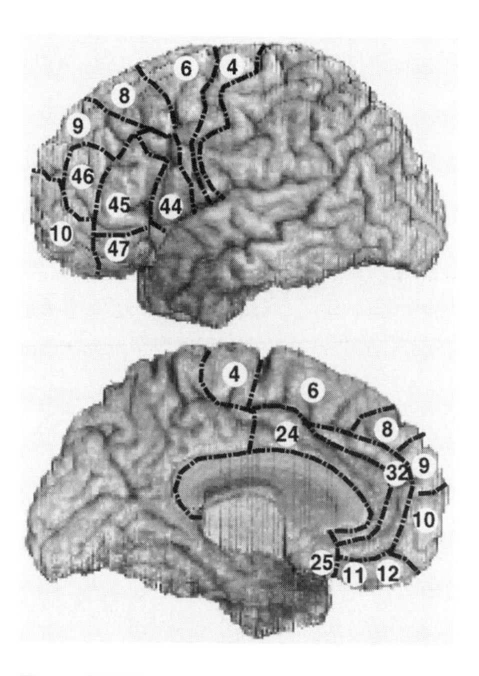

Figure 3.3 Diagramatic representation of frontal lobe cortical areas as seen on a 3-D reconstuction based of Magnetic Resonance images obtained from a living human brain. The lateral view of the left hemisphere is shown above and the mesial view of the same hemisphere is shown below (courtesy of Hanna Damasio).

phonemic representation of word forms, and are often designated as the perisylvian structures, which include auditory, somatosensory, and motor cortices. The classic Broca and Wernicke areas are part of this group. Other regions (in the left temporal lobe) represent additional neuronal sites that mediate between those that support conceptual knowledge and those that implement language production or comprehension (Damasio & Damasio, 1992). It was recently proposed that these language areas are not a single, uniform mediational site for all words, but rather separable regions within a large network which preferentially assists with the processing of words denoting entities that belong to distinct conceptual categories (Damasio et al., 1996).

Among the limbic cortices contained within the frontal lobes are the posterior regions of the orbital surface (areas 13, 47), which when damaged lead to disturbances of social behavior, as for instance the inability to recognize the social value of real-life situations and the inability to plan future actions to the advantage of the subject. Lesions in the orbital sectors of the prefrontal cortex produce major changes in personality and emotional states. Irritabiltiy, euphoria, intolerance, sudden depression, impaired social judgment and facetiousness with sexual content, are some of the manifestations described over the last century on the subject (as described in Damasio & Van Hoesen, 1983). Individuals with damage to ventromedial frontal cortices exhibit sociopathic behavior with an inablility to activate mechanisms that are linked to punishment and reward (Damasio, Tranel, & Damasio, 1990).

Higher-order prefrontal association cortices (areas 46, 9, 10, 11, 12) have been involved in decision-making, planning, creativity, and the regulation of affect and emotion. On the other hand, lesions in the dorsolateral portion of the prefrontal cortex are associated with impairment of working memory (Goldman-Rakic, 1984), as well as with impairment of higher-cognitive abilities that allow the extraction of meaning from ongoing experiences, the organization of mental contents that control creative thinking and language, as well as artistic expression and the planning of future actions (Damasio, 1985). Small lesions in the prefrontal cortices may produce minor impairments that are difficult to detect and tend to recover. Large bilateral prefrontal lesions, however, although they leave motor, perceptual, and language functions intact, are incompatible with maintaining a socially adapted, fully self-conscious and creative life.

The frontal lobes of the great apes were studied experimentally during the first part of the century using ablations and electrophysiology, as well as histology of post-mortem brain tissue for cyto- and myeloarchitecture. Most studies addressed one of the great apes (usually the chimpanzee) and rarely included representatives from other species.

The motor cortex of the orangutan, gorilla, and chimpanzee was studied experimentally by electrical stimulation in the beginning of the century by Leyton and Sherrington (1917). According to these authors, the motor cortex "embraces almost all of the free surface and a large part of the sulcal surfaces of gyrus centralis anterior; it also extends over the mesial border upon gyrus marginalis for a distance about half-way toward sulcus cinguli" (p. 218). Several individuals were included in their study and interindividual variability was noted in the motor responses to the stimulation of identical sites. These authors also found differences between the right and left motor areas of the same individual. "The motor area for face and tongue movements seems, relatively to the rest of the motor area, more extensive in orang than in chimpanzee... The largest and most highly developed brain we examined was that of a gorilla, and the motor area in that specimen appeared to be, on the whole, the most extensive and differentiated of those experimented upon" (Leyton & Sherrington, 1917, p. 219). Apart from the above distinctions, there seemed to be no clear difference between the motor area from different species of the anthropoids examined. Furthermore "stimulation of the middle and posterior parts of the inferior frontal convolution of the left hemisphere failed in chimpanzee, orang, or gorilla to evoke any vocalisation. Ablation of a large portion of that area in one chimpanzee, chosen because it was a noisy and vociferous animal, produced no obvious impairment or change in vocalisation" (1917, p. 220).

Two other chimpanzees, Becky and Lucy, were the protagonists of an experiment on the functional attributes of the prefrontal cortex, which led to the development of frontal leucotomy in humans with psychotic disturbances (Moniz, 1936). Becky was an affectionate female chimpanzee, while Lucy was described as a crotchety old-maid who bit anyone who tried to approach her. Both were tame and seemed well acclimated to captivity, but had different temperaments (Jacobsen, Wolfe, & Jackson, 1935). Both

Becky and Lucy had one and later both frontal association areas removed during different stages of the experiment. Before the operations, they were both capable of performing several tests (delayed response, problem box, stick-and-food). After bilateral removal of the prefrontal cortices, although they exhibited no paresis, they were likely to suddenly stop in the middle of a movement sequence and engage in something else, before they would return and open the box or obtain the food. Becky's and Lucy's temperament also changed strikingly. They did not show their usual excitement, but rather knelt quietly before the cage or walked around. Whenever they made a mistake during an experiment, they showed no emotional disturbance, but quietly waited for the next trial. "[i]f the animal failed, it merely continued to play or to pick over its fur . . . It was as if the animal had joined the 'happiness cult of Elder Micheaux,' and had placed its burdens on the Lord" (Jacobsen et al., 1935, p. 10). After extirpation of frontal association areas in the two chimpanzees, "a marked reduction in complexity of behavior" was recognized (Jacobsen et al., 1935, p. 2).

No comparative studies are available addressing lesions of frontal cortices in the orangutan or the gorilla. Such information has to be extrapolated from studies on monkey species or humans and chimpanzees and then be related to the underlying organization of the brain for each species.

EXTERNAL MORPHOLOGY OF THE FRONTAL LOBE

Sulcal patterns

The external morphology of the brain and of the frontal lobe of the orangutan and the gorilla was described by several investigators in the earlier parts of the twentieth century. Selected works include those by Papez (1929), Tilney and Riley (1928) who described the brain of both apes, and Bolk (1902), Mingazzini (1928) and Papez (1929) who studied the brain of the orangutan. Retzius (1906) provided a photographic documentation of the different views of all ape brains, including the orangutan and the gorilla. A larger comparative study of the external morphology of the brain of the gorilla and the orangutan was conducted by Conolly (1950) on a large number of individuals, while more recently Zilles and Rehkaemper (1988) described the orangutan brain.

According to Conolly (1950), the orangutan brain has a distinct shape that includes a high frontal lobe, a prominent rostral keel and a deeply excavated orbital region. Furthermore, "There is an extensive variability in the frontal lobe of the orang-utan, ranging from schematic simplicity in some, generally young specimens, to a great complexity and apparent instability – features which make it perhaps the most interesting of the great ape brains" (Conolly, 1950, p. 77). Zilles and Rehkaemper (1988) also describe the orangutan brain as having a remarkable height on the lateral view that reaches nearly three-quarters of its total length. They add that a typical feature of the orangutan brain is the keel-like form of the orbital sector of the frontal lobe which can easily be observed when the brain is viewed from the front (Figure 3.4). According to them, the orangutan brain appears to be more similar to the gorilla brain than to the chimpanzee brain.

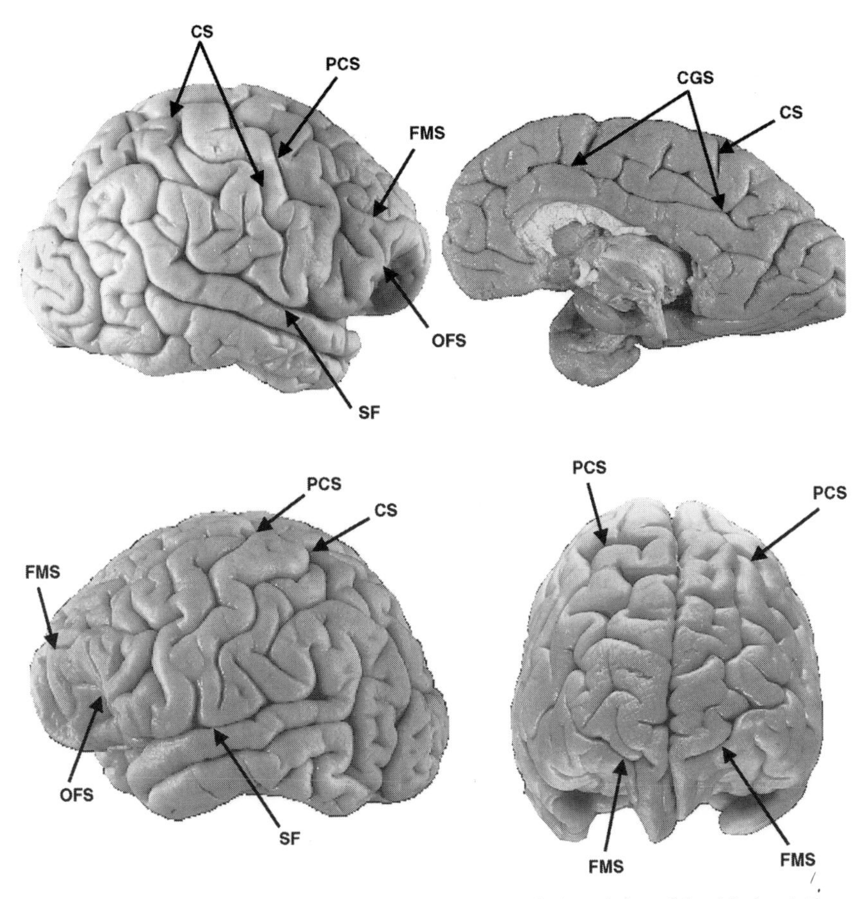

Figure 3.4 Photographs of the brain of an orangutan showing the lateral view of the right hemisphere (upper left), the lateral view of the left hemisphere (lower left), and the frontal view of both hemispheres of the same specimen (lower right). The mesial view of the right hemisphere of another specimen is seen in the upper right. PCS: precentral sulcus; CS: central sulcus; FMS: frontomarginal sulcus; OFS: orbitofrontal sulcus; SCA: subcentral anterior sulcus; SF: Sylvian fissure; OLS: olfactory sulcus; ORS: orbital sulci. Note that here the precentral sulcus is continuous in both hemispheres.

On the lateral view of the gorilla brain (Figure 3.5), we can identify the central sulcus as extending from the most dorsal part of the hemisphere to the lower parts of the lateral surface, close to the Sylvian fissure. There is variability in the sulcal pattern of each hemisphere in both the gorilla and the orangutan, and in some cases the central sulcus notches to the mesial surface (as in most human brains), while in other cases it is present only on the lateral surface (as in the brains of most gibbons and macaques).

The precentral sulcus, is located anterior to the central sulcus and runs parallel to it on the lateral surface. In some of the gorilla as well as the orangutan hemispheres, there is a rather long, continuous precentral sulcus (Figures 3.4 and 3.5), while in others this

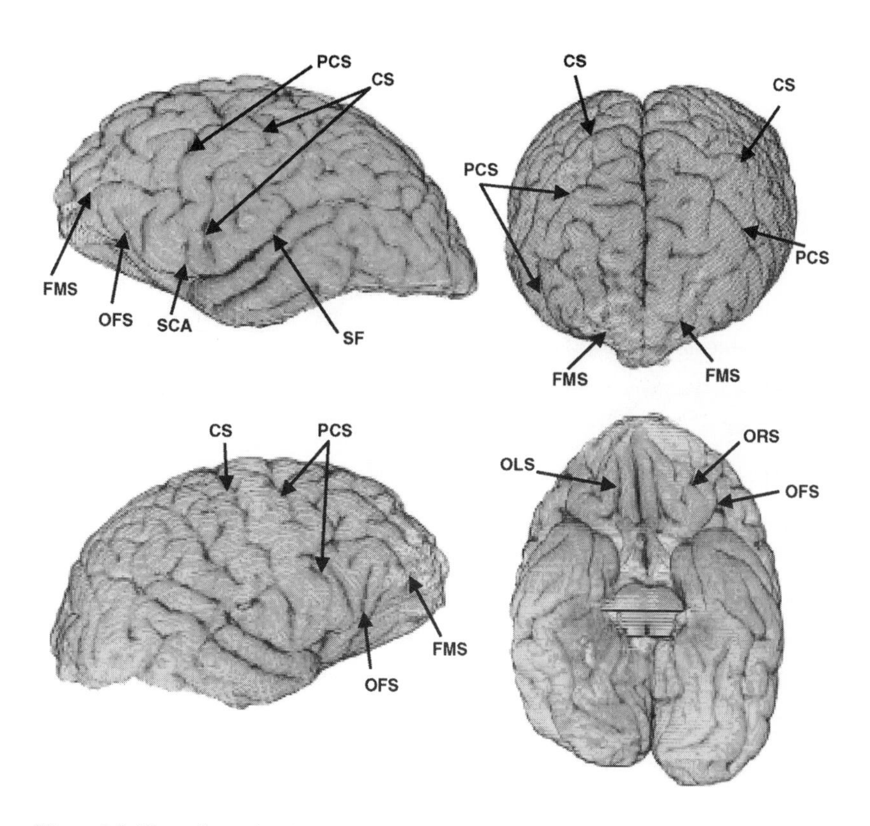

Figure 3.5 Three-dimensional reconstructions of a Magnetic Resonance scan of a gorilla post-mortem specimen showing the lateral view of the left hemisphere (upper left), the frontal view of both hemispheres (upper right corner), the lateral view of the right hemisphere (lower left), and the orbital view of both hemispheres (lower right). Abbreviations for the sulci are as in Figure 3.4. Note that the PCS is continuous in the left hemisphere but in the right hemisphere it has two discontinuous segments.

sulcus is segmented into smaller segments that run more or less parallel to the central sulcus (Figure 3.5). Such findings are also present in human brains, where in many instances the precentral sulcus is composed of two segments, a superior and an inferior, while in others it is long and uninterrupted (Ono, Kubik, & Abernathey, 1990). Neither the gibbon nor the macaque brains have a sizeable precentral sulcus, but instead, exhibit a couple of small sulci anterior to the central sulcus on the higher and lower sectors of the lateral surface of the hemisphere (Figure 3.2). The sulcal pattern of orangutans and gorillas is similar to the pattern seen in chimpanzees.

The cortex anterior to the precentral sulcus on the lateral surface of the hemisphere is characterized by a complex fissural pattern in all hominoids. The fissural pattern of the prefrontal cortex of the great apes is more similar to the very complex human pattern than to the rather simple pattern seen in other primates. It certainly distinguishes the great apes from the other primates, placing them together with humans in a group with a unique hominoid fissural pattern. Some of the most prominent frontal sulci in the human brain include the superior, intermediate, and inferior frontal sulci, as well as the sulci in the frontal operculum, the horizontal ramus of the Sylvian fissure, the ascending ramus of the Sylvian fissure and the diagonal sulcus, which separate the operculum into three separate segments (pars orbitalis, pars triangularis and pars opercularis). The subcentral anterior sulcus also belongs to the frontal lobe and is to be found anterior to the central sulcus on the lower part of the lateral surface, while in the frontal pole we find the frontomarginal sulcus. On the mesial surface is the cingulate sulcus, and on the orbital surface we have the olfactory sulcus and a complex set of orbital sulci (see, for instance, Duvernoy, 1991; Damasio, 1995).

The gorilla and the orangutan present on the lateral surface of the frontal lobe two or three major sulci that run in an anterior–posterior direction, connecting in most cases with the precentral sulcus and its complex ramifications. There is also an orbitofrontal sulcus to be found, a rather long ascending sulcus that starts on the orbital surface of the brain and extends onto its lateral surface contributing to the formation of an operculum in some individuals among the gorillas and the orangutans. This sulcus is typically present in the brain of the apes and is well known in the anthropological literature. Absence of this sulcus, but presence of the inferior frontal sulcus (representative of the human pattern) in hominid fossil endocasts has been suggested to mark the presence of Broca's area and consequently the first appearance of language abilities (Falk, 1983). Close to the orbitofrontal sulcus is the subcentral anterior sulcus. A frontomarginal sulcus is present in the frontal pole region in the great apes and is not to be confused with the sulcus "rectus" or "principalis" typical of the brains of monkeys and gibbons. This was erroneously suggested by Conolly (1950) who based the homology simply on the similar location of the sulci, but did not take into account the cortical morphology underlying the frontal pole. However, it was recently shown that on the basis of the underlying cytoarchitecture this sulcus traversing the frontal pole in the hominoids is homologous to the human frontomarginal sulcus, which is situated in the frontopolar gyri formed by cortical area 10 (Semendeferi, 1994).

On the mesial surface of the gorilla and orangutan brains there is, as in the human brain, a long, continuous sulcus, the cingulate sulcus, which may show side branches in some orangutans and gorillas (as it does in many humans). On the orbital surface, along with the ever present olfactory sulcus, both species present a complex set of orbital sulci, running in an anterior–posterior direction with some crossings in a mesial to lateral direction. Several smaller, secondary sulci are present in both species and their location, size, and orientation varies from one individual to another.

Unlike the complex prefrontal fissurization of the hominoid brain, in gibbons and macaques we see a prominent long sulcus rectus (also called principalis), a V-shaped arcuate sulcus (which in the gibbon is more straight), a small precentral superior sulcus, a subcentral anterior sulcus, a small frontal superior sulcus and an orbitofrontal sulcus.

The above description of the external morphology of the frontal lobe in the orangutan and the gorilla focused on the similarities rather than the differences in the configuration of the gyri and the complex sulcal patterns of the two great apes. A discussion of the differences in the external morphology between the two species requires some understanding of interindividual variability that we know is present among the great apes, just as it is known to be present in the human brain (Ono et al., 1990). Non-invasive techniques (3-D reconstructions based on Magnetic Resonance scans) are presently being used to address many of these questions on the living brain of all great apes (Semendeferi & Damasio, 1997).

Asymmetries of the frontal lobe structures

In humans, the anteriormost portion of the right hemisphere was found to be wider and protrude farther anteriorly than its counterpart (Le May, 1976; Galaburda, LeMay, Kemper, & Geschwind, 1978). These findings are in agreement with results from another study based on endocasts of hominoid skulls (Holloway & De La Costelareymondie, 1982), in which all human endocasts examined (total of 14) were also found to be wider in the right sector.

In the apes, LeMay, Billig, and Geschwind (1982) found some asymmetries in the width and in the length of the frontal lobes of some, but not all, of the specimes studied. Regarding the maximum *width* of the frontal lobes, most of the gorillas (4 out of 6 specimens) presented a right preponderance, with the remaining specimens showing no difference between the two sides. In the case of the orangutans, asymmetry in the width of the frontal lobes was present in half of the specimens examined, but absent in the other half. In contrast to the gorillas, most of the chimpanzees examined (4 out of 6) did not present any asymmetry in the width of the frontal lobes. Regarding asymmetries in the *length* of the frontal sector of the hemisphere (also called petalias), most gorillas (4 out of 6), most orangutans (5 out of 6) and half of the chimpanzees showed greater length on the right, a right petalia.

Holloway and De La Costelareymondie (1982) also found that in contrast to humans, the gorilla specimens could be split into two groups. Half of them (out of a total of 40) had no asymmetry in the width of the frontal lobe and the other half showed a right petalia. In a fashion very similar to the gorillas, more than half of the twenty orangutan specimens studied had no frontal petalia, while the remaining showed a right petalia (there was one exception with a left frontal petalia) . Regarding the chimpanzees, most of the bonobos had a right frontal petalia, while most of the common chimpanzees did not.

These two studies (Le May et al., 1982; Holloway & De La Costelareymondie, 1982) suggest that, while most humans have asymmetries in the width of the frontal lobes favoring the right hemisphere, apes present larger variability in regard to this feature. The gorilla and bonobo patterns are closer to the human pattern with more than half of the specimens studied presenting a right frontal advantage. When comparing humans and apes regarding asymmetries in the *length* of the frontal lobe (Le May et al., 1982) all hominoids present a more homogeneous picture, in that half or more of all specimens show a right frontal petalia.

Anatomical asymmetries are also present in other sulcal patterns of the right and left hemispheres. The ends of the central sulcus present some variability mostly pertaining to the upper end. LeMay & Geschwind (1975) found that the lower end of the central sulcus lies more posteriorly in the left than in the right hemispheres in most orangutan brains (9 of 12 brains) suggesting a larger left frontal lobe. More studies on asymmetries of the brain of the great apes are necessary to settle the questions still remaining and such attempts are now being undertaken (Semendeferi, Rilling, Insel, & Damasio, 1996; Semendeferi, Damasio, Rilling, & Insel, 1997b).

SIZE OF THE BRAIN AND THE FRONTAL LOBE

The weight of the brain has been documented by many investigators, and several reports exist including post-mortem specimens of gorillas and orangutans, as well as chimpanzees that date as far back as the last century (Bischoff, 1876; Fick, 1895a,b; Kennard & Willner, 1941). Since the weight of the brain tissue can vary considerably depending on whether the sample is obtained from fresh tissue or after fixation (as discussed extensively by Blinkov & Glezer, 1968; Stephan, Baron, & Frahm, 1991), meaningful comparisons of values across studies can prove to be a challenging task. Furthermore, there seems to be a certain degree of individual variation and sexual dimorphism in the size of the brain of the apes, which may in part explain why, according to some investigators, the gorilla has the largest brain, followed by the orangutan or the chimpanzee as the second brainiest of the apes, while others place them in a different order.

More recent measurements have also included weights of the brain tissue of the gorilla (Stephan, Frahm, & Baron, 1981) and the orangutan (Zilles & Rehkaemper, 1988) and volumes of cranial capacities (Tobias,1971) of all great apes. Since the relationship between the two measures is thought to be very close (Falk, personal communication), values are directly comparable. Stephan et al. (1981) reported on the brain volume for the gorilla (500 cc) and also for the chimpanzee (405 cc), gibbon (102 cc) and human (1330 cc). The same year, Zilles & Rehkaemper summarized a variety of previous reports on the brain volume for the orangutan (11 males and 14 females) and came up with a mean fresh brain weight of 333 g, where the males had a mean brain weight of 359 ± 13.9 g and the females 306 ± 14 g. Pilbeam and Gould had summarized earlier (1974) values on cranial capacities from the existing published data at the time for all apes. The mean value they established for the male gorillas was 550 cc and for the female gorillas 460 cc. In the case of the orangutans, males had a mean brain volume of 415 cc and females 370 cc. Furthermore, in their review they include values for the chimpanzee (males 410 cc and females 380 cc) and the bonobo (males 355 cc and females 339 cc), which exhibit less sexual dimorphism than gorillas and orangutans.

Even before the Pilbeam and Gould study, Tobias (1971) had measured the cranial capacities of a large number of individuals in all hominoids; he reported that gorillas had a mean size of cranial capacity of 534.5 cc, orangutans 434.4, chimpanzees 398.5 cc, and, as expected, gibbons only 104 cc, while the cranial capacity for humans was 1345 cc. These values were obtained from the measurements of large cohorts of gorillas (414), orangutans (203), chimpanzees (163), gibbons (95), and humans (1000).

One of the reasons driving early interest in the overall size of the brain was the attempt to identify factors responsible for the cognitive and behavioral differences among the species and in particular among primates. Later, it became understood that mere comparisons of brain to body size are not an adequate measure of intellectual complexity and that more sophisticated and cumbersome approaches are necessary, involving the search for differences, even in size, of subdivisions of the brain at the macroscopic level (absolute or relative sizes of, e.g., lobes, cortex, white matter), or the search for differences of microscopic components (e.g. numbers and size of neurons, density of fibers, distribution and frequency of neurotransmitters etc.).

Going back to the middle of the century, Blinkov and Glezer (1968) had provided one of the most comprehensive attempts to describe qualitatively and quantitatively the complexity of the nervous system of humans and to compare some of their human data with certain mammalian species including orangutans and chimpanzees. The parameters investigated by these authors include the surface of the cortex of the cerebral hemisphere and of its subdivisions. They specifically reported the neocortex of the occipital, temporal, inferior parietal, limbic, and frontal lobes of the brains of humans and orangutans, as well as of the chimpanzee. Blinkov and Glezer (1968) mentioned in particular the frontal lobe, stating that the surface area of the "frontal region" (prefrontal cortex) and of the precentral region (motor and premotor cortex) in relation to total surface of the hemisphere was: 32.8% in the human, 22.1% in the chimpanzee, 21.3% in the orangutan, and 21.2% in the gibbon.

Studies of the size of the frontal lobes in gorillas and orangutans are rare. However, as early as 1928, Tilney estimated the surface area of the frontal lobe in relationship to the neocortex. He found it to be: 32% for the gorilla, 33% for the chimpanzee, and 47% for the human. Sixty years later, Uylings and Van Eden (1990) readdressed the question of the volume of the prefrontal cortex and found its relative size to be similar in the human and the orangutan (approximately 29% of the total volume of the isocortex in each species), while for the macaque it was 18% and the marmoset a mere 13%.

Contemporary applications of imaging techniques

Let us turn now to current studies of these same questions. In a recent study performed in the Department of Neurology at the University of Iowa, Semendeferi et al. (1997a) used three-dimensional reconstructions of Magnetic Resonance (MR) brainscans of post-mortem specimens to compare the size of the frontal lobe in all extant hominoids (humans, great apes, lesser apes, and the macaque). The great ape brain specimens were obtained after the natural death of the adult animals due to non-neurological causes and included a female chimpanzee, a male gorilla, and a male orangutan. Measurements involving brainscans of post-mortem specimens have the advantage of being free of the effects of shrinkage associated with tissue processing for histological slices (even if correction factors have been established to compensate for these effects, and are being used in several studies; Frahm, Stephan, & Stephan, 1982; Semendeferi, 1994, 1995). *In vivo* scans that are currently being analyzed (Semendeferi et al., 1996; 1997b; Semendeferi & Damasio, 1997) have the further advantage of being free of shrinkage

Table 3.1 *Volume of hemisphere, frontal lobe and its sectors*[a]

	Hemisphere	*Frontal Lobe*	*Dorsal*[b]	*Mesial*[b]	*Orbital*[b]
Orangutan	268 553	94 705	43 964	23 512	8 059
(male)		35.3%	58%	31%	11%
Gorilla	348 336	112 912	50 603	27 497	15 671
(male)		32.4%	54%	29%	17%
Chimpanzee	305 521	109 800	51 274	23 199	14 142
(male)		35.9%	58%	26%	16%

[a] Values were obtained from MR scans of postmortem specimens fixed in formalin. They are given in mm^3 and include both hemispheres.
[b] Sectors include cortex and immediately underlying white matter.

related to autolysis time and preservation method, which the study of post-mortem specimens does not eliminate.

In the above mentioned study, the frontal lobe was subdivided into its traditional anatomical subdivisions, dorsal, mesial, and orbital (Figure 3.6), because they are known to be involved, to a greater or lesser extent, in the functions described earlier. The measurements also included the whole brain as represented by the two hemispheres, and differences among species were addressed in both absolute and relative terms. All measurements were obtained using Brainvox, a suite of programs designed to study the brain, both qualitatively and quantitatively (Damasio & Frank, 1992; Frank, Damasio, & Grabowski, 1997).

With respect to the size of the hemispheres, the gorilla brain was larger than the orangutan brain (348 cm^3 and 268 cm^3 respectively; Table 3.1). The size of the frontal lobes was somewhat larger (in absolute terms) in the gorilla brain (113 cm^3) than in the orangutan brain (95 cm^3). In relative terms however, the frontal lobe of the orangutan was found to occupy a larger percentage of the hemisphere than the frontal lobe of the gorilla. I should point out that, as this study was carried out in only one specimen of each species, the finding will have to be replicated in a larger sample before we can generalize these differences. Preliminary results of identical measurements obtained in the living brain of the orangutan and the gorilla, however, confirm the above mentioned differences (Semendeferi et al., 1997b).

The three sectors of the frontal lobe "cortex" (a region including the cortex and immediately underlying white matter) were also larger in the gorilla than in the orangutan in absolute terms, but the differences were modest in 2 of the 3 sectors (Table 3.1). The dorsal sector was 50 cm^3 in the gorilla and 44 cm^3 in the orangutan, while the mesial sector was 27 cm^3 and 24 cm^3 respectively. The largest difference was found in the orbital sector. Here, the orangutan had a particularly small size with 8 cm^3 versus 16 cm^3 for the gorilla.

The percentages that these sectors occupy in the frontal lobe also vary somewhat between the two species, with the dorsal sector in the gorilla occupying only 54% of the total frontal lobe, while in the orangutan it occupies 58%. For the mesial sector of the gorilla brain, the percentage was 29% and for the orangutan 31%, again larger. However, when it came to the orbital sector, the percentage for the orangutan was only 11%, while for the gorilla it was 17%.

On the basis of such a small sample as this, it appears that the frontal lobe as a whole is somewhat smaller in the gorilla than it is in the orangutan or the chimpanzee. On the other hand, in the orangutan the small size of the orbital sector in absolute and relative terms stands out when this ape is compared to any other hominoid (see Table 3.1).

INTERNAL ORGANIZATION OF THE FRONTAL LOBE

Following the very early studies of the gross anatomy and the cyto- and myeloarchitecture of the orangutan brains (Tiedemann, 1827; Bischoff, 1876; Campbell, 1905; Mauss, 1908, 1911; Tilney & Riley, 1928), there have been only a few studies addressing aspects of the internal organization of the brain of the orangutan (Zilles & Rehkaemper, 1988), and of the gorilla (Stephan et al., 1981; Armstrong, 1982). Armstrong (1982) has addressed the organization of the thalamus of the gorilla, along with that of the chimpanzee, gibbon, and human, and has estimated the number of neurons that form the major thalamic nuclei. In none of the thalamic complexes studied (including the mediodorsal nucleus of the thalamus that is known to be heavily connected with the prefrontal cortex) did she find a major difference between the gorilla and the chimpanzee. The orangutan brain was not part of the study.

Numerous studies have investigated the frontal lobe in several monkey species (Barbas & Pandya, 1989; Preuss & Goldman-Rakic, 1991; Morecraft, Geula, & Mesulam, 1992; Carmichael & Price, 1994), but very little is known about this part of the brain in the apes. Our major sources of information about the overall cortical organization of the orangutan brain remain the works by Campbell and Mauss on major cytoarchitectonic and myeloarchitectonic subdivisions of the cortex at the beginning of this century. Campbell (1905) identified four major regions within the frontal lobe across these species: the precentral, the intermediate precentral, the frontal, and the prefrontal cortices. The extent of these areas and their precise topographic location in relationship to the complex sulcal pattern of the frontal cortices varies between the two species.

The frontal lobe is a mosaic of many cytoarchitectonic fields (Figure 3.3). The prefrontal cortex, the area rostral to the motor and premotor cortices, is also called the frontal higher-order association cortex or the frontal granular cortex, depending on whether we refer to its functional or structural attributes, respectively. Comparative information on the intrinsic architecture of the frontal lobe of the extant hominoids is necessary if we are to attempt to relate structural adaptations to species-specific behaviors. Different cortical areas changed to a different extent over time under different selection pressures for each species, resulting in distinct neural organizations in each extant species (Armstrong, 1990), and thus knowledge of the organization and the size of individual cortical areas is important.

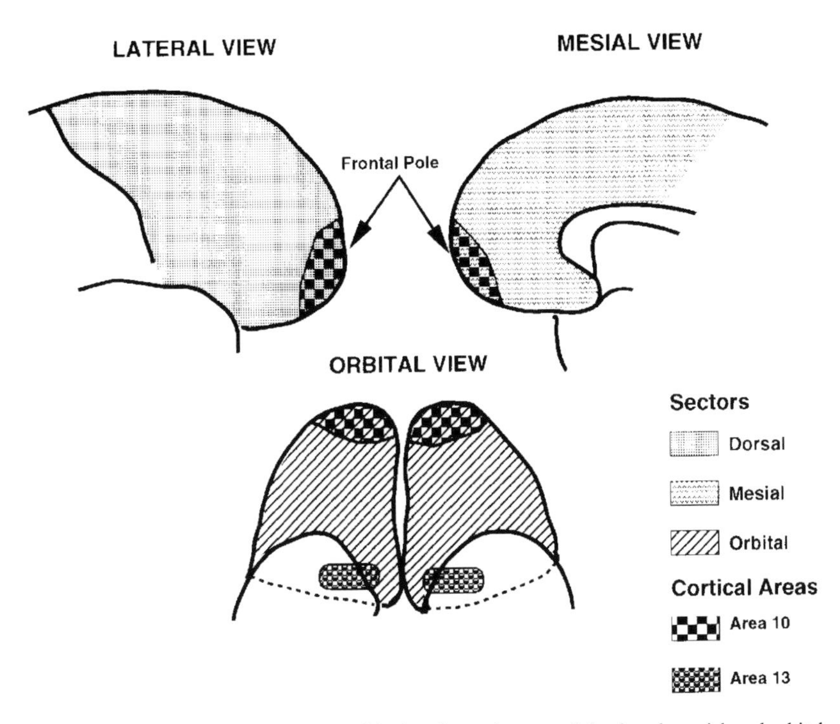

Figure 3.6 Diagramatic representation of the location and extent of the dorsal, mesial, and orbital sectors of the frontal lobe, as well as cortical areas 10 and 13.

Areas 13 and 10 of the prefrontal cortex

The only study addressing the cytoarchitectonic organization of parts of the prefrontal cortex of the gorilla and other apes is that of Semendeferi (1994). In the face of the almost complete absence of information on the hominoid prefrontal cortex, two areas were chosen from different parts of the frontal lobe in this study (Figure 3.6). Area 13 was chosen as part of the limbic related cortices of the orbital surface of the frontal lobe and area 10 as part of the higher-order association cortices in the frontal pole.

The frontal pole in the human brain is the most anterior sector of the frontal lobe. It is thought to be part of the structures important in planning and decision-making; and cortical area 10 is a large component of this sector of the frontal lobe. Cortical area 13 has been known to be part of the posterior sector of the orbitofrontal cortex in the macaque brain, but was not included by Brodmann in his map of the human brain. Today, area 13 is considered to be part of a circuit relevant to emotion, particularly in connection with social stimuli. Are areas 10 and 13 present in the gorilla and the orangutan? Is their internal organization similar among hominoids? And how large are they in the two species?

Area 13 has been known to be present in the cortex of the macaque, and damage to this area has been associated with changes in emotional states and disinhibition of emotional reactions. Removal of this cortex has been found to enhance aversive reactions and reduce aggressive reactions in several threatening situations. These emotional alterations have been interpreted on the basis of the close relationships that the posterior orbital cortex has with the limbic structures, especially the mediodorsal nucleus of the thalamus and the amygdala (Butter & Snyder, 1972). Studies performed in the wild, as well as in laboratory social settings, on monkeys with orbitofrontal ablations, have shown significant reductions and losses of behaviors that are considered important for the maintenance of social bonds (Kling & Steklis, 1976).

A distinct area 13 was identified and described in all great apes and humans in the posterior sector of the orbitofrontal cortex (Semendeferi, 1994). Its structural organization (cytoarchitecture, size of cortical layers, neuronal density, and space for connections) was found to be similar across hominoids, in general terms, but subtle differences were present between species. The size of area 13 is similar in the orangutan and the gorilla.

Area 13 is characterized by a cytoarchitectonic organization that involves the presence of six cortical layers. As in other parts of the cortex, layer I (superficial layer) is acellular, while layers II and III (supragranular layers) are known to have short (layer II) and long association (layer III) connections. Granular layer IV has intrinsic connections, while layers V and VI (infragranular layers) have connections with subcortical structures. All species have a medium layer I and a thin layer II including small to medium granular cells. Layer III is thick and includes pyramidal cells that tend to increase in size towards the border with layer IV. Layer IV is incipient. Layer Va is more densely packed with prominent pyramidal cells, while Vb is more sparsely populated. Layer VI (lower layer, bordering with the underlying white matter) includes pyramidal and fusiform cells and its border with the white matter is not sharp. Area 13 has two particular features that distinguish it from the surrounding cortex in all species: the presence of an incipient layer IV and a horizontal striation of the cells in layers V and VI.

In regard to the distribution of the relative size of the cortical layers, the orangutan presents a slightly distinct pattern from that of the other great apes with a larger granular layer IV and similarly sized supra- and infragranular layers. In the gorilla (as well as in the other African hominoids), area 13 has a more typical limbic appearance in the sense that it includes features that have been known to characterize this part of the cortex in several monkey species (e.g., a smaller layer IV and infragranular layers that are larger than the supragranular ones). In other words, it seems that the orangutan not only has a smaller and less diversified orbitofrontal cortex, but it also includes a limbic area (area 13) that is in some respect less limbic than the homologous area of the other hominoids. This can only suggest a decreased representation of the limbic cortices, at least in the frontal region, in the orangutan.

An interesting finding is that the density of neurons in area 13 of the orangutan is the lowest among the great apes, while that of the gorilla is the highest (see Table 3.2). Furthermore, the space available for connections as determined by the Grey-Level Index (GLI) (Schleicher & Zilles, 1989) is also lowest in the orangutan and highest in the gorilla (GLI estimates the ratio of space occupied by cell bodies versus space available for connections). Low neuronal density in the orangutan could be expected to be followed by increased space for connections. Fewer neurons should allow for more intracellular space available for connections unless fewer neurons would coincide with larger size of neurons. This possible explanation for the odd combination mentioned above still remains to be investigated. Space for postsynaptic connections is particularly increased, especially in the supragranular layers in the gorilla's area 13.

The absolute volume of area 13 (Table 3.2) has no major differences between the orangutan (317 mm^3) and the gorilla (273 mm^3), although such differences can be found in other species (e.g., the bonobo has a small area 13). The same similarity of size is seen when the relative volume of area 13 is considered: the two species have almost identical values of the percentage of area 13 in relation to the total volume of the brain, 0.08% for the gorilla and 0.09% for the orangutan.

It should be noted, however, that although the volume of area 13 in relation to the total hemisphere does not stand out in either species, the relative anterior–posterior extent of area 13 within the orbitofrontal cortex varies considerably between the two species. In the gorilla, only a small part of the posterior orbitofrontal cortex is occupied by area 13, whereas, in the orangutan, a large segment of the orbital surface of the frontal lobe is occupied by a single cortical area (namely area 13). This difference is not surprising since the orbitofrontal cortex of the orangutan is very small. Its total length is 2.4 cm, while in the gorilla it is 3.6 cm. This difference, however, cannot be due to a difference in the total length of the hemisphere, since this measure is very similar in the two species, namely 9.1 cm in the orangutan and 9.7 cm in the gorilla. Furthermore, the extent of area 13 in the anterior–posterior direction in the orbitofrontal cortex is the same in both species (0.54 cm).

Area 10 (Figure 3.6) is thought to be part of the structures important in planning, decision-making, and undertaking of initiatives. The behavior of patients with damage to the frontal cortices including area 10, loses "its goal-oriented character and becomes a series of isolated, immediate reactions to particular stimuli, subject to the influence of an external field or perseverative factors" (Luria & Homskaya, 1962). Such patients were evaluated as maintaining the capacity of anticipating a course of events, but as unable to picture themselves in relation to those events as potential agents (Teuber, 1962).

Area 10 has been known to be present in humans, chimpanzees, and several monkey species, and was recently investigated also in the orangutan and the gorilla (Semendeferi, 1994). The cytoarchitectonic organization of area 10 in the orangutan has a homogeneous appearance and resembles closely that of area 10 in the human and the two chimpanzee species. It is an area characterized by the presence of all six cortical layers and a distinct, well-formed layer IV that includes small pyramidal cells. The supragranular layers (II and III) include medium to large pyramidal cells, that have a homogeneous distribution. These cells become larger, the closer they are to the borders with layer IV. The infragranular layers (V and VI) are wide and include large pyramidal cells that vary in their distribution within the sublayers, with a lower packing density in Vb than in Va. The borders with the white matter are well defined.

Table 3.2 *Size and internal organization of areas 13 and 10*[a]

Species[d]	Volume[b]			Relative Size of Cortical Layers[c]			Density of neurons per mm		Grey – Level Index (GLI)			
	Brain (cm^3)	A 13 (mm^3)	A10 (mm^3)	Cortical Layers	A 13	A 10	A 13	A 10	A Mean	13 Layers	A Mean	10 Layers
Orangutan (male)	356.2	316.6	1611.1				42 400	78 182	18.6		20.1	
				LI	7.0	8.5						
				LII, III	43.0	37.0				17.9		19.8
				LIV	6.0	7.5				19.7		21.5
				LV, VI	44.0	47.0				18.1		19.7
Gorilla (female)	362.9	273.2	1942.5				54 783	47 300	14.6		15.9	
				LI	8.0	10.0						
				LII, III	35.0	34.0				12.9		16.4
				LIV	3.5	6.5				16.7		17.9
				LV, VI	53.5	49.5				15.3		15.1
Chimpanzee (unknown)	393.0	269.9	2239.2				50 686	60 468	18.6		17.5	
				LI	9.0	11.5						
				LII, III	32.5	44.5				17.6		17.7
				LIV	3.5	8.0				19.4		18.5
				LV, VI	55.0	36.0				18.9		18.0
Bonobo (female)	378.4	110.5	2804.9				44 111	55 690	17.0		18.2	
				LI	8.5	10.5						
				LII, III	30.5	46.5				16.7		19.0
				LIV	3.0	8.0				18.1		21.6
				LV, VI	58.0	35.0				16.7		16.2

[a] Values were obtained from histological sections
[b] The figures on the brain refer to total brain structure and on areas 10 and 13 to measurements of the cortex in one hemisphere
[c] As a percentage of total cortical width
[d] The individuals included here are different from the ones included in Table 3.1

Area 10 in the orangutan occupies the frontal pole extending on its mesial and orbital surfaces, as well as along the length of the frontomarginal sulcus on the lateral surface. In the anterior–posterior direction within the walls of the frontomarginal sulcus, area 10 is replaced by another neighboring area with different cytoarchitectonic features including a very prominent layer II. On the mesial side, another cortical area is identified as neighboring area 10 that is characterized by prominent layers II, Va, and VIa. On the orbital side, the cortex succeeding area 10 has a wider layer IV and the pyramidal cells in layer IIIa are larger than those in layer III of area 10.

The gorilla frontal pole cortex has a distinct appearance and, although area 10 (unlike area 13) presents some slight variations among all hominoids, it is in the gorilla that we find the most distinctive architecture. Most frontal pole cytoarchitectonic features described in the orangutan are also present in the gorilla. Some aspects of its organization differ however, including both qualitative and quantitative features. In the gorilla, layers II and Va are very prominent and, as a consequence the cortex does not appear as homogeneous as it does in the other hominoids. Also, layer IV is thinner in the gorilla frontal pole than it is in the other hominoids. There are also differences in the relationships to the neighboring areas. It is on the mesial surface of the frontal pole of the gorilla, and not on the lateral surface (as in the orangutan), that we have the first appearance of a neighboring cortical area. The cortex of the frontal pole in the gorilla extends farther on the dorsal surface of the hemisphere, in contrast to the other hominoids, where it extends the farthest on the orbital surface.

The density of the neurons in area 10 of the orangutan (78 000 neurons per mm^3) is distinct from that of the gorilla (47 000 neurons per mm^3) and the two species have the extreme values found for this area among hominoids, with the chimpanzee values falling in between. It is worth noticing that the density of neurons in all hominoid species is higher for area 10 and lower for area 13, but the gorilla is an exception with exactly the opposite relationship. The GLI values further support the decreased cellular density for the gorilla frontal pole, with a uniquely low value among the great apes (Table 3.2). The density is particularly low in the supragranular layers, indicating more space for synapses relating cortical information.

The volume of area 10 in the gorilla and the orangutan is not very different in absolute terms (1942.5 mm^3 in the gorilla and 1611.1 mm^3 in the orangutan), but it is smaller than in the two chimpanzee species (Table 3.2). In relative terms, as a percentage of the volume of the entire brain, the values are similar. Both the orangutan and the gorilla have area 10 occupying 0.5% of the total volume of the brain, which is slightly less than that found in the chimpanzee (0.6%) or the bonobo (0.7%).

The overall organization of the cortex forming the frontal pole is similar among all hominoids, including the gorilla. One feature of this cortex, the distribution of the relative size of the cortical layers, sets the orangutan and the gorilla apart from the other hominoids. In both species, the supragranular layers are thinner than the infragranular layers, whereas the reverse is the case in the other African hominoids. Nevertheless, if we go beyond the major similarities in the overall organization, and consider all quantitative measurements and qualitative observations, then the suggestion of a unique organization in the gorilla frontal pole cortex appears. Once again I would note that these observations are based on one individual only (both hemispheres) and that more specimens need to be examined before definitive conclusions can be generalized about area 10 in this species. Until then it is most parsimonious to refer to area 10 in the gorilla brain as peculiar or distinct.

CONCLUDING REMARKS

As we have seen, early studies on the frontal lobe of the great apes concentrated for the most part on descriptions of the external morphology and of total size of post-mortem specimens. It has been only recently that computerized imaging techniques have been used to study parts of the brain of the great apes and selected macroscopic and microscopic components of the frontal lobe have been analyzed.

Gorillas and orangutans have large brains, a complex morphology in the frontal lobe, and sulcal patterns that are very distinct from those of any monkey species and similar to those of the other hominoids, including humans. Nevertheless, it is clear that differences are present between the species both in the size as well as in the organization of the frontal lobe.

On the basis of a very small sample, the gorilla appears to have a slightly smaller frontal lobe (in relation to the hemisphere) than the other apes. Nevertheless, due to a complete lack of information regarding intraspecific variation of the brain of the spes, such a finding can be interpreted only with great caution.

If a larger sample were to show that gorillas have smaller frontal lobes, then what could such a finding suggest? Thoughts that come to mind pertain to differences observed in the behavior of the gorillas that, according to some investigators, may set them apart cognitively from the rest of the apes. Can a smaller overall frontal lobe be associated with the suggested differences in self-recognition, mental-mapping abilities used in food acquisition, technical skills in tool use, or even social sophistication? Did older claims involving such behavioral differences at the expense of the gorillas (e.g., Suarez & Gallup, 1981) finally meet their neuroanatomical substrate? Our current understanding of the complexity of neural systems does not allow us to simply accept such oversimplification. Even if all of the above attributes were shown to rely heavily on frontal lobe cortices (which may not necessarily be the case), several considerations should prevent us from accepting such an association. First, behavioral differences in the mental abilities of the apes as documented by field and laboratory studies remain controversial (Patterson & Cohn, 1994; Byrne, 1996). Second, we have to address what exactly it means to have a smaller frontal lobe – is it the cortex, the white matter, or the subcortical structures that underlie frontal lobe cortices that are smaller? If it is the cortex that is smaller, what does *that* mean? Does less cortex mean less computing power, less integrative power? What if it is the white matter that is smaller? Would not this mean decreased interconnectivity? What is functionally more important, more basic computing power or better possibilities of crosstalk between different brain regions? Which segments of the cortex are smaller (dorsal, orbital, mesial)? Are there specific cortical areas that differ in size, and which of their components can be identified as distinct when one species is compared to another?

Some sectors of the gorilla frontal lobe are indeed specialized, while others are closer to the hominoid pattern. The dorsal sector of the frontal lobe is somewhat smaller in the gorilla when compared to that of the orangutan or the chimpanzee, while the mesial and orbital sectors do not stand out. The dorsal segment of the frontal lobe cortices is known to be involved in several functions (in humans), as described in the beginning of the chapter. Some of them, like working memory, extraction of meaning from ongoing experiences, and organization of mental contents that control planning of future actions, can be recognized in the apes as well. Are any of these abilities different or less prominent in the gorilla, when compared to the orangutan or the chimpanzees?

Area 10, involved in planning, decision-making, and undertaking of initiatives, is a highly specialized area in this species with several distinguishing features. For example, the layer involved in intrinsic (local) connections (layer IV), as well as the layers involved in long and short association connections (layers II and III) are smaller in the gorilla than in the other species, but layers V and VI, known to be involved in connections with subcortical structures, are larger. Does this mean decreased connectivity between this area and other higher-order association areas, but more output to subcortical structures?

Emotional reactions to social stimuli and behaviors important for the maintenance of social bonds are altered if limbic cortices of the orbitofrontal cortex along with area 13, are damaged. In contrast to area 10, this area of the gorilla brain has an appearance which is very similar to that seen in the other hominoids, both in its internal organization as well as in its size. The only feature that stands out is an increased space between cells in the upper layers (possibility of increased interconnectivity for limbic cortices?).

In the orangutan, the dorsal and mesial sectors of the frontal lobe, as well as area 10, do not stand out in the orangutan when compared to the other apes, but the orbital sector of the frontal lobe, including area 13, differs considerably.

The orangutan has a unique organization in the orbital sector of the frontal lobe among the apes. This sector's overall size is smaller in both absolute and relative terms than in any other hominoid. Its posterior parts are more homogeneous in the sense that a single cortical area (area 13) occupies a relatively larger segment. Furthermore, this limbic area appears to be "less limbic" in its organization when compared to the other hominoids, by having a more robust granular layer (layer IV) and a similar size ratio of supra- to infragranular layers (both features are more representative of prefrontal association cortices). In other words, the orangutan seems to have a more restricted frontal limbic cortex. The uniqueness in the orbitofrontal sector of the orangutan has been seen in several specimens and is thus a rather robust finding.

How can this uniqueness in the orbital sector of the orangutan frontal lobe be interpreted? Recent evidence from lesion studies in humans, associating the orbital and mesial frontal sector with a variety of deficits in behaviors in the social domain, make it reasonable to believe that this sector is important for the survival of members of complex social groups. On the other hand, orangutans are known to have a more solitary life style and less complex social organization than other apes. It is tempting to relate the uniqueness of the social organization of the orangutans with the smaller and more homogeneous frontal limbic cortex underlying behaviors associated with social stimuli. Further analysis of the limbic cortex in this ape (Semendeferi et al., 1996, 1997b) in association with current studies of the orangutan social system (Van Schaik & Van Hoof, 1996) can help shed more light on this issue.

Variation in the size and the organization of the frontal lobes among the great apes is a fact, and differences might reflect species–specific adaptations, functional specializations, and/or major evolutionary events relating to changes in the organization of the hominoid brain. Mere consideration of relationships such as total brain size to body size are inadequate for the understanding of species-specific adaptations and their underlying neural circuitry. Instead, the study of specific neural circuits or cortical areas have to be targeted and compared across closely related taxa using a variety of approaches at the macroscopic and microscopic levels. Continuing develoments in the neurosciences and our current understanding of the extreme complexity of the primate brain demand sophisticated approaches at multiple levels in the study of the hominoid brain and oversimplification cannot be accepted.

As shown above, noninvasive techniques available for the study of the living human brain have just started to be applied to the apes in combination with novel techniques for the analysis of post-mortem brain tissue. Such efforts in combination with findings from behavioral studies of the apes can address questions involved in the understanding of what makes each hominoid species unique and what set hominids apart from our closest relatives.

ACKNOWLEDGMENTS

The author thanks Hanna Damasio for her generous contribution to this chapter, as well as Gary van Hoesen, Este Armstrong, and Karl Zilles for their helpful comments.

REFERENCES

Armstrong, E. (1982). Mosaic evolution in the primate brain: Differences and similarities in the hominid thalamus. In E. Armstrong & D. Falk (eds.), *Primate brain evolution* (pp. 131–161). New York: Plenum Press.

Armstrong, E. (1990). Evolution of the brain. In G. Paxinos (ed.), *The human nervous system* (pp. 1–16). New York: Academic Press.

Barbas, H. & Pandya, D. N. (1989). Architecture and intrinsic connections of the prefrontal cortex in the rhesus monkey. *Journal of Comparative Neurology*, **286**, 353–375.

Bischoff, T. L. W. (1876). Uber das Gehirn eines OrangOutan. *Sitzungsberichte der Mathematisch-physikalischen Klasse der bayrischen Akademie der Wissenschaften, Munchen*, **6**, 193–205.

Blinkov, S. M. & Glezer, I. I. (1968). *The human brain in figures and tables.* New York: Plenum Press.

Bolk, L. (1902). Beitrage zur Affenanatomie. 2. Uber das Gehirn von Orang-Utan. *Peterus Camper Nederlandsche Bijdragen tot se Anatomie*, **1**, 25–84.

Butter, C. M. & Snyder, D. R. (1972). Alterations in aversive and aggressive behaviors following orbital frontal lesions in rhesus monkeys. *Acta Neurobiologiae Experimentalis*, **32**, 525–565.

Byrne, R. W. (1996). The misunderstood ape. In A. E. Russon, Bard, K. A., & Parker, S. T. (eds.), *Reaching into thought: The minds of the great apes* (pp. 111–130). Cambridge University Press.

Campbell, A. W. (1905). *Histological studies on the localization of cerebral function*. Cambridge University Press.

Carmichael, S. T. & Price, J. L. (1994). Architectonic subdivision of the orbital and medial prefrontal cortex in the macaque monkey. *Journal of Comparative Neurology*, **346**, 366–402.

Connolly, C. J. (1950). *External morphology of the primate brain*. Springfield, IL: Bannerstone House,.

Damasio, A. R. (1985). The frontal lobes. In K. Heilman and E. Valenstein (Eds.), *Clinical neuropsychology* (pp. 339–375). New York: Oxford University Press.

Damasio, H. (1995). *Human brain anatomy in computerized images*. New York: Oxford University Press.

Damasio, H. & Damasio, A. R. (1992). Memory, language and decision-making: contributions from the lesion method in humans. In Y. Christen & P. Churchland, (eds.), *Neurophilosophy and Alzheimer's disease* (pp. 108–122). Berlin: Springer-Verlag.

Damasio, H. & Frank, R. (1992). Three-dimensional in vivo mapping of brain lesions in humans. *Archives of Neurology*, **49**, 137–143.

Damasio, A. R. & van Hoesen, G. W. (1983). Emotional disturbances associated with focal lesions of the frontal lesions of the frontal lobe. In K. Heilman & P. Satz (eds), *Neuropsychology of human emotion: recent advances* (pp. 85–110). New York: Guilford Press.

Damasio, A. R., Tranel, D., & Damasio, H. (1990). Individuals with sociopathic behavior caused by frontal damage fail to respond autonomically to social stimuli. *Behavioral Brain Research*, **41**, 81–94.

Damasio, H., Grabowski, T. J., Tranel, D., Hichwa, R. D., & Damasio, A. R. (1996). A neural basis for lexical retrieval. *Nature*, **380**, 499–505.

Duvernoy, H. (1991). *The human brain*. Vienna: Springer-Verlag.

Falk, D. (1983). Cerebral cortices of East African early hominids. *Science*, **221**, 1072–1074.

Fick, R. (1985a). Vergleichend-anatomische Studien an einem erwachsenen Orang-Utang. *Archiv fur Anatomie und Physiologie, Leipzig*, 1–100.

Fick, R. (1895b). Beobachtungen an einem zweiten erwachsenen Orang-Utang und einem Schimpansen. *Archiv fur Anatomie und Physiologie, Leipzig*, 289–318.

Frahm, H. D., Stephan, H., & Stephan, M. (1982). Comparison of brain structure volumes in insectivora and primates. I. Neocortex. *Journal fuer Hirnforschung*, **23**, 375–380.

Frank, R. J., Damasio, H., & Grabowski, T. J. (1997). Brainvox: An interactive, multimodal, visualization and analysis system for neuroanatomical imaging. *Neuroimage*, **5**, 13–30.

Galaburda, A., LeMay, M., Kemper, T., & Geschwind, N. (1978). Right–left asymmetries in the brain. *Science*, **199**, 852–856.

Goldman-Rakic, P. S. (1984). The frontal lobes: Uncharted provinces of the brain. *Trends in Neurosciences*, **7**, 425–429.

Holloway, R. L. & De La Costelareymondie, M. C. (1982). Brain endocast asymmetry in pongids and hominids: some preliminary findings on the paleontology of cerebral dominance. *American Journal of Physical Anthropology*, **58**, 101–110.

Jacobsen, C. F., Wolfe, J. B., & Jackson T. A. (1935). An experimental analysis of the functions of the frontal association areas in primates. *Journal of Nervous and Mental Disease*, **82**(1), 1–14.

Kennard, M. & Wilner, M. (1941). Findings at autopsies of seventy anthropoid apes. *Endocrinology*, **28**, 967–970.

Kling, A. and Steklis, H. D. (1976). A neural substrate for affiliative behavior in nonhuman primates. *Brain, Behavior & Evolution*, **13**, 216–238.

LeMay, M. (1976). Morphological cerebral asymmetries of modern man, fossil man, and nonhuman primates. *Annals of New York Academy of Science*, **280**, 349–366.

LeMay, M. & Geschwind, N. (1975). Hemispheric differences in the brains of great apes. *Brain, Behavior & Evolution*, **11**, 48–52.

LeMay, M., Billig, M. S., & Geshwind, N. (1982). Asymmetries of the brains and skulls of nonhuman primates. In E. Armstrong & D. Falk (eds.), *Primate brain evolution: Methods and concepts* (pp. 263–277). New York: Plenum Press.

Leyton, A. S. F. & Sherrington, C. S. (1917). Observations on the excitable cortex of the chimpanzee, orang-utan, and gorilla. *Quarterly Journal of Experimental Physiology*, **11**, 135–221.

Luria, A. R. & Homskaya, E. D. (1962). Disturbance in the regulative role of speech with frontal lobe lesions. In J. M. Warren and K. Akert (eds.), *The frontal granular cortex and behavior* (pp. 353–371). New York: McGraw Hill.

Mauss, T. (1908). Die faserarchitektonische Gliederung der Grosshirnrinde bei den niederen Affen. *Journal fur Psychologie und Neurologie*, **13**, 263–325.

Mauss, T. (1911). Die faserarchitektonische Gliederung der Cortex cerebri von den anthropomorphen Affen. *Journal fur Psychologie und Neurologie*, **18**, 410–467.

Mingazzini, G. (1928). Beitrag zur Morphologie der ausseren Grosshirn-hemispharenoberflache bei den Anthropoiden (Schimpanse und Orang). *Archiven fuer Psychiatrie*, **85**, 1–219.

Moniz, E. (1936). *Tentatives operatoires dans le traitement de certaines psychoses.* Paris: Masson.

Morecraft, R. J., Geula, C., & Mesulam, M.-M. (1992). Cytoarchitecture and neural afferents of orbitofrontal cortex in the brain of the monkey. *Journal of Comparative Neurology*, **323**, 341–358.

Ono, M., Kubik, S., & Abernathey, C. D. (1990). *Atlas of the cerebral sulci.* New York: Thieme Medical Publishers.

Papez, J. W. (1929). *Comparative neurology: Manual and text for the study of the nervous system of vertebrates.* New York: Hafner Publishing Co.

Patterson, F. & Cohn, R. (1994). Self-recognition and self-awareness in lowland gorillas. In S. T. Parker, R. W. Mitchell, & M. L. Boccia, M. L. (eds.), *Self-awareness in animals and humans: Developmental perspectives* (pp. 273–290). New York: Cambridge University Press.

Pilbeam, D. & Gould, S. J. (1974). Size and scaling in human evolution. *Science*, **186**, 892–901.

Preuss, T. & Goldman-Rakic, P. S. (1991). Myelo- and cytoarchitecture of the granular frontal cortex and surrounding regions in the strepsirhine primate galago and the anthropoid primate macaca. *Journal of Comparative Neurology*, **310**, 429–474.

Retzius, G. (1906). *Das Affenhirn.* Jena: Gustav Fischer.

Schleicher, A. & Zilles K. (1989). A quantitative approach to cytoarchitectonics: analysis of structural inhomogeneities in nervous tissue using an image analyzer. *Journal of Microscopy*, **157**(3), 367–381.

Semendeferi, K. (1994). Evolution of the hominoid prefrontal cortex: A quantitative and image analysis of areas 13 and 10. Dissertation thesis, University of Iowa, Iowa City, Iowa.

Semendeferi, K. (1995). Comparative imaging analysis of the hominoid prefrontal cortex: Implications for the evolution of areas involved in emotional (area 13) and cognitive (area 10) states. *American Journal of Physical Anthropology*, Supplement, **20**, 403.

Semendeferi, K. & Damasio, H. (1997). Comparison of sulcal patterns in the living brain of the great apes. *Society for Neuroscience Abstracts*, **23**.

Semendeferi, K., Damasio, H., Frank, R., & Van Hoesen, G. W. (1997a). The evolution of the frontal lobes: a volumetric analysis based on three-dimensional reconstructions of magnetic resonance scans of human and ape brains. *Journal of Human Evolution*, **4**(32), 375–388.

Semendeferi, K., Damasio, H., J. Rilling, & Insel, T. (1997b). The volume of the cerebral hemispheres, frontal lobes and cerebellum in living humans and apes using in vivo Magnetic Resonance mor-

phometry. *American Journal of Physical Anthropology*, Supplement, **24**, 208–209.
Semendeferi, K., Rilling, J., Insel, T., & Damasio, H. (1996). Brain volume and its components in living apes and humans. *Society for Neuroscience Abstracts*, **22**, 675.
Stephan, H., Baron, G., & Frahm, H. D. (1991). *Insectivora. Comparative Brain Research in Mammals.* New York: Springer-Verlag.
Stephan, H., Frahm, H., & Baron, G. (1981). New and revised data on volumes of brain structures in isectivores and primates. *Folia Primatologica*, **35**, 1–29.
Suarez, S. & Gallup, G. G., Jr. (1981). Self-recognition in chimpanzees and orangutans, but not gorillas. *Journal of Human Evolution*, **10**, 175–188.
Teuber, H. L. (1962). The riddle of frontal lobe function in man. In J. M. Warren & K. Akert (eds.) *The Frontal granular cortex and behavior* (pp. 410–477). McGraw-Hill Company.
Tiedemann, F. (1827). Das Hirn des Orang-Outangs mit dem des Menschen verglichen. *Zeitschrift fur Physiologie*, **2**, 17–28.
Tilney, F. & Riley, H. A. (1928). *The brain from ape to man*. New York: Paul B. Hoeber, Inc.
Tobias, P. V. (1971). The distribution of cranial capaciity values among living hominoids. *Proceedings of the 3rd International Congress of Primatology*, **1**, 18–35.
Tuttle, R. H. (1986). *Apes of the world.* New Jersey: Noyes Publications.
Uylings, H. & Eden, C. G. van. (1990) Qualitative and quantitative comparison of the prefrontal cortex in rat and in primates, including humans. *Progress in Brain Research*, **85**, 31–62.
Van Schaik, C. P. & Van Hoof, J. A. R. A. M. (1996). Toward an understanding of the orangutan's social system. In McGrew, W. C., Marchant, L. F., & Nishida, T. (eds.) *Great ape societies* (pp. 3–15). New York: Cambridge University Press.
Zilles, K. & Rehkaemper, G. (1988). The brain, with special reference to the telencephalon. In J. Schwartz (ed.), *Orangutan biology* (pp 151–176). New York: Oxford University Press.
Zilles, K., Schlaug, G., Matelli, M., Luppino, G., Schleicher, A., Dabringhaus, A., Seitz, R., & Roland P. E. (1995). Mapping of human and macaque sensorimotor areas by integrating architectonic, transmitter receptor, MRI and Pet data. *Journal of Anatomy*, **187**, 515–537.

PART II

Cognition and tool use in gorillas and orangutans

4

Intelligent tool use in wild Sumatran orangutans

ELIZABETH A. FOX, ARNOLD F. SITOMPUL, AND CAREL P. VAN SCHAIK

INTRODUCTION

Until recently, there was limited evidence of tool use (*sensu* Beck, 1980) by wild orangutans (*Pongo pygmaeus*), and no evidence of tool manufacture, despite several long-term field studies in Borneo (Ulu Segama: MacKinnon, 1974; Tanjung Puting: Galdikas, 1982; Kutai: Rodman 1973; Mitani 1985; Suzuki, 1989; Gunung Palung: Mitani et al., 1991) and Sumatra (Ketambe, in Gunung Leuser: Rijksen, 1978; Sugardjito 1986; Utami & Mitrasetia, 1995). Orangutans had been observed using only "found objects" as tools (Byrne, 1995): simple, unmodified raw materials such as leaves to wipe off feces (MacKinnon, 1974), a pad of leaves for holding spiny durian fruit (S. Utami, personal communication), a leafy branch for a bee swatter (E. Fox, personal observation), a bunch of leafy branches held together as an "umbrella" while traveling in the rain (Rijksen, 1978; E. Fox, personal observation), a single stick as a backscratcher (Galdikas, 1982), and a branch or tree trunk as a missile (all studies). Because observations of flexible tool behavior in wild primates had been limited to chimpanzees (*Pan troglodytes*), it had often been suggested that the manufacture and flexible use of tools arose only in the chimpanzee–hominid clade (McGrew, 1992, 1993).

Such limited tool use by wild orangutans was inconsistent with observations of captive orangutans. In contrast to their wild counterparts, captive and rehabilitant orangutans demonstrated a rich array of flexible tool behavior, at least as complex as that demonstrated by chimpanzees under similar conditions (Lethmate, 1982; Russon & Galdikas, 1993, 1995). Why did captive and rehabilitant orangutans demonstrate such complex tool behavior, while wild orangutans remained relatively unsophisticated?

Two hypotheses were proposed to explain this inconsistency. The first hypothesis presumed that wild orangutans had all the necessary cognitive abilities to make and use tools for foraging. However, because orangutans possessed strong hands, as well as strong jaws with large, thick-enameled teeth, they were able to access almost all foods without using tools (Lethmate, 1982; Galdikas, 1982). Therefore, although as a legacy from the common ancestor of the great apes, orangutans possessed the cognitive capacities to make and use tools, these capacities remained unexpressed (Parker, 1996).

The second hypothesis was that the cognitive capacities for tool use were not developed in wild orangutans. Rather, the technological skills displayed in captivity reflected a high capacity for imitation or emulation in a situation of persistent exposure to human models and abundant suitable materials. This capacity might, in part, be potentiated by early social interaction with humans (Miles, Mitchell, & Harper, 1996), as was suggested for the development of extraordinary cognitive skills such as language comprehension and production, and attribution of mental states to others (Povinelli, 1993; Tomasello, 1994; Rumbaugh, Savage-Rumbaugh, & Sevcik, 1994). Under this hypothesis, no advanced cognitive skills (expressed in tool manufacture and flexible use) were expressed in the wild because wild orangutans lacked exposure to potentiating factors.

Recently, researchers have observed tool manufacture and flexible tool use by wild orangutans (*Pongo pygmaeus abelii*) at the Suaq Balimbing Research Station in northwestern Sumatra (Van Schaik and Fox, 1994; Van Schaik, Fox, & Sitompul, 1996). These observations prove that wild orangutans both possess and express the cognitive capacities necessary to manufacture and use tools. By doing so, they join chimpanzees and humans as the only animals to demonstrate intelligent tool manufacture and flexible tool use in their natural environment.

The data presented in this chapter represent three years of continuous field research. New evidence for a *tool kit* is presented, as well as evidence for *task-dependent adjustment of tools*. These observations indicate intelligent tool behavior (Parker & Gibson, 1977; Beck, 1980; Byrne, 1995), including comprehension of object–object relations, schematic anticipation, and advanced representational capacities (Bard, 1993; Gibson, 1993; Mitchell, 1994). In addition, anecdotal evidence of at least program-level imitation (*sensu* Byrne, 1995) is presented. Finally, three hypotheses are proposed to explain the geographical distribution of tool behavior in wild orangutans.

STUDY AREA AND METHODS

The Suaq Balimbing Research Station (3° 04' N, 97° 26' E) is located within the Kluet portion of Gunung Leuser National Park. It consists of lowland swamp forest located in the flood plain of the Krueng Lembang River, near sea-level and approximately 10 km inland from Sumatra's west coast. The forest grades from seasonally flooded forest on levees immediately adjacent to the river, to seasonal or permanent freshwater swamps, to peat swamp forest. Annual rainfall is *c.* 3500 mm (1994–1996), with two dry and two wet seasons annually. Fruit production in these swamp and riverine forests is approximately tenfold that of the adjacent hill areas (Van Schaik, unpublished). Suaq Balimbing lies approximately 70 km south-southwest of Ketambe, the other long-term orangutan study site in Gunung Leuser National Park.

Local orangutan population density at Suaq Balimbing is higher than that recorded from any other research site. In the approximately 1 km^2 around the field station, annual mean density is 6.9 individuals/km^2 (van Schaik, et al., 1996), although local orangutan abundance fluctuates dramatically, roughly in proportion to the availability of orangutan preferred fruits (Van Schaik & Fox, 1996). By comparison, orangutan population densities in protected areas of Borneo range from 0.5 to 3.0/km^2 (Rijksen, et al., 1996), and in Ketambe, the only other protected orangutan study site in Sumatra, maximum orangutan population density is 5.2/km^2 (Van Schaik, Priatna, & Priatna, 1995). As of July 1996, fifty adult orangutans were individually recognized in the 4.5 km^2 study area of Suaq Balimbing.

Orangutan research at Suaq Balimbing has been ongoing since February 1994. Ranging and behavioral data are recorded using focal animal sampling and all-occurrence methods (Altmann, 1974) during full day, nest-to-nest follows (10–13 hour days). For this study, follows were conducted by EAF, AFS, CvS, and four research assistants. Interobserver reliability was 97% or higher.

Initially, tool behavior data were collected *ad libitum*. By mid-1995, however, the data collection procedure was systematized in an attempt to ensure that the following data were recorded for all observed tool behavior sessions: (1) date, (2) time session started and stopped, (3) species of tree in which tool was used, (4) height in tree of tool-use session, (5) food obtained with tool, (6) visual estimate of tree-hole width and length (for insect extractive foraging), (7) manner of tool modification (stripped of bark, shortened during session, chewed or beveled at end), (8) manner of tool manipulation (tool held by mouth, hand, or both), (9) manner of tool use (e.g., hammering, poking, probing/scraping), and (10) *ad libitum* notes. Field conditions sometimes prevented us from collecting all of the desired information. Tools were collected whenever possible. Using calipers, tool width was measured at both ends and at the midpoint. Tool length was measured using a standard tape measure.

The distributions of both tool width and tool length departed from normality and were skewed to the right. Prior to statistical analysis, logarithmic transformations were therefore applied to improve the fit to a normal distribution (Sokal & Rohlf, 1981). In all cases, statistical tests were performed on both raw and transformed data. Non-parametric tests were performed to monitor the sensitivity of the differences to aspects of the distribution. The results of the different tests were highly convergent, however, so only the outcomes of tests performed on the logarithmically transformed data are presented.

RESULTS

Contexts of tool use

All manufactured tools were used in two foraging contexts: extracting insects or honey from tree holes, or prying seeds from hard-husked fruit. All tool use occurred in trees. All tools were made from live branches from which twigs and leaves were removed. In the case of insect/honey extraction (extractive foraging: *sensu* Parker and Gibson, 1977), these branches were usually taken less than one meter from the foraging site. In contrast, tools used for frugivory were frequently manufactured before entering the fruit tree. Most tools were fairly straight and unbranched; only one of 201 tools recovered was forked. Orangutans variably modified tools by removing the bark with their teeth ("stripping"), chewing the tip ("fraying"), or splitting the tip ("beveling"). In most cases, the tool was held between the teeth when used and transferred to the hand between uses. However, in 21% of 38 systematically recorded cases of insect-extractive tool behavior, the tool was held in the hand, although a precision grip was never used. Rather, the tool was grasped using a pad-to-side hold (*sensu* Marzke & Wullstein, 1996).

For insect-extractive foraging, orangutans used tools to exploit tree holes inhabited by termites, ants, bees, or stingless bees. Adult insects, eggs/larvae, and/or honey were eaten. Tools were inserted into the tree hole and manipulated with variable force. Although sometimes difficult to distinguish, it was possible to discriminate between three tasks: (1) *hammering*, thereby breaking open termite or bee nests; (2) *poking*, which also ruptured insect nests and prompted adult ants or bees to exit; and/or (3) *probing/scraping*, to extract honey from stingless bees' nests or to obtain ants or termites (both adults and eggs/larvae). Of the 115 insect-extraction tools with known contexts, most (72%) were used for stingless bees' nests, whereas ants and termites each accounted for 14% of tools. Not all tool use was successful. Often, animals were observed to probe a tree hole with a tool but then leave the site without eating.

Orangutans often used multiple tools at a single tree hole. Tools sometimes broke during a tool-use session, or were seemingly inadvertantly dropped, necessitating the manufacture of another tool. Other times, a second tool was manufactured after the first tool was seemingly purposefully discarded after preliminary use, perhaps because the first tool was incorrectly manufactured for the intended purpose. Tools were also discarded after prolonged use, as if the tool had worn out or no longer met the requirements for the task at hand.

For seed extraction, orangutans used tools to obtain the edible seeds of ripe *Neesia sp.* fruits. These large fruits (*ca.* 15 × 10 cm) dehisce when ripe, but remain difficult to break open. Adult males are sometimes able to break open unripe *Neesia*, and some large subadult males and females can occasionally open ripe *Neesia* using hands, feet, and teeth, but most animals are usually unsuccessful in such attempts. In addition, the seeds are embedded in a mass of irritant stinging hairs, so that the seeds are difficult to access. Orangutans circumvented these obstacles by picking the fruit, positioning themselves in a stable position on a branch (this frequently necessitated moving away from the site from which the fruit was picked), and grasping the picked fruit with one or more hands/feet. Holding the tool in their mouths, orangutans then inserted the tool into the crack between two valves of the fruit and repeatedly scraped toward the apex (Figure 4.1). The irritant hairs were thereby scraped up and out of the fruit. All cracks were systematically subjected to the same treatment. By the same scraping motion with the tool, the nutritious arillate seeds were pushed toward the open top of the fruit and picked out with the lips or fingers. A single seed-extraction tool was frequently used to process many fruits; orangutans often traveled between feeding sites (in the same or a different tree) with tools held in their mouths.

Figure 4.1 Dehisced *Neesia* fruit with tool inserted into the crack between two valves.

Evidence for a tool kit

Although many animals use tools (Beck, 1980), most tool behavior is limited to the restricted use of an unmodified tool in a single context. Increased flexibility in tool behavior usually entails using an unmodified tool for multiple contexts, or different types of unmodified tools to accomplish the same task (Beck, 1980; Savage, 1995). A *tool kit*, however, comprises different manufactured tools, each used in a specific context (Boesch and Boesch, 1990; McGrew, 1992, 1993). Until now, manufacture and use of a tool kit in nonhumans has been observed only in chimpanzees (McGrew, 1992). According to Byrne (1995), the ability to create a tool kit indicates the ability to: (1) understand the physical properties needed in a tool in order to accomplish the task at hand, (2) modify a raw material so that it meets the requirements for the task at hand, and (3) relate each tool to a particular task and manufacture the tool to meet the corresponding demand.

The orangutan tool kit: seed-extraction and insect-extraction tools

Orangutans at Suaq Balimbing – like many populations of chimpanzees (Boesch & Boesch, 1990) – have a tool kit, comprised of insect-extraction and seed-extraction tools. Insect-extraction tools are long and wide ($n = 124$: *length*: mean 23.5 cm, median 21.5 cm, range 9.5–51.5 cm; *width*: mean 8.5 mm, median 8.3 mm, range 2.7–15 mm). In contrast, seed-extraction tools are relatively thin and short ($n = 69$: *length*: mean 13.4 cm, median 13.0 cm, range 9.5–20.5 cm; *width*: mean 7.7 mm, median 7.0 mm, range 4.7–12 mm). Insect- and seed-extraction tools differ significantly in both length and width (*ln*-transformed data: $t[191] = 11.42$, $P < 0.0001$, and $t[191] = 2.48$, $P < 0.05$, respectively), as well as in their frequency of being stripped (73% of 116 tools, vs. 90% of 70 tools; $Chi^2_{[1]} = 6.52$, $P < 0.05$ [cont. corr.]) (Figurte 4.2). It was almost always possible to accurately differentiate seed- and insect-extraction tools by eye.

Recent preliminary field research at Trumon (Sitompul, et al., in preparation), a peat swamp forest approximately 40 km southeast of Suaq Balimbing, has yielded the discovery of another population of orangutans that uses seed-extraction tools for *Neesia*. The seed-extraction tools collected from Trumon are quite similar in size to those collected from Suaq Balimbing ($n = 19$: *length*: mean 13.01 cm, median 12.9 cm, range 9.6–17.0 cm; *width*: mean 7.68 mm, median 7.3 mm, range 4.4–10.6 mm). These two research sites were connected by continuous forest until approximately 1990. Such recent separation does not eliminate the possibility that seed-extraction tool size is a cultural artifact of orangutans that until recently comprised a single population. None the less, a more probable explanation is that tool size serves a functional purpose.

Variation among insect-extraction tools

Among insect-extraction tools, there was significant variation among termite, ant, and stingless-bee tools (excluding the one tool used on a bees' nest). Although they did not show significant variation in length, their width varied significantly ($F_{[2, 112]} = 3.15$, $P < 0.05$). Unplanned contrasts among means (Scheffé test, Sokal & Rohlf, 1981) revealed that ant tools were significantly thinner than termite tools ($F_{[1, 30]} = 3.14$, $P < 0.05$). The three types of insect-extraction tools also differed in the tendency to have the bark stripped off ($Chi^2_{[2]} = 11.33$, $P < 0.01$) (Figure 4.3). Pairwise repetitions of the contingency test showed that ant *vs.* termite tools, as well as stingless-bee *vs.* termite tools, differed significantly in the tendency to have their bark stripped ($Chi^2_{[1]} = 6.30$, 6.25 respectively; $P \approx 0.01$ for both). Technically, then, one can distinguish between three different kinds of insect-extraction tools: termite tools, ant tools, and stingless-bee tools. Termite tools are relatively wide and are often unstripped. Stingless-bee tools are of medium width and are more often stripped. Ant tools are the thinnest of the insect-extraction tools and are almost always stripped.

By themselves, however, these statistically significant differences do not yet warrant classifying ant, termite, and stingless-bee tools into three distinct components of a tool kit. These tools have no real qualitative difference, and, unlike seed- *vs.* insect-extraction tools, it is not possible to visually distinguish ant, termite, and stingless bee-tools (Figure 4.4). Furthermore, while it is relatively easy to account for the functional variation in seed- *vs.* insect-extraction tools (see below), this task is not easy for ant *vs.* termite *vs.* stingless-bee tools. Rather, there is a gradient of tool features across tool type which may well be functionally relevant to an orangutan, but if so, this relevance is not immediately apparent to human observers. For the present then, the tool kit is limited to seed-extraction and insect-extraction tools. The significant variation among insect-extraction tools currently remains unexplained.

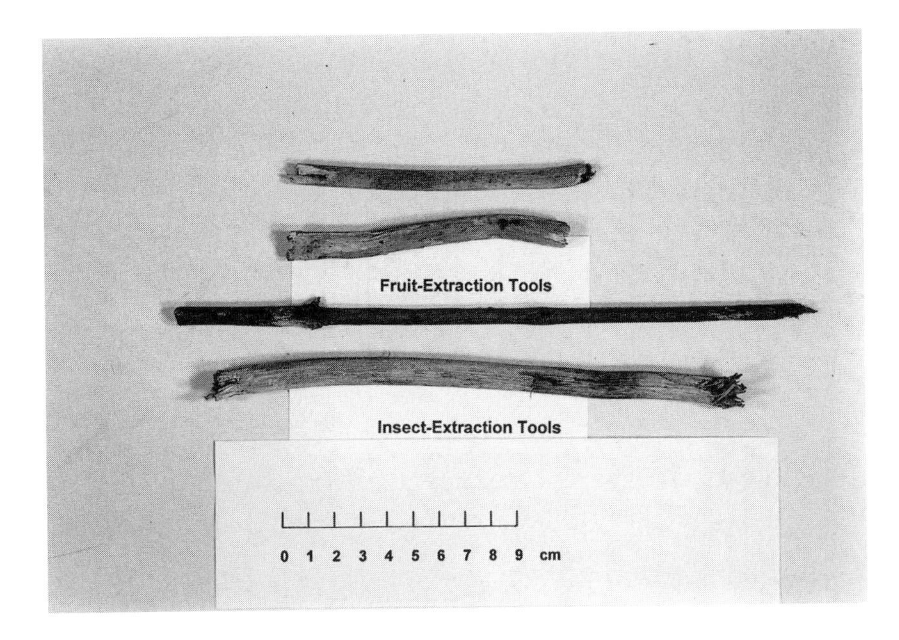

Figure 4.2 Insect- and seed-extraction tools are easily diastinguished by eye. Insect-extraction tools are longer, wider, and less frequently stripped of bark than seed-extraction tools.

Task-dependent adjustment of tools

While all insect-extraction tools are used in tree holes, not all tree holes are the same. They may be of different depths or widths, and insect nests may vary in hardness. Such variation in tree hole conditions likely creates situations in which differently constructed tools would be more useful than those of different manufacture. In order to meet these varying conditions most efficiently, then, an efficient tool user should modify a tool to fit the immediate task. Preliminary evidence suggests that orangutans are indeed capable of such task-dependent adjustment of tools.

Insect-extraction tools

When orangutans manufactured insect-foraging tools, they sometimes used the tool momentarily, paused, visually or manually inspected the tree hole, modified the tool further, and then used the same tool again. They also occasionally used multiple tools in a single session at a tree hole. For example, one adult female (Butet) used four tools in succession to obtain honey from a single tree hole. The first two tools were made 20 and 22 cm long, respectively. The third tool was initially 43.5 cm long, but was reduced (by biting) 16.5 cm, creating a tool of 27 cm. The fourth tool again started out longer, at 30.5 cm, but was soon reduced to 22.5 cm, very close to the length of the first two tools. Although it was not possible to measure the tree-hole depth, Butet's behavior suggests that shorter tools were more functional than longer tools at this particular extractive foraging site.

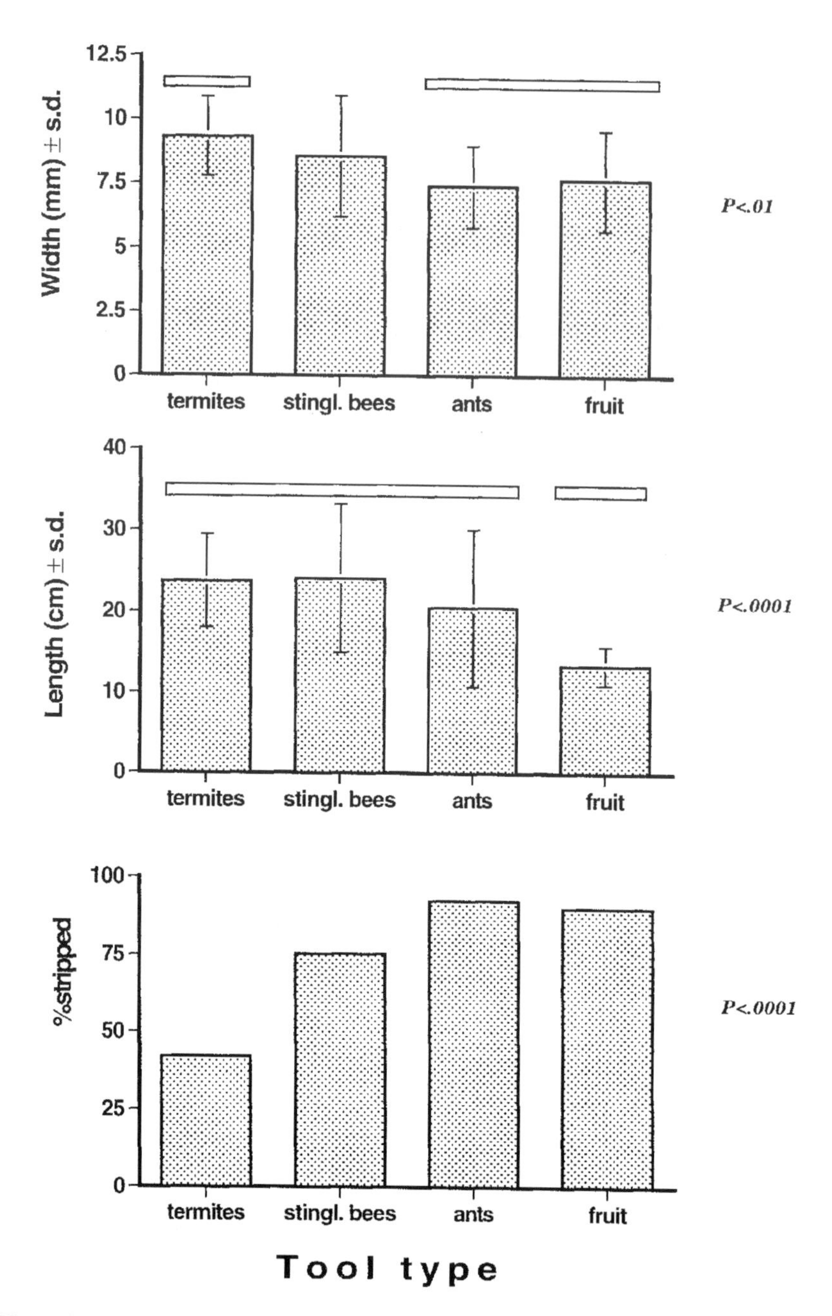

Figure 4.3 Variation among insect-extraction tool types.

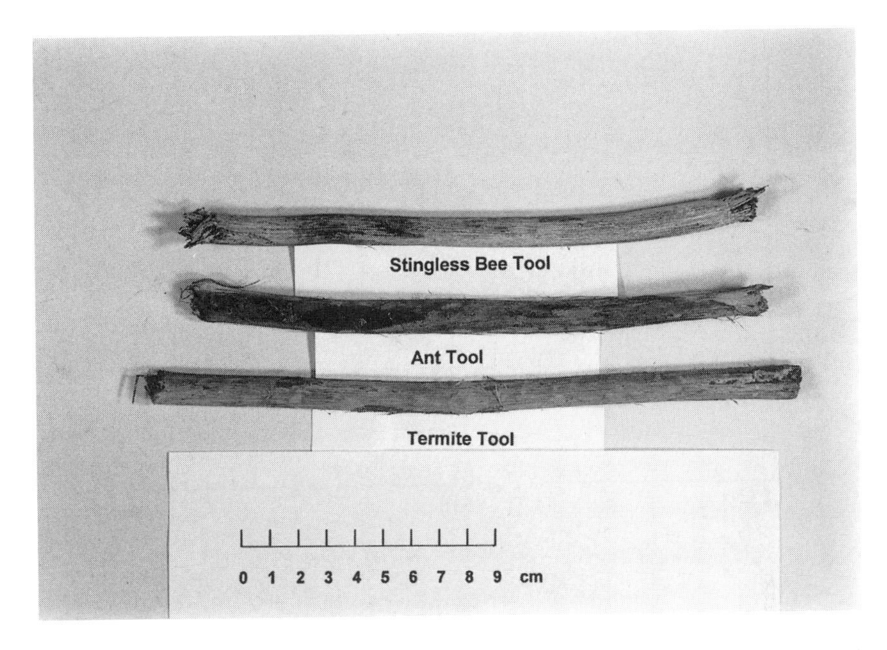

Figure 4.4 Termite, ant, and stingless bee tools. Despite statistically significant variation of physical properties among some tool types, it is not possible to visually distinguish types of insect-extraction tools.

If orangutans manufacture tools appropriate to a tree-hole's dimensions or to fit the task requirements (e.g., probing vs. hammering), variation in tool dimensions should be greater between rather than within tool-behavior sessions. This hypothesis is best tested within both individuals and tool context in order to control for interindividual differences in tool-making style and for variation among tool types. Sufficient data were available for one adult female (Ani), the most prolific tool-user at Suaq Balimbing. In fifteen tool-behavior sessions in which multiple tools were used, Ani's stingless bee tools showed variation between sessions in both length ($F_{[14,28]} = 4.83$, $P < 0.001$) and width ($F_{[14,28]} = 2.66$, $P < 0.05$). Between multiple-tool tool-behavior sessions for termite extraction, Ani had a (nonsignificant) tendency to vary tool length ($F_{[2,3]} = 8.63$, $P = 0.06$), but not width ($F_{[14,28]}$ 1.59, n.s.).

Albeit limited, the data suggest that orangutans do adjust tool dimensions – particularly tool length – between sessions. While tool width varied less than tool length, tool width did show allometric scaling with tool length ($n = 124$, $r = +0.530$, $P < 0.001$) (Figure 4.5). Such allometry makes functional sense, since longer tools must be stronger in order to withstand leverage force.

Seed-extraction tools

Orangutans rarely used multiple tools to process a single *Neesia* fruit; in fact, they frequently used one tool to process many fruits. It is therefore not relevant to test for

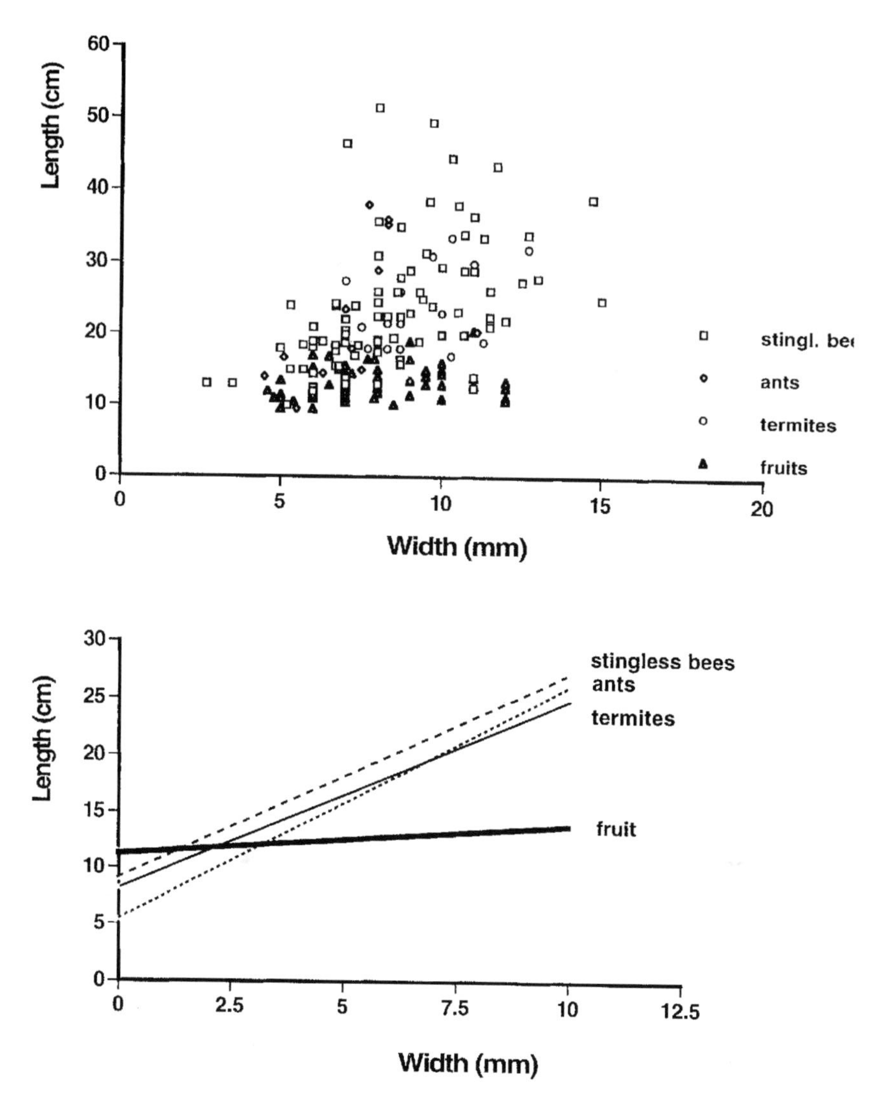

Figure 4.5 Allometry of tools. Insect-extraction tools are allometrically scaled for length and width, while seed-extraction tools are not.

variation across fruit-extraction tool-use sessions. Additionally, the lengths and widths of seed-extraction tools were not allometrically scaled (Figure 4.5). This lack of allometry is not surprising, since the depth of a *Neesia* fruit is not particularly variable, and seed-extraction tools must always be thin enough to fit between the cracks of the fruit husk.

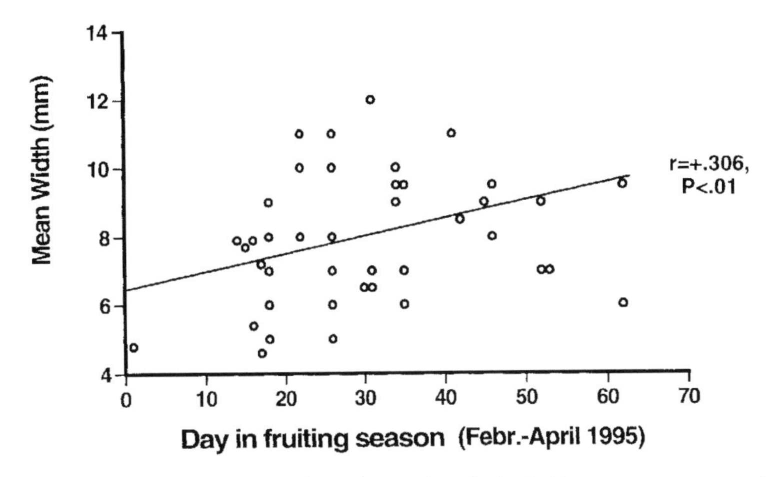

Figure 4.6 Width of seed-extraction tools over time. As the fruiting season progresses and *Neesia* fruits dehisce, seed-extraction tools are manufactured increasingly wider.

However, as *Neesia* fruits ripen, the cracks in the fruit husk gradually widen. Therefore, tool width may change over time, even while length is expected to remain constant. This relationship proved true: tools widened as the three-month *Neesia* season progressed (Figure 4.6). We conclude that orangutans adjusted the width of seed-extraction tools to meet the seasonal changes in the task of seed extraction.

Learning tool behavior

Although generally considered solitary, orangutans at Suaq Balimbing spent more time in social parties (interindividual distance < 50 meters) than orangutans at other sites (Fox & Van Schaik, unpublished). Thus, there may be greater opportunity for horizontal transfer of information, such as tool behavior, at Suaq Balimbing. During adult female/subadult male consortships, for example, consort partners frequently observed each other manufacturing and using tools. Typically, one orangutan initiated tool behavior. The consort partner then approached the tool user, positioned her/his face less than 20 cm from the tree hole, and intently observed the tool-behavior session in progress. The observing orangutan then attempted to pick at the tree hole with her/his fingers during the tool-behavior session. When the tool-user abandoned the tree hole, the observing orangutan either: (a) also left the tree hole, or (b) picked at the tree hole with her/his fingers. Adult orangutans were never seen to use a tool after the tool-user abandoned the tree hole.

Interestingly, observing tool behavior by consort partners occurred during both forced and unforced consortships. An adult female that was typically intolerant of spatial proximity with a forcibly consorting subadult male would approach this same subadult male when he used tools to obtain insects, as well as tolerate the subadult male's proximity while she exploited a tree hole. Food sharing during tool-behavior was never observed, although the observing orangutan not infrequently picked at the tree hole during a tool-behavior session or ate insects from the site after the tool-user abandoned the tree hole. After the tool-behavior session ended, an adult female resumed her attempts to avoid a forcibly consorting subadult male.

Opportunities for learning through vertical transfer – from mother to offspring – were also common. As with consorting individuals, infants closely scrutinized their mothers' tool-behavior sessions. Infants frequently inserted their fingers into the tree hole during the tool-behavior session or after their mother abandoned the tree hole. Infants also played with the tool after the mother discarded it.

The ontogeny of tool behavior in one female infant, Andai, was documented in some detail. Andai is the offspring of Ani, a frequent tool-user. While still quite dependent on her mother (approximately 3–4 years old, i.e., occasionally nursing and being carried by Ani, still sleeping in Ani's night nest), Andai frequently observed her mother use tools. As Andai grew older (5–6 years old, i.e., weaned, no longer carried by Ani, most frequently building own night nest), she made her own tools and probed tree holes after Ani finished her tool-behavior session. Not long afterwards, Andai initiated her own tool-behavior sessions without succeeding Ani at a tree hole. Now independent of her mother (7–8 years old), Andai is a frequent and competent tool-user.

These observations suggest that orangutans learn tool behavior from conspecifics rather than independently invent tool-behavior skills. However, even if orangutans learn tool behavior, it is not possible to discriminate between the types of learning: stimulus enhancement, response facilitation, emulation, or impersonation/imitation (Byrne, 1995). Some observations strongly suggest that at least program-level imitation (Byrne & Byrne, 1991, 1993; Russon, this volume) occurs. For example, during one tool-behavior session with Ani and Andai, Ani manufactured her tool in a rare way: She broke off a branch, stripped it, then shortened it *from both ends*. Ani made two tools in exactly this way. When she abandoned the tree hole, Andai approached the tree hole and also made a tool in the same way: breaking, stripping, and shortening *from both ends*. Like her mother, Andai also made two tools in this manner. Other than this observation, orangutans have never been observed to shorten a tool from both ends at once.

DISCUSSION

Tool manufacture and use in wild orangutans at Suaq Balimbing is both *habitual* (*sensu* McGrew 1993) and *natural*. Individuals of all sex and age classes (except very young infants) manufacture and use tools. It is highly unlikely that tool behavior was transferred from humans, either directly or indirectly through the release of rehabilitant orangutans. First, Suaq Balimbing is far from any established human habitation, and orangutan rehabilitation programs have never been conducted in the vicinity. Moreover, if rehabilitant-to-wild transfer was important, tool manufacture should have been observed in several other sites where wild orangutans have been in contact with released rehabilitants (e.g., Ketambe: Rijksen 1978; Tanjung Puting: Galdikas 1982). However, this is not the case. In fact, there are no reports of ex-rehabilitant orangutans at Ketambe manufacturing and using tools after intense contact with humans has ended and they are entirely dependent on their natural habitats. (Ex-rehabilitants at Tanjung Puting maintain contact with humans and are never completely free of human influence.) Therefore, the advanced cognitive capacities demonstrated by orangutans at Suaq Balimbing are not artifacts of human influence.

Implications for orangutan intelligence

From observations at other study sites, it is already known that wild orangutans can flexibly use "found objects" as tools (Galdikas, 1982), indicating that orangutans understand the physical properties needed in a tool in order to accomplish a task. Wild orangutans at Suaq Balimbing, however, demonstrate more complex tool behavior, with resulting important implications for intelligence. First, by modifying raw materials to manufacture a tool, they demonstrate the ability to maintain a mental representation of a tool's necessary physical properties. Second, by having a tool kit, the orangutans at Suaq Balimbing associate and maintain multiple mental representations of tools and relate each to its particular use (Byrne, 1995). Finally, by manufacturing seed-extraction tools before they enter the fruit tree, orangutans demonstrate forethought, or schematic anticipation, of the task to be performed (Goodall, 1986; Mitchell, 1994; Byrne, 1995). Thus, wild orangutans demonstrate intelligent tool use (Parker & Gibson, 1977) on a parallel with their captive counterparts as well as with chimpanzees.

Furthermore, orangutans at Suaq Balimbing use multiple tools in a single session. From observational data, two possible reasons for such behavior can be deduced. First, an orangutan may accidentally drop a tool while picking at the tree hole with her/his fingers. In this case, multiple tools in a single session should be uniformly made.

At other times, however, the orangutan dropped the tool while at the same time reaching for a branch to make another tool. In this second case, it seems probable that dropping the tool was intentional because the dropped tool was unsuitable for the task. Tools may be unsuitable because the orangutan either miscalculates some aspect of the task or is doing exploratory probing with the first tool to determine the requirements of the situation. In addition, tools may become unsuitable because the task requirement changes as the session progresses. For example, an orangutan may have to hammer with a thick, unstripped tool before she/he can probe with a thinner, stripped tool. In short, tools within sessions may vary as orangutans perfect the tool for the task.

For this second type of multiple-tool use, in which tools within a session vary, to be considered intelligent rather than repeated trial-and-error learning such as that shown by *Cebus* monkeys (Visalberghi & Trinca, 1989), it must be shown that tools are modified in a systematic way. For example, tools within sessions should become increasingly longer or shorter, wider or thinner. Currently, our data are insufficient for this type of analysis.

Origin and maintenance of tool behavior at Suaq Balimbing

The preceding observations clearly demonstrate that flexible tool manufacture and tool use in wild orangutans at Suaq Balimbing is a robust phenomenon. It is displayed by many individuals in a variety of contexts of extractive foraging on insect and plant foods. While resolving the apparent paradox about the lack of tool use in wild orangutans, these observations pose another question: Why do orangutans at Suaq Balimbing manufacture and use tools, while no comparable behavior has been reported from established, long-term study sites (MacKinnon 1974; Rijksen 1978; Galdikas 1982; Rodman 1973)? Below, three hypotheses are suggested to explain population differences in tool behavior. These three hypotheses are currently being investigated by ongoing research (Sitompul, Van Schaik, & Fox, in preparation).

Necessity hypothesis

The high orangutan density at Suaq Balimbing may produce severe scramble competition for food, compelling the usually highly frugivorous orangutans to exploit food sources from which they would otherwise refrain. In short, orangutans at Suaq Balimbing invented tools, and now maintain food-extractive tool behavior, in order to meet their sustenance needs. This hypothesis presupposes that many orangutan foods in swamp habitats are distributed in a manner leading to scramble competition: trees with few ripe fruits per crown, liana shoots, bark, insects, understory herbs, palm hearts, etc. If the necessity hypothesis holds, several predictions follow. First, compared to orangutans at other sites, orangutans at Suaq Balimbing should have less favorable energy budgets, as expressed in day journey lengths and/or length of active day per food consumed. Second, insectivory should be a replacement of other foods, particularly fruit. Therefore, insectivory should be negatively correlated with frugivory, rather than as a supplement of fruit. Third, subordinate individuals should show higher rates of insectivory than dominant individuals as they are diplaced from fruit trees and must resort to alternative food sources.

Opportunity hypothesis

The opportunity hypothesis argues that the swampy environment of Suaq Balimbing creates propitious circumstances for tool invention. In a swamp forest, social insects can not build nests on the ground; therefore, there is a higher density of social insects inhabiting tree holes than that found in nonswamp habitats. This hypothesis predicts that orangutans at Suaq Balimbing show a constant, high degree of insectivory – higher than orangutans at other sites – especially extractive foraging on insects. Second, Suaq Balimbing orangutans that travel into the adjacent hill forest should not use tools there because the density of social insects in trees is lower than that found in the swamp. While the opportunity hypothesis may apply to Suaq Balimbing, it fails to account for the absence of tool behavior in the seemingly similar swamp habitats of Kalimantan (Indonesian Borneo). A more detailed comparison of the Suaq Balimbing and Kalimantan swamp habitats is necessary to account for the observed differences in tool behavior.

Limited invention hypothesis

The necessity and opportunity hypotheses posit strict ecological bases for the evolution of tool behavior at Suaq Balimbing. While ecology almost certainly plays a role, such a strictly focused approach ignores the cognitive and social mechanisms of invention and the maintenance of tool behavior. The third hypothesis, limited invention, thus states that tool behavior may be invented only rarely and/or may have difficulty spreading and being maintained. This idea can not truly be tested; rather it is supported only by eliminating all plausible alternatives. None the less, it does predict that tool behavior should occur in a limited geographic area bounded by dispersal barriers, but without any obvious ecological correlates. The limited invention hypothesis is consistent with the use of tools in processing ripe *Neesia* fruits, which are present in at least two other sites, but eaten less frequently (Tanjung Puting: Galdikas 1988; Gunung Palung: C. Knott, personal communication). However, this hypothesis is inconsistent with the absence of tool behavior in Ketambe, which retains a continuous forest connection with Suaq Balimbing.

CONCLUSION

As the strict ecological hypotheses ignore cognitive/social aspects of invention, the limited invention hypothesis ignores ecological factors which make the invention of tool behavior possible. It is most likely that both ecology and the limitations of invention play a role in the geographic distribution of tool behavior in orangutans. Once tool behavior is invented, its maintenance in an individual is a function of the success of tool-foraging effort per tool-manufacturing effort. Success of tool-foraging effort is, of course, largely determined by ecology: the occupation rate of tree holes by social insects, or by the number of *Neesia* trees in a given area. Second, if tool-behavior is learned from conspecifics (see above), maintenance of tool behavior in a population is related to the rate of the transfer of knowledge. Learning requires at least one exposure to the learned behavior per lifetime and may require multiple exposures within a limited time frame. As a function of the frequency of tool behavior per social time, the transfer of tool-behavior knowledge may be more easily maintained in a high density, highly social population such as that at Suaq Balimbing.

ACKNOWLEDGMENTS

We thank the University of Indonesia for sponsoring the work of Elizabeth Fox and Carel van Schaik, the Indonesian Institute of Sciences (LIPI) for permission to conduct research in Indonesia, and the Director General of Forest Protection and Nature Conservation (PHPA) for permission to work in Gunung Leuser National Park. The Suaq Balimbing Research Station is supported by the Wildlife Conservation Society (WCS, founded as the New York Zoological Society). Elizabeth Fox's research was funded by the Wenner-Gren Foundation for Anthropological Research, the L. S. B. Leakey Foundation, Wildlife Preservation Trust International, and Primate Conservation, Incorporated. Carel van Schaik received funding from WCS and the Arts & Sciences Research Council of Duke University. We thank Abdussamad, Azhar, Bahlias, Ibrahim, Idrusman, Samsuar, and Tantanis for their able assistance with fieldwork. We thank M. Griffiths, K. Monk, S. Poniran, A. Selch, and J. Supriatna for support in the field; and H. Preuschoft and M. Riley for discussion and comments on the manuscript.

REFERENCES

Altmann, J. (1974). Observational study of behavior: Sampling methods. *Behaviour*, **49**, 227–267.

Bard, K. A. (1993). Cognitive competence underlying tool use in free-ranging orangutans. In A. Berthelet & J. Chavaillon (eds.), *The use of tools by humans and non-human primates* (pp. 356–378). Oxford University Press.

Beck, B. B. (1980). *Animal tool behavior: The use and manufacture of tools by animals.* New York: Garland STPM Press.

Boesch, C. & Boesch, H. (1990). Tool use and tool making in wild chimpanzees. *Folia Primatologica*, **54**, 86–99.

Boesch, C. & Boesch, H. (1993). Diversity of tool use and tool-making in wild chimpanzees. In A. Berthelet & J. Chavaillon (eds.), *The use of tools by humans and non-human primates* (pp. 158–174). Oxford University Press.

Byrne, R. W. (1995). *The thinking ape: Evolutionary origins of intelligence.* Oxford University Press.

Byrne, R. W. & Byrne, J. M. E. (1991). Hand preferences in the skilled gathering tasks of mountain gorillas (*Gorilla g. beringei*). *Cortex*, **27**, 521–546.

Byrne, R. W. & Byrne, J. M. E. (1993). Complex leaf-gathering skills of mountain gorillas (*Gorilla g. beringei*): Variability an standardization. *American Journal of Primatology*, **31**, 241–261.

Fox, E. A. & van Schaik, C. P. Unpublished data.

Galdikas, B. M. F. (1982). Orang-utan tool use at Tanjung Puting Reserve, Central Indonesian Borneo (Kalimantan Tengah). *Journal of Human Evolution*, **10**, 19–33.

Galdikas, B. M. F. (1988). Orangutan diet, range, and activity at Tanjung Puting, Central Borneo. *International Journal of Primatology*, **9**, 1–35.

Gibson, K. R. (1993). Animal minds, human minds. In K. R. Gibson & T. Ingold (eds.), *Tools, language and cognition in human evolution* (pp. 3–19). Cambridge University Press.

Goodall, J. (1986). *The chimpanzees of Gombe.* Cambridge, MA: Harvard University Press.

Kummer, H. & Goodall, J. (1985). Conditions of innovative behaviour in primates. *Philosophical Transactions of the Royal Society of London, Series B*, **308**, 203–214.

Lethmate, J. (1982). Tool-using skills of orang-utans. *Journal of Human Evolution*, **11**, 49–64.

MacKinnon, J. (1974). The behaviour and ecology of wild orang-utans (*Pongo pygmaeus*). *Animal Behaviour*, **22**, 3–74.

Marzke, M. W. & Wullstein, K. L. (1996). Chimpanzee and human grips: a new classification with a focus on evolutionary morphology. *International Journal of Primatology*, **17**(1), 117–139.

McGrew, W. C. (1992). *Chimpanzee material culture: Implications for human evolution.* Cambridge University Press.

McGrew, W. C. (1993). The intelligent use of tools: Twenty propositions. In D. R. Gibson & T. Ingold (eds.), *Tools, language and cognition in human evolution* (pp. 151–170). Cambridge University Press.

Mitani, J. C. (1985). Mating behaviour of male orangutans in the Kutai Game Reserve, Indonesia. *Animal Behaviour*, **33**, 392–402.

Mitani, J. C., Grether, G. F., Rodman, P. S., & Priatna, D. (1991). Associations among wild orang-utans: sociality, passive aggregations or chance? *Animal Behaviour*, **42**, 33–46.

Mitchell, R. W. (1994). The evolution of primate cognition: simulation, self-knowledge, and knowledge of other minds. In D. Quiatt & J. Itani (eds.), *Hominid culture in primate perspective* (pp. 177–232). Boulder: University Press of Colorado.

Parker, S. T. (1996). Apprenticeship in tool-mediated extractive foraging: The origins of imitation, teaching, and self-awareness in great apes. In A. E. Russon, K. A. Bard, & S. T. Parker (eds.), *Reaching into thought: The minds of the great apes* (pp. 348–372). Cambridge University Press.

Parker, S. T. & Gibson, K. R. (1977). Object manipulation, tool use and sensorimotor intelligence as feeding adaptations in cebus monkeys and great apes. *Journal of Human Evolution*, **6**, 623–641.

Povinelli, D. J. (1993). Reconstructing the evolution of mind. *American Psychologist*, **48**, 493–509.

Rijksen, H. D. (1978). *A field study on Sumatran orangutans* (Pongo pygmaeus abelii, *Lesson 1827*): *Ecology, behaviour, and conservation*. Mededelingen Landbouwhogeschool Wageningen. Wageningen, BV: H. Veenman & Zonen.

Rijksen, H. D., Ramono, W., Sugardjito, J., Lelana, A., Leighton, M., Karesh, W., Shapiro, G., Seal, U. S., Traylor-Holzer, K., & Tilson, R. (1996). Estimates of orangutan distribution and status in Borneo. In R. D. Nadler, B. F. M. Galdikas, L. K. Sheeran, & N. Rosen (eds.), *The neglected ape* (pp. 117–122). New York: Plenum Press.

Rodman, P. S. (1973). Population composition and adaptive organisation among orang-utans of the Kutai Reserve. In R. P. Michael & J. H. Crook (eds.), *Comparative ecology and behaviour of primates* (pp.171–209). London: Academic Press.

Rumbaugh, D. M., Savage-Rumbaugh, E. S., & Sevcik, R. A. (1994). Biobehavioral roots of language: a comparative perspective of chimpanzee, child, and culture. In R. W. Wrangham, W. C. McGrew, F. B. M. de Waal, & P. G. Heltne (eds.), *Chimpanzee culture* (pp. 319–334). Cambridge, MA: Harvard University Press.

Russon, A. E. & Galdikas, B. M. F. (1993). Imitation in free-ranging rehabilitant orangutans. *Journal of Comparative Psychology*, **107**, 147–161.

Russon, A. E. & Galdikas, B. M. F. (1995). Constraints on great apes' imitation: Model and action selectivity in rehabilitant orangutan (*Pongo pygmaeus*) imitation. *Journal of Comparative Psychology*, **109**(1): 5–17.

Savage, C. (1995). *Bird brains: The intelligence of crows, ravens, magpies, and jays*. San Francisco: Sierra Club.

Sitompul, A. F., van Schaik, C. P., & Fox, E. A. (in preparation). Why don't all wild orangutans use tools? An environmental hypothesis.

Sokal, R. R. & Rohlf, F. J. (1981). *Biometry*, 2nd edn. New York: W. H. Freeman & Company.

Sugardjito, J. (1986). Ecological constraints on the behaviour of Sumatran orang-utans (*Pongo pygmaeus abeli*) in the Gunung Leuser National Park, Indonesia. Utrecht: Drukkerjj Pressa Trajectina.

Suzuki, A. (1989). Socio-ecological studies of orangutans and primates in Kutai National Park, East Kalimantan in 1988–1989. *Kyoto University Overseas Research Report of Studies on Asian Nonhuman Primates*, **7**, 1–42.

Tomasello, M. (1994). The question of chimpanzee culture. In R. W. Wrangham, W. C. McGrew, F. B. M. de Waal & P. G. Heltne (eds.), *Chimpanzee cultures* (301–318). Cambridge, MA: Harvard University Press.

Utami, S. S., & Mitrasetia, T. (1995). Behavioral changes in wild male and female Sumatran orangutans (*Pongo pygmaeus abelii*) during and following a resident male take-over. In R. D. Nadler, B. M. F. Galdikas, L. K. Sheeran, & N. Rosen (eds.), *The neglected ape*, (pp. 183–190). New York: Plenum Press.

Van Schaik C. P. & Fox, E. A. (1994). Tool use in wild Sumatran orangutans (*Pongo pygmaeus*). Paper presented at the International Primatological Society meetings, Kuta, Bali, Indonesia, August 1996.

Van Schaik C. P. & Fox, E. A. (1996). Temporal variability on orangutan gregariousness revisited. Paper presented at the International Primatological Society meetings, Madison, Wisconsin, USA, August 1996.

Van Schaik, C. P., Fox, E. A., & Sitompul, A. F. (1996). Manufacture and use of tools in wild Sumatran orangutans. *Naturwissenschaften*, **83**, 186–188.

Van Schaik, C. P., Priatna, A., & Priatna, D. (1995). Population estimates and habitat preferences of orangutans based on line transects of nests. . In R. D. Nadler, B. F. M. Galdikas, L. K. Sheeran, & N. Rosen (eds.), *The neglected ape*, (pp. 129–147). New York: Plenum Press.

Visalberghi, E. & Trinca, L. (1989). Tool use in capuchin monkeys: distinguishing between performing and understanding. *Primates*, **30**, 511–521.

5

Orangutans' imitation of tool use: a cognitive interpretation

ANNE E. RUSSON

Imitation and tool use have both been portrayed as key abilities shaping the evolution of the human mind. The two combined – imitation of tool use – may have even greater significance. Imitation of tool use may represent a more advanced ability than imitation *per se* (Mitchell, 1994) and imitation may have been instrumental in promoting the acquisition and spread of tool use in hominids (e.g., Parker & Gibson, 1979; Visalberghi, 1993b). Whether any nonhuman species can imitate tool use, then, has implications for models of intelligence within the primate order, particularly evolutionary ones. The possibility that nonhumans can has begun to loom large with the increasing number of studies showing that great apes can imitate and use tools to unexpectedly sophisticated levels of complexity. This raises the credibility of the many claims of imitative tool use in great apes that have accumulated through the twentieth century.

Some newer work on imitation since the late 1980s has challenged the credibility these claims. First, great apes have independently acquired the types of tool use that were ostensibly imitatively learned (e.g., Nash, 1982; Paquette, 1992). Resemblances between great apes' tool strategies and demonstrated tool strategies could, then, derive from similarities in the affordances of the tools and the demands of the problem, that is, environmental contingencies that guide individual learning along similar paths (e.g., Galef, 1990, 1992; Whiten & Ham, 1992; Nagell, Olguin, & Tomasello, 1993; Call & Tomasello, 1994, 1995; Tomasello, 1996). Second, criticisms have been lodged against most of the studies behind claims of imitative tool use in great apes, for their lack of one or another of the methodological controls deemed necessary to eliminate the influence of all other social learning mechanisms other than imitation (e.g., Whiten & Ham, 1992; Heyes & Galef, 1996; Russon, Mitchell, Lefebvre, & Abravanel, 1998). Third, great apes' imitative responses to tool demonstrations, tested under conditions devised with these challenges in mind, have been unimpressive (Tomasello, Dasilva, Camak & Bard, 1987; Nagell et al., 1993; Call & Tomasello, 1994; Whiten et al., 1995; Whiten & Custance, 1996; Call, this volume).

Disagreement continues, however, because other recent studies have elicited more impressive performances, suggesting that these doubts do not do justice to the whole body of findings and that they underrate great apes' accomplishments (e.g., Byrne &

Russon, 1998; Meinel & Russon, 1996; Miles, Mitchell, & Harper, 1996; Parker, 1996a; Russon, Mitchell, Lefebure, & Abravnal, 1998). One problem that perhaps fuels disagreements is our lack of an explicit model of imitative tool use in great apes. Great ape imitative tool use may differ from human imitative tool use because great apes and humans differ in the cognitive systems that govern their abilities. Recent theoretical work on primate cognition, especially great ape cognition, is helping clarify the cognitive processes used (e.g., Gibson, 1990; Greenfield, 1991; Visalberghi, 1993a,b; Byrne, 1994; Matsuzawa, 1996; Russon, Bard, & Parker, 1996, in press; Visalberghi & Limongelli, 1996; Byrne & Russon, 1998). This offers potential for resolving some disagreement, insofar as better models of great ape cognition may better indicate what sorts of tool tasks to demonstrate and how to interpret great apes' responses to them.

This chapter attempts to establish what characteristics great apes' imitative tool use should have, based on the nature of great ape cognition, of tool use, and of imitation. These characteristics then serve as criteria to reassess the credibility of existing cases of "imitative" tool use in orangutans and other great apes.

TOOL USE AND IMITATION GOVERNED BY GREAT APE COGNITION

In primatology, the currently accepted definitional standard for tool use was set by Beck (1980) and, for imitation, by scholars like Thorndike (1898) and Thorpe (1963); for recent updates, see McGrew (1992) and Heyes & Galef (1996). Beck's tool use is "the external employment of an unattached, environmental object to alter more efficiently the form, position, or condition of another object, another organism, or the user itself when the user holds or carries the tool during or just prior to use and is responsible for the proper and effective orientation of the tool" (1980, p. 10). The best-known definitions of imitation are Thorndike's "learning an act by seeing it done rather than doing it" (1898, p. 50) and Thorpe's "copying of a novel or otherwise improbable act or utterance, or some act for which there is clearly no instinctive tendency" (1963, p. 135).

Under these definitions, all great apes can use tools and they can probably imitate. Evidence for their tool-using abilities is extensive (e.g., Beck, 1980; Galdikas, 1982; Jordan, 1982; Lethmate, 1982; Wood, 1984; Goodall, 1986; Gomez, 1988; reviewed by McGrew, 1992; Van Schaik, Fox, & Sitompul, 1996). Evidence for their imitative abilities is less solidly established but growing (e.g., Byrne & Byrne, 1993; Russon & Galdikas, 1993a; Tomasello, Savage-Rumbaugh, & Kruger, 1993; Custance & Bard, 1994; Whiten et al., 1995; Meinel & Russon, 1996; Miles et al., 1996; Visalberghi & Limongelli, 1996; Whiten & Custance, 1996). The question that remains is whether they can imitate tool use or whether, as argued by Call (this volume), they cannot.

Great apes' tool use and imitation are likely governed by various mental processes, of which cognition is likely a major contributor (e.g., Köhler, 1927; Parker & Gibson, 1977, 1979). A sizable body of research suggests that great ape cognition differs from cognition in other nonhuman primates or in humans (e.g., Gallup, 1983; Parker &

Gibson, 1990; Mitchell, 1993; Byrne, 1995; Parker, 1996b; Russon et al., 1996). For these reasons, great apes' tool use and imitation likely have their own *look* distinct from the tool use and imitation of other primates, in the sense of having distinctive behavioral characteristics. Evidence concurs.

Great apes, humans, and some monkeys share a particularly flexible and generalized, or "intelligent," form of tool use compared with many other tool-using species (e.g., Parker & Gibson, 1977, 1979). Given the same tool to solve the same problem, however, great apes' tool-using attempts often show characteristics of insightful understanding of the causal relations essential to the tool use, whereas monkeys' attempts rarely, if ever, do: monkeys achieve causal understanding to the extent of mastering simple tool use, such as using a stick to rake in an object, but their acquisition of this ability often shows the characteristics of trial-and-error guided, successive approximation (e.g., Klüver, 1933; D'Amato & Salmon, 1984; Antinucci, 1989; Parker & Poti', 1990; Visalberghi, Fragaszy & Savage-Rumbaugh, 1995; Anderson, 1996; Langer, 1996; Visalberghi & Limongelli, 1996). Chimpanzees' versus cebus' responses to one version of the tube task illustrate this difference. In the tube task, food is inserted into a tube so that it can be obtained only by pushing it out with a probe; the version of the task used provided reeds bundled and taped together as potential probes. Cebus often tried inappropriate probes; one, for instance, bit the tape off the bundle, discarded the reeds, and inserted the tape into the tube (Visalberghi, 1993a). Chimpanzees often chose appropriate probes and rarely inserted inappropriate items into the tube (Visalberghi et al., 1995). Great apes' tool use differs from humans' in its smaller repertoire of basic tool-object manipulations available to build complex tool operations and in attaining lower levels of tool use complexity (e.g., Reynolds, 1983; McGrew, 1992; Gibson, 1993b; Hewes, 1994; Langer, 1996).

Similarly for imitation, both great apes and humans have shown the ability to imitate but other nonhuman primates have not; and great ape imitation is less readily elicited and achieves lower degrees of fidelity than human imitation (e.g., Anderson, 1996; Moore, 1992, 1996; Whiten & Ham, 1992; Mitchell & Anderson, 1993; Nagell et al., 1993; Whiten et al., 1995; Meltzoff, 1996; Visalberghi & Limongelli, 1996; Whiten & Custance, 1996). For example, after observing a rake tool being used to rake in out-of-reach food, with the rake head either on its back edge or on its tines, many children copied both features (raking and rake-head orientation) while great apes copied the first feature (raking) and ignored the second (rake-head orientation) (Nagell et al., 1993; Call & Tomasello, 1994).

Hierarchization

One quality of great ape cognition that may account for these particular characteristics in great ape tool use and imitation is hierarchization. Hierarchization creates more complex cognition through the creation of higher-level processes, by integrating combinations of lower-level cognitive units into unified, higher-level ones (Piaget, 1937/1954; Gibson, 1993b; Langer, 1993). Its power comes from its capacity to recycle its own products, cognitive units, as cognitive input for reuse as building blocks in yet further combinations. In conjunction with the combinatorial mechanisms that allow use of several cognitive units at a time, hierarchization can generate a flexible range of high-level cognitions from a relatively small repertoire of basic cognitive units. Hierarchization appears to distinguish great ape from other nonhuman primate cognition; *per se*, hierarchization unites great ape with human cognition, but great apes' hierarchical ceiling may separate the two (e.g., Gibson, 1990; Byrne, 1994; Langer, 1996; Matsuzawa, 1996; Russon et al., 1998; Byrne & Russon, 1998). Their ceiling appears to be *second-order* cognition, cognition that generates integrated combinations of *first-order* cognitions; first-order cognitions directly order simple sensori-motor phenomena like individual motor actions, objects, or a set of objects (e.g., Whiten & Byrne, 1991; Langer, 1993, 1996). In causal cognition, for instance, the means–ends cognition that underpins tool use, first-order cognitions directly manipulate single sets of objects to create simple causal effects, such as using one object to push or bang another or using a stick to get an object out of reach; second-order causal cognitions integrate several first-order causal cognitions, such as using one object to push a second object so that it moves then using the first object to block the moving object so that it stops (Langer, 1996). Second-order cognition is rudimentary in comparison with mature human achievements, corresponding to the Piagetian preoperational cognition characteristic of human children 1.5–6 years old (e.g., Piaget, 1962; Case, 1985); and great apes have achieved only levels like those seen in children 3–3.5 years old (Premack, 1988).

Hierarchically organized behavior does differ observably from associatively governed, chained behavior, making for its distinct look. It is characterized by behavioral *routines* or *programs*. Like chained behavior, routines comprise combinations of behavioral components; unlike it, routines organize the combinations so that their components are interco-ordinated and subordinated to higher-level structures. At least five behavioral features have been used to detect the presence of routines (e.g., Miller, Galanter, & Pribram, 1960; Dawkins, 1976; Case, 1985; Reynolds, 1991; Byrne & Russon, 1998; Russon, submitted). A first is *iteration* of combinations of behaviors *to criterion*, that is, re-enacting the combination until a predetermined criterion is achieved. The second and third concern patterns across performances – *organizational stability* of combinations, in the sense of the arrangement of the components, in conjunction with *flexibility in components themselves*, like motor action detail. Low-level flexibility serves the need to adjust behavioral strategies to local contingencies as well as to interco-ordinate, or smooth the transition between, components. Fourth and fifth are *tolerance for interruption* (e.g., correct errors or handle disruption, like fix a tool or repulse an intruder) and *use of optional and/or alternative subroutines* (e.g., *sometimes* use a probe in addition to a hammer to eat palm nuts, or use different techniques to make one tool), in the sense of recovering control of behavior and resuming operation from the point of interruption or insertion. These five features are observable and they are diagnostic of routines, or behavior governed by hierarchical cognition. On this basis, cognitive hierarchization is evident in both the imitation and tool use of great apes (e.g., Byrne & Russon, in press).

Great apes' usage of tools indicates second-order hierarchical cognition in the realm of means–end or causal reasoning (e.g., Reynolds, 1983; Goodall, 1986; Boesch & Boesch, 1990; McGrew, 1992; Boesch, 1993; Langer, 1996). Second-order causal cognition is characterized by the manipulation of object–object relationships to effect specific outcomes (e.g., Inhelder & Piaget, 1964; Langer, 1996; Piaget & Inhelder, 1967; Parker & Gibson, 1979; Reynolds, 1982, 1983; Case, 1985; Whiten & Byrne, 1991; Visalberghi, 1993b; Visalberghi & Limongelli, 1996). Raking in an item is a simple example of manipulating an object–object relationship; it entails understanding that rake and item must be manipulated so that the rake head is *behind* the item and remains there while the rake is pulled, in order to move the item within reach; great apes handle this deftly. Great ape imitation has also been ascribed to hierarchical cognition (Byrne & Byrne, 1993; Russon & Galdikas, 1994; Byrne & Russon, 1998) at the second order (Whiten & Byrne, 1991; Whiten, 1996) or relational level (e.g., Russon & Galdikas, 1994; Russon, 1995; Russon et al., 1998). Great apes show second-order cognition beyond threshold levels in both abilities: They can use tool sets that can challenge 6–7–year-old humans (Toth et al., 1993; Matsuzawa, 1994, 1996; *contra* Greenfield, 1991) and they show some capacity for abilities considered to entail advanced second-order cognition, like pretense, self-awareness, and demonstration teaching, that are considered to be based on imitation (e.g., Gallup, 1983; Boesch, 1991; Mitchell, 1994; Parker, Mitchell, & Boccia, 1994; Parker, 1996a).

GREAT APES' IMITATION OF TOOL USE

If tool use and imitation each has its own distinct look in great apes, their imitative tool use may have its own look as well. Hierarchical and other characteristics of great apes' flexible form of tool use, of the process of imitation, and of great ape cognition allow prediction of what we could, and should not, expect if great apes were to imitate tool use.

Tool use

Flexible tool use – hereafter, tool use – is instrumental behavior that aims for specific ecological outcomes; it is of little use unless it achieves them. To achieve outcomes, both the aims *and* the behavioral means of attaining them must be understood and built. Fully acquiring tool use by imitation would then entail copying all the manipulations of tool–object relationships, their arrangement, their application, and all the fine action adjustments needed to achieve the outcome. I will argue that this is virtually impossible.

Tool use constitutes complex behavior composed of multiple components. This has implications for the novelty criterion in imitation. Novelty is not all-or-none in complex behavior because some components can be novel while others are not (Visalberghi & Fragaszy, 1990; Visalberghi, 1993b; Whiten & Custance, 1996). In complex behavior that is hierarchically organized, including tool use, novelty more likely inheres in the applications and arrangements of components than in the components themselves (e.g., Piaget, 1952); components tend to be simple behavioral units that are already known and so available as building blocks for constructing more complex arrangements (e.g., in tool use, motor actions, and basic object–object relationship manipulations). Because components tend to be already known, there is seldom need to acquire them imitatively, although novel components could be acquired this way (Byrne & Russon, 1998; Visalberghi, 1993b). Imitation of tool use is then likely to focus on the arrangements or applications of relationship manipulations more than on motor action detail, and imitative tool use is likely to show a mix of novel and known features.

To effect desired outcomes, manipulations of tool–object relationships must be flexibly fine-tuned to the ecological contingencies of the problem, not to the social contingencies associated with imitation (e.g., how to adjust a rake's movement to draw in a ball, not how to match the bend in the model's knees). The inevitable variation in a given type of problem from one instance to another would otherwise be devastating. If tool use is to be functional and efficient, it makes better sense to relegate the acquisition of local motor action detail to individual experience than to imitation (Byrne & Byrne, 1991, 1993).

Tool use commonly involves nonobservable components like what a probe does once inserted in a termite nest, how hard to hammer a nut with a rock to crack it, or what happens when sucking a siphon. If components are nonobservable, they cannot be imitated and must be acquired independently – a constraint recognized almost a century ago (Sherrington, 1906). Tool use, therefore, is likely to be acquired only partly by imitation and its acquisition should almost invariably entail multiple cognitive and learning mechanisms.

This makes it unlikely that tool use would, should, or even could be acquired by imitation alone. Imitation likely guides or enhances acquisition of some but not all features of tool use; other mechanisms likely contribute as well. Hierarchization in great ape cognition makes it possible that the imitation used in acquiring instrumental ecological skills evolved to work in conjunction with other processes, not independently (Byrne & Russon, 1998; Russon et al., 1998). Specific implications for the imitation of flexible tool use are:

- imitative tool use should be an inexact replica of demonstrated tool use;
- imitative tool use should admit low-level deviation from demonstrated tool use;
- imitative tool use should neglect nonobservable facets of demonstrations
- imitation may be only one of several mechanisms used in acquiring tool use;
- imitative tool use may include known as well as novel components;
- imitation of tool use should tend to target the arrangements or applications of relational manipulations more often than motor actions or individual relational manipulations.

Human imitation

Even human imitation has its limits, especially when it aims at complex forms of instrumental, ecological behavior like tool use.

Human imitation of complex behavior commonly involves multiple trials, not a single one. First trials tend to produce flawed approximations and later trials to improve on them (e.g., Piaget, 1937/54; Meltzoff, 1996); each imitative trial advances skills a little.

Human instrumental imitation is selective: It tends to focus on elements of demonstrated behavior that are *relevant* to effective performance and to discount elements that are not, and to focus on novel demonstrations, or novel facets of them, that *challenge* imitators' current abilities (e.g., Piaget, 1937/54; Vygotsky, 1962; Yando, Seitz, & Zigler, 1978; Abravanel & Gingold, 1985; Bauer & Mandler, 1989). This pertains to tool use because some elements of tool-using behavior are relevant, in the sense of being essential to solution, while others are not, and some features will challenge an observer's abilities, while others will not. Implications for human imitation of tool use are:

- imitative acquisition of tool use may require multiple trials;
- imitation may selectively reproduce relevant features of demonstrated tool use;
- imitation may selectively act upon components of tool use that challenge observers' current abilities and selectively ignore others.

Similar multitrial attempts to imitate demonstrations have been observed in rehabilitant orangutans, along with similar selectivity with respect to both relevance and challenge (Russon & Galdikas, 1995; Russon, 1996). This suggests these implications likely apply to great apes' imitation of tool use as well, and to greater, not lesser, degrees (Russon & Galdikas, 1995).

Great ape cognition

Great apes likely experience constraints to their imitative tool use beyond those seen in humans, owing to the more limited power of their cognition. Their cognitive development progresses more slowly, more narrowly, and in fewer modalities than it does in humans (e.g., Parker, 1996a). Internal constraints, like limited combinatorial capacity, are clearly one source for these limits (e.g., Langer, 1993, 1996) but so are experiential ones: great apes do not normally experience the extensive cognitive scaffolding that humans do, like intense and repetitive imitative practice. Both constraints limit the building blocks great apes construct for producing or parsing behavior. Among the consequences of these limits are that great apes achieve second-order cognitive abilities, including imitation, later than humans (beyond 3–4 years of age vs. 14 months – e.g., Dumas & Doré, 1986; Mitchell, 1994; Langer, 1996) and that their ultimate achievements appear limited to second-order cognition. Their potential for understanding is thereby reduced in comparison with humans.

Some evidence suggests that the limited modifiability of great apes' cognition affects their imitation. The clearest examples concern imitating gestures, not tool use; the former entails imitating motor actions. For instance, I trained an enculturated juvenile female orangutan, Daidai, to copy gestures, including touching one forefinger to the center of the other palm. Once she was proficient, I demonstrated a new gesture, touching one forefinger to the tip of the other forefinger. Daidai watched the new gesture then touched one forefinger to her other palm at the base of its forefinger –

directly between the center of the palm and the forefinger's tip. In response to two redemonstrations, she successively touched her first forefinger to the middle of the other forefinger then finally its tip. Her responses do *not* suggest random trial-and-error learning because they changed *only* in the direction of more closely matching the demonstration and no information other than redemonstration was provided to cue change. Daidai clearly registered the gesture change but she had difficulty or even resisted modifying her already known gesture. Chimpanzees have shown similar resistance to modifying gestures that were previously correct (Custance & Bard, 1994; Kojima, personal communication). Since imitating tool use is considered more difficult than imitating gestures, limited modifiability should constrain great apes' imitative tool use in similar fashion.

Some evidence also suggests that great apes' limited combinatorial capacity could constrain their imitation to very few features of demonstrated behavior per attempt. Daidai, for instance, was trained to copy touching one forefinger to one eye and then a new gesture was demonstrated, touching each forefinger to each eye. She responded correctly in touching with both forefingers, but she touched both to one eye. Other apes have similarly seized only a few of the demonstrated features at a time (e.g., Rensch, 1973; Custance & Bard, 1994; Meinel & Russon, 1996; Miles et al., 1996).

Great apes' understanding and encoding of the behavior they observe may be further limited to undifferentiated, basic forms; human children initially do the same (e.g., Piaget, 1937/54; Abravanel & Gingold, 1985). One example involved a demonstrator joining interlocking sticks by *inserting* the end of one into a socket in the end of another then *screwing* them together; an orangutan observer joined these sticks by *butting* their ends together, in undifferentiated contact, then *twisting* back and forth (Meinel & Russon, 1996). Other apes have shown similar patterns: The chimpanzees and orangutans tested by Tomasello's group copied the demonstrated *raking* but neglected to copy the particular *orientation* of the rake's head's (e.g., Tomasello et al., 1987; Nagell et al., 1993; Call & Tomasello, 1994) and the chimpanzees tested by Whiten et al. (1995) used *twist-like* manipulations that mixed *twist* with *pull* to remove a bolt from its hole, rather than distinguishing *twist* from *pull* as the demonstrator had.

For tool use, if great apes' overall cognitive ceiling limits their causal reasoning to second-order cognition, which focuses on manipulating object–object relationships, challenge-based selectivity would suggest that they may preferentially imitate patterns associated with object–object relationships (e.g., Visalberghi, 1993b; Russon, 1995; Russon & Galdikas, 1995). Some research concurs: Byrne and Byrne (1991, 1993), Russon and Galdikas (1995), and Byrne and Russon (1998) have argued that, when mountain gorillas and rehabilitant orangutans imitate complex ecological behavior, they tend to replicate arrangements or applications, that is, the *routines* or *programs* that organize combinations of relational components, more than individual motor actions or even individual relationship manipulations. Byrne termed the imitation of arrangements *program-level imitation* to distinguish it from imitation of motor actions (Byrne & Byrne, 1993).

Specific implications for great apes' imitation of tool use are:

- great ape imitation of tool use may be slow because of resistance to change;
- great ape imitation of tool use may be constrained by the small number of components that can be handled at a time;
- great apes may tend to imitate the tool–object relationship manipulations demonstrated in undifferentiated forms;
- in imitating tool use, selectivity guided by challenge will likely lead great apes to focus on reproducing arrangements of the tool–object relational manipulations demonstrated and to neglect lower-level components demonstrated, those that concern motor action details, particular tools, and individual tool–object relational manipulations.

These characteristics predicted for great ape imitative tool use are consistent with two long-standing views – that great apes achieve limited gains from a few demonstrations or attempts at imitative learning (e.g., Köhler, 1927), and that imitation found in free-ranging apes stems from years of observation and many imitative attempts (e.g., Lawick-Goodall, 1973; McGrew, 1992). If great apes are constrained in the range of behavior they can understand, the number of components they can combine at once, the depth and precision of their encoding, and the rate at which they accommodate change, they will need many imitative "drafts" to generate a close copy of a demonstration.

IMITATION OF TOOL USE IN ORANGUTANS

Orangutans may seem unlikely targets for investigations of imitative tool use because they are known for neither in the wild: their renown instead is for their lack of tool use and their pronounced asociality (but see Fox, Sitompul, & Van Schaik, this volume). Despite this unlikely state of affairs, wild orangutans are now known to engage in many manipulations of object–object relationships governed by the same cognitive ability, causal reasoning, at the same order of complexity, relational reasoning, as that involved in classic tool use (e.g., Bard, 1993; Russon & Galdikas, 1994). That this sort of cognition is the key to tool use is supported by evidence that orangutans may be the *most* sophisticated tool users of the great apes when the need or opportunity arises (e.g., Yerkes & Yerkes, 1929; Beck, 1980; Galdikas, 1982; Lethmate, 1982; McGrew, 1992; Van Schaik et al., 1996). Their apparent asociality likewise masks imitative capabilities equal to those of the other great apes; provided favorable conditions, they imitate as well as or better than other great apes (e.g., Russon & Galdikas, 1993a; Meinel & Russon, 1996; Miles et al., 1996).

If orangutans imitate tool use, their imitative tool use should show the distinctive features that stem from the demands of flexible tool use, the process of imitation itself, and the limited form of hierarchization that characterizes great ape cognition. Enough allegations of orangutan imitative tool use exist to check for these features. A considerable portion of this evidence is from studies conducted in the early part of the twentieth century (e.g., Haggerty, 1910, 1913; Yerkes, 1916; Hornaday, 1922; Köhler, 1927; Yerkes & Yerkes, 1929); a small amount comes from studies conducted over the middle of the century, when interest in imitation lagged (e.g., Harrisson, 1962, 1969; Wright, 1972; Horr, 1977; Galdikas, 1982); and a substantial amount has appeared in the wave of imitation research that arose in the late 1980s. Because much of the older evidence was challenged in the recent wave of imitation research, I focused on recent studies as less subject to methodological criticisms, even though they are by no means free of problems. I selected cases from each research program that has studied orangutan imitation and assessed them for the features that should distinguish great apes' imitative tool use.

Experimental evidence

Several researchers have recently attempted to elicit imitation of tool use in orangutans experimentally: Tomasello, Russon, Meinel, and Miles. As each presented different tasks, their findings are discussed individually.

1. The Tomasello group

Tomasello's group conducted one of its experimental tests of imitative tool use on captive orangutans (Call & Tomasello, 1994). They demonstrated raking in out-of-reach food with a rake tool to orangutan observers then offered orangutans the opportunity to obtain the food themselves, making an identical tool available. Many of the orangutans did use the tool to rake in food but did not copy one facet of the demonstrated technique, flipping the rake head from its tines onto its back edge before raking. Call and Tomasello concluded that the orangutans were emulating but not imitating – i.e. copying demonstrated outcomes, but not behavioral techniques (Köhler, 1927; Tomasello, 1990). This interpretation, however, ignores the fact the orangutans did copy the basic technique they observed, *raking* food with the tool, and that they did not try techniques with the tool that great apes have used when trying to solve this task independently, like poking, tapping, or throwing (Parker, 1969) (for detailed argument, see Byrne & Russon, 1998; Russon, 1995; Russon et al., 1998). Raking represents a behavioral technique (the technique of raking a target object with a tool object), not merely the inevitable product of affordances or dynamic properties of environmental entities (behavior determined by a rake tool) nor merely an outcome (food raked in), albeit a behavioral technique organized at relational rather than motor action levels. And the behavioral technique for raking entails manipulation of object–object relationships, the predicted target of great ape imitation. That orangutans ignored some components of the demonstrated technique and that they may have acquired some components by individual, trial-and-error learning are also consistent with the predicted characteristics of great ape imitative tool use – that it tends to match relational manipulations at undifferentiated levels and that it may relegate the acquisition of lower-level components to experientially guided adjustment. A third pattern consistent with predictions is that some orangutans, given further opportunities to obtain food with the tool, showed a transition phase in which they used both tine and edge techniques. They were successful with both techniques and both techniques constitute behavioral routines (see Byrne & Russon, 1998), so persisting with both shows a tolerance for alternative techniques embedded within the overall behavioral technique; this is one index of hierarchically organized behavior.

2. Russon

In my work with Daidai, I conducted a formal "do what I do" imitation experiment starting in 1994, when she was about 6 years old, and continuing during my 1995 and 1996 visits. I followed standard procedures (e.g., Hayes & Hayes, 1952; Custance & Bard, 1994). I first trained her to perform any of a set of gestures on cue, where training used standard operant procedures like molding, shaping, and reinforcement and the cue was my demonstrating a chosen gesture to her. Once Daidai reliably copied training gestures, I introduced novel items intermittently into the stream of demonstrations. Some novel demonstrations involved manipulating multiple object–object relationships in specific arrangements (e.g., put one object *on* another *then* put that object combination *in* a box). As a baseline control against her inventing similar combinations alone, I used only objects I had seen in her possession but that she had discarded (fruit seeds, coconut shells).

The task most relevant here was putting a jackfruit seed (like a large lima bean) ***into*** a hole in the center of 20 cm length of garden hose, then putting the seed-in-hose ensemble ***onto*** a coconut shell dish (Russon et al., 1998). This is the sort of sequencing task that Whiten and Custance (1996) advocate as a good test of hierarchical imitation. I had never observed Daidai or any other rehabilitants spontaneously manipulate fruit seeds other than to remove them from food and discard them. I had observed others but not Daidai spontaneously arrange pieces of coconut shells into piles or bang them, and manipulate longer lengths of hose like rope (e.g., loop hose over bars, weave it through bars, flap). In line with the view that great ape imitation focuses on novel arrangements of known relational manipulations, I built towards this key task by first establishing each relational component separately, i.e. I demonstrated three simple relational manipulations – put a jackfruit seed *onto* a shell "dish," put a tube or piece of hose *onto* a shell, and put a jackfruit seed *into* a tube or piece of hose. Daidai imitated each relational manipulation correctly and immediately, and was rewarded. I tested each relational manipulation (demonstration–response) twice, for reliability; the 6 trials took a total of 3 minutes. I then demonstrated the key task – put a seed ***into*** the hose's hole ***then*** put the combination ***onto*** the shell. Daidai imitated immediately after receiving the items, that is, on the first trial and within seconds. She put the seed ***into*** the hose hole, pushing it in repeatedly because it tended to slip out. Seed in hose accomplished, she ***then*** carried her seed-in-hose ensemble to the coconut-shell dish and dropped it ***on***. The seed slipped out of the hose as she carried it and she did not retrieve it, either not noticing or intent on the next step, so she ultimately placed an empty hose ***onto*** the dish. She none the less reproduced the novel arrangement of these relational manipulations demonstrated and she corrected some errors in lower-level components to achieve each (e.g., correct poor insertion by pushing the seed *into* the hose). Also striking is that she did *not* attend to copying the demonstrated outcome – the very opposite of what Tomasello's emulation hypothesis predicts.

3. Meinel

Meinel (1995; Meinel & Russon, 1996) tested non-enculturated captive orangutans at the Metropolitan Toronto Zoo for their abilities to imitate object use. Tasks were designed to require manipulating relations between objects (e.g., *hook* one object with another, *rub* one object with another); some of the tasks involved tool use. All the tool tasks demonstrated required manipulating and intercoordinating several object–object relationships (e.g., *join* sticks by *inserting* and *screwing*, then use *the joined* stick to *rake* in a reward).

One of the tool tasks is analyzed here; others showed similar patterns. The task was obtaining a banana placed on a boat. The boat was set afloat on the moat surrounding the orangutans' enclosure and out of their reach. Obtaining the banana required a tool to hook the boat and pull it within reach. The demonstration was visible to all orangutans. It was a priori expected that Juara, an adolescent male, would be the most likely to imitate it because in baseline observations he most often poked at out-of-reach items with single sticks; the prediction proved true. Figure 5.1 shows the demonstrated routine (left side) and Juara's response (right side), reading from top to bottom. The object–object relations manipulated are shown in *italics*. The elements Juara imitated are shown in **bold** characters. In real time, the whole of the demonstrated routine preceded the whole of Juara's response; the two routines have been realigned in Figure 5.1 to highlight comparable features.

The demonstrator stood on the visitor side of the moat, ***near*** the boat, holding three interlocking sticks. She joined stick 1 to stick 2 (***join*** → 2) by inserting one end of stick 1 into a socket in one end of stick 2 with exaggerated screwing movements (***insert – screw***), as if joining the sticks required screwing them together (it did not). She then added stick 3 to the stick 1–2 unit (***join*** 12 → 3), using the same insert-and-screw technique. With the long stick she reached for the boat (***reach***), hooked it (***hook***), and pulled it within reach (***pull***). She demonstrated this repeatedly over a half hour then gave the orangutans several sets of three interlocking sticks.

The orangutans first tried other techniques to get the food, unsuccessfully, then lost interest and discarded the sticks around the enclosure. After about 10 minutes Juara left the edge of the moat, where he had been intently watching the banana boat, to search for and collect three interlocking sticks from around the enclosure (*collect*). He brought his collection to the edge of the moat near the boat (collection ***near*** boat), then "joined" two of the sticks (***join*** 1 → 2) by butting their ends together and repeatedly twisting them back and forth against one another (*together – twist*). This failed to join the sticks securely, possibly because the joining sockets had been destroyed by the orangutans. Probably because this joining technique failed, Juara modified it: he again "joined" the two sticks by butting their ends together but then grabbed them in his right fist, across the point where they were butted together (*together – hold*). His fist held them together; in other words, it acted as the joint. He reached for the boat with this joined stick (***reach***).

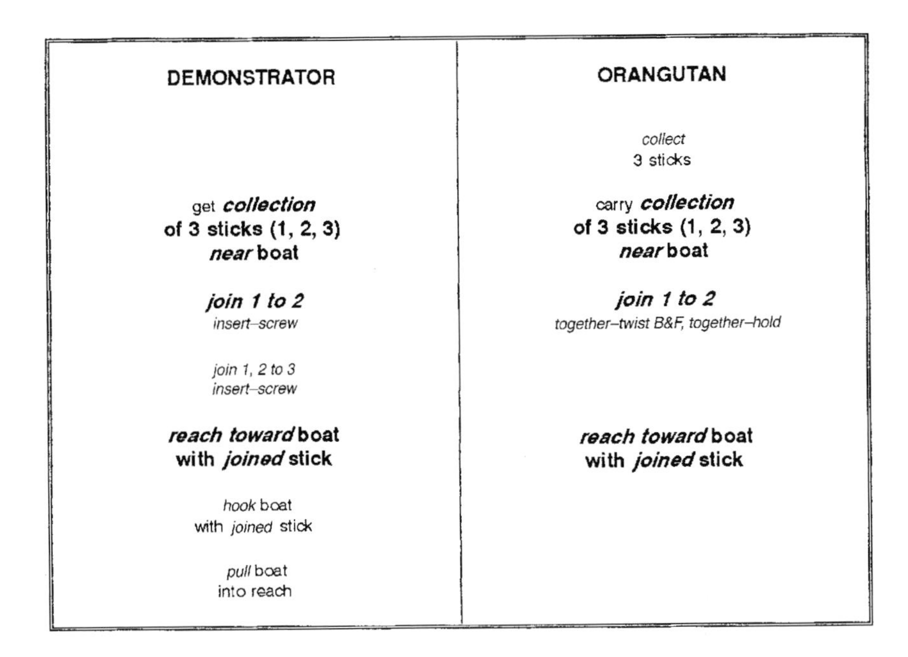

Figure 5.1 Banana boat afloat on a moat.

Although Juara did not hook or even touch the boat this way, what he reproduced from the demonstration is consistent with the view that he understood and encoded the multiple relations involved in the demonstrated tool strategy: *join* sticks, then *reach* boat with *joined* stick. He reproduced each of the individual relational manipulations in correct causal sequence; and he added, substituted, and corrected components, as needed, to render the strategy functional (add *collect* sticks initially; correct ineffective, imitative *join*ing technique by substituting an idiosyncratic *join*ing technique). Juara also reproduced undifferentiated versions of the relations needed to realize the demonstrated joining technique (*together* versus *insert*, *twist* versus *screw*). His two attempts at joining show his ability to adjust his techniques to local contingencies.

4. Miles

In her long-term developmental study of Chantek, a sign-using enculturated orangutan, Miles observed and elicited a range of imitative tool use. Chantek's imitations elicited in a DO THE-SAME-THING game have been reported; one tool task described in detail is drawing (Miles et al., 1996). The drawing task is interesting because it involves manipulating several tool–object relationships (Case, 1985) – that between pencil and paper, including the pencil's orientation, and that between demonstrator's and imitator's marks on the page (e.g., parallel, perpendicular). For Chantek, a caregiver demonstrated drawing a circle or a straight line on a piece of paper then signed to Chantek to DO THE-SAME-THING. In all the cases reported, Chantek copied the basic pencil–paper relationship manipulation demonstrated: He *drew* rather than doing other things with the pencil like chewing it or stabbing the paper. He also replicated additional demonstrated relationships. Following four circle demonstrations, Chantek equally often drew over the demonstrated circle and drew partial circles on his own. Following line demonstrations, he most often drew lines by swiping the marker across the page or pressing it down hard and moving it about on the page; he also drew a line parallel to the demonstrated line several times, traced over the demonstrated line twice, and made several short lines separate from and perpendicular to the demonstrated line once. Two of his parallel line responses followed partial molding by the caregiver and one followed the caregiver's request to do another line after he had just made one on the paper. In quality, his imitations often show relatively undifferentiated versions of demonstrated relationships, like partial circles or lines and circles inconsistently oriented to the model. He also made more precise copies, like complete circles, showing a capacity for imitating relational manipulations in complex rather than undifferentiated form; this may reflect his enculturation, especially learning not to ignore the fine points of demonstrations (Mitchell, personal communication).

Chantek also spontaneously imitated caregiver behaviors like "cooking." He cooked cereal, putting it in a pot and placing that on top of the stove, and his camera, putting it in a pot of water (Miles et al., 1996). Here again, he correctly reproduced relational components (put items *into* a pot *then* put pot-with-contents *onto* the stove, put water *into* a pot *then* put items *into* water-in-pot) while neglecting some critical details (heat is needed to cook, cameras do not cook well). One feature ignored, heat, may not be readily observable.

Observational data

The widest range of allegedly imitative tool use comes from an observational study of free-ranging rehabilitant orangutans by Russon and Galdikas (1993a). These orangutans were all ex-captives; a few were partially enculturated but most had been rescued from pitifully abusive and depriving conditions. This study identified 54 incidents in which imitation was implicated; 41 of them involved tools, for a total of 23 distinct types of tool use apparently influenced by imitation.

Not all 23 types of tool use were amenable to the analysis proposed here. First, some of the tool use was simple, making it difficult to show convincingly, on the basis of behavioral characteristics, that matching owed to imitation rather than to parallel experience. The focus predicted for great apes' imitative tool use, arrangements of tool–object relational manipulations, is most evident in complex, multistage tool use. I therefore selected those types of tool use that required a specific arrangement of at least *three* relational manipulations to render the technique effective. I established this from the basic form of the technique demonstrated, that is, the simplest set and arrangement of the tool–object manipulations necessary to achieve the desired outcome. In most cases, I established the basic technique from ad lib observation of camp staff and consideration of the physical exigences involved. Second, these data are observational so the methodological controls that support inferences of imitation differ from those typical of experiments. I chose cases for which controls against nonimitative processes were strongest: (a) orangutans copied a demonstrated technique when evidence showed they knew alternative techniques, especially ones that were equally effective, easier, and more common than the demonstrated one (in line with Dawson and Foss' two-action method, one of the better controls against nonimitative matching; Heyes, 1996) or (b) orangutans acquired a novel demonstrated technique extremely rapidly (in the rare circumstance that acquisition was directly observed). Evidence of alternative techniques came from a larger observational database I collected on object and tool use in these same orangutans (800 h, collected 1990–1991) and literature on tool use in orangutans; these data allowed establishing a baseline of the abilities and proclivities of particular individuals and orangutans as a species.

Allowing for these considerations reduced the sample to 12 cases of tool use that could be examined for the influence of imitation:

- make fire
- weed paths
- sweep paths
- make bridges
- apply and remove paint
- put baby in cage and wipe cage
- brush teeth
- hang hammock
- sand blowgun dart
- sharpen axe blade
- hammer nails into wood
- siphon fuel

Earlier analyses already revealed that several of these cases incorporated similar basic manipulations of object–object relationships, but organized them into novel arrangements and applications (Russon & Galdikas, 1995): sharpen an axe blade (***dip*** a stone ***in*** water ***then rub*** the wet stone back and forth ***on*** the blade's surface), apply and remove paint (***dip*** a brush ***in*** paint ***then rub*** it back and forth ***on*** various surfaces, then ***dip*** a rag ***in*** water ***then rub*** it back and forth ***on*** painted surfaces), and sand a blowgun dart (***rub*** the dart's base ***on*** a hairy leaf). Space precludes reanalyzing all 12 cases in detail, so 3 cases are selected to illustrate; 3 others are considered elsewhere (make fire, weed paths, and hammer nails into wood; see Byrne & Russon, in press; Russon, 1996). For each of the 3 selected cases, the demonstration and the orangutan's copy are described in terms of their manipulations of tool–object relationships (e.g., *probe*, *join*, *rub*, *hammer*, *insert*) as well as all behavior outside the tool use, retaining the exact sequence recorded in field notes.

Sharpening an axe blade

I arrived with Supinah near Inas, one of the camp staff, after he had started sharpening an axe blade. Supinah was a young adult female orangutan with extensive human experience. Inas was sitting on a log with an axe blade, a sharpening stone, and a scoop of water. Supinah watched him sharpening the blade then did so herself.

Figure 5.2 details the incident. The left column shows Inas' demonstration and the right column, Supinah's behavior. Behavior is spaced vertically to indicate interactive timing (top to bottom). Inas often interrupted or elaborated the basic technique to correct errors, insert optional activity, or handle interference so, to facilitate comparisons, relational manipulations that were components of the basic technique are shown in ***bold italics***, nonrelational manipulations are in *italics*, and peripheral or irrelevant behavior is indented. Only one set of materials was available, so periods when each actor had possession of the set are boxed.

Inas' demonstration involved the following behavioral manipulations. His basic sharpening technique was rubbing a wet stone against the blade until it was sharp. Object–object relational manipulations fundamental to this technique include one core complex operation, a manipulation co-ordinating two relationships – repeatedly rubbing the stone back and forth *on* the blade and *along* the blade's edge (***rub stone along edge of axe***), interspersed with three periodic operations, wetting the stone when it became dry – commonly after 30–40 rubs – by dipping it in a scoop of water (***dip***), reorienting the blade relative to the stone to enable working its other side (***turn***), and testing the blade's sharpness (***test***) by chopping with it or feeling its edge. After any of these periodic manipulations, rubbing is resumed unless the blade tests sharp enough. Disturbances interrupting Inas' task, and that were not relevant to it, were offering and giving tools to Supinah, puffing on a cigarette and tapping its ashes off, retrieving items Supinah stole, and acknowledging Supinah's request for tools.

Supinah watched Inas' sharpening closely and twice, Inas gave her the stone and axe blade. On the first occasion, she fumbled the stone, sucked the (wet) stone, then rubbed the stone 3–4 times on the face of the axe blade; on the second, she fumbled the stone, retrieved the scoop of water, dipped the stone in the water, and rubbed the stone 3–4 times on the face of the axe blade.

Superficial inspection of Supinah's behavior could give the impression that she merely performed a couple of actions that the situation happened to afford and that happened to match the demonstrator's. Inas clearly demonstrated a multitude of behaviors that Supinah did not copy in contrast to very few that she did and the situation limited the range of manipulations afforded. However, the particular set of manipulations she matched tells a more interesting story. The behaviors she matched were the relational manipulations of the basic technique – rub the stone on the face of the blade (***rub stone on face of axe***), and wet the stone by dipping it in the scoop of water (***dip***) then rub it on the face of the blade (***rub stone on face of axe***). Other observational data on orangutan tool using suggests that she did not simply discover these manipulations independently: She already knew them and she deliberately matched them. She did not perform alternative manipulations that the axe afforded and that were known to her, like hitting or chopping with it. Other data revealed that she already knew the individual relational manipulations, ***rub*** one item ***on*** another and wet an item by ***dipping*** it ***in*** liquid, and she used the arrangement ***dip then rub*** in other routines

INAS	SUPINAH
Rub stone along edge of axe	
turn	
dip	
Rub stone along edge of axe	
test	watch axe sharpening intently
Rub stone along edge of axe	
offer Supinah stone	extend hand to receive stone
withdraw stone	
Rub stone along edge of axe	move to sit closer to INAS
give Supinah stone and axe, comment	receive stone and axe; suck stone
point & simulate rubbing near blade	place blade on log
	move stone towards blade
	pick up axe
	Rub stone on face of axe (3-4 times)
request axe & stone back	taste blade & stone, try to chew
	give stone and axe back
dip	
place blade on log for support	watch axe sharpening intently
Rub stone along edge of axe	
turn	dip hand in scoop, drink from hand
Rub stone along edge of axe	
puff cigarette, tap ash off	
Rub stone along edge of axe	
dip	
Rub stone along edge of axe	
retrieve scoop, replace near self	take scoop of water, drink from scoop
dip	
Rub stone along edge of axe	
turn	
Rub stone along edge of axe	
dip	
dip blade, wipe clean	
test	
dip blade, wipe clean, re-orient stone	
Rub stone along edge of axe	pick up cigarette, puff cigarette
retrieve cigarette	dip hand in scoop, drink from hand
Rub stone along edge of axe	request gesture for stone, blade
acknowledge Supinah (tap her hand)	
Rub stone along edge of axe	request gesture for stone, blade again
give Supinah stone, blade	receive stone, blade; fumble them
	move to sit beside water scoop
	dip
	Rub stone on face of axe
	change hold on stone, blade
	Rub stone on face of axe
	suck blade, put blade down

Figure 5.2 Sharpening axe blade.

(e.g., wash clothes, eat soap). However, she was unlikely to have tried this application: This was the only occasion in over 1000 hours of observation, collected during three 3–6 month sampling periods that spanned three years, on which I observed camp staff sharpening blades, and she was unlikely to have invented this technique independently because this forest had no stones. Her rendition of one relational manipulation was undifferentiated compared with Inas'; she *rubbed* the stone, imprecisely placed *on the face* of the blade, *3–4 times* – he *rubbed* it, precisely placed *along the edge* of the blade, *30–40 times*. Finally, she ignored elements of Inas' behavior that were irrelevant to the blade-sharpening technique – except for smoking the cigarette, a preferred but normally illicit activity. These characteristics are consistent with those predicted for great apes' imitative tool use. They also exemplify what Tomasello (1996) advocated as the crux of true imitation, selection of relevant aspects of a demonstration with respect to its intent.

Siphoning

Supinah attempted to siphon from a fuel drum into a jerry can in a way that closely matched the camp technique (see Figure 5.3; demonstration and copy realigned to highlight comparable features). This incident was discussed in Russon and Galdikas (1993a) but not in terms of relational manipulations. No demonstration directly preceded her attempt, so I established the basic technique from other observations of staff siphoning fuel. When Supinah tried to siphon, the fuel drum was empty; this was the only circumstance in which orangutans' manipulating the fuel drum was tolerated.

The basic technique in camp used three tools, a fuel drum, a jerry can, and a length of hose to serve as the siphon; also involved were lids of both drum and can, and fuel in the drum at the outset. The technique is to open both containers (*unscrew then remove* their lids), place the can near the drum with its mouth near the drum's hole (*near*), insert the first end of the siphon into fuel in the drum (*insert*), suck second end of the siphon until fuel flows (fuel *through* siphon), then quickly insert the second end into the can (*insert*, so fuel flows between the two – *out* of the drum *and through* the siphon *and into* the can).

Supinah matched most of the basic relational manipulations in siphoning fuel – open both containers (*unscrew then remove* lids), place the can *near* the drum, *insert* the first end of the siphon into the drum, suck the second end of the siphon while the first end is in fuel drum (missing the relational reason for this, fuel *through* siphon), and *insert* the second end of hose into the can (again missing the complex relational patterns required). Evidence from the object-manipulation database confirmed that she already knew these basic relational manipulations. She also, however, matched their necessary arrangement except for the timing of placing the can near the drum and the critical timing between sucking the hose and inserting it into the can. Because of the complexity of the arrangement and the fact that, for her, it could not have been experientially shaped (orangutans were not allowed to fiddle with the fuel drum *unless* it was empty), it is highly unlikely that she invented it on her own. Her other major functional error stemmed from the fact that the fuel drum was empty, a circumstance over which she had no control. Her errors could have stemmed from not understanding what happens in siphoning when fuel is present – a factor that is not observable and that must be worked out by direct experience (humans also learn much of this by experience, after distasteful mouthfuls of fuel caused by hearty sucking or spilled fuel caused by unskilled transfer of the siphon from mouth to can). Supinah's other deviations from the demonstrated routine look like error corrections (e.g., reinsert or retrieve the fallen

unscrew then remove both lids	***unscrew then remove*** both lids
place can ***near*** drum	
Insert one end of siphon into fuel in drum	***Insert*** first end of hose in drum (empty)
put second end of siphon in mouth	*put second end of hose in mouth*
suck second end so fuel flows ***through***	*suck (or blow on) second end* of siphon
	insert first end of siphon farther in drum
	extract first end of siphon and taste it
	try to drink from can (empty)
	Insert first end of siphon into drum
	hold first end of siphon in place in drum
	put second end of siphon in mouth
	suck second end of hose
	place can ***near*** drum
	drop second end siphon, pick up again
quickly ***Insert*** second end of siphon in can	***Insert*** second end of siphon in can

Figure 5.3 Supinah siphons.

hose), investigations of unexpected outcomes (e.g., extract the hose after sucking, perhaps having tasted fuel fumes), or pauses for preferred activities (e.g., try to drink from the can). All resemble deviations that humans can insert into tool use that is hierarchically organized.

Sweeping

Princess, a young adult female wise to human ways, swept paths using the same special twig broom and special sweeping technique that camp staff used. No demonstration preceded this incident so I identified the basic technique from other observations of staff sweeping paths.

Camp staff and other Indonesians use twig brooms to sweep debris from paths, different from the brooms they use in houses. A twig broom is a bunch of long, straight twigs whipped together by vine or rattan "twine"; users hold it by its whipping "handle". Each sweeping stroke involves scraping the broom's bristles across the ground in front of the body so that debris is pushed off the side of the path (broom *sweep* debris *and* debris *off* path), then lifting the broom (broom *off* path) and returning it near the last stroke's starting-point (broom *near* start of last stroke) before lowering it (broom *on* path) and beginning a new stroke. A back-and-forth stroke could equally well be used, but it is not. While sweeping, the broom's twigs often become misaligned because the bundle is bound only around its outside. Indonesians repair the broom by tapping its butt end repeatedly against an available surface like the ground, the palm of their hand, or a wooden surface (*invert* broom, butt *perpendicular* to surface, *iterate* (butt *tap* surface) *until* twigs are *together*). Once twigs are realigned, sweeping resumes. Such repairs occur every few minutes during a sweeping bout.

Princess obtained a twig broom by luck, spotting an abandoned one near the camp staff quarters. She picked it up and, holding it appropriately by its whipping "handle," swept 5–6 strokes across the ground. Her stroke involved scraping the broom's bristles across the ground in front of her body (broom *sweep* path; there was no evident debris to remove), then lifting the broom (broom *off* path), returning it near that stroke's starting point (broom *near* start of last stroke), lowering it (broom *on* path), then beginning another stroke. She did *not* use the back-and-forth stroke that orangutans typically used when rubbing/scrubbing one object against another, but rather the same controlled sweeping strokes that Indonesians use with twig brooms. Princess paused, swept two more strokes, then inverted the broom and absently tapped its butt on the ground 9 times and against her knuckles 6 times, so that its twigs became better aligned (*invert* broom, butt *perpendicular* to surface, *iterate* (butt *tap* surface) *until* twigs are *together*). She changed hands, swept 5 strokes, tapped the butt on the ground 3 times, swept 9 strokes, paused, swept 12 strokes, paused, swept 3 strokes, paused, swept 3 strokes, tapped the butt 5 times. She finally unravelled the whipping binding the twig bundle, the broom fell apart, and she left.

Princess replicated all the relational manipulations essential to the camp technique for sweeping paths in their proper arrangement, except removing debris, and she appropriately inserted the routine for repairing the broom into the main routine when the broom's twigs became misaligned. This example may come the closest to matching motor action detail as well as arrangements of relational manipulations, because Princess used the same sweeping stroke as camp staff. Interestingly, though, she used idiosyncratic motor action detail to realign twigs: she sometimes tapped the broom's butt against her knuckles whereas, in my experience, staff used their palms when they tapped against their hands.

These reanalyses of existing candidates for orangutans' imitative tool use suggest that, in at least some cases, orangutans did indeed imitate tool-using routines in the distinctive great ape way. They did not commonly imitate tool use at the level of motor actions; they commonly adjusted motor actions flexibly, in line with local problem contingencies and in order to effect the desired relationship between tool and target and its desired outcome. Neither did they commonly imitate basic manipulations on object–object relationships; my own data on object manipulation and tool use in these rehabilitant orangutans and the literature both suggest that orangutans are quite capable of understanding and manipulating a wealth of simple object–object relationships independently (e.g., Galdikas, 1982; Lethmate, 1982; Russon & Galdikas, 1993b). In line with many extant views, what they appeared to imitate most commonly were novel arrangements or applications of these components to solve physical world problems. Orangutans' copies did tend to show low fidelity with respect to the original because they tended to neglect some essential relationships or to encode undifferentiated versions of them; both are consistent with predictions.

ORANGUTANS COMPARED WITH OTHER GREAT APES

This approach to identifying imitative tool use is based on cognitive qualities that are hypothesized to be shared by all great apes. Evidence allows exploring whether other great apes' alleged imitative tool use has similar qualities only for chimpanzees, because they alone have been systematically enough studied. I know of only one study on gorilla imitation of complex ecological behavior (Byrne & Byrne, 1993) and it focused on manipulative, not tool-based, food processing in wild mountain gorillas. The Byrnes argued that these gorillas imitated hierarchical *programs* for behavior, i.e. the structures that organize behavior into complex patterns, rather than the motor actions; this work is one recent source of the hypothesis that great ape imitative tool use targets relational-level rather than action-level operations. Moore's (1992) review found older reports of gorillas using powder-puffs, facecloths, latches, and keys but these reports are from early- to mid-century studies that have since been criticized.

I know of only one study specifically on bonobos, Toth et al.'s (1993) study on the imitative acquisition of stone tool use and manufacture by Kanzi, an enculturated language-trained bonobo; other studies discussed below have tested chimpanzees and bonobos together. In Toth's study, a human demonstrated knapping stone flake tools from a core, then using the flake tool to cut a cord, releasing the door to a box holding a food reward. Kanzi acquired the ability to make stone tools, and to use them to get the food, by imitation. He did, however, devise his own knapping technique of throwing the knapping stone at the core versus striking with it, and his technique may have failed to produce acute angle flakes systematically (Gibson, 1993b). Both limitations show effective imitation of the basic object–object hitting relationship but imprecision in how to hit, a refinement necessary for systematic stone tool making – possibly due to ignoring demonstrated relational subtleties. Twenty years earlier, Wright (1972) taught stone tool knapping to an orangutan using demonstration combined with various operant procedures. Like Kanzi, the orangutan acquired a basic technique that effectively detached a flake but did not always produce a functional cutting tool, and he refined the details of his technique on the basis of direct experience. Both patterns are consistent with the proposal that great apes can acquire the arrangement of basic relational operations imitatively, but that they often work out the technique's refinements and details independently.

The comparison with chimpanzees is particularly intriguing, because chimpanzees are the only habitual tool users among the great apes (e.g., McGrew, 1992) and they have been treated as the most likely nonhuman primates to imitate (e.g., Moore, 1992;

Whiten & Ham, 1992). Comparisons of imitative tool use between chimpanzees and orangutans – species that ostensibly face such different ecological and social pressures – should offer important insights into the cognitive abilities they share.

With some variation, recent major reviews of nonhuman imitation found claims for chimpanzees imitating the following range of tool use (Moore, 1992; Whiten & Ham, 1992; Mitchell, 1994): use screwdrivers or drills, dig with spades or trowels, hammer stakes or nails, make and use ladders (including stack boxes), open locks, use sandpaper, open paint cans, paint items with brush and pot of paint, sweep or scrub floors, dust, vacuum, spray with a spray bottle, regulate water flow by lever, wash laundry or dishes or dolls, towel-dry washed dolls, sew, "thread" a nail, sharpen pencils, use eyebrow tweezers, brush hair or teeth, apply makeup, wipe nose with hankie or perineal area with leaves, and, in the wild, cracks nuts or probe. These reviews primarily cover work conducted before the 1980s revival of imitation research, so many of the studies behind these cases have been challenged. What is none the less striking is that this purportedly imitative tool use by chimpanzees is close to identical in range and quality to that claimed for orangutans studied under more stringent methodological conditions. The fine-grained data needed to assess the organization and details of this "imitative" chimpanzee tool use are often unavailable, but these cases do involve applying a few from a limited set of relational manipulations to a variety of problems (e.g., ***rub*** one object ***on*** another's surface, ***insert*** an object ***into*** or ***thread*** it ***through*** another). Chimpanzees' use of these relational manipulations in a variety of tasks (e.g., sweep floor, wipe nose) shows they must be able to adjust manipulative detail to the contingencies posed by particular tasks.

Researchers studying wild chimpanzees have recently turned their attention away from imitation *per se* in favor of the more general process of social learning because of the methodological difficulty of distinguishing imitation from other social learning mechanisms on the basis of observation (e.g., McGrew, 1992; Boesch, 1996). New candidates for imitative tool use still appear, however. An interesting case comes from Boesch's (1991, 1993) work of teaching stone nut-cracking: A female observer maintained the *same orientation of her hammer stone relative to her nut* that had been demonstrated, but repeatedly *changed her own posture and motor actions independently*, often using a very poor hammering technique (Gibson, 1993a). This sounds very like an effort to imitate specific object–object relationships and manipulations upon them while varying their constituent motor actions.

Most recent work on chimpanzee imitation has used captive subjects. Tomasello's group is among the most active in studying their imitative tool use. Tomasello et al. (1993) tested six enculturated and nonenculturated chimpanzees and bonobos on imitating complex object manipulations, six of which involved simple tool use (e.g., flatten Play-Doh with a roller, open a can with a stick, paint foam on a floor with a brush, wipe foam off with a squeegee). Most of these tool manipulations have been imitated by orangutans. Both chimpanzees and bonobos imitated some of the tool-use tasks; the range of tasks they imitated was restricted and enculturated subjects were more adept than nonenculturated ones. There were no clear chimpanzee–bonobo differences, although species comparisons are badly compromised by the distribution of subject ages and the small sample size. Details of manipulations were not provided, but subjects' successful imitation of tool use alone implies their effective imitation of relational manipulations. Results suggest low fidelity reproduction in the form of "partial" copies (only means or ends). Tomasello's group also experimentally tested chimpanzees' imitating rake tool use (Tomasello et al., 1987; Nagell et al. 1993); in fact, Call and Tomasello's (1994) experiment on orangutans is a replication of this chimpanzee work. The results on the two species look basically the same: chimpanzees, like orangutans, tended to copy the raking techique to obtain out-of-reach food with the rake tool, but not the rake head's orientation; and some of the chimpanzees likewise showed a transition phase in which they used both head positions alternatively. In terms of success with the tool, orangutans actually did a bit better than chimpanzees or children (Nagell et al., 1993). The same imitative pattern seems to apply, replicating arrangements or applications of simple object-object relational manipulations.

DISCUSSION/CONCLUSION

To begin, I identified behavioral features that should characterize imitative tool use in great apes and serve to detect it. The proposal derives from a model of great ape cognition as hierarchical, but constrained to rudimentary levels and limited combinatorial capacity. This type of cognition likely affects the targets and power of great apes' imitation; considered in conjunction with qualities of human instrumental imitation and constraints operative in the causal cognition governing flexible tool use, this points to behavioral indices that should distinguish great ape imitative tool use. Great apes' cognitive strengths may incline them to imitate arrangements or applications of simple object–object relational manipulations used in demonstrations, more than motor action detail; their cognitive limitations may lead to imprecise copies with respect to both motor action detail and the differentiation of relational manipulations; the characteristics of flexible tool use may require great apes to acquire some components of demonstrated tool use independently rather than by imitation; and, because of great apes' cognitive constraints, they can accommodate few modifications to their existing behavior at one time.

The cases of alleged imitative tool use that I assessed show these behavioral indices and so offer evidence that orangutans, and likely the other great apes, can imitate tool use. Not all cases that have been reported were assessed this way, but the prediction was for a tendency, and these assessments show that such a tendency exists.

Predictions were explicitly stated as tendencies so as *not* to imply that great apes cannot or do not imitate motor actions: Studies on gestural imitation in great apes clearly show that they can (e.g. Hayes & Hayes, 1952; Custance & Bard, 1994; Miles et al., 1996). Rather, given great apes' cognitive strengths and limitations, the novelty and the challenge essential to spontaneous instrumental imitation commonly occur elsewhere. Predictions may similarly *not* extend to other uses of imitation, like interpersonal ones. Interpersonal and instrumental imitation may differ in humans (e.g. Yando et al.,

1978; Nadel, 1986; Užgiris, 1981). Hints of this possibility are now surfacing for great apes, including the possibility that these different functions are served by different cognitive mechanisms. One view proposes distinguishing *emulation* (imitating changes in the environment, like goals) from *impersonation* (imitating behavior) (Tomasello, 1990, after Wood, 1989); another suggests that instrumental imitation favors the use of *program-level* imitation (replicating programs, or arrangements and applications of behavioral complexes) whereas interpersonal imitation favors *action-level* imitation (closely matching the demonstrator's behavior), although the difference may be only one of degree (Byrne & Russon, 1998). Great apes appear capable of both; which they use may reflect the demands of the task at hand.

Two broader implications issue from this exercise. First, the exercise illuminates problems with attempts to portray imitation of outcomes and imitation of behavioral techniques as separate processes, *emulation* versus *impersonation,* in the context of tool use; similar problems have been identified elsewhere (e.g., Whiten & Ham, 1992; Whiten & Custance, 1996). Tool use entails means–end reasoning, that is, generating behavioral means that *link to* specific ends. Insofar as, and to the extent that, this implies the intertwining of means and ends, the two are not separable in the way suggested. The term "raking," for instance, covers the behavioral means as well as the behavioral end. Raking as outcome versus raking as means may reflect, at most, different faces of the relational manipulation that links the two. Second, this exercise suggests an alternative method of identifying imitation than the standard one of excluding all other social learning processes. Exclusion has been used because of the belief that imitation leaves no distinctive behavioral trace; it has greatly affected progress in the study of imitation because of the difficulty of controlling convincingly against input from all other possible learning processes. The current view suggests that, contrary to standard belief, imitation can produce a distinctive behavioral trace. This claim is not novel – cognitive developmental psychologists long ago articulated the qualities of imitative products that indicate the nature of their governing cognitive mechanisms (e.g., for a review, see Mitchell, 1987) – but it deserves more attention in the study of nonhuman imitation.

ACKNOWLEDGMENTS

The research associated with this chapter was supported by grants from the Natural Sciences and Engineering Council of Canada, and from Glendon College and York University, Toronto, Canada. I am extremely grateful to the agencies and individuals that made field work on rehabilitant orangutan imitation possible: The Orangutan Research and Conservation Project, The Wanariset Orangutan Reintroduction Project, the Indonesian Department of Forestry and its Nature Protection Agency (PHPA), the Indonesian Institute of Sciences (LIPI), and many outstanding Indonesian field staff, research volunteers, and Canadian research assistants. Thanks also for comments that contributed to improving the chapter, to the book's editors, and to Richard Byrne.

REFERENCES

Abravanal, E. & Gingold, H. (1985). Learning via observation during the second year of life. *Developmental Psychology*, **21**, 614–623.

Anderson, J. R. (1996). Chimpanzees and capuchin monkeys: Comparative cognition. In A. E. Russon, K. A. Bard & S. T. Parker (eds.), *Reaching into thought: The minds of the great apes* (pp. 23–56). Cambridge University Press.

Antinucci, F. (ed.) (1989). *Cognitive structure and development in nonhuman primates.* Hillsdale, NJ: Erlbaum.

Bard, K. A. (1993). Cognitive competence underlying tool use in free-ranging orang-utans. In A. Berthelet & J. Chavaillon (eds.), *The use of tools by humans and non-human primates* (pp. 103–113). Oxford University Press.

Bauer, P. & Mandler, J. (1989). One thing follows another: Effects of temporal structure on 1– and 2–year-olds' recall of events. *Developmental Psychology*, **25**, 197–206.

Beck, B. B. (1980). *Animal tool behavior: The use and manufacture of tools by animals.* New York: Garland STPM Press.

Boesch, C. (1991). Teaching among wild chimpanzees. *Animal Behaviour*, **41**, 530–532.

Boesch, C. (1993). Aspects of transmission of tool-use in wild chimpanzees. In K. R. Gibson & T. Ingold (eds.), *Tools, language and cognition in human evolution* (pp. 171–183). Cambridge University Press.

Boesch, C. (1996). Three approaches for assessing chimpanzee culture. In A. E. Russon, K. A. Bard, & S. T. Parker (eds.), *Reaching into thought: The minds of the great apes* (pp. 404–429). Cambridge University Press.

Boesch, C. & Boesch, H. (1990). Tool use and tool making in wild chimpanzees. *Folia Primatologica*, **54**, 86–99.

Byrne, R. W. (1994). The evolution of intelligence. In P. J. B. Slater & T. R. Halliday (eds.), *Behaviour and evolution* (pp. 223–265). Cambridge University Press.

Byrne, R. W. (1995). *The thinking ape: Evolutionary origins of intelligence.* Oxford University Press.

Byrne, R. W. & Byrne, J. M. E. (1991). Hand preferences in the skilled gathering tasks of mountain gorillas (*Gorilla gorilla berengei*). *Cortex*, **27**, 521–546.

Byrne, R. W. & Byrne, J. M. E. (1993). The complex leaf-gathering skills of mountain gorillas (*Gorilla g. berengei*): Variability and standardization. *American Journal of Primatology*, **31**, 241–261.

Byrne, R. W. & Russon, A. E. (1997). Learning by imitation: A hierarchical approach. *Behavioural and Brain Sciences.* (1998, 21, 667–721).

Call, J. & Tomasello, M. (1994). The social learning of tool use by orangutans *(Pongo pygmaeus)*. *Human Evolution*, **9**, 297–313.

Call, J. & Tomasello, M. (1995). The use of social information in the problem-solving of orangutans and human children. *Journal of Comparative Psychology*, **109**, 301–320.

Case, R. (1985). *Intellectual development: Birth to adulthood.* New York: Academic Press.

Custance, D. M. & Bard, K. A. (1994). The comparative and developmental study of self-recognition and imitation: The importance of social factors. In S. T. Parker, R. W. Mitchell, & M. L. Boccia (eds.), *Self-awareness in animals and humans: Developmental perspectives* (pp. 207–226). Cambridge University Press.

D'Amato, M. R. & Salmon, D. P. (1984). Cognitive processes in cebus monkeys. In H. L. Roitblat, T. G. Bever, & H. S. Terrace (eds.), *Animal cognition*, (pp. 149–168). Hillsdale, NJ: Erlbaum.

Dawkins, R. (1976). Hierarchical organization: A candidate principle for ethology. In P. P. G. Bateson & R. A. Hinde (eds.), *Growing points in ethology*. Cambridge University Press.

Dawson, B. V. & Foss, B. M. (1965). Observational learning in budgerigars. *Animal Behaviour*, **13**, 470–474.

Dumas, C. & Doré, F.-Y. (1986). *Intelligence animale: Recherches Piagetiennes* [Animal Intelligence: Piagetian Studies]. Sillery, Québec, Canada: Presses de l'Université du Québec.

Galdikas, B. M. F. (1982). Orang-utan tool use at Tanjung Puting Reserve, Central Indonesian Borneo (Kalimantan Tengah). *Journal of Human Evolution*, **10**, 19–33.

Galef, B. G. Jr. (1990). Tradition in animals: Field observations and laboratory analyses. In M. Bekoff & D. Jamieson (eds.), *Interpretations and explanations in the study of behaviour: Comparative perspectives* (pp. 3–28). Hillsdale, NJ: Erlbaum.

Galef, B. G. Jr. (1992). The question of animal culture. *Human Nature*, **3**, 157–178.

Gallup, G. G. Jr. (1983). Towards a comparative psychology of mind. In R. L. Mellgren (ed.), *Animal cognition and behavior* (pp. 473–510). New York: North Holland.

Gibson, K. R. (1990). New perspectives on instincts and intelligence: Brain size and the emergence of hierarchical mental construction skills. In S. T. Parker & K. R. Gibson (eds.), *"Language" and intelligence in monkeys and apes: Comparative developmental perspectives* (pp. 97–128). New York: Cambridge University Press.

Gibson, K. R. (1993a). Generative interplay between technical capacities, social relations, imitation and cognition. In K. R. Gibson & T. Ingold (eds.), *Tools, language and cognition in human evolution* (pp. 131–137). Cambridge University Press.

Gibson, K. R. (1993b). Tool use, language and social behavior in relationship to information processing capacities. In K. R. Gibson & T. Ingold (eds.), *Tools, language and cognition in human evolution* (pp. 251–269). Cambridge University Press.

Gomez, J. C. (1988). Tool-use and communication as alternative strategies of problem solving in the gorilla. *Primate Report*, **19**, 25–8.

Goodall, J. (1986). *The chimpanzees of Gombe*. Cambridge, MA.: Harvard University Press.

Greenfield, P. (1991). Language, tools and the brain: The ontogeny and phylogeny of hierarchically organized sequential behavior. *Behavioral and Brain Sciences*, **14**, 531–595.

Haggerty, M. E. (1910). Preliminary studies on anthropoid apes. *Psychological Bulletin*, **7**, 49.

Haggerty, M. E. (1913, August). Plumbing the minds of apes. *McClure's Magazine*, **41**, 151–154.

Harrisson, J. (1962). *Orang-utan*. London: Collins.

Harrisson, J. (1969). The nesting behaviour of semi-wild juvenile orang-utans. *Sarawak Museum Journal*, **17**, 336–384.

Hayes, K. J. & Hayes C. (1952). Imitation in a home-reared chimpanzee. *Journal of Comparative and Physiological Psychology*, **45**, 450–459.

Heyes, C. M. (1996). Introduction: Identifying and defining imitation. In C. M. Heyes & B. G. Galef, Jr. (eds.), *Social learning in animals: The roots of culture* (pp. 211–220). New York: Academic Press.

Heyes, C. M. & Galef, B. G. Jr. (eds.) (1996). *Social learning in animals: The roots of culture*. New York: Academic Press.

Hewes, G. W. (1994). The baseline for comparing human and nonhuman primate behavior. In D. Quiatt & J. Itani (eds.), *Hominid culture in primate perspective* (pp. 59–93). Niwot, CO: University Press of Colorado.

Hornaday, W. (1922). *The minds and manners of wild animals*. New York: Scribners.

Horr, D. A. (1977). Orang-utan maturation: Grouping up in a female world. In S. Chevalier-Skolnikoff & F. E. Poirier (eds.), *Primate bio-social development* (pp. 289–322). New York: Garland.

Inhelder, B. & Piaget, J. (1964). *The early growth of logic*. New York: W. W. Norton.

Jordan, C. (1982). Object manipulation and tool-use in captive pygmy chimpanzees (*Pan paniscus*). *Journal of Human Evolution*, **11**, 35–39.

Klüver, H. (1933). *Behavior mechanisms in monkeys*. University of Chicago Press.

Köhler, W. (1927). *The mentaity of apes*. (2nd revised edn). New York: Vintage Books.

Langer, J. (1993). Comparative cognitive development. In K. R. Gibson & T. Ingold (eds.), *Tools, language and cognition in human evolution* (pp. 300–313). Cambridge University Press.

Langer, J. (1996). Heterochrony and the evolution of primate cognitive development. In A. E. Russon, K. A. Bard, & S. T. Parker (eds.), *Reaching into thought: The minds of the great apes* (pp. 257–277). Cambridge University Press.

Lawick-Goodall, J. van (1973). Cultural elements in a chimpanzee community. In E. W. Menzel (ed.), *Precultural primate behavior* (pp. 144–184). Basle: S. Karger.

Lethmate, J. (1982). Tool-using skills of orang-utans. *Journal of Human Evolution*, **11**, 49–64.

Matsuzawa, T. (1994). Field experiments on use of stone tools in the wild. In R. W. Wrangham, W. C. McGrew, F. B. M. de Waal & P. G. Heltne (eds.), *Chimpanzee cultures* (pp. 351–370). Cambridge MA: Harvard University Press.

Matsuzawa, T. (1996). Chimpanzee intelligence in nature and in captivity: isomorphism of symbol use and tool use. In W. C. McGrew, L. F. Marchant, & T. Nishida (eds), *Great ape societies* (pp. 196–209). Cambridge University Press.

McGrew, W. C. (1992). *Chimpanzee material culture: Implications for human evolution*. Cambridge University Press.

Meinel, M. (1995). Eliciting true imitation of object use in captive orangutans. Unpublished BA thesis, Glendon College, York University, Toronto, Canada.

Meinel, M. & Russon, A. E. (1996). Eliciting true imitation of object use in captive orangutans. Presented at the XVIth Congress of the International Primatological Society, Madison WI, August 11–16.

Meltzoff, A. N. (1996). The human infant as imitative generalist: A 20–year progress report on infant imitation with implications for comparative psychology. In C. M. Heyes & B. G. Galef, Jr. (eds.), *Social learning in animals: The roots of culture* (pp. 347–370). New York: Academic Press.

Miles, H. L., Mitchell, R. M., & Harper, S. (1996). Simon says: The development of imitation in an enculturated orangutan. In A. E. Russon, K. A. Bard, & S. T. Parker (eds.), *Reaching into thought: The minds of the great apes* (pp. 278–299). Cambridge University Press.

Miller, G. A., Galanter, E., & Pribram, K. (1960). *Plans and the structure of behavior*. New York: Holt, Rinehart & Winston.

Mitchell, R. W. (1987). A comparative-developmental approach to understanding imitation. In P. P. G. Bateson & P. H. Klopfer (eds.), *Perspectives in ethology* (vol. 7, pp. 197–227). New York: Plenum.

Mitchell, R. W. (1993). Mental models of mirror self-recognition: Two theories. *New Ideas in Psychology*, **11**, 295–325.

Mitchell, R. W. (1994). The evolution of primate cognition: Simulation, self-knowledge, and knowledge of other minds. In D. Quiatt & J. Itani (eds.), *Hominid culture in primate perspective* (pp. 177–232). Boulder, CO: University Press of Colorado.

Mitchell, R. W. & Anderson, J. R. (1993). Discrimination learning of scratching, but failure to obtain imitation and self-recognition in a long-tailed macaque. *Primates*, **34**, 301–309.

Moore, B. R. (1992). Avian movement imitation and a new form of mimicry: Tracing the evolution of a complex form of learning. *Behaviour*, **122**(3–4), 231–263.

Moore, B. R. (1996). The evolution of imitative learning. In C. M. Heyes & B. G. Galef, Jr. (eds.), *Social learning in animals: The roots of culture* (pp. 245–266). New York: Academic Press.

Nadel, J. (1986). *Imitation et Communication entre Jeunes Enfants* [Imitation and communication between young children]. Paris: Presses Universitaires de France.

Nagell, K., Olguin, K., & Tomasello, M. (1993). Processes of social learning in the tool use of chimpanzees and human children. *Journal of Comparative Psychology*, **107**, 174–185.

Nash, V. J. (1982). Tool use by captive chimpanzees at an artificial termite mound. *Zoo Biology*, **1**, 211–221.

Paquette, D. (1992). Discovering and learning tool-use for fishing honey by captive chimpanzees. *Human Evolution*, 7, 17–30.

Parker, C. (1969). Responsiveness, manipulation and implementation behavior in chimpanzees, gorillas, and orang-utans. *Proceedings of the Second International Congress of Primatology*, 1, 160–166. New York: S. Karger.

Parker, S. T. (1996a). Apprenticeship in tool-mediated extractive foraging: The origins of imitation, teaching and self-awareness in great apes. In A. E. Russon, K. A. Bard, & S. T. Parker (eds.), *Reaching into thought: The minds of the great apes* (pp. 348–370). Cambridge University Press.

Parker, S. T. (1996b). Using cladistic analysis of comparative data to reconstruct the evolution of cognitive development in hominids. In E. P. Martins (ed.), *Phylogenies and the comparative method in animal behavior* (pp. 361–398). New York: Oxford University Press.

Parker, S. T. & Gibson, K. R. (1977). Object manipulation, tool use and sensorimotor intelligence as feeding adaptations in Cebus monkeys and great apes. *Journal of Human Evolution*, **6**, 623–641.

Parker, S. T. & Gibson, K. R. (1979). A model of the evolution of language and intelligence in early hominids. *Behavioral and Brain Sciences*, **2**, 367–407.

Parker, S. T. & Gibson, K. R. (eds.), (1990). *"Language" and intelligence in monkeys and apes*. Cambridge University Press.

Parker, S. T. & Potì, P. (1990). The role of innate motor patterns in ontogenetic and experiential development of intelligent use of sticks in cebus monkeys. In S. T. Parker & K. R. Gibson (eds.), *"Language" and intelligence in monkeys and apes* (pp. 219–246). Cambridge University Press.

Parker, S. T., Mitchell, R. W., & Boccia, M.L. (eds.) (1994). *Self-awareness in humans and animals: Developmental perspectives*. Cambridge University Press.

Piaget, J. (1937/1954). *The construction of reality in the child*. New York: Basic Books.

Piaget, J. (1952). *The origins of intelligence in children*. New York: International Universities Press.

Piaget, J. (1962). *Play, dreams and imitation in children*. New York: W. W. Norton.

Piaget, J. & Inhelder, B. (1967). *The child's conception of space*. New York: W. W. Norton.

Premack, D. (1988). "Does the chimpanzee have a theory of mind?" revisited. In R. W. Byrne & A. Whiten (eds.), *Machiavellian intelligence* (pp. 160–179). Oxford, U.K.: Clarendon Press.

Rensch, B. (1973). Play and art in monkeys and apes. In E. Menzel, Jr. (ed.), *Precultural primate behavior*. Symposium of the IVth International Congress of Primatology (vol. 1). Basle: Karger.

Reynolds, P. C. (1982). The primate constructional system: The theory and description of instrumental tool use in humans and chimpanzees. In M. van Cranach & R. Harré (eds.), *The analysis of action* (pp. 243–385). Cambridge University Press.

Reynolds, P. C. (1983). Ape constructional ability and the origin of linguistic structure. In De Grolier, E. (ed.), *Glossogenetics: The origin and evolution of language* (pp. 185–200). Amsterdam: Harwood Academic Publishers.

Reynolds, P. C. (1991). Structural differences in intentional action between humans and chimpanzees – and their implications for theories of handedness and bipedalism. In M. Anderson & Fl. Merrell (eds.), *Semiotic modeling* (pp. 19–46). Berlin: Walter Gruyter & Co.

Russon, A. E. (1995). Aping imitation. Contribution to *Imitation in comparative evolutionary perspective*, invited symposium (A. Russon, organizer), 25th annual Jean Piaget Society Symposium, Berkeley, June 1–3.

Russon, A. E. (1996). Imitation in everyday use: Matching and rehearsal in the spontaneous imitation of rehabilitant orangutans (*Pongo pygmaeus*). In A. E. Russon, K. A. Bard, & S. T. Parker (eds.),

Reaching into thought: The minds of the great apes (pp. 152–176). Cambridge University Press.

Russon, A. E. (1998). The nature and evolution of orangutan intelligence. *Primates*, **39**, 485–503.

Russon, A. E. & Galdikas, B. M. F. (1993a). Imitation in free-ranging rehabilitant orangutans (*Pongo pygmaeus*). *Journal of Comparative Psychology*, **107**(2), 147–161.

Russon, A. E. & Galdikas, B. M. F. (1993b). The complexity of orangutan tool use. Symposium contribution to Cognitive Ethology of the Pongids, organized by A. Russon, at the XXIII International Ethological Conference, Torremolinos, Spain, September 1–9.

Russon, A. E. & Galdikas, B. M. F. (1994). Hierarchical organization in rehabilitant orangutan object manipulation. Symposium contribution to Hierarchical Organization in Primate Intelligence, R. W. Byrne & A. E. Russon, at the XVth Congress of the International Primatological Society, Kuta, Bali, August 3–8.

Russon, A. E. & Galdikas, B. M. F. (1995). Constraints on great apes' imitatiion: Model and action selectivity in rehabilitant orangutan (*Pongo pygmaeus*) imitation. *Journal of Comparative Psychology*, **109**, 5–17.

Russon, A. E., Bard, K. A., & Parker, S. T. (eds.) (1996). *Reaching into thought: The minds of the great apes*. Cambridge University Press.

Russon, A. E., Mitchell, R. E., Lefebvre, L., & Abravanel, E. (1998). The comparative evolution of imitation. In J. Langer & M. Killen (eds.), *Piaget, evolution & development* (pp. 103–143). Hillsdale, NJ: Erlbaum.

Sherrington, C. S. (1906). *The integrative action of the nervous system* (pp. 388–392). New Haven, CT: Yale University Press.

Thorndike, E. L. (1898). Animal intelligence: An experimental study of the associative process in animals. *Psychological Review Monograph*, **2**(8), 551–553.

Thorpe, W. H. (1963). *Learning and instinct in animals*. (2nd edn). London: Methuen.

Tomasello, M. (1990). Cultural transmission in the tool use and communicatory signalling of chimpanzees? In S. T. Parker & K. R. Gibson (eds.), *"Language" and intelligence in monkeys and apes* (pp. 274–311). New York: Cambridge University Press.

Tomasello, M. (1996). Do apes ape? In C. M. Heyes & B. G. Galef, Jr. (eds.), *Social learning in animals: The roots of culture* (pp. 319–346). New York: Academic Press.

Tomasello, M., Savage-Rumbaugh, E. S., & Kruger, A. C. (1993). Imitative learning of actions on objects by children, chimpanzees, and enculturated chimpanzees. *Child Development*, **64**, 1688–1705.

Tomasello, M., Davis-Dasilva, M., Camak, L., & Bard, K. (1987). Observational learning of tool use by young chimpanzees. *Human Evolution*, **2**, 175–185.

Toth, N., Schick, K. D., Savage-Rumbaugh, E. S., Sevcik, R., & Rumbaugh D. M. (1993). Pan the tool maker: investigations into the stone tool-making and tool-using abilities of a bonobo (*Pan paniscus*). *Journal of Archaeological Science*, **20**, 81–91.

Užgiris, I. (1981). Two functions of imitation during infancy. *International Journal of Behavioral Development*, **4**, 1–12.

Van Schaik, K., Fox, E. A., & Sitompul, A. F. (1996). Manufacture and use of tools in wild Sumatran orangutans: Implications for human evolution. *Naturwissenschaften*, **83**, 186–188.

Visalberghi, E. (1993a). Tool use in a South American monkey species. An overview of characteristics and limits of tool use in *Cebus apella*. In A. Berthelet & J. Chavaillon (eds.), *The use of tools by humans and nonhuman primates* (pp. 118–131). Oxford University Press.

Visalberghi, E. (1993b). Capuchin monkeys: A window into tool use in apes and humans. In K. R. Gibson & T. Ingold (eds.), *Tools, language and cognition in human evolution* (pp. 138–150). Cambridge University Press.

Visalberghi, E. & Fragaszy, D. M. (1990). Do monkeys ape? In S. T. Parker & K. R. Gibson (eds.),

"Language" and intelligence in monkeys and apes (pp. 247–273). New York: Cambridge University Press.

Visalberghi, E. & Limongelli, L. (1996). Acting and understanding: Tool use revisited through the minds of capuchin monkeys. In A. E. Russon, K. A. Bard & S. T. Parker (eds.), *Reaching into thought: The minds of the great apes* (pp. 57–79). Cambridge University Press.

Visalberghi, E., Fragaszy, D. M., & Savage-Rumbaugh, S. (1995). Performance in a tool-using task by common chimpanzees (*Pan troglodytes*), bonobos (*Pan paniscus*), an orangutan (*Pongo pygmaeus*), and capuchin monkeys (*Cebus apella*). *Journal of Comparative Psychology*, **109**, 52–60.

Vygotsky, L. (1962). *Thought and language*. Cambridge, MA: MIT Press.

Whiten, A. (1996). Imitation, pretense, and mindreading: Secondary representation in comparative primatology and developmental psychology? In A. E. Russon, K. A. Bard, & S. T. Parker (eds.), *Reaching into thought: The minds of the great apes* (pp. 300–324). Cambridge University Press.

Whiten, A. & Byrne, R. W. (1991). The emergence of metarepresentation in human ontogeny and primate phylogeny. In A. Whiten (ed.), *Natural theories of mind* (pp. 267–282). Oxford: Blackwell.

Whiten, A. & Custance, D. M. (1996). Studies of imitation in chimpanzees and children. In C. M. Heyes & B. G. Galef, Jr. (eds.), *Social learning in animals: The roots of culture* (pp. 291–318). New York: Academic Press.

Whiten, A. & Ham, R. (1992). On the nature and evolution of imitation in the animal kingdom: Reappraisal of a century of research. In P. J. B. Slater, J. S. Rosenblatt, C. Beer, & M. Milinski (eds.), *Advances in the Study of Behaviour* (vol. 21, pp. 239–283). New York: Academic Press.

Whiten, A., Custance, D. M., Gomez, J.-C., Teixidor, P., & Bard, K. (1995). Imitative learning of artificial fruit processing in children (*Homo sapiens*) and chimpanzees (*Pan troglodytes*). *Journal of Comparative Psychology*, **110**(1), 3–14.

Wood, D (1989). Social interaction as tutoring. In M. H. Bornstein & J. S. Bruner (eds.), *Interaction in human development* (pp. 59–80). Hillsdale, NJ: Erlbaum.

Wood, R. J. (1984). Spontaneous use of sticks by gorillas at Howletts Zoo Park, England. *International Zoo News*, **31**(3), 13–18.

Wright, R. V. S. (1972). Imitative learning of a flaked stone technology – The case of an orangutan. *Mankind*, **8**, 296–306.

Yando, R., Seitz, V., & Zigler, E. (1978). *Imitation: A developmental perspective*. Hillsdale, NJ: Erlbaum.

Yerkes, R. M. (1916). The mental life of monkeys and apes. *Behavior Monographs* (Vol. 3). New York: Holt.

Yerkes, R. M. & Yerkes, A. W. (1929). *The great apes*. New Haven, CT: Yale University Press.

6

Object manipulation and skill organization in the complex food preparation of mountain gorillas

RICHARD W. BYRNE

NATURAL HISTORY

In his monograph on the mountain gorilla of the Virunga Volcanoes, Schaller describes how gorillas eat over thirty different foods (Schaller, 1963, pp. 156–165). The accounts are brief, and sometimes based on only a very few observations. Nevertheless, it is immediately clear to the reader that interestingly different techniques are involved with each plant.

Some, such as the vine *Droquetia iners*, are "merely pushed into the mouth." Almost as little care is taken with small ferns *Polypodium* sp.: to eat them a gorilla "reaches below the branch and without looking grabs a handful of the hanging ferns which it pulls in. After severing and discarding the roots with one bite, it stuffs the greens into its mouth." Biting off encasing material is a method used in several different ways to detach inedible, contaminated, or, occasionally edible parts. Bark from the tree fern *Cyathea deckenii* is bitten off and discarded in small piles, and only the tender inside is eaten; taproots of the herbs *Cynoglossum amplifolium* and *C.geometricum* are hauled out of the ground and their tough bark is bitten off before consumption; in contrast, the dry bark of *Hagenia* and *Hypericum* trees is bitten off, but consumed.

Other plants evidently present more challenge to preparation. Often, this is a consequence of the relatively greater hardness and toughness of the encasing material compared with the fragile but edible interior. Schaller noted that feeding remains of bamboo *Arundinaria alpina* suggest that the tough and hairy outer layers are "peeled back to expose the pith, much as a human prepares a banana." Elephant grass *Pennesetum purpureum*, and wild ginger *Afromomum* sp., are also peeled in somewhat similar ways in order to eat their soft interiors. The pith of the tree-like *Vernonia adolfi-frederici* is a tougher proposition than these giant herbs and grasses. A gorilla must first detach a branch and shorten it to manageable proportions, then it bites into a section of the bark and wood and "jerks backward with the head while at the same time pushing the distal end of the branch with one hand in the opposite direction." This technique is elaborated in consumption of one of the commonest gorilla foods, the umbelliferous herb *Peucedanum linderi*. Schaller describes an individual who "holds the stalk horizontally in both hands and rapidly detaches slivers of bark by biting and tearing them off with abrupt sideways jerks of the head. Finally she eats the tender center."

The most interesting descriptions, however, are those for plants which are physically defended. The simplest method, of course, is to gingerly pick out the edible portions; for instance, gorillas pick the ripe fruit of blackberry *Rubus* sp. from among the long spines. Nettles are defended by stings rather than spines. While the virulence of these nettles was such that Schaller found "my knees were swollen and red welts covered my face," he considered the gorillas insensitive to the stinging hairs. However, his observations that a blackback (adolescent) male ate the top of the nettle *Laportea alatipes*, and an infant ate the flowers by picking them off individually, suggests that this may have been a premature deduction: in all nettles, the flowers and tip of the plants specifically lack active stings. It seems even less likely that gorillas are insensitive to the spines of thistles *Carduus* sp., for which Schaller describes several methods of feeding.

This range of elaborate manual skills is not inconsistent with what is known of gorilla hand anatomy. Among the great apes, the gorilla is unusual in having fingers relatively short compared to the thumb, in fact theirs is closest in formation to the human hand (Napier, 1961). Although the human "pad-to-pad" precision grip – thumb and index finger – is impossible for nonhuman primates, the gorilla's ability to use a "tip-to-tip" precision grip with the conical fingertips of these digits (as well as the "pad-to-side" grip more general in Haplorhine primates) allows a potentially wide range of deft manual actions (Christel, 1993).

In texts on human evolution, it is axiomatic to stress the importance of manual dexterity and control, in particular the retention of the primitive pentadactyl hand in primates and the enhanced flexibility of control in great apes. Primates in the wild show skills of manual precision almost entirely in food preparation and grooming. The latter is general in Haplorhines, but food preparation beyond the most rudimentary actions is certainly not. Thus Schaller's discovery that the mountain gorilla shows a wide range of varied and often deft methods of food preparation has obvious significance for understanding the evolution of manual skills in hominids. Given the complexity of some of the methods he described, and his explicit mention that "gorillas showed great dexterity in combining the use of hands and mouth, with the result that the palatable parts of each plant were rapidly exposed and eaten" (Schaller, 1963, p. 156), it is surprising that no work on gorilla food preparation was attempted again until 1989. By that time, however, a great deal more was known about the ecological circumstances of gorilla feeding.

ECOLOGY

In the Virunga Volcanoes of Rwanda and eastern Zaïre, the herbaceous vegetation[1] is generally low in "secondary compounds," plant poisons adapted to deter consumption

[1] "Meadows" of large herbs occur in the hagenia–hypericum zone of the Virungas, and most likely formed originally after tree-destruction by elephants or gorillas. Gorillas do pull down trees to get access to arboreal leaves, and this benefits the plants on which they primarilysubsist, a sort of unintended "farming."

by animals (Waterman, Choo, Vedder, & Watts, 1983). These plant poisons are almost ubiquitous in tropical vegetation, where leaves tend to be long-lived structures; animals like gorillas and humans, lacking digestive specialization, cannot safely consume most tropical leaves, let alone subsist on them. In contrast, the herbs of the Virungas are temperate plants: fast-growing, short-lasting, and chemically unprotected. Moreover, they are often high in protein (Waterman et al., 1983). And, whereas for many primates the edible plant items make up only a minute fraction of the total biomass in the habitat, among the herb meadows of the Virungas the plants which the gorillas actually do eat are almost as abundant as those which they do not (Watts, 1984). Finally, the location of these plants is predictable, in a rather simple way; in fact, the great majority of the gorillas' diet consists of plants which are very uniformly distributed in time and space (Watts, 1984). Only the seasonally shooting bamboo, found in a narrow monospecies zone, and the giant Afro-Alpine plants (notably *Senecio johnstonii* and *Lobelia wollostonii*), found in a distinct zone on the higher regions of the volcanoes, show any significant variation in spatio-temporal patterning.

It would seem from this picture that the foraging task confronting mountain gorillas is a trivial one, compared with the demands made on most primate species: their foodstuffs are largely unvarying in time and space, nutritious and chemically unprotected, and common, both in relative and absolute terms. That is indeed how the gorilla's foraging has been characterized in popular literature (Attenborough, 1979). However, this is to forget the significance of the processing skills described by Schaller. Food plants are easy to find, but are hard to process. The main feeding challenge confronting mountain gorillas is, rather than any difficulty of finding foods to eat, the technical problems in food preparation – dealing with painful stings, sturdy spines, and delicate foods embedded in tough outer cases.

Gorillas must therefore be expected to compete by employing skill. Unlike many species of primate, gorillas seldom eat foods for which overt contest competition would be worthwhile: each plant is of relatively small value, and easily replaced. (In fact, female gorillas frequently dispute with each other, causing many minor injuries [Watts, 1994]. The function of this agonism is uncertain, but the disputes are not centered on those few high-quality foods which gorillas eat and so do not seem to reflect food competition.) However, the large body size of gorillas, combined with the high water and cellulose content of their major foods (Waterman, et al., 1983), means that gorillas must process a considerable bulk of food merely to stay alive. For a young male to gain body mass and strength in order to increase his power, or for a nursing female to provide nutrition to a developing baby, requires considerably more. Most of this bulk food consumption requires significant preparation: over 80% of the diet of Watts' study group was made up of just four plants, all difficult ones to process (Watts, 1984): nettle *Laportea alatipes*, bedstraw *Galium ruwensoriense*, thistle *Carduus nyassanus* and *Peucedanum linderi*, locally called celery. Efficient food preparation – maximizing input of digestible material while minimizing damage to hands and ingestion of indigestible matter – is at a premium for mountain gorillas.

In 1989, a study of mountain gorilla food-preparation techniques was begun at Karisoke, Rwanda. The first task was to study the methods used by adult and independently ranging immature gorillas, and 510 hours of detailed individual observations were collected (Byrne & Byrne, 1991, 1993). Focal animal sampling (Altmann, 1974) was used, to amass at least 400 minutes for each of the 38 individuals over 3 years age (in many cases, far more data were obtained). Two observers worked together to clarify details of technique, employing video and hand-held computer recording where these methods could increase the quality and quantity of data. The 1989 study forms the basis for the rest of this chapter; future work will concentrate on unweaned immatures and the development of their manual skills.

COGNITIVE ANALYSIS

Complexity and laterality

Six distinct techniques were identified, but the 4 food plants that make up the bulk of these gorillas diet accounted for 5 of them – the complex ones – whereas the sixth was used for eating a wide variety of different leaves and vines, and was a relatively simple, unimanual process (Byrne & Byrne, 1991). This simpler technique involved only the repeated picking of items to accumulate a bundle for putting in the mouth, and was used for plants lacking physical defenses of any kind. Individuals typically showed only weak behavioral lateralization on this technique, with little preference for one hand or the other. Simple manual actions in the repertoires of great apes and even humans are known to show little lateralization (Marchant, McGrew, & Eibl-Eibesfeldt, 1995; McGrew & Marchant, 1996), so this finding was to be expected. However, a much more complicated and bimanual[2] technique, that for dealing with the unwieldy stems of thistle *Carduus nyassanus* which possess spiny flanges along much of their length, showed even less lateralization, for an interesting reason. Most gorillas dealt with thistle by alternating between preparing mouthfuls of food left- and right-handed, apparently in order to avoid having to move the unwieldy – and prickly – stem around them (or move themselves around to the other side of the same stem). Thus, although the overall technique was rather complex (see Figure 1e in Byrne & Byrne, 1991), strongly lateralized processing would be a disadvantage. Similar use of both mirror forms of a single technique in alternation was seen when the stems of the nettle *Urtica massaica* were eaten: the stem is gingerly held by the tip, and bent over the side of a finger to break it. The tough "skin" of the nettle remains attached at one side, so the stem can be pinched delicately and lifted, detaching the skin (and its powerful stings) of one side. Then the maneuvre is reversed, breaking the stem over the other finger, again lifting to remove the rest of the skin, and the peeled stem eaten. In both cases, the utility of being able to alternate the two mirror forms of the process evidently outweighs any advantage to be had from lateralized processing. For a similar reason, we would predict that any

[2] With processes in which both hands are used, "isomers" or "mirror forms" are better terms for the two chirally different ways of carrying out the process, rather than "left-handed" and "right-handed." In the original publications, the two isomers were arbitrarily characterized as L and R, on the basis of the hand used to pass the prepared food bundle into the mouth. L corresponds in general to right-handed precision manipulation, with left-handedpower-grip support.

technique for wholly arboreal foods would be unlateralized and unimanual, because of the advantage of being able to use whichever hand was free from climbing.

The other four techniques showed very strong lateralization, at individual level. Of 156 cases for which there were enough data for analysis, significant hand preference was found for 132 (cf. the 8 "expected" merely from repetitious statistical testing), and many individuals used exclusively one mirror form. All these techniques were multi-stage processes, and both hands were necessarily used in complementary roles at several stages. The ubiquity and strength of manual laterality may be related to bimanual action. In laboratory tasks with chimpanzees, Hopkins has found that bimanual action elicits strongly lateralized behaviour (Hopkins, 1994). With the gorillas' bimanual techniques, it was not always obvious which hand was the "dominant" one. At each stage, one hand was usually acting as a vice, employing some sort of power grip, thus enabling the other hand to perform a manipulation by using some sort of precision grip. Thus, one criterion that might be used to decide dominance is the type of grip applied: the dominant hand would be expected to be used for precision roles. Consideration of guitar or violin playing suggests that the distinction between dominant and subordinate hand is not always so easily made: right-handers use fingers of their left hand for skilled manual actions of fretting and stopping, and it may be the precise timing required from plucking and bowing that prescribes use of the dominant hand. In any case, the need to keep hold of a half-processed bundle of food evidently constrains gorillas' procedures at times, so that the same – or sometimes, the opposing – hand has to be used for consecutive stages. Sometimes, first a left-hand and then a right-hand precision grip are used successively in a single process. Despite these potential difficulties, it was possible to identify the hand used for the more precise and delicate manipulative actions for each of the four techniques. For three of them a small but significant population bias towards right-handedness[3] was found. On a scale where 50% represents equal numbers of left- and right-handers, the population bias ranged from 58% for nettle (Kolmogorov–Smirnov one-sample test for deviation from symmetry, $D=0.11$, $p=0.05$) to 60% for bedstraw and 64% for thistle leaf (both $D=0.15$, $p=0.01$). Not only does thistle-leaf preparation show the strongest trend towards right-handedness, but this technique evoked the highest number of 100% exclusive hand users (16 individuals right-handed for precision manipulations, 5 left-handed, again a significant population effect). McGrew and Marchant single out this as the only good evidence of exclusive hand use in any study of a great ape (McGrew & Marchant, 1996).

As well as giving evidence for right-handedness, these same three techniques were strongly correlated with each other in the direction and strength of preference across individuals ("hand specialization," in McGrew and Marchant's terms). In contrast, celery stem-processing correlated with none of the leaf-processing tasks in either

[3] "Handedness" is generally used whenever there is a significant population trend towards left or right in individual hand preferences. However, McGrew and Marchant (1996) reserve the term handedness for cases where a population level effect ("task specialization") emerges in a consistent direction on, arbitrarily, more than six independent tasks. Only three significant task specializations were found for mountain gorillas.

direction or strength of preference, and – while it evoked just as powerful individual laterality – it produced exactly equal left- and right-hand preferent individuals. However, hand preference for celery did correlate with the mirror form of the technique used for dealing with bamboo. It would appear that there are at least two *families* of manual techniques: groupings within which the related techniques differ in organization but share behavioral laterality. The family resemblance suggests that each group develops from a single, more generalized technique used during infancy. Between the families, direction and strength of laterality is unconnected, and only the more delicate manipulations involved in leaf-processing evoke population right-handedness. The reason for psychologists' interest in findings of manual laterality in apes is, of course, the hope for insight into the origins of human handedness. It is possible that the gorillas' right-handedness for leaf-preparation is only coincidentally related to human right-handedness, but assuming a real link (whether of homology or analogy) immediately highlights the significance of these leaf-processing skills. In contrast, the absence of any trace of population laterality for stem-processing techniques would tend to link these skills to those of other nonhuman primates, suggesting they are more primitive traits. The possibility of a group of human manual skills, related to gorilla stem-processing, with no population trend in laterality despite strong individual hand preferences, is an interesting one that deserves further investigation.

There are other reasons for thinking that stem-processing skills are more primitive. Although in monkeys nothing like the organized, deliberately structured approach of the gorillas has been found, their method of eating pith from the inside of tough stems is not dissimilar. Harrison, presenting wild vervet monkeys *Cercopithecus aethiops* with pieces of cultivar sugar cane, found that the monkeys' basic approach was to hold the cane in both hands, bite into the outer case with the incisors, and tear back with a jerking movement of the head against a forward push with the hands (Harrison, 1996), much as Schaller described in mountain gorillas. This action peels back and sometimes detaches pieces of outer stem, revealing the pith. Some aspects of the process may be found in a wide range of primates, part of a genetically coded repertoire of basic primate feeding methods (Glickman & Sroges, 1966; Parker & Gibson, 1977; Torigoe, 1985). Even in strepsirhines there are close similarities. The hapalemurs are specialist feeders on bamboo, like giant pandas. The details of food preparation have been studied in captive *Hapalemur griseus* (Stafford, Milliken, & Ward, 1993). The lemurs remove the outer casing in a similar way to monkeys, and moreover their actions are highly structured into an organized sequence, in a way very reminiscent of gorillas eating celery. In the case of gorilla techniques, the variety and flexibility of methods, and the very local usefulness of many of them, argues against innate specification in general. In contrast, *H. griseus* subsists entirely on one species of bamboo, and even its relatives at generic level (*H. simus*, *H. aureus*) vary only in specializing on different species of bamboo. Given the basic structural similarity of all bamboos, an essentially identical processing technique would be effective in all cases. It is therefore possible that, for lemur bamboo-processing, not only the actions sequenced, but also the organization of the skill, come under rather tight genetic control.

Even for stem-processing, such "hard-wiring" is perhaps implausible in the case of a great ape, but it should be noted that consumption of pith from terrestrial herb vegetation is also characteristic of lowland gorillas (*Gorilla g. gorilla*) (Rogers, Williamson, Tutin, & Fernandez, 1988), pygmy chimpanzees (*Pan paniscus*) (Kano, 1983; Badrian & Malenky, 1984), and some populations of chimpanzees (*P. troglodytes*) (Wrangham, Conklin, Chapman, & Hunt, 1991). These species rely heavily on ripe fruit, a food source generally unavailable to mountain gorillas, but regularly fall back on pith consumption in periods of fruit shortage. Great ape stem-handling techniques may depend on rather general primate abilities.

Organization and control

A very different pattern of results is found when the leaf-processing techniques of the Virungas gorillas are examined. Each is adapted to the defenses of a specific herb, and, since these temperate plants are not found even a few thousand feet lower on the mountains, the techniques cannot be primitive, hard-wired skills, but must instead be extensively dependent on learning. Fine details, such as the particular grip best suited to holding a loose bundle to enable debris to be picked out, are learnt individually. This is clear, because these details show much idiosyncratic variation, and the variation shows no sign of running in families or groups, as they would if imitation were involved (Byrne & Byrne, 1993). (In wild gorillas, other individuals are not permitted nearby when feeding, with the exceptions that mothers allow their infants to remain with them, and silverbacks tolerate infants in general.) Just the same applies to individual manual laterality: no trace of any association between hand preference of mother and offspring, or silverback and infants, was found (Byrne & Byrne, 1991). A striking contrast is found when the overall organization of each technique is analyzed. Techniques of adults and weaned juveniles are highly standardized across the population. This contrast, plus the implausibility of some of the techniques being effectively shaped by trial-and-error or computed individually, has been used to argue for "program-level imitation" as the means of acquisition (Byrne, 1993; Byrne & Byrne, 1993).

The standardized aspects of leaf-processing techniques include not only the *sequential order* of the several stages, but also the "loops" of control where some stages of a process can be iterated a number of times as *subroutines* to build up a larger handful of half-processed food, and the presence of extra *optional stages*, triggered, for instance, by a need to remove contaminating debris. While minor details vary from plant to plant and individual to individual, the control architecture of each process is the same. Taken as a whole, these details "parse" the flow of control, and reveal the hierarchical structure of the techniques (Figure 6.1) (Byrne & Russon, 1998).

While this structure would be expected for human skills, nothing similar has been reported in any other nonhuman primate population. Among animals in general, many skills have hierarchical organization (Dawkins, 1976), but these show no signs of acquisition by learning. Each is a species-typical pattern, quite at variance with other aspects of the species' behavioural complexity: nest building in weaver birds or termites, the courtship displays of many amphibians, the constructions of bower birds.

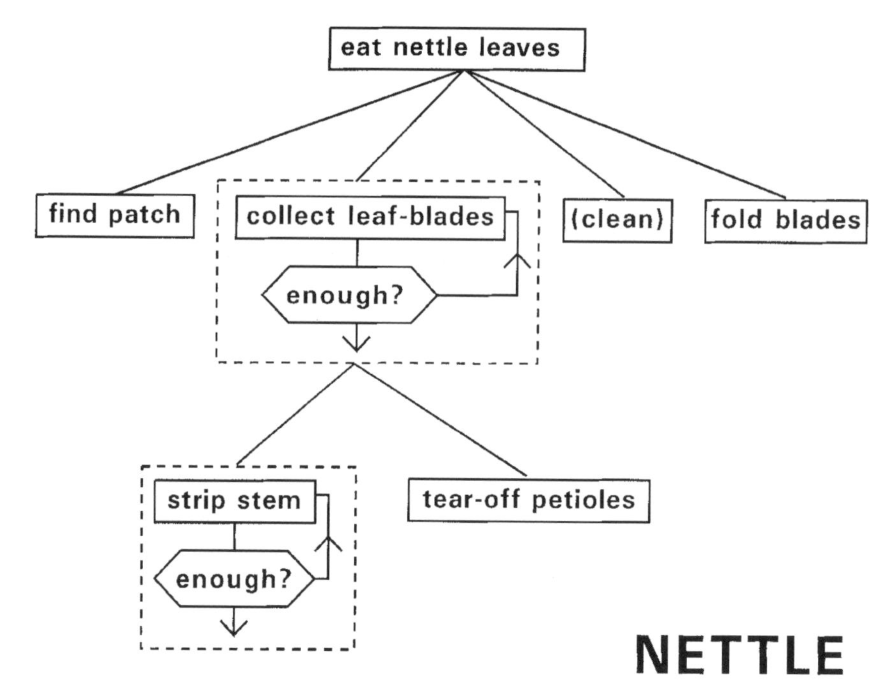

Figure 6.1 *Architecture of control for nettle preparation.* The hypothetical program organization, of minimum hierarchical complexity needed to produce the actual behavior observed in a mountain gorilla preparing nettle leaves to eat. The "top goal" is placed at the top of the figure, indicating its control of the goals below it, and this may occur recursively, increasing hierarchy "depth."

Even among other species of great apes, it is not easy to find skills of comparable logical complexity to those of the mountain gorillas. Chimpanzee insect-gathering with tools is one likely candidate. As generally presented, the skill is a matter of delicate probing in and out of holes, but, in fact, neither the probe nor the hole comes ready for the chimpanzee (McGrew, 1992). Probes need to be fashioned, by detaching leaves from a stem or leaf-blades from a leaf-midrib. In some cases, no suitable material is available nearby to a *Macrotermes* termite mound or *Campanotus* ant nest, and then chimpanzees have been observed to make the tool in advance (Goodall, 1986). In termite-fishing, the mound is hard until near the time when winged alates emerge. Only then do the termites soften the baked mud above each emergence tube with water, and chimpanzees can evidently identify the precise sites of tubes and pick off the soft mud with their fingernails. The logical organization of insect-fishing has not been studied from the same perspective as gorilla plant preparation, but all descriptions suggest a systematic, organized, and elaborated activity; hierarchical organization would be expected. Of the component parts of the activity, only tool manufacture requires bimanual co-ordination, and, while insect-fishing evokes high degrees of laterality, there is no population bias to right or left (Marchant & McGrew, 1996). Hammering hard nuts – on a larger, flat "anvil" stone or tree-root by using a smaller, round "hammer" stone – is certainly a more difficult task for chimpanzees than insect-fishing, since it is acquired later in development and not by all individuals (Boesch & Boesch, 1983). However, the difficulty is apparently one of fine motor control and power, more than logical complexity. For hammering, and also for the technically simpler task of leaf-sponging for water, high degrees of hand preference are found but no consistent population bias (Boesch, 1991; Sugiyama, Fushimi, Sakura, & Matsuzawa, 1993).

McGrew and Marchant conclude similarly that chimpanzee tool use and gorilla plant-food processing demand equivalent motor skills (McGrew & Marchant, 1996, p. 268). Their careful review of all sizeable studies of laterality and handedness in great apes found few signs of population-level handedness, and they rightly dismissed most of these as a spurious consequence of apparently huge but actually invalid sample sizes. The problem is that the need for *independent* data has not yet been fully appreciated in this area: all too often, thousands of actions are treated as independent samples for statistical analysis, when they are clearly nothing of the sort. Those studies that met proper criteria for independence and yet achieved statistical significance were, interestingly, all bimanual tasks: the present study of gorillas, and two studies of captive chimpanzees (Hopkins, 1994, 1995). It seems that the need for right-handed action, generally assumed to increase efficiency through left-hemisphere neural specializations, is greatest in tasks that require co-ordination of complementary actions. Presumably the fact that such tasks necessarily cannot be performed while hanging one-handed in a tree, so ambidextrality is of no benefit, is also influential.

DISCUSSION

Although evidence of hierarchically organized learned activities is at present restricted to humans, gorillas, and perhaps chimpanzees, it would be premature to rule definitively on other great apes. Regular tool-using in wild orangutans has only recently been described from a single population (Van Schaik, this volume), and no adequate and sizeable studies of laterality have yet been carried out in either orangutans or pygmy chimpanzees (McGrew & Marchant, 1996). It would not be surprising to find, as several times before, that all the great apes share a cognitive potential originally identified in only one species. However, the ability to learn elaborate, hierarchically organized programs of action may be unique to great apes (Byrne & Russon, 1998).

The potential to construct novel, organized programs of action is fundamental to what in humans is called "problem-solving," "thinking" or "planning" (Lashley, 1951; Newell, Shaw, & Simon, 1958; Miller, Galanter, & Pribram, 1960). In every analysis of human problem solving, tasks are tackled by hierarchically splitting a problem into a series of related subgoals (Newell & Simon, 1972; Byrne, 1977; Young, 1978). Thus, a complex problem is usually solved by means of a hierarchical structure of nested subroutines. Keeping track of this organizational structure requires some kind of working memory capacity (not necessarily the same as the sequential short-term memory capacity tested in serial learning), and this may be more restricted in nonhuman primates than in people (Byrne & Russon, 1998; Gibson, 1990). Nevertheless, the exciting thing about recent analyses of manual skills employed by wild great apes for food procurement and preparation is that they may point to the evolutionary origin of computational planning and thought (Byrne, 1998; Parker & Milbrath, 1993).

The implications of the distribution of manual laterality in great ape activities are that – except where there are good reasons favoring ambidextrality, as in thistle-stem eating by mountain gorillas – any acquired, complex manual skills will be highly lateralized. Moreover, where close co-ordination between the two hands, each with a different role complementing the other, is required, it would seem that a right-hand preference is favored. Unfortunately, it is not easy to get comparable data for humans. In western populations, the prevalence of writing distorts the picture, and in any case the standard "psychology text book" story of 10:90 bias to the right is largely based on questionnaire results and tests of fine-motor tool use. The only ethological study of manual laterality in traditional human populations was restricted to unimanual activities, but found a strong association between laterality, population handedness, and tool use (Marchant et al., 1995). Common non-tool-use activities were unlateralized in three cultures (G/wi hunter gatherers, Himba pastoralists, Yanomamö horticulturalists), whereas tool use with power grips showed a significant bias to population handedness; only skilled tool use with precision grips showed a trend to exclusive use of one hand. Thus, only for tool use was a strong population trend found, 16:84 towards right handedness. Although considerably less than the degree of right-handedness usually assumed to typify humans, this is more striking than the trends found for bimanual tasks in great apes: 36:64 for gorilla thistle-leaf preparation (Byrne & Byrne, 1991); 35:65 for chimpanzee bimanual probing (Hopkins, 1995).

There are two possibilities that could explain right-handedness in gorillas, chimpanzees, and humans: essentially, these may be described as convergence or common descent. Both depend on some link between manual laterality and cortical laterality, but note that this link is not a simple one (Corballis, 1991). On the one hand, if there is some strong ergonomic reason to specialize in representing motor skills of a particular sort within the left hemisphere of the brain, then analogous motor systems will develop at times in various species, whenever they tackle tasks of the particular sort. The left-hemisphere specialization found for vocalization in singing canaries (Nottebohm, 1977), Japanese monkeys (Petersen et al., 1978) and humans is presumably of this sort. On the other hand, in a group of close relatives, a shared pattern of handedness may result from their possession of homologous brain structures, and the ancestral bias to one side or the other may have been weak or coincidental. Given the very short phylogenetic distance between the African apes and humans, this is perhaps more likely in this case (but note the eclectic distribution of claims of right-handed manipulation in more distantly related primates (Bradshaw & Rogers, 1993)). The bimanual tasks which in great apes evoke handedness are most likely to be homologous with those that underwrite human tool using skills.

ACKNOWLEDGMENTS

For support of the field work on which the chapter is based, I thank the National Geographic Society and the Carnegie Trust for the Universities of Scotland; and for permission and general assistance, I thank the Dian Fossey Gorilla Fund and L'Office Rwandaise du Tourisme et Parcs Nationaux.

REFERENCES

Altmann, J. (1974). Observational study of behaviour: sampling methods. *Behaviour*, **49**, 227–265.

Attenborough, D. (1979). *Life on earth*. London: Collins.

Badrian, N. L. & Malenky, R. K. (1984). Feeding ecology of *Pan paniscus* in the Lomako Forest, Zaire. In R. L. Susman (ed.), *The pygmy chimpanzee: evolutionary biology and behaviour* (pp. 275–299). New York: Plenum Press.

Boesch, C. (1991). Handedness in wild chimpanzees. *International Journal of Primatology*, **12**, 541–558.

Boesch, C. & Boesch, H. (1983). Optimisation of nut-cracking with natural hammers by wild chimpanzees. *Behaviour*, **26**, 265–286.

Bradshaw, J. & Rogers, L. (1993). *The evolution of lateral asymmetries, language, tool use, and intellect*. San Diego: Academic Press.

Byrne, R. W. (1977). Planning meals: Problem-solving on a real data-base. *Cognition*, **5**, 287–332.

Byrne, R. W. (1993). Hierarchical levels of imitation. Commentary on M. Tomasello, A. C. Kruger, and H. H. Ratner "Cultural learning." *Behavioural and Brain Sciences*, **16**, 516–517.

Byrne, R. W. (in 1997). The technical intelligence hypothesis: an additional evolutionary stimulus to intelligence? In A. Whiten & R. W. Byrne (eds.), *Machiavellian intelligence II: evaluations and extensions* (pp. 289–311). Cambridge University Press.

Byrne, R. W. & Byrne, J. M. E. (1991). Hand preferences in the skilled gathering tasks of mountain gorillas (*Gorilla g. beringei*). *Cortex*, **27**, 521–546.

Byrne, R. W. & Byrne, J. M. E. (1993). Complex leaf-gathering skills of mountain gorillas (*Gorilla g. beringei*): Variability and standarization. *American Journal of Primatology*, **31**, 241–261.

Byrne, R. W. & Russon, A. (1998). Learning by imitation: a hierarchical approach. *Behavioral and Brain Sciences*, **21**, 667–721.

Christel, M. (1993). Grasping techniques and hand preferences in apes and humans. In H. Preuschoft & D. J. Chivers (eds.), *Hands of primates* (pp. 91–108). New York: Springer Verlag.

Corballis, M. C. (1991). *The lopsided ape*. Oxford University Press.

Dawkins, R. (1976). Hierarchical organisation: a candidate principle for ethology. In P. P. G. Bateson & R. A. Hinde (eds.), *Growing points in ethology* (pp. 7–54). Cambridge University Press.

Gibson, K. R. (1990). New perspectives on instincts and intelligence: brain size and the emergence of hierarchical mental construction skills. In S. T. Parker & K. R. Gibson (eds.), *"Language" and intelligence in monkeys and apes* (pp. 97–128). Cambridge University Press.

Glickman, S. E. & Sroges, S. R. (1966). Curiosity in zoo animals. *Behaviour*, **26**, 151–158.

Goodall, J. (1986). *The chimpanzees of Gombe: patterns of behavior*. Cambridge, MA: Harvard University Press.

Harrison, K. E. (1996). Skills used in food processing by vervet monkeys, Cercopithecus aethiops. Dissertation, University of St Andrews.

Hopkins, W. D. (1994). Hand preferences for bimanual feeding in 140 captive chimpanzees (*Pan troglodytes*): rearing and ontogenetic determinants. *Developmental Psychobiology*, **217**, 395–407.

Hopkins, W. D. (1995). Hand preferences for a coordinated bimanual task in 110 chimpanzees (*Pan troglodytes*): cross-sectional analysis. *Journal of Comparative Psychology*, **109**, 291–297.

Kano, T. (1983). An ecological study of the pygmy chimpanzees (*Pan paniscus*) of Yalosidi, Republic of Zaire. *International Journal of Primatology*, **4**, 1–31.

Lashley, K. S. (1951). The problem of serial order in behaviour. In L. A. Jeffress (ed.), *Cerebral mechanisms in behaviour: the Hixon Symposium*. New York: Wiley.

Marchant, L. F. & McGrew, W. C. (1996). Laterality of limb function in wild chimpanzees of Gombe National Park: comprehensive study of spontaneous activities. *Journal of Human Evolution*, **30**, 427–443.

Marchant, L. F., McGrew, W. C., & Eibl-Eibesfeldt, I. (1995). Is human handedness universal? Ethological analyses from three traditional cultures. *Ethology*, **101**, 239–258.

McGrew, W. C. (1992). *Chimpanzee material culture: implications for human evolution*. Cambridge University Press.

McGrew, W. C. & Marchant, L. F. (1996). On which side of the apes? Ethological study of laterality of hand use. In McGrew, W. C., Marchant, L. F., & Nishida, T. (eds.) *Great ape societies* (pp. 255–272). Cambridge University Press.

Miller, G. A., Galanter, E., & Pribram, K. (1960). *Plans and the structure of behaviour*. New York: Holt, Rinehart and Winston.

Napier, J. R. (1961). Prehensility and opposability in the hands of primates. *Symposia of the Zoological Society of London*, **5**, 115–132.

Newell, A. & Simon, H. A. (1972). *Human Problem Solving*. New York: Prentice-Hall.

Newell, A., Shaw, J. C., & Simon, H. A. (1958). Elements of a theory of human problem solving. *Psychological Review*, **65**, 151–166.

Nottebohm, F. (1977). Asymmetries in neural control of vocalization in the canary. In S. Harnad, R. W. Doty, L. Goldstein, J. Jaynes, & G. Krauthamer (eds.), *Lateralization in the nervous system* (pp. 23–44). New York: Academic Press.

Parker, S. T. & Gibson, K. R. (1977). Object manipulation, tool use, and sensorimotor intelligence as feeding adaptations in early hominids. *Journal of Human Evolution*, **6**, 623–641.

Parker, S. T. & Milbrath, C. (1993). Higher intelligence, propositional language, and culture as adaptations for planning. In K. R. Gibson & T. Ingold (eds.), *Tools, language and cognition in human evolution* (pp. 314–333). Cambridge University Press.

Petersen, S., Beecher, M., Zoloth, S., Moody, D., & Stebbins, W. (1978). Neural lateralization of species-specific vocalizations by Japanese macaques (*Macaca fuscata*). *Science*, **202**, 324–327.

Rogers, M. E., Williamson, E. A., Tutin, C. E. G., & Fernandez, M. (1988). Effects of the dry season on gorilla diet in Gabon. *Primate Reports*, **22**, 25–33.

Schaller, G. B. (1963). *The mountain gorilla*. Chicago University Press.

Stafford, D. K., Milliken, G. W., & Ward, J. P. (1993). Patterns of hand and mouth lateral biases in bamboo leaf shoot feeding and simple food reaching in the gentle lemur (*Hapalemur griseus*). *American Journal of Primatology*, **29**, 195–207.

Sugiyama, Y., Fushimi, T., Sakura, O., & Matsuzawa, T. (1993). Hand preference and tool use in wild chimpanzees. *Primates*, **34**, 151–159.

Torigoe, T. (1985). Comparison of object manipulation among 74 species of non-human primate. *Primates*, **26**, 182–194.

Waterman, P. G., Choo, G. M., Vedder, A. L., & Watts, D. (1983). Digestibility, digestion-inhibitators and nutrients and herbaceous foliage and green stems from an African montane flora and comparison with other tropical flora. *Oecologia*, **60**, 244–249.

Watts, D. P. (1984). Composition and variability of mountain gorilla diets in the central Virungas. *American Journal of Primatology*, 7, 323–356.

Watts, D. P. (1994). Social relationships of immigrant and resident female gorillas. 2. Relatedness, residence, and relationships between females. *American Journal of Primatology*, **32**, 13–30.

Wrangham, R. W., Conklin, N. L., Chapman, C. A., & Hunt, K. D. (1991). The significance of fibrous foods for Kibale Forest chimpanzees. *Philosophical Transactions of the Royal Society of London, Series B*, **334**, 171–178.

Young, R. (1978). Strategies and structure of a cognitive skill. In G. Underwood (ed.), *Strategies of information processing*. New York: Academic Press.

7

Development of sensorimotor intelligence in infant gorillas: the manipulation of objects in problem-solving and exploration

JUAN C. GÓMEZ

INTRODUCTION

This chapter is about the development of intelligent manipulations of objects in hand-reared infant gorillas. It is situated within the framework of what Köhler (1921) and Piaget (1936) identified as "practical" or "sensorimotor" intelligence. This kind of intelligence is nonlinguistic and nonreflective. It consists of the ability to produce novel coordinated sequences of actions adapted both to the physical and the social environment. In previous work (Gómez, 1990, 1991), I have explored the practical intelligence of young gorillas in the social domain. In this chapter I will explore their practical intelligence in the physical domain.

In his pioneering work, Köhler (1921) proposed that chimpanzees showed intelligent behavior when confronted with practical problems whose solving involved carrying out physical displacements or manipulations of objects. Chimpanzees were capable of producing actions whose organization was adapted to challenging environmental conditions. For example, when confronted with a goal hanging from the ceiling, the chimpanzees would drag a box under it, or make it fall with a stick, or simply find a roundabout way to approach the goal. Köhler thought that these and other comparable behaviors demonstrated some kind of intelligence or understanding in the apes, different from mere trial-and-error learning *à la* Thorndike (1898). He suggested that this practical intelligence could be explained within the framework of *Gestalt* concepts, but he never developed a detailed theoretical account of it.

It was Piaget (1936) who some years later provided a comprehensive theory of practical intelligence. He called it "sensorimotor" intelligence to emphasize the lack of underlying symbolic representations supporting the organization of the actions. His theory was based not upon the study of chimpanzees (although he knew and made extensive use of Köhler's studies), but of human babies. According to him, one could trace the development of this kind of intelligence from birth to the end of the second year of life. It would start as a mere collection of automatic reflexes, but by means of several developmental mechanisms (notably, assimilation and accommodation) it would progress through a series of stages culminating in the development of interiorized representations of actions. Piaget (1936, 1937) provided an elaborate taxonomy of sensorimotor actions organized in a framework based upon a diachronic distinction of six stages and a synchronic differentiation into several domains (problem-solving, object concept, notion of space, notion of causality, etc.). Each domain would develop along those six stages, and each stage would be characterized by certain milestone behaviors.

Infancy research in the last 20 years has suggested important modifications to the Piagetian account of sensorimotor development. Infants are now thought to have a more sophisticated perceptual understanding of the world than Piaget – focused mainly upon infants' actions – assumed. However, his taxonomy of developmental milestones has been largely confirmed and remains a powerful tool for comparative research (see, for example, Parker & Gibson, 1990). Furthermore, a concept of practical intelligence, although in need of further theoretical exploration and identified with other terms such as procedural or implicit knowledge, remains an essential component of developmental theories (Karmiloff-Smith, 1992). In this chapter I will make use of some Piagetian concepts but without a commitment to Piaget's theory of sensorimotor development as a whole.

One important point in which I will be following Piaget is the differentiation between exploratory and instrumental manipulations of objects.[1] Usually, practical intelligence is identified with problem-solving, especially tool use. However, an important component of Piagetian sensorimotor intelligence is the exploratory mechanisms known as "circular reactions." These involve the manipulation of objects outside a "serious" problem-solving context, and, as argued by Parker (1993), they might constitute a fundamental way of adaptation in primates. In this chapter I will examine both aspects of object manipulation in gorillas – for problem solving and as exploratory actions.

SENSORIMOTOR INTELLIGENCE IN NONHUMAN PRIMATES

Piaget's model of sensorimotor intelligence has been applied to nonhuman primates by a number of researchers who believed that it could be specially useful for the comparative study of intelligence in monkeys and apes (Jolly, 1972; Parker, 1977; Chevalier-Skolnikoff, 1976, 1977; Redshaw, 1978; see Doré & Dumas, 1987, and Vauclair, 1996, for overviews). Typically, apes were found to develop through the same sequence of stages in domains such as the "object concept" or the co-ordination of actions in means–ends structures. However, the results are controversial as to the degree and/or speed of achievement of infant apes in the highest stages of certain domains of development, particularly those involving the spontaneous manipulation and exploration of objects.

For example, Mathieu and Bergeron (1981) concluded that nursery-reared chimpan-

[1] This distinction was also made by Köhler (1917), who provided extensive descriptions of the spontaneous object manipulations of chimpanzees. However, he did not elaborate upon it.

zee infants complete all six stages of sensorimotor development, but they reported important deficits in their understanding of causality and an almost total absence of tertiary circular reactions. Mignault (1985) emphasized that combinations of two or more objects were observed only "rarely" and that the chimpanzees "seemed to interact with objects not so much in an attempt to discover their properties (exploration) as just for the sake of manipulating them" (p. 754). Vauclair and Bard (1983) found that chimpanzees were limited to "simple and repetitive manipulations of a single object," such as simple holding or moving on a substrate. Finally, Potí and Spinozzi (1994) reported that, although many features of sensorimotor intelligence (object permanence and means–ends co-ordinations of schemes) were present in infant chimpanzees, they were poor at object exploration and object combinations. All in all, in his recent overview, Vauclair (1996) feels justified to conclude that in the chimpanzee there is a "relative absence of object manipulation and object–object combination" and that, therefore, one could speak of a "human specificity" in complex object manipulation.

This alleged departure from the human pattern of object manipulation would seem to be even more pronounced in the case of gorillas. In a series of preliminary reports Chevalier-Skolnikoff (1976, 1977, 1983) claimed to have found evidence of all 6 stages of sensorimotor intelligence in a sample of 10 gorillas ranging from infants to adults. However, examples of highest stages of tool use and object manipulation were seen only in the adults. Three subsequent systematic studies reported an absence of complex object manipulations in young gorillas. Redshaw (1978), working with 4 baby gorillas from birth to 18 months, found that they followed the same course of development as humans, but they failed to reach the higher levels in some domains of sensorimotor intelligence. In constructive play, the gorillas did not combine objects in an organized way, and they failed to use implements such as rakes. Furthermore, the gorillas were unable to recruit the help of humans to manipulate objects as external causal agents.

Similar results were found by Perinat and Dalmáu (1989) in a longitudinal study of 2 baby gorillas between 21 and 34 months of age. The subjects showed very little motivation to manipulate objects, specially in a constructive and explorative way, and almost no ability to use other individuals as agents. Their manipulations of objects reached only the level of secondary circular reactions and were subordinated to the satisfaction of primary needs. Spinozzi and Natale (1989), working with a single gorilla infant – Romina – failed to find even secondary circular reactions. Tertiary circular reactions were never seen and Romina never engaged in manipulations involving object–object relations, except for some elementary actions like hitting towards the end of the second year (Antinucci, 1990). In contrast, the object concept did develop almost in parallel to human children. In relation to the use of objects as tools, Natale (1989) found that Romina's performance with a stick was always very poor. Although, during her third year of life she was capable of systematically establishing contact between the stick and the goal, her manipulations are described as "clumsy" and demonstrating a "lack of understanding of the relation between force, angle and movement." By 36 months, Romina discovered a procedure of her own, which proved to be very success-

Table 7.1 *Subjects*

Name	Gender	Estimated age (in months) upon arrival	Arrival date at zoo
Muni	Female	5–10	30–5–80
Bioko	Male	8–12	15–7–81
Guinea	Female	6–10	20–11–81
Nadia	Female	5–8	30–10–81
Arila	Female	16–20	30–4–82

ful: throwing a blanket over the goal and retrieving it. However, the authors consider this to be a procedure that requires a lesser level of intelligence.

In summary, the data about these seven gorilla infants show an uneven development of sensorimotor intelligence. Gorillas complete the object permanence series at about the same rate as human infants and other apes, but they appear as clumsy and uninsightful manipulators of elementary tools, and they almost never show any interest in exploring and manipulating objects in themselves. In these studies, infant gorillas appear to be even more limited than chimpanzees in object manipulation, confirming the early impression of Yerkes and Yerkes (1929) that gorillas "rank low" in their ability to use objects.[2] This would also contribute to consolidate the picture of divergence between ape and human infants when their manipulatory abilities are tested within the Piagetian framework (Vauclair, 1996).

In the next sections I will present data from a longitudinal study with infant gorillas which show a different picture of their manipulatory skills.

THE PRESENT STUDY

The subjects were 5 young captive gorillas (see Table 7.1). Since they had been originally captured in the wild, their ages are only an estimate, and their development has been punctuated by the traumatic experience of capture and transportation to Europe, the exact circumstances of which are not known. The gorillas were subject to a hand-rearing procedure in a zoo nursery environment. One of them (Muni) enjoyed hand-rearing with several human companions as an "only" gorilla baby for over 1 year. Then 3 more gorillas arrived: Bioko, Nadia, and Guinea. These were kept together with several hours of daily contact with Muni until eventually (about 1 year after their arrival) the four gorillas formed a single group.

During her first year as an "only child," Muni enjoyed daily contact with a small group of human caretakers and researchers, as well as occasional visitors. She had access

[2] None the less, Yerkes and Yerkes (1929) warned against concluding from this inferiority in manipulative skills that gorillas had an inferior level of intelligence.

to a variety of rooms and outdoor areas, including most facilities of the nursery area of the zoo. This means that she could manipulate objects that are not usually part of artificial captive environments, and extensively interact with a variety of people.

After the arrival of the other gorillas and especially 1 year later, when the four animals were housed in cages, the variety of objects, environments, and people available decreased, although Muni was still allowed to be taken out of her new cage for an additional year. This process of environmental impoverishment continued when three of the gorillas (Guinea having been transferred to another zoo) were moved to a public exhibition area with greatly limited access to objects. This happened about 2 years after the move to the new cage. However, during this period of environmental degradation, the gorillas spent a part of their weekly time (about 10 hours on the average) in contact with humans who took with them some manipulable objects. For further details about the hand-rearing conditions of the subjects see Gómez (1992).

The fifth subject mentioned in this chapter (Arila) was never fully integrated into the group. Although she spent some time every week in the company of humans, she never experienced the amount of interaction of the other animals, neither was she exposed to the same variety of objects and physical environments.

Procedure

The data analyzed in this chapter are part of an extensive database about the psychological development of young gorillas compiled during several years of study of the above subjects. The more complete and systematic longitudinal record corresponds to Muni. The data consist of hand-written descriptions and videotapes. They correspond to spontaneous behaviours and reactions to experimental and semi-experimental situations. By "experimental" I mean situations prepared beforehand with controlled variables and predictions about possible outcomes; for example, the administration of object permanence tests in a controlled environment and with a scheduled sequence of presentations. By "semi-experimental" I refer to situations in which the observer took advantage of a particular circumstance or interest of the subject to administer an informal test of a target ability (for example, tool-use).

All notes were taken on the spot by one or two observers. On the whole, this procedure is similar to that used by Piaget in his sensorimotor intelligence observations with the addition of formal experiments. It can also be compared to the diary method used in the longitudinal study of early language development (Braunwald & Brislin, 1979). In this chapter I will report data upon the use of objects as tools in problem solving situations and the exploratory manipulations of objects. Data about object permanence and the understanding of external causality are also available in the database. The object permanence data essentially confirm the results presented in previous studies. Results showing the gorillas' ability to understand external causal agents appear in Gómez (1990, 1991).

Results: tool-use and problem-solving

For the purposes of this chapter I will classify problem-solving skills into two main categories. On the one hand, finding roundabout ways towards goals that cannot be

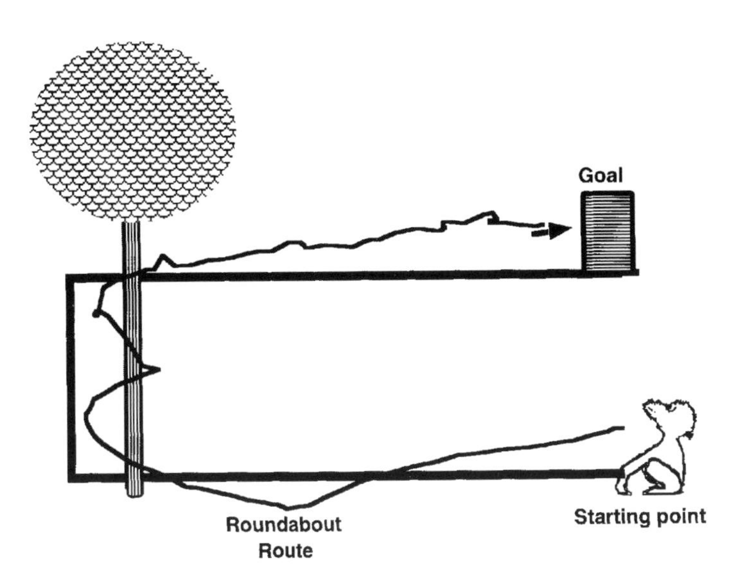

Figure 7.1 A problem-solving situation in which Muni – a 10-month-old gorilla – reaches a goal through an elaborate roundabout route that implies going away from the goal and climbing up a tree.

directly reached, and, on the other, the more complex skill of creating new means to reach a goal. This can be further subdivided into: (a) Modifications of the environment to create conditions that render a roundabout solution possible; and (b) Using objects as a hand extension or substitute to manipulate the goal itself.

Roundabout behaviors

The five gorillas of this study exhibited the ability to reach goals by means of roundabout detours almost from the very beginning of the observations. Their first successful attempts at finding alternative roundabout routes were observed by the second half of their first year of life. The onset of this skill is indeed facilitated by the early locomotor maturity of gorillas, but roundabout behaviors are not just a matter of "approaching a goal." Detours may involve a coordination of locomotor actions in complex means–ends structures, as in the following example (Figure 7.1):

Muni (at 10 months of age) is trying to get to a goal situated on a platform that is too high for her to climb onto it. She solves the problem by using a tree as a roundabout way to the goal.

The young gorilla is faced here with a very complex roundabout route that forces her initially to go away from the goal, perform an elaborate and uncertain climb (with the goal out of sight) and only then approach the goal.

Thus, although this early manifestation of problem-solving does not require object manipulation, the gorillas are already engaging in complex means–ends co-ordinations. Moreover, in our gorillas, roundabout behaviors were soon combined with the second kind of problem-solving behaviors: manipulations of environmental objects to create new detours. The development of the skills to create "bridges" towards unreachable goals was longitudinally studied in Muni. Figure 7.2 shows the development of these behaviors in comparison with ordinary roundabouts strategies. The first – clumsy and unsuccessful – attempts at creating new roundabout ways were first recorded at 10 months of age. However, in the next months this strategy progressively increased in frequency replacing the mere use of already disposed means. This can be interpreted as a preference for shorter and easier roundabout routes, which are usually those created by the gorilla. For example, in the platform situation depicted in Figure 7.1, Muni would now prefer to move a nearby object (e.g., a box) under the goal, instead of performing the long detour up the tree. By the end of the first year of life, instrumental strategies had become her preferred way of problem solving.[3].

Displacing and adequately adjusting the position of an object to be used as a stool or ladder is not a straightforward action. It is an intelligent behavior that, depending upon the conditions of each particular problem, may require a great deal of skill. As shown in Figure 7.3, it took Muni several months to pass from clumsy, mostly unsuccessful attempts to skillful and successful manipulations of objects to be used as roundabout ways. These objects included a variety of boxes, crates, plastic bins, a toy tricycle, a collection of brooms and poles (used for climbing purposes), etc. At the same time as these skills were developing (end of the first year of life), the first instances of objects used as hand extensions were recorded.

Tools as hand extensions

The simplest manifestation of this ability is the use of already disposed extensions such as a string attached to the goal. Redshaw (1978) found this behavior in her gorillas during their first year of life. This is confirmed by my own observations. But the behavior that seemed to be more problematic for Redshaw's gorillas was the creation of previously unexisting connections to extend the manipulatory power of their hands.

At 12 months, Muni was observed hitting the floor and the walls with sticks or randomly raking dead leaves on the floor, but there was no evidence that she was actually intending to produce external modifications in the environment. She seemed to be essentially concentrated on the activity itself in a ludic mood, without visually attending to the effects. Muni's definite attempts at using objects as intermediaries to manipulate other objects were seen at 15 months: she started to use tissue or straw to take mucus out of her nose. At 22 months, Muni would use sticks to manipulate excrement on the floor. At 28 months, she used sticks or similar objects for a variety of purposes, including the manipulation of spiders and other obnoxious targets, picking up food from a container, reaching the latch of a door, trying to poke objects out of a hole, etc. She could use pieces of cardboard, straw, and other materials for similar functions.

[3] The subsequent abrupt fall in the frequency of this strategy that can be seen in Figure 7.2 is due to the emergence of an even more popular strategy: recruiting the help of humans as external agents; see Gómez, 1990, 1992.

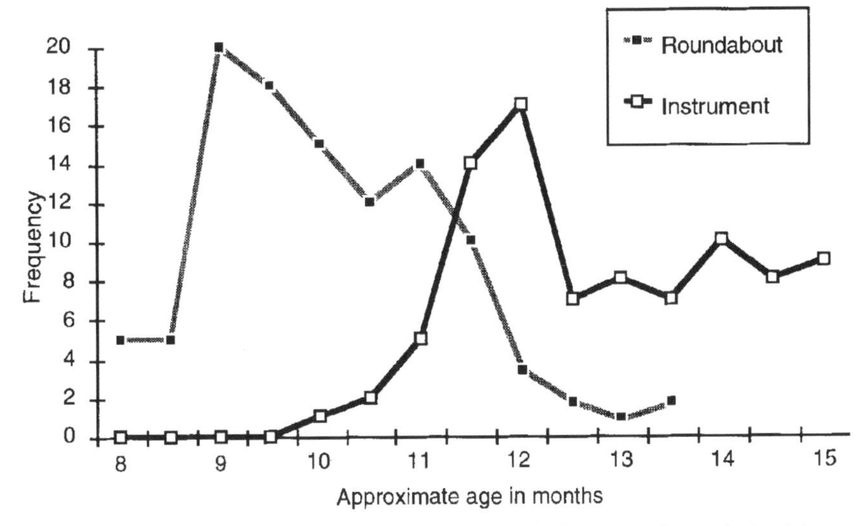

Figure 7.2 Development of roundabout strategies compared with instrumental strategies involving object manipulation. Data from July 1980 to March 1981. Note that instrumental strategies are progressively substituted for roundabout ones.

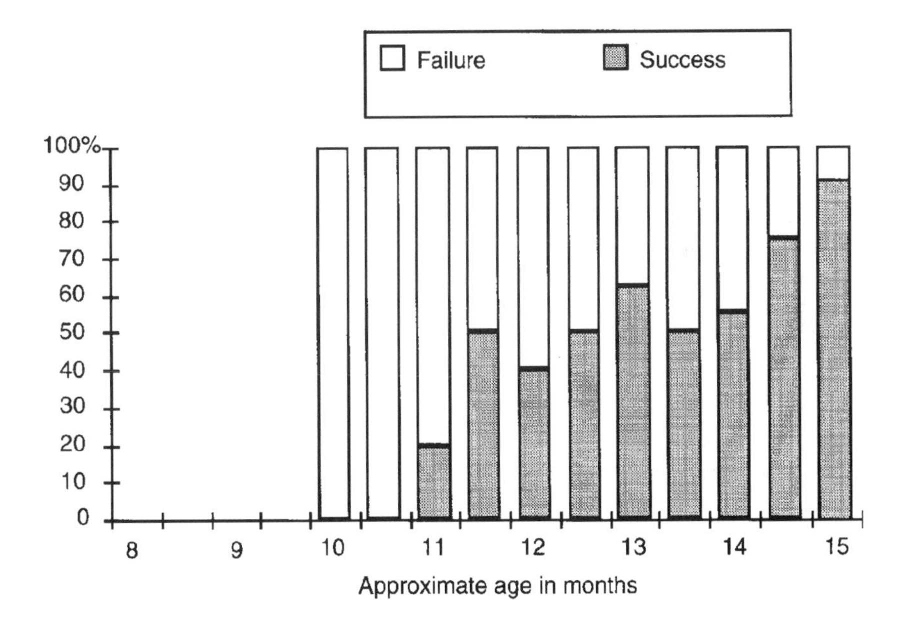

Figure 7.3 Percentage of instrumental strategies in which Muni succeeded in solving the problem. Data from July 1980 to March 1981.

By 3.5, years Muni was capable of engaging in complex co-ordinations of objects, such as using a container to take water from a water dispenser (which involved a bimanual co-ordination: one hand to press the dispenser; the other to hold the container in the right position). It was at around this same time that Muni developed a favorite procedure to attract out of reach objects. She had been moved to a new cage with vertical bars in its door allowing for the retrieval of outside objects. She developed a very skillful use of straw bundles that were thrown over the object and then retrieved. This procedure involved a certain amount of repetition since it was very rare that the object could be successfully displaced in a single trial. This throwing–retrieving–throwing cycle was eventually executed at great speed and with a high degree of accuracy. Sometimes it was also possible to observe some preparation of the straw bundle before its use. This consisted mainly of pressing together the straw into a more consistent bundle. Objects other than straw bundles were also used occasionally (e.g., cardboard pieces, plastic pieces, ropes).

It is important to remark that there was always straw available in the cages during the later period of observations (when the gorillas were confined in a highly impoverished environment), whereas sticks and branches were relatively rare objects. Straw became the most important manipulandum at their disposal: it was used in play sequences, as an instrument to clean or wipe themselves, as material for nest building, and as a versatile retrieval tool.

When formally tested for the use of sticks as instruments in similar situations, Muni was at first clumsier with this instrument than with the straw. However, she very soon reached a comparable level of mastery in terms of her success rate. This is apparently similar to the use of a blanket by Romina, which Spinozzi and Natale (1989) considered to be an inferior kind of tool use. However, it is not clear why this strategy should be cognitively less sophisticated than the use of a stick. The use of straw bundles – a loose and unreliable material – by Muni seemed to require a high degree of skill. She had to achieve a complex combination of accuracy in the positioning of the straw upon or just beyond the goal, the dragging movement with the straw and the systematic repetition of the operation. Of course, the subjective impressions of complexity or simplicity caused by the manipulations of the subjects should not be the final criterion with which to analyze their actions. More objective and theoretically based analyses of the microstructure of problem solutions should be used. However, in the meantime the solutions that depart from the classical stick-use strategy should not be dismissed as simple or unsophisticated. It could be the daily availability of the straw in the case of Muni and the blanket in the case of Romina that explains their readiness and skill in using them, and not any intrinsic difference in difficulty. In favor of this interpretation would be the speed with which Muni learned to use sticks when provided with them.

A similar phenomenon was observed in one of our other subjects: Arila. She was first formally tested for her ability to use sticks when she was about 4–5 years old. Her first responses were highly clumsy and mostly ineffective, although in several attempts she worked with the apparent aim of establishing contact between the stick and the goal. However, the conditions of the bars through which she had to operate seemed to be especially difficult and the sticks she had too short. When 3 days later she was tested in better conditions (the bars were devoid of additional wire mesh) her performance was almost perfect.

Two other gorillas – Nadia and Bioko – also showed spontaneous tool manipulations. At around 2–2.5 years they were using straw or tissue to clean themselves or other people or simply to manipulate unpleasant material such as excrement. They would also use pieces of cardboard or wood to make contact with insects such as ants. These manipulations do not require the execution of complex trajectories imposed upon the target with the tool, but they already show the basics of tool use as a hand extension or substitute.

Manipulations requiring more complex co-ordinations between the targets and the tools were also attempted by these gorillas, initially with a lesser degree of success. For example, in his attempts at using containers to procure water from the water dispensers, Bioko very soon mastered the basics of pushing the dispenser with a finger while holding the container under it with the other hand. However, it took several months for him to learn to hold the container with its upper part well oriented to the water jets. Because water could be obtained simply by pressing the dispenser with the mouth (the device was specifically designed for that) the gorillas tended to do this when they were really thirsty. This easier method could explain why the right procedure with the container was not immediately discovered by a trial-and-error process, but in the course of manipulations that were closer to exploratory actions (see next section).

In conclusion, these findings show that young gorillas are more sophisticated users of objects in problem-solving situations than suggested by the data of previous studies. They have an early preference for procedures involving the creation of roundabout ways to the goals, but they also use objects as extensions of their hands in a sophisticated way. This fits better with the emerging picture of gorillas as complex manipulators of food in the wild (Byrne, 1996) and the earlier observations of Chevalier-Skolnikoff (1976, 1983) on the well-developed sensorimotor intelligence of adult captive gorillas.

EXPLORATORY MANIPULATIONS OF OBJECTS

Previous studies are almost unanimous in pointing out the poverty of the exploratory behaviors of gorillas in relation to objects, reporting an almost total absence of tertiary circular reactions and object combinations[4]. Some were unable even to observe simple secondary circular reactions.

The concept of "circular reactions," originally coined by Baldwin (1895), was developed by Piaget (1936) as a way of systematizing the study of the exploratory manipulation of objects (see Table 7.2). A circular reaction (from now on CR) is, first of all, a repeated action. Primary CRs involve actions carried out upon the body of the subject itself (e.g., thumb-sucking). Secondary CRs involve actions performed upon the

[4] The exception is Chevalier-Skolnikoff's (1976, 1977). However, she includes tool use and tertiary circular reactions under the same category. Here I distinguish between these two kinds of manipulations, reserving the term "tertiary circular reactions" to the exploratory handling of objects.

Table 7.2 *Object exploration in a Piagetian framework: definitions and criteria of identification*

Stage	Definitions and criteria
Stage 1 *REFLEX*	Automatic assimilation of objects to pre-existing, stereotyped schemes. No proper exploration.
Stage 2 *PRIMARY CIRCULAR REACTIONS*	*Definition*: Repetition of acquired behaviors usually focused upon the subject's own body. *Criteria*: As above, but also including actions upon objects in which the subject does not look at the external effects.
Stage 3 *SECONDARY CIRCULAR REACTIONS*	*Definition*: New actions upon objects that are repeatedly carried out and whose effects are contemplated by the subject. *Criteria*: The subject looks at the object when the effects are visible
Stage 4 *DERIVED SECONDARY CIRCULAR REACTIONS*	*Definition*: Known actions sequentially applied to new objects as if in search of which schemes are good for it. *Criteria*: Any sequential application of two or more Secondary CR as defined above.
Stages 5 & 6. *TERTIARY CIRCULAR REACTIONS*	*Definition*: New actions are tried with one or more objects as if in search of what novelties they can offer. *Criteria*: Variations in the actions and visual attention to the effects of those variations.

environment provoking effects that arouse the interest of the infant. The aim is to repeat the same action, although innovations may be found by chance and retained in new secondary CR. Derived Secondary CRs happen when subjects apply a set of secondary reactions already in their repertoire to a new object: several actions are applied one after the other as if trying to discover which *known* things the new object can do. Finally, Tertiary CRs involve the application of repeated actions upon the same object but with a systematic and intentional variation of the structure of the action (e.g., changing the force, orientation, height, etc. with which an object is thrown) as if the subject is "experimenting" to see what *new* things an object can do.

Piaget's reference to the attention and intentions of the infants as a criterion to differentiate between different kinds of circular reactions introduces some degree of ambiguity in their identification. To circumvent this problem, primatologists have tried to use more objective criteria to define the different kinds of CRs. For example, it is usual to consider as Secondary CR any action performed upon an external object, and as Tertiary CR any action involving the combination of two or more objects. I suggest, however, that more stringent criteria should be used. For example, the presence of objects in the actions of an organism does not guarantee that a reaction is secondary: if

Table 7.3 *A description of the incidence of exploratory behaviors in the gorillas*

	Primary CR (stage 2)	Secondary CR (stage 3)	Derived Second. CR (stage 4)	Tertiary CR (stage 5–)	Constructive Object Comb. (stage 5–)
MUNI	very frequent	very frequent	very frequent	very frequent	very frequent
NADIA	very frequent	very frequent	common	Occasional	common
GUINI	very frequent	occasional	rare		
BIOKO	very frequent	frequent	frequent	occasional	occasional
ARILA	very frequent	common	common	rare	occasional

the attention of the subject appears to be on its own proprioceptive sensations, then the reaction should better be considered as "primary" (e.g., hitting upon an object may be a primary reaction if the attention of the subject is upon his or her own proprioceptive sensations). In secondary CR, subjects must focus their visual attention upon the environmental effects of their actions. Therefore, the criterion I will be using in the analysis of my data is that the subject *watches* the effects of his/her actions upon the object under manipulation. This is operationalized as "looking at the object when the effect is visible." Similarly, in a strict Piagetian definition, in Tertiary CRs the attention of the subject must be upon the *anticipation* of changes in the effects when changes in the actions are introduced. I operationalize this as the occurrence of variations in the gorilla's manipulation of the object(s) accompanied by looks at their visible effects.

Another important dimension to be considered in the analysis of object manipulation is the complexity of the manipulative sequences and the number of objects combined in any manipulation. As pointed out before, object combinations are usually identified with at least the 5th Piagetian stage of development. However, what is relevant is the *constructive* nature of the combination and the introduction of *variations* in the manipulation: simply banging two objects together may be closer to a Primary or a Secondary CR if the attention of the subject is concentrated upon its own proprioceptive sensations or upon the noise provoked, and no variations are introduced in the way the objects are handled together.

A characteristic of all CRs is that they are not subordinated to other goals. They are repeated for the sake of exploration. This feature can also be described as their being *intrinsically motivated* (Perinat & Dalmáu, 1989). This is what differentiates them from intentional groping and insightful problem-solving. Infants need not be pursuing a particular goal when exploring an object in a secondary or tertiary way, although new actions discovered during problem solving may be subsequently explored with CRs.

Results

Table 7.3 shows a summary of the preliminary results of analyzing the database of observations of the 5 gorillas included in this study.

Object manipulations that could not be differentiated from Primary CRs were observed in Muni at 6–7 months of age. For example, she would take a piece of cardboard, place it on her face and start pounding gently on herself. Or she would take a ball, put it on her belly, and start rolling it while displaying playfaces. There was no evidence that her attention was on the effects her actions were provoking upon the world. Rather she seemed to be enjoying bouts of self-stimulation in which objects were incorporated. Similar incorporations of objects into what I suggest should be interpreted as a complex kind of primary circular reactions were frequently observed in all the gorillas from the beginning of the observations.

Secondary CRs were also frequently seen in all gorillas. They could be simple actions like hitting objects or scratching upon surfaces, or more complex actions like making noise by hitting two objects together, sprinkling water by obstructing the water dispenser with a finger, opening and closing doors, turning a lights-switch on and off in a room, etc. It is important to emphasize that, especially in the case of Muni, Nadia, and Bioko, this did not involve just a limited repertoire of favorite actions. The list of objects they explored was virtually as long as the list of objects available to them.

The exploration of objects by sequentially applying known actions (Derived Secondary CRs) was also common in four of the subjects studied. For example, at 12 months Nadia explored a plastic toy barrel by mouthing it, looking at it while holding it in front of her eyes, rubbing it upon her chest, hitting with it on the floor, and again mouthing it, always looking at the object after acting upon it. At 10 months, Muni took a toy shovel and hit the floor with it, then dragged it, bent its handle, etc., observing her own manipulations.

Tertiary CRs and constructive object combinations were also frequently seen in 3 of the gorillas. One of them – Muni – produced this kind of action very frequently and with a high degree of complexity. She started to produce Tertiary CRs early in her second year of life. For example, after contemplating how a human lifted the lid of a box, she started a series of opening and closing actions with the same objects. The lid fitted very narrowly, so that it was difficult to close it again. She tried different actions until she succeeded; however, she reopened the box immediately and started the cycle again. Her actions were not simple repetitions: she seemed to be adapting her movements in search of effectiveness, but then this effectiveness was further explored because the result – sometimes achieved after considerable effort – was immediately undone.

At 2.5 years, a human showed Nadia how to throw an apple against the glass panel of her cage. The human threw it with enough strength to make it bounce back and catch it with his hand again. Nadia watched the demonstration. When the apple fell down by her side, Nadia took it and threw it against the glass. She repeated this action several times. Then she started throwing the apple down against the floor. After repeating this a few times, she threw it against a wall. She then threw the apple up in the air watching how it fell. She used different throwing procedures for this. She spent the whole day carrying out actions like these. Her final modification consisted of incorporating the apple-throwing procedure into a sequence of social play, as a way of teasing her partner into action.

Some of the manipulations produced by Muni and the other gorillas involved the combination of two or more objects in constructive ways. These combinations preceded even the onset of clear-cut Tertiary CRS. For example, at 11 months, after finding a button and putting it in her mouth, Muni took an empty water bottle and holding the button with her mouth she dropped it inside the bottle. She then tried vainly to take it out again with her fingers. Eventually she changed her strategy, leaning the bottle into her mouth and taking the button with the mouth. Immediately after that, she let it drop into the bottle again, and again tried she to take it out with her fingers, and then with the mouth. She finally kept to the mouth method, repeating the action half a dozen times. Some minutes later, she tried to introduce the same button through a hole in the floor, first directly with her mouth, then with her fingers.

The exploratory actions we have just described are intrinsically motivated. The gorillas explore the objects' properties in themselves, not because they are important to solve any primary need (unless we consider "exploration" to be a primary drive, as some people suggest [White, 1959]). Even some food manipulations are directed at exploring its physical properties, as in the case of Nadia's manipulation of the apple. Many of these exploratory actions are organized around a particular goal, like closing the lid of the box in the above example of Muni. However, the repeated undoing of the target state as soon as it is achieved indicates that this is not a simple problem-solving activity, but an exploration of the procedures and the objects themselves

In summary, the results reported here show that manipulations showing the features of Secondary, Derived, and Tertiary circular reactions are present in young gorillas reared in contact with humans and human objects. These manipulations involve not only individual objects, but also object–substrate and object–object combinations. They are frequent, varied, constructive, and intrinsically motivated. As in the case of the problem-solving behaviors analyzed in the previous section, these observation are consistent with the emerging picture of gorillas as potentially sophisticated manipulators (Byrne, 1996).

OBJECT EXPLORATION AND COGNITIVE DEVELOPMENT

There is little question about the usefulness of tool-using behaviours. Knowing how to organize complex detour displacements or how to modify the environment to reach a goal is clearly adaptive. But what is the usefulness of exploratory manipulations? There is an extensive literature on the subject of exploratory behavior in animals and humans that is relevant to this question (White, 1959; Bruner, Jolly, & Sylva, 1976; Archer & Birke, 1983; Görlitz & Wohlwill, 1987). The consensus seems to be that exploration, as well as play, may be a way of promoting behavioral flexibility and variability. In the realm of object manipulation, the exploratory handling of objects could enrich the repertoire of actions at the disposal of organisms when later they have to confront serious problem-solving situations. Experiments have demonstrated a connection between exploratory play with objects and subsequent success in problem-solving with those objects, both in animals and humans (Birch, 1945; Sylva, 1977). However, little attention has been paid to the cognitive processes that may underlie this beneficial effect of exploratory manipulations. In this section I would like to discuss one such process.

In a recent attempt at reconciling Piagetian constructivism with modern findings about the early cognitive abilities of human babies, Karmiloff-Smith (1992) suggests that an important cognitive difference between humans and animals, including anthropoids, may lie in the humans' ability to go beyond successful action in their dealings with the world. Human infants and children are not only interested in discovering how to achieve goal X with action Y, but also in *understanding* how this happens. According to Karmiloff-Smith, humans do this by *redescribing* their practical or procedural knowledge – implicit in nature – into more explicit representations. Human cognitive development is viewed as a process of progressive re-representation of knowledge into more explicit and flexible formats (but not necessarily linguistic formats). This redescription of knowledge happens not under the pressure of environmental challenges (for example, when children are unable to solve a problem and have to find a new method), but as a consequence of success. After achieving mastery of a problem, children re-represent their successful procedures in more explicit terms. For example, young children who are capable of balancing blocks on a rod by "feeling" (combining visual and tactile information), go beyond their success and create a "theory" of why blocks balance – for instance, the theory that blocks always balance on their geometrical centre. In this way, they do not have to find the right solution for each block anew: they directly place the blocks on their centre, thereby achieving the solution in a more efficient way. But, paradoxically, these – initially inaccurate and nonverbalizable – "theories" may now make them fail with blocks whose weight is not symmetrically distributed, which they were able to balance with their earlier strategy (see Karmiloff-Smith, 1992; chapter 3). Only later, when their theories are refined as a consequence of further redescription, can children deal with the odd cases becoming much better problem solvers.

My suggestion is that exploratory circular reactions may be an early form of re-representation, and perhaps an important cognitive tool in preparing implicit representations for explicit redescription. The practical exploration of objects by means of what Piaget called Tertiary Circular Reactions involves going beyond successful instrumental actions. It seems to be a practical exploration of the actions themselves, or, in Piagetian terms, of the underlying *schemes* that organize the actions. Remember those elaborate manipulations produced by Muni until she succeeded in closing the lid of a box, only to immediately undo what she had achieved and start again. Could these "tertiary" explorations act as a way of gaining and storing information about the procedures themselves, perhaps as a way of analyzing them into their schematic components? Could this information be represented in different – more explicit – formats in the minds of gorillas and young children (for example, in the form of more abstract schemes), or would it remain procedural and implicit but, none the less, constitute a more articulate practical "re-description" of the objects, the actions, and the relationships between them?

Whatever the nature of the knowledge stored by humans or gorillas performing circular reactions, it is clear that these behaviors allow them to go beyond successful action. Their actions do not stop once they have solved a particular problem: they engage in an active exploration whose aim might be the achievement of some further "understanding" of what they are already capable of doing. This interpretation opens the possibility of applying Karmiloff-Smith's concept of representational redescription – originally conceived as a specifically human mechanism of development – to non-human intelligence. It also offers an interesting new avenue to understand exploratory manipulations from a cognitive-developmental point of view.[5]

SUMMARY AND CONCLUSIONS

In this chapter I have shown data demonstrating that young gorillas are capable of achieving the highest levels of sensorimotor intelligence that some previous studies had failed to identify. I have shown that their early tool-using abilities are far from simple, and that, besides being sophisticated problem solvers, they are also capable of going beyond practical success to engage in complex explorations of objects, including object combinations. These findings concur with those of Byrne and Byrne (1993; Byrne, this volume) in wild gorillas and Parker (this volume) suggesting a higher level of complexity in gorilla sensorimotor cognition than had been assumed. The presence of complex object manipulations in gorillas does not rule out the existence of possible qualitative differences between apes and humans *within* that complexity. The nature of these differences is beyond the scope of this chapter. In any case, whatever the results of future research about the differential properties of cognitive development in gorillas and humans, it is important not to confuse alleged divergences from the human (or for that matter chimpanzee; see Byrne, 1996) pattern with lack of sophistication.

An important unresolved question is: what is the proper theoretical characterization of circular reactions and exploratory manipulations from the point of view of cognitive development and evolution? I have suggested that Karmiloff-Smith's theory of representational redescription may provide an avenue to explore this most important aspect of early cognitive development.

Finally, although this chapter has been devoted mainly to the problem of physical cognition, I want to emphasize that the sophisticated manipulations shown by the gorillas in this study occurred in a social context. It is not only that the physical environment of the gorillas, including the objects they had access to, was designed by humans, but also that humans demonstrated, or provided clues about, the manipulation of the objects. Independently of the degree of "genuine" imitation showed by the apes (Byrne, 1995), other people's actions upon particular objects acted as a powerful mechanism in the elicitation of the gorillas' explorations of these objects (Gómez, 1992). The study of cognitive development in nonhuman primates will have to take into account the importance of this "human factor" in their cognitive achievements (see Tomasello, Kruger, & Ratner, 1993; and Gómez, 1993).

[5] See Inglis (1983) for an earlier attempt at characterizing the cognitive dimension of exploratory behaviors.

ACKNOWLEDGMENTS

The data used in this chapter were analyzed while enjoying a DGICYT (Direccion General de Investigación Cientifica y Tecnica) grant with Fernando Colmenares (PB95–0377). I am grateful to Madrid Zoo for the permission given to work with the subjects in their facilities. My thanks to the editors of the volume for their valuable comments and advice, especially to Bob Mitchell for a most careful and detailed reading of early versions of the manuscript which improved not only the chapter but also my own thinking about the problems addressed in it. Needless to say, the editors are not responsible for any remaining shortcomings.

REFERENCES

Antinucci, F. (1990). The comparative study of cognitive ontogeny in four primate species. In S. T. Parker & K. R. Gibson (eds.), *"Language" and intelligence in monkeys and apes: Comparative developmental perspectives* (pp. 157–171). Cambridge University Press.

Archer, J. & Birke, L. (eds.) (1983). *Exploration in animals and humans* . Wokingham: Van Nostrand Rheingold.

Baldwin, J. M. (1895). *Mental development in the child and the race*. N. York: Macmillan.

Birch, H. G. (1945). The relation of previous experience to insightful problem-solving. *Journal of Comparative Psychology*, **38**, 267–383.

Braunwald, S. R. & Brislin, R. W. (1979). The diary method updated. In E. Ochs & B. B. Schieffelin (eds.), *Developmental pragmatics* (pp. 21–42). New York: Academic Press.

Bruner, J., Jolly, A., & Sylva, K. (eds.). (1976). *Play: its role in development and evolution.* Harmondsworth: Penguin.

Byrne, R. W. (1995). *The thinking ape*. Oxford University Press.

Byrne, R. W. (1996). The misunderstood ape: cognitive skills of the gorilla. In A. Russon, K. Bard, & S. Parker (eds.), *Reaching into thought: The minds of the great apes* (pp. 111–130). Cambridge University Press.

Byrne, R. W. & Byrne, J. M. E. (1993). Complex leaf-gathering skills of mountain gorillas (*Gorilla g. beringei*): variability and standardization. *American Journal of Primatology*, **31**, 241–261.

Chevalier-Skolnikoff, S. (1976). The ontogeny of primate intelligence and its implications for communicative potential: A preliminary report. In S. Harnad, H. Steklis, & J. Lancaster (eds.), *Origins and evolution of language and speech* (vol. 280, pp. 173–211). New York: New York Academy of Sciences.

Chevalier-Skolnikoff, S. (1977). A Piagetian model for describing and comparing socialization in monkey, ape, and human infants. In S. Chevalier-Skolnikoff & F. E. Poirier (eds.), *Primate biosocial development: Biological, social, and ecological determinants* (pp. 159–187). New York: Garland.

Chevalier-Skolnikoff, S. (1983). Sensorimotor development in orangutans and other primates. *Journal of Human Evolution*, **12**, 545–561.

Doré, F. & Dumas, C. (1987). Psychology of animal cognition: Piagetian studies. *Psychological Bulletin*, **102**(2), 219–233.

Gómez, J. C. (1990). The emergence of intentional communication as a problem-solving strategy in the gorilla. In S. T. Parker & K. R. Gibson (eds.), *"Language" and intelligence in monkeys and apes: Comparative developmental perspectives* (pp. 333–355). Cambridge University Press.

Gómez, J. C. (1991). Visual behavior as a window for reading the minds of others in primates. In

A. Whiten (eds.), *Natural theories of mind: Evolution, development and simulation of everyday mindreading* (pp. 195–207). Oxford: Blackwell.

Gómez, J. C. (1992) *El desarrollo de la comunicación intencional en el gorila*. Ph.D. Dissertation, Universidad Autónoma de Madrid.

Gómez, J. C. (1993). Intentions, agents and enculturated apes. *Behavioral and Brain Sciences*, **16**(3): 520–521.

Görlitz, D. & Wohlwill, J. F. (eds.). (1987). *Curiosity, imagination, and play*. Hillsdale, NJ: LEA.

Inglis, I. R. (1983). Towards a cognitive theory of exploratory behaviour. In J. Archer & L. Birke (eds.), *Exploration in animals and humans* (pp. 72–116). Wokingham: Van Nostrand Rheingold.

Jolly, A. (1972). *The evolution of primate behavior. First edition*. New York: MacMillan.

Karmiloff-Smith, A. (1992). *Beyond modularity: A developmental perspective on cognitive science*. Cambridge, MA: MIT Press.

Köhler, W. (1917). Aus der Anthropoidenstation auf Teneriffa. III: Intelligenzprüfungen an Anthropoiden. *Abhandlungen der Preussische Akademie der Wissenschaften. Physikalische-Mathematische Klasse*. N° 1).

Köhler, W. (1921). *Intelligenzprüfungen an Menschenaffen* . Berlin: Springer. [English translation: *The mentality of apes*. New York: Vintage, 1927].

Mathieu, M. & Bergeron, G. (1981). Piagetian assessment on cognitive development in chimpanzee (*Pan troglodytes*). In B. Chiarelli & R. S. Corrucini (eds.), *Primate behaviour and sociobiology* (pp. 142–147). Berlin: Springer.

Mignault, C. (1985). Transition between sensorimotor and symbolic activities in nursery-reared chimpanzees (*Pan troglodytes*). *Journal of Human Evolution*, **14**, 747–758.

Natale, F. (1989). Causality II. The stick problem. In F. Antinucci (eds.), *Cognitive structure and development in nonhuman primates* (pp. 121–133). Hillsdale, NJ: Erlbaum.

Parker, S. T. (1977). Piaget's sensorimotor series in an infant macaque: a model for comparing unstereotyped behaviour and intelligence in human and nonhuman primates. In S. Chevalier-Skolnikoff & F. E. Poirier (eds.), *Primate biosocial development: biological, social, and ecological determinants* (pp. 43–112). New York: Garland.

Parker, S. T. (1993). Imitation and circular reactions as evolved mechanisms for cognitive construction. *Human Development*, **36**, 309–323.

Parker, S. T. & Gibson, K. R. (eds.). (1990). *"Language" and intelligence in monkeys and apes: Comparative developmental perspectives*. Cambridge University Press.

Perinat, A. & Dalmáu, A. (1989). La comunicación entre pequeños gorilas criados en cautividad y sus cuidadoras. *Estudios de Psicología*, **32–34**, 11–29.

Piaget, J. (1936). *La naissance de l'intelligence chez l'enfant*. Neuchatel: Delachaux et Niestlée.

Piaget, J. (1937). *La construction du réel chez l'enfant*. Neuchâtel: Delachaux et Niestlé.

Potì, P. & Spinozzi, G. (1994). Early sensorimotor development in chimpanzees (*Pan troglodytes*). *Journal of Comparative Psychology*, **108**, 93–103.

Redshaw, M. (1978). Cognitive development in human and gorilla infants. *Journal of Human Evolution*, 7: 133–141.

Spinozzi, G. & Natale, F. (1989). Early sensorimotor development in gorilla. In F. Antinucci (ed.), *Cognitive structure and development in nonhuman primates* (pp. 21–38). Hillsdale, NJ: Erlbaum.

Sylva, K. (1977). Play and learning. In B. Tizard & D. Harvey (eds.), *Biology of play* (pp. 59–73). London: Heinemann.

Thorndike, E. L. (1898). Animal intelligence: an experimental study of the associative processes in animals. *Psychological Review: Series of Monograph Supplements*, 2(4), 1–109.

Tomasello, M., Kruger, A. C., & Ratner, H. H. (1993). Cultural learning. *Behavioral and Brain Sciences*, **16**(3): 495–552.

Vauclair, J. (1996). *Animal cognition*. Cambridge, Mass.: Harvard University Press.

Vauclair, J. & Bard, K. A. (1983). Development of manipulations with objects in ape and human infants. *Journal of Human Evolution*, **12**, 631–645.

White, R. W. (1959). Motivation reconsidered: the concept of competence. *Psychological Review*, **66**(5): 297–333.

Yerkes, R. M. & Yerkes, A. W. (1929). *The great apes*. New Haven: Yale University Press.

8

Tool use in captive gorillas

SARAH T. BOYSEN, VALERIE A. KUHLMEIER,
PETER HALLIDAY, AND YOLANDA M. HALLIDAY

Ethological and experimental investigations of tool use have been explored across a wide range of mammalian and avian species, with the extensive review by Beck (1980) still the penultimate volume. The initial field observations of tool use in wild chimpanzees (*Pan troglodytes*) came from the pioneering work of Goodall (1964, 1968), and were followed by meticulous reports from McGrew (e.g., 1974, 1987; Brewer & McGrew, 1990; for overview, see 1992), detailing two different types of tool use, including termite-fishing and ant-dipping. Subsequent observations of tool use, modification, and transport, including stone tool use by chimpanzees, have provided a window toward understanding cultural transmission among populations of chimpanzees now isolated throughout eastern and western Africa (Boesch & Boesch, 1990; McGrew, 1992; Matsuzawa, 1994). As numerous field workers continue to collect long-term data and behavioral observations of wild chimpanzee populations, new tool types and functions are regularly reported (Sugiyama, 1985, 1995; McGrew, 1992; Suzuki, Kuroda, & Nishihara, 1995; Yamakoshi & Sugiyama, 1995).

The rich archival literature now available on primate tool use, particularly for chimpanzees, coupled with increased awareness of the positive effects of environmental enrichment for captive primates, have resulted in a wide range of creative and innovative efforts at providing captive great apes opportunities for tool use in both zoo and laboratory settings. Despite this history of research, opinions of what can be classified as a tool or tool use vary. In accordance with Goodall (1970), Parker & Gibson (1977), and, more recently, Boesch & Boesch (1990), the present study defines a tool as an object that is held in the hand, foot, or mouth and employed in a functional manner that enables the user to attain a specific goal. This is a narrower definition than that presented in Beck (1980) and excludes activities previously considered to be tool use in other studies, such as an orangutan draping leaves over its head (Galdikas, 1982), or a chimpanzee making a nest (Nishida & Hiraiwa, 1982).

Among the apes, chimpanzees alone have been shown to be the documented tool-users of the African forest, with no reports of tool use of any type from either lowland (*Gorilla gorilla gorilla*) or mountain gorillas (*Gorilla gorilla beringei*), or from the bonobo. (*Pan paniscus*). Among the Asian apes and lesser apes, neither captive nor wild gibbons or siamangs have been observed to use tools, although McGrew (1992) noted that the arboreal habitat of these species make observation in the wild difficult. Although rehabilitant orangutans (*Pongo pygmaeus*) have shown remarkable abilities to imitate and otherwise replicate complex behavioral sequences, including utilization of tools as observed from their human caregivers (Russon & Galdikas, 1993), only recently has tool manufacture and use for wild orangutan populations been reported (Fox, Sitompul, & Van Schaik, this volume; Van Schaik & Fox, 1996), despite long-term studies of orangutans in several Indonesian sites (e.g., Galdikas,1979, 1982; Galdikas & Teleki, 1981). Prior to these recent observations, object manipulations related to nesting and covering, particularly of the head, have been discussed relative to tool use (Galdikas, 1982). Ex-captive and captive orangutans, however, display a large tool-use repertoire. For example, ex-captive orangutans use sticks to dig, pry loose objects, and disturb ant nests (Galdikas, 1982). They have also been observed to use sticks to obtain objects (Lethmate, 1982), and use a ladle-like tool to retrieve food from a cylindrical feeder (King, 1986). Among other apes, Jordan (1982) also reports tool use among captive bonobos at European zoos, and bonobos at the San Diego Wild Animal Park use sticks to obtain honey from small holes in a log (V. Kuhlmeier, personal observations, 1994).

Despite the recent observations of orangutan tool use in the wild, the common chimpanzee (*Pan troglodytes*) remains the most prolific tool-user in both the wild and captivity. Ant-dipping, during which a slender probe is used to extract ants from small openings in an ant-hill, has been observed in all three subspecies of common chimpanzee (*P. troglodytes*, Tutin & Fernandez, 1992; *P. t. schweinfurthi*: Goodall, 1964; Nishida, 1973; McGrew, 1979; Nishida & Hiraiwa, 1982; and *P. t. verus*: Boesch & Boesch, 1990). Termite- fishing has been documented in *P. t. troglodytes* (Fay & Carroll, 1994) and *P. t. schweinfurthi* (McGrew, 1979). Chimpanzees also create probes to fish for honey from beehives (*P. t. verus*: Boesch & Boesch, 1990; Brewer & McGrew, 1990; *P. t. troglodytes*: Merfield & Miller, 1956; Izawa & Itani, 1966; Tutin & Fernandez, 1992), and also use sticks to dig for bees (*P. t. troglodytes*: Fay & Carroll, 1994; *P. t. schweinfurthi*: Yamagiwa, Yumoto, Ndunda, & Maruhashi, 1988). If breaking the hive is desired, chimpanzees will do so using strong sticks (*P. t. troglodytes*: Fay & Carroll, 1994). Goodall (1968) and Nishida (1980) have both reported the use of leaves as sponges to hold liquid for drinking. Recently, Yamakoshi & Sugiyama (1995) have described a "mortar and pestle" technique of obtaining sap among *P. t. verus*. The chimps pound the crown of an oil palm tree with a stalk until a hole is formed, then withdraw handfuls of pulp from the hole. *P. t. verus* has also been the only subspecies in the wild to use branches or stones as hammers and anvils to crack nuts (Boesch & Boesch, 1990).

Gorillas show a similar wild/captive tool-use dichotomy as the orangutan. Mountain gorillas (*G. g. beringei*) have not been seen to use tools in the wild, despite long-term observation (e.g., Fossey & Harcourt, 1977), and Tutin & Fernandez (1992) report that lowland gorillas (*G. g. gorilla*), although living sympatrically with tool-using chimpanzees at Lope Reserve, do not show tool-use behavior. In captivity, however, young

Table 8.1. *Study group composition*

Subjects ($N = 15$)	Sex	Age in years
Group 1		
Djala	Male	10
Sangha	Female	6
Sounda	Female	6
Kouillou	Male	7
Group 2		
Bitam	Male	20
Biju	Male	4
Tamtam	Male	8
Tamba	Female	5
Shamba	Female	33
Baby Doll	Female	30
Boulas	Male	5
Mouila	Female	31
Group 3		
Kijo	Male	16
Founa	Female	20
Tebe	Female	11

lowland gorillas have used rake-like tools to obtain goal objects (Chevalier-Skolnikoff, 1977), and have even shown behavioral flexibility by switching from a stick to blanket when the latter proved to be a more effective tool with which to drag a goal object into reach (Natale, Potí, & Spinozzi, 1988).

In comparison to the other great ape species, tool-use data on adult lowland gorillas even in captivity is surprisingly sparse (Fontaine, Moisson, & Wickings, 1995), given their numbers held in zoos and other collections around the world (McGrew, 1992). Wood (1984) reports the use of sticks to rake in food that was placed beyond reach in a captive gorilla group at Howletts Zoo Park, Canterbury, UK, but few other observations have been reported. Since the Wood (1984) report, an additional enrichment apparatus was built for the Howletts gorillas. In this chapter, we describe observations of the gorillas' use of the sites, using sticks to obtain semi-liquid foods from the baited sites. We present observations of tool modification, innovative techniques using stick tools, social interaction and regulation of access at the tool site, and age and sex differences.

METHODS

The study group consisted of 15 captive lowland gorillas, including 8 females and 7 males, housed in 3 social groups at Howletts Zoo Park (Table 8.1). The animals ranged in age from 4 to 33 years. Each group was housed in an indoor/outdoor enclosure equipped with resting shelves, deep straw bedding, and a variety of large climbing structures. The gorillas were videotaped over a 7–month period (June–December), in 1–2 hour blocks, for a total of 17 hours of tool use at a tool site baited with peanut butter. The animals were also on public display throughout all videotaping. The tool site was a dome-shaped apparatus that was attached to the wire mesh of the gorillas' outdoor enclosure. The side protruding into the enclosure was convex, with holes drilled through the surface which allowed sticks to be inserted. The back side of the tool site was accessible to the outside walkway, permitting caretakers to bait a shallow tray inside the site with either peanut butter or honey, both preferred foods of the gorillas. Tree branches were placed in the animals' outdoor habitat which could potentially be used as dipping tools. Thus, the gorillas obtained the peanut butter by selecting (and often modifying) a stick, inserting it through the holes in the tool site, dipping into the peanut butter without actually being able to see the food or the process themselves, then withdrawing the stick in order to lick the treat off the tool.

Tool insertions were continuously recorded during each videotaping session, with a session beginning following baiting of the site, when the first gorilla picked up a stick and began using it to obtain the peanut butter. A session was considered ended when the last gorilla left the food site following depletion of the bait. Consequently, sessions varied in number and duration, although at least one session per day was recorded. However, since the gorillas were maintained in their social group during filming, participation by individual animals was wholly a function of a variety of social factors, some of which will be described as results.

From the videotapes, analyses of each session were completed, with all instances of site sharing, observation of tool use by another gorilla, novel tool-use methods, and tool modification recorded for each animal. Tool sharing was defined as the simultaneous use of the tool site by two or more gorillas, while observation of tool use was defined as the immediate presence of a nonparticipating gorilla watching the tool-user. Novel tool use was any method of tool use that deviated from the typical insert/dip/retract method. Tool modifications consisted of any manipulation of the stick which changed the original physical condition of the stick prior to using it to obtain food. Three observers evaluated all videotaped sessions, with significant interrater reliability (> 95%).

RESULTS

Throughout the 17 total hours of observation, 283 sessions at the tool site were evaluated. Session length varied across the subjects from > 1 minute to 34 minutes. Similarly, the number of individual tool insertions at the site varied per session, from a low of 16 insertions to 1743 tool insertions by a single subject during one session. Simultaneous sharing of the tool site was observed in only 25% of the total observed sessions, and thus the vast majority of observations of tool use involved a single gorilla using the tool site at one time. There was no correlation between the percentage of sessions that the tool site was shared (between two or more gorillas) and the age of the subjects (Spearman's correlation, $r_s = 0.205$); males and females also did not significantly differ in sharing the tool site (Mann–Whitney, $U = 41$). Thus, observation of a solitary tool-user was seen much more frequently than simultaneous sharing of the tool site.

Observation of tool use by one or more gorillas other than the tool-user was seen in 47% of the sessions. Typically, a second gorilla watched as the tool-user dipped for peanut butter. In the case of observers, there was a strong negative correlation between the age of the observer and the number of observations of a tool-user ($r_s = -0.816$, $p < 0.001$). That is, younger gorillas were much more likely to participate as passive observers of a tool-use session than were older gorillas. With respect to sex, males and females did not differ in how frequently they were seen as passive observers of another tool-user (Mann–Whitney, $U = 35$). The wide variation in session length and number of tool insertions among individuals was mostly due to the frequent displacements (63 total) of tool users by more dominant individuals.

A variety of novel tool use strategies were observed across the various gorillas who participated. A total of 15 novel food retrieval methods were seen throughout the study period, with 29% of the sessions involving at least 1 novel method (Table 8.1). There was, however, no correlation between age and novel tool use ($r_s = 0.2871$). Similarly, males and females showed no significant difference in displaying innovation in tool use (Mann–Whitney, $U = 35$). During the study period, 99 instances of tool modification were observed. Modifications included changing the length of the stick by breaking it with the hands or mouth, or changing the diameter of the stick by peeling off bark, or removing leaves and small twigs. Again, males and females did not differ significantly in number of sessions observed during which at least one observed instance of tool modification occurred (Mann–Whitney, $U = 25$) and there was no relationship between tool modification and age of the gorilla. None of the gorillas used sticks in a consistent, stereotyped manner during each session.

DISCUSSION

The observed tool modifications and methods for novel use closely resemble descriptions of those observed in the wild for chimpanzees' ant-dipping (Nishida & Hiraiwa, 1982; Boesch & Boesch, 1990) and termite-fishing (McGrew, 1979; Nishida & Hiraiwa, 1982). Also consistent with the data for wild chimpanzee tool use was the fact that there were no significant correlations between age of the animal, and the percentage of tool modification or novel use. The youngest gorilla (4 years) was observed using sticks for dipping in a manner which procedurally did not differ from the techniques seen in the oldest members of the group. Similarly, both Goodall (1970) and Nishida (1973) report that, among chimpanzees, adult-like tool use began no earlier than four years of age, and was similarly indistinguishable from adult tool use.

Unlike the different trends reported among male and female chimpanzees in the wild (e.g., Boesch & Boesch, 1981, 1990), the absence of significant differences between male

Table 8.2 *Novel retrieval methods*

Method	% of Sessions method used	No. of gorillas observed using method
Insert with one hand, remove with other	15	9
Insert, move from side to side, remove	11	10
Hold one stick in each hand	3	5
Hold more than one stick in hand	2	6
Insert forcefully	1	2
Use mouth to guide stick	1	2
Insert with one hand, push down into pot with other	< 1	1
Insert end of stick in hole and drop stick in hole	< 1	1
Insert with both hands, first one hand, and adding the other	< 1	1
Rotate stick while inserting	< 1	1
Use one hand to guide stick into hole, then use other hand	< 1	1
Hold stick between thumb and first finger, with last finger up	< 1	1
Remove with mouth	< 1	1
Insert, dip twice in hole, remove	< 1	1
Use stick to bring peanut butter up, then use finger to extract	< 1	1

and female captive gorillas in all aspects of observed tool use is notable. For wild chimpanzees, in both ant-dipping and termite-fishing, female chimpanzees predominate in tool use, performing the insect extraction at much higher rates than males and persisting for longer periods of time (Nishida, 1973; McGrew, 1979; Boesch & Boesch, 1990). For stone tool use among wild chimpanzees, female chimpanzees, but not males, crack open Coula nuts with clubs and stones in trees, and are more successful than males at cracking open the nuts when on the ground (Boesch & Boesch, 1981).

The use of novel probing techniques by some of the adult animals may provide an opportunity for examining the potential for cultural transmission of tool use between generations of gorillas. Of the 6 novel methods that were demonstrated by more than 1 gorilla (Table 8.2), only 3 appear to confer any advantage to retrieving food from the tool sites. These include moving the stick from side to side while inserting it in the hole, rotating the stick while inserting it, and dipping the tool twice into the food before withdrawing it. All 3 methods would likely result in a larger amount of food obtained during each individual tool insertion. The other 3 more frequently observed novel methods, such as: (1) inserting the stick with one hand and removing with another, (2) inserting the stick forcefully, or (3) the gorillas' using their mouth to guide the stick, would not result in the retrieval of a greater amount of peanut butter over time. In fact, both a forceful insertion and a mouth-guided probe appeared to result in less-accurate probing through the small holes of the tool site. None the less, more than one gorilla was observed using these strategies, suggesting that observational learning and/or emulation of a particular novel technique might not necessarily involve a functional assessment by the observing gorilla(s) of the appropriateness or efficiency of a specific tool-use strategy.

In summary, these findings, along with additional observations of other captive gorilla groups (Fontaine et al., 1995) indicate that gorillas have the requisite cognitive capacities and motor facility to modify and use tools to extract food from artificially embedded sources. As noted by Van Schaik & Fox (1996), reflecting upon the previously observed disparity between captive and wild orangutan tool use, perhaps the tool activities observed in captive gorilla groups unfold under combinations of certain key circumstances, including: the potential for observing human tool use in adjacent food-preparation areas in zoo facilities; a predisposed capacity for emulation by all great ape species; available leisure time under captive conditions (since subsistence foraging is not necessary); ready access to tool production materials, coupled with the gorillas' motivation to explore and manipulate the environment. It remains to be seen whether further study of wild gorilla populations, despite numerous long-term studies, might also yield new observations of tool use comparable to the recent discovery of tool use by wild orangutans, including a tool kit for insect foraging and frugivory (Van Schaik & Fox, 1996). Clearly, the cognitive capacities are readily available for tool use by gorillas, although the question remains whether habitat conditions for any of the gorilla populations that remain are such that the energy expended to extract food from embedded sources through tool use would confer a significant advantage.

REFERENCES

Beck, B. B. (1980). *Animal tool behavior: the use and manufacture of tools by animals.* New York: Garland.

Boesch, C. & Boesch, H. (1981). Sex differences in the use of natural hammers bywild chimpanzees: a preliminary report. *Journal of Human Evolution*, **10**, 585–593.

Boesch, C. & Boesch, H. (1990). Tool use and tool making in wild chimpanzees. *Folia Primatologica*, **54**, 86–99.

Brewer, S. M. & McGrew, W. C. (1990). Chimpanzee use of a tool set to get honey. *Folia Primatologica*, **54**(1–2), 100–104.

Chevalier-Skolnikoff, S. (1977). A Piagetian model for describing and comparing socialization in monkey, ape, and human infants. In S. Chevalier-Skolnikoff & F. E. Poirier (eds.). *Primate bio-social development: Biological, social & ecological developments.* (pp. 159–187). New York: Garland.

Fay, J. M. & Carroll, R. W. (1994). Chimpanzee tool use for honey and termite extraction in Central Africa. *American Journal of Primatology*, **34**, 309–317.

Fontaine, B. Moisson, P. Y., & Wickings, E. J. (1995). Observations of spontaneous tool making and tool use in a captive group of western lowland gorillas (*Gorilla gorilla gorilla*). *Folia primatologica*, **65**, 219–223.

Fossey, D. & Harcourt, A. H. (1977). Feeding ecology of free-ranging mountain gorilla (*Gorilla gorilla beringei*). In T. H. Clutton-Brock (ed.). *Primate ecology* (pp. 415–447). London: Academic Press.

Galdikas, B. M. F. (1979). Orangutan adaptation at Tanjung Puting Reserve: Mating and ecology. In D. A. Hamburg & E. R. McCown (eds.). *The great apes.* (pp. 195–233). Menlo Park, CA: Benjamin/Cummings Publishing Co.

Galdikas, B. M. F. (1982). Orang-utan tool use at Tanjung Puting Reserve, Central Indonesian Borneo (Kalimantan Tengah). *Journal of Human Evolution*, **11**, 19–33.

Galdikas, B. M. F. & Teleki, G. (1981). Variation of subsistence activities of male and female pongids. *Current Anthropology*, **22**, 241–256.

Goodall, J. van Lawick (1964). Tool-using and aimed throwing in a community of free-living chimpanzees. *Nature*, **201**, 1264–1266.

Goodall, J. van Lawick (1968). The behavior of free-living chimpanzees in the Gombe Stream Reserve. *Animal Behavior Monographs*, **1**, 161–311.

Goodall, J. van Lawick (1970). Tool-using in primates and other vertebrates. In D. S. Lehrman, R. A. Hinde, & E. Shaw (eds.). *Advances in the Study of Behavior* (vol. 3, pp. 195–249). London: Academic Press.

Izawa, K. & Itani, J. (1966). Chimpanzees in Kasakati Basin, Tanganyika (I.) Ecological study in the rainy season 1964–1964. Kyoto University *African Studies*, **1**, 73–156.

Jordan, C. (1982). Object manipulation and tool-use in captive pygmy chimpanzees (*Pan paniscus*). *Journal of Human Evolution*, **11**, 35–39.

King, B. J. (1986). Individual differences in tool-using by two captive orangutans(*Pongo pygmaeus*). In B. J. King & A. Taub (eds.). *Current perspectives in primate social dynamics* (pp. 469–475). New York: Van Nostrand Reinhold.

Lethmate, J. (1982). Tool-using skills of orangutans. *Journal of Human Evolution*, **11**, 49–64.

Matsuzawa, T. (1994). Field experiments on use of stone tools in the wild. In R. W. Wrangham, W. C. McGrew, F. B. M. de Waal, & P. G. Heltne (eds). *Chimpanzee cultures* (pp. 351–370). Cambridge, MA: Harvard University Press.

McGrew, W. C. (1974). Tool use by wild chimpanzees in feeding upon driver ants. *Journal of Human Evolution*, **3**, 501–508.

McGrew, W. C. (1979). Evolutionary implications of sex differences in chimpanzeepredation and tool use. In D. Hamburg & E. McCown (eds.). *Perspectives on human evolution* (vol. 5, pp. 441–463). Menlo Park, CA: Benjamin/Cummings Publishing Co.

McGrew, W. C. (1987). Tools to get food: The subsistants of Tasmanian aborigines and Tanzanian chimpanzees compared. *Journal of Anthropological Research*, **43**(3), 247–258.

McGrew, W. C. (1992). *Chimpanzee material culture.* Cambridge University Press.

Merfield, F. & Miller, H. (1956). *Gorilla hunter.* New York: Farrar, Straus, and Cudahy.

Natale, F., Potí, P., & Spinozzi, G. (1988). The development of tool use in a macaque and a gorilla. *Primates*, **29**, 413–416.

Nishida, T. (1973). The ant-gathering behavior by the use of tools among wild chimpanzees of the Mahali Mountains. *Journal of Human Evolution*, **2**, 357–370.

Nishida, T. (1980). The leaf-clipping display: a newly-discovered expressive gesture in wild chimpanzees. *Journal of Human Evolution*, **9**, 117–128.

Nishida, T. & Hiraiwa, M. (1982). Natural history of a tool-using behavior by wild chimpanzees in feeding upon wood-boring ants. *Journal of Human Evolution*, **11**, 73–99.

Parker, S. T. & Gibson, K. R. (1977). Object manipulation, tool use, and sensorimotorintelligence as feeding adaptations in Cebus monkeys and great apes. *Journal of Human Evolution*, **6**, 623–641.

Russon, A. E. & Galdikas, B. M. F. (1993). Imitation in free-ranging rehabilitant orangutans (*Pongo pygmaeus*). *Journal of Comparative Psychology*, **107**, 147–161.

Sugiyama, Y. (1985). Brush-stick of chimpanzees found in south-west Cameroon and their cultural characteristics. *Primates*, **26**, 361–374.

Sugiyama, Y. (1995). Drinking tools of wild chimpanzees at Bossou. *American Journal of Primatology*, **37**, 263–269.

Suzuki, S., Kuroda, S., & Nishihara, T. (1995). Tool-set for termite-fishing by chimpanzees in the Ndoki Forest, Congo. *Behavior*, **132**(3–4), 219–235.

Tutin, C. E. G. & Fernandez, M. (1992). Insect-eating by sympatric lowland gorillas (*Gorilla g. gorilla* and chimpanzees (*Pan t. troglodytes*) in the Lope Reserve, Gabon. *American Journal of Primatology*, **28**, 29–40.

Van Schaik, C. & Fox E. A. (1996). Manufacture and use of tools in wild Sumatran orangutans. *Naturwissenschaften*, **83**, 186–188.

Wood, R.J. (1984). Spontaneous use of sticks by gorillas at Howletts Zoo Park, England. *International Zoo Yearbook*, **31**, 13–18.

Yamagiwa, J., Yumoto, T., Ndunda, M., and Maruhashi, T. (1988). Evidence of tool-use by chimpanzees (*Pan troglodytes schweinfurthi*) for digging out a bee nest in the Kahuzi-Biega National Park, Zaire. *Primates*, **29**, 405–411.

Yamakoshi, G. & Sugiyama, Y. (1995). Pestle-pounding behavior of wild chimpanzees at Bossou, Guinea: A newly observed tool-using behavior. *Primates*, **36**, 489–500.

9

A survey of tool use in zoo gorillas

SUE T. PARKER, MARY KERR, HAL MARKOWITZ,
AND JAY GOULD

INTRODUCTION

Chimpanzees have long been famous for their tool use in the wild (Goodall, 1964), as well as in captivity (Köhler, 1927). Orangutans are now known to use tools in the wild (Van Schaik, Fox & Sitompul, 1996) as well as in captivity (Lethmate, 1982). Even bonobos, who are not known to use tools in the wild (Kano, 1986), are widely known for their tool use in captivity (Savage-Rumbaugh & Lewin, 1994). Of the great apes, only gorillas have been described as rarely using tools in captivity (McGrew, 1992), as well as never using tools in the wild (Schaller, 1963). In his review of tool use in great apes, McGrew (1992) cites only one report of tool use – use of a stick to rake in food – in captive gorillas. The idea that gorillas lack problem-solving capacities traces back to Yerkes' study of the gorilla Congo (Yerkes, 1927). Recently, this idea was reinforced by a study of the development of tool use in an infant gorilla (Natale, 1989)

On the other hand, at least two factors suggest that gorillas should be capable of tool use. First, gorillas display 5th- and 6th-stage sensorimotor cognitive abilities similar to those of other great apes (Chevalier-Skolnikoff, 1977; 1983), and tool use – at least, purposeful use of a variety of tools to solve problems – is an index of 5th-stage sensorimotor intelligence similar to that of 2–year-old human children (Piaget, 1952). Second, orangutans, gorillas, chimpanzees, bonobos, and humans are closely related sister species that share a common ancestor roughly dating back 14 to 18 million years ago (see Begun, this volume).

This chapter reports on a survey undertaken to assay the varieties and frequencies of tool use reported in captive lowland gorillas living in zoos in the US and Canada. The survey was designed to elicit data on what kinds of tool use and object manipulation, if any, had been observed in gorillas. Fifty-two zoos were contacted in 1994, of which 12 returned questionnaires. Respondents from these 12 zoos sent reports on a total of 56 gorillas. Of these, 52 were reported to use tools. Two of the non-tool users were less than 18 months old. See Table 9.1 for a list of participating zoos and numbers of gorillas.

Table 9.1. *Zoos reporting tool use in gorillas*

Name of zoo	Location	Tenure of group	Number of gorillas in group	Number of tool-users
Brookfield Zoo	Chicago, IL	12 years	9 (+ 1 solitary)	9
Calgary Zoo	Calgary, ON	11 years	6	4
Cheyenne Mt. Zoo	Colorado Springs, CO	?	3	3
Cincinnati Zoo	Cincinnati, OH	16	7	7
Como Zoo	St. Paul, MI	10	3	3
DuMond Jungle	Miami, FL	?	1	1
Gulf Breeze Zoo	Gulf Breeze, FL	6	3	2
Little Rock Zoo	Little Rock, AK	6	4	3
Metro Toronto Zoo	Toronto, ON	20	5	5
National Zoo	Washington, DC	15	4	4
San Francisco Zoo	San Francisco, CA	35	7	7
Woodland Park Zoo	Seattle, WA	25	4	4
Totals 12 Zoos			56 gorillas	52 tool-users

NATURE OF THE SURVEY

Respondents were asked to provide the following information on the gorilla exhibit: number of group members, number of adults and immatures, number of males, length of time in exhibit of oldest animal, and the nature of the objects in the exhibit (from a check sheet). They were also asked to give the following information on each gorilla in their zoo: name, sex, birthdate, mother and fathers name, place of birth, rearing, health, time in group, rank, tool-user or not, object modifier or not, mode of acquisition of tool use, age of first observation. They were asked to check off the categories of tool use observed.

Ten categories of tool use were listed: probing, scratching or rubbing with an object, raking, levering, striking or breaking with an object, aimed throwing, sponging, using a ladder or stool, and using an object as a bridge. The questionnaire also asked respondents to report other forms of tool use that were not listed. In addition, respondents were asked to report such forms of complex object manipulation as stacking objects, placing objects inside other objects, tying objects, and bouncing and rolling balls.

The questionnaire defined tool-users in the following terms: "To qualify as a tool user, the animal must move a detached object for the purpose of changing the condition and/or position of another object or organism (Beck, 1980; Parker & Gibson, 1977; Van

Table 9.2 *Overall summary of tool use in zoo gorillas*

Categories of tool use on questionnaire	Proportion of zoos reporting		Proportion of tool-using gorillas	
Throwing missiles	10/12	(83%)	31/52	(60%)
Sponging	10/12	(83%)	25/52	(48%)
Probing with stick	7/12	(58%)	24/52	(46%)
Grooming self	7/12	(58%)	18/52	(35%)
Hitting with weapon	9/12	(75%)	15/52	(29%)
Using ladder/stool	8/12	(75%)	15/52	(29%)
Raking	7/12	(58%)	14/52	(27%)
Hammering	5/12	(42%)	9/52	(17%)
Using bridge	3/12	(25%)	9/52	(17%)
Levering	3/12	(25%)	8/52	(15%)
Additional categories of tool use reported				
Digging with stick	1/12	(08%)	2/52	(04%)
Using container for water	3/12	(25%)	4/52	(08%)
Using stick as a fork to hold food	1/12	(08%)	2/52	(04%)
Using stick to prop up own sore limb	1/12	(08%)	2/52	(04%)
	Total no. of Kinds of tool use 14 kinds		Proportion of gorillas exhibiting tool use 52/56 (93%)	

Two of the 56 individuals were infants less than 18 months old.

Lawick-Goodall, 1970) or move and use a detached object for the purpose of reaching a goal, e.g., to reach an object or bridge a gap."

RESULTS OF THE SURVEY

Ninety-six percent of the gorillas surveyed lived in social groups. All of these were long-standing groups, most of which included both mature and immature individuals, parents and offspring, and both sexes. The largest group had 9 members, the smallest, 3 members. Of the 52 gorillas whose origins were specified, 15 were wild born. Of the 37 captive born, only 4 had been hand-raised. Few respondents ventured to say whether forms of tool use had been acquired through observation or not. Eleven and a half months was the youngest age reported for first tool use, but this question was answered by only about half the respondents.

Ninety-three percent of the gorillas used tools. Fifty-two percent of the 52 individ-

Figure 9.1 Gorillas from San Francisco Zoo using sticks as probes.

uals who used tools, modified objects for use as tools. At least 15% and as many as 60% of tool-using individuals used each of the 10 kinds of tools listed on the questionnaire. Aimed throwing (60%), sponging (48%), and probing with a stick (46%) were the most commonly reported forms of tool use, hammering and using a bridge (both 17%), and use of a lever (15%) were the rarest forms of tool use. In addition to the 10 categories listed on the questionnaire, 4 other categories of tool use were noted by respondents: use of a stick to dig, use of a cup to drink water, use of a stick to hold food as a fork, use of a stick to prop up a sore limb. Both the variety and the goal-directed nature of gorilla tool use suggest that it is intelligent. Use of tools to procure out-of-reach objects, for example, is characteristic of 5th-stage sensorimotor intelligence in human infants (Piaget, 1952) See Table 9.2 for a list of the number of individual gorillas engaging in various kinds of tool use, and number of zoos reporting them. See photograph of gorillas from the San Francisco Zoo using sticks as probes.

In addition to these 14 forms of tool use, respondents reported 9 forms of object manipulation that, like tool use, are indicative of 5th-stage sensorimotor level understanding of space and causality (Chevalier-Skolnikoff, 1977; 1982). Fifty-four percent of the gorillas (30 individuals) engaged in at least 1 of the following categories of complex object manipulation (in order of decreasing frequency): rolling and bouncing balls, stacking objects, placing objects inside other objects, and using an object as a sled. See Table 9.3 for a list of frequencies.

Table 9.3 *Categories of object manipulation in zoo gorillas*

Categories of object use reported	Number of zoos reporting	Proportion of individuals reported using tools
Rolling balls/tires	6/12	9/56
Bouncing balls	4/12	4/56
Nesting objects inside	3/12	3/56
Stacking objects	2/12	2/56
Holding ball overhead to get seeds out	1/12	3/56
Covering head with object to avoid rain	1/12	2/56
Trading objects with keeper	1/12	2/56
Using tub as sled	1/12	2/56
Covering other with cloth	1/12	1/56
	Total no. of Categories	Total no. of Individuals
	9	30

The absolute numbers and percentages of tool-users and the variety of forms of tool use strongly suggest that tool use and complex object manipulation are common among zoo gorillas. This suggests that, contrary to the popular view, gorillas do not differ significantly from other great apes in these tool-using abilities. It remains to be seen whether they display more complex forms of tool use seen in wild chimpanzees. Complex tool use involves sequential use of two or more implements in service of a given task, as, for example, use of an anvil to hold a nut while it is being hammered, or a wedge to stabilize the anvil, or a pick to remove nut meat from nut that has been hammered open (Sugiyama, 1997).

ACKNOWLEDGMENTS

We thank the following individuals for collecting and reporting the data used in this brief report listed alphabetically by zoo: Rich Bergl, Brookfield Zoo; Rob Sutherland, Calgary Zoo; Michael Burton, VMD, Cheyenne Mountain Zoo; Ron Evans, Cincinnati Zoo; James Hauge, Como Zoo; Caroline Bettinger, DuMond Conservancy; Pat Quinn, Gulf Breeze Zoo, Ann Rademacher, Little Rock Zoo; Vanessa Phelan, Metropolitan Toronto Zoo; Rob Shumaker, National Zoo; Mary Kerr and Joanne Tanner, San Francisco Zoo; Dawn Prince, Woodland Park Zoo. We also thank Joanne Tanner for the photograph.

REFERENCES

Beck, B. (1980). *Animal tool behavior*. New York: Garland.

Chevalier-Skolnikoff, S. (1977). A Piagetian model for comparing the socialization of monkey, ape, and human infants. In S. Chevalier-Skolnikoff & F. Poirier (eds.), *Primate biosocial development* (pp. 159–188). New York: Garland.

Chevalier-Skolnikoff, S. (1983). Sensorimotor development in orangutans and other primates. *Journal of Human Evolution*, **12**, 545–561.

Goodall, J. (1964). Tool using and aimed throwing in a community of free-living chimpanzees. *Nature*, **201**, 1264–1266.

Kano, T. (1986). *The last ape: Pygmy chimpanzee behavior and ecology* (Evelyn Ono Vineberg, trans.). Stanford University Press.

Köhler, W. (1927). *The mentality of apes*. New York: Vintage.

Lethmate, J. (1982). Tool-using skills of orangutans. *Journal of Human Evolution*, **11**(49–64).

McGrew, W. (1992). *Chimpanzee material culture*. Cambridge University Press.

Natale, F. (1989). Causality II: The stick problem. In F. Antinucci (ed.), *Cognitive structure and development in nonhuman primates* (pp. 121–125). Hillsdale, NJ: Erlbaum.

Parker, S. T. & Gibson, K. R. (1977). Object manipulation, tool use, and sensorimotor intelligence as feeding adaptations in cebus monkeys and great apes. *Journal of Human Evolution*, **6**, 623–641.

Piaget, J. (1952). *The origins of intelligence in children*. New York: International Universities Press.

Savage-Rumbaugh, S. & Lewin, R. (1994). *Kanzi: The ape at the brink of the human mind*. New York: John Wiley & Sons.

Schaller, G. (1963). *The mountain gorilla: Biology and behavior*. University of Chicago Press.

Sugiyama, Y. (1997). Social tradition and the use of tool-composites by wild chimpanzees. *Evolutionary Anthropology*, **6**, 23–27.

Van Lawick-Goodall, J. (1970). Tool-using in primates and other vertebrates, *Advances in the study of behavior* (vol. 3, pp. 195–249).

Van Schaik, C. P., Fox, E. A., & Sitompul, A. F. (1996). Manufacture and use of tools in wild Sumatran orangutans: Implications for human evolution. *Naturwissenschaften*, **83**, 186–188.

Yerkes, R. M. (1927). The mind of a gorilla: I and II *Genetic Psychology Monographs*, **2**, 1–193; 337–551.

PART III

Communication in gorillas and orangutans

10

Symbolic communication with and by great apes

H. LYN MILES

CAN AN APE LEARN LANGUAGE?

In 1887, Müller challenged Darwin to demonstrate the truth of human evolution by teaching language to an ape. More than 100 years later, scientists are no longer worried about supporting claims of human evolution, but they are still teaching apes aspects of human language and discerning what apes understand. With more than 30 great apes from all four species exposed to some level of language learning and human culture, it is possible to make preliminary comparisons among the great apes. The abilities of chimpanzees and bonobos have been extensively reviewed (Desmond, 1979; Gardner, Gardner, & Van Cantfort, 1989; Savage-Rumbaugh, 1991; Wallman, 1992), but less is known about gorillas and orangutans (Ristau, 1996). In this chapter, I summarize the ape language studies, compare the language abilities of all great ape species, and discuss specific issues such as methodology, vocabulary, symbolic capacity, communicative competence, reference, and grammar. To begin, I describe briefly three types of language-learning studies: those based on speech, those based on gesture, and those based on arbitrary visual symbols.

SPEECH STUDIES

Perhaps the first study to teach an ape some aspect of human language was Garner's (1900, p. 138) attempt to teach the chimpanzee Moses four words with "a bribe of corned beef": the English "mamma," the French "feu," the German "wie," and the Nkami "nkgwe." Moses learned to say "feu," and imitated the mouth movements of "mamma," but was less successful with the other two sounds. A few years later, Witmer (1909) described a performing chimpanzee Peter, who had been trained to say "mama," though with great effort and unwillingness. Witmer himself taught Peter to make the English "peh" sound for the letter p in 5 minutes, and compared Peter's difficulties to those of a child with speech difficulties.

Another chimpanzee, imported for Witmer, eventually stayed with Furness (1916), who tried to teach her, another chimpanzee, and two orangutans to speak. He taught one orangutan to say "papa" after 6 months of daily training, which involved imitation of appropriate facial movements; Furness hoped to help the orangutan to control her lip movements better by repeating the sound and lip movements while both he and she were facing a mirror, thereby giving her feedback on the appearance of her lip movements. After 6 months she spontaneously said "papa," for which Furness praised her, and she eventually used it as his name. He next taught her "cup" by placing a spatula at the back of her throat and blocking off her nose (to force her to breathe through her mouth) and quickly removing the spatula, all the while imitating her mouth and tongue movements. Eventually the orangutan herself placed Furness' finger over her own nose to help her make the sound, and no longer required the spatula. More trials with variations led to her learning to use the word "cup" to refer to a cup as well as to thirst. Unfortunately, the orangutan died soon after Furness taught her the sound "th" by holding her tongue against her teeth while forcing her to breathe through her mouth. Furness then taught a chimpanzee to say "mama," which she said poorly; the chimpanzee also failed to learn to say "cup."

Winthrop and Luella Kellogg (1933/1967) sought to examine the influence of human upbringing on a chimpanzee by raising one, Gua, starting at 7.5 months of age, with their own slightly older child, Donald, for 9 months. The Kelloggs tried but failed to teach Gua to say "papa" by facing her while saying the word and moving her lips in the appropriate way. Like her predecessors, Gua reacted correctly to many words and phrases, but she never learned to speak. Similarly, Ladygina-Kots (1935/1982) raised the chimpanzee Joni from 6 months to 4 years of age, and noted that he did not learn speech. Another couple, Keith and Cathy Hayes (1955; see also Hayes, 1951; Hayes & Nissen, 1971) more explicitly set about to teach speech to a chimpanzee they named Viki, using methods similar to those used by Furness. Through patient reinforcement of sounds, and physical manipulation of Viki's face, they taught Viki to say four words with great difficulty: "mamma" and "papa," which referred to her human parents; "cup," which referred to a drink; and "up," which was a request for a piggyback ride. They also reinforced three other sounds Viki initially made spontaneously in play, and then imitated when her parents made them: clicking teeth, as a request for a ride in a car; "tsk," for a cigarette; and a clicking sound to be allowed out in the yard.

A more detailed understanding of the process of speech production by an ape comes from the more recent research of Laidler (1980) who conducted a developmental study of the speech and cognitive ability of Cody, an infant male orangutan raised in Laidler's home and at the London Zoo. Laidler failed to get Cody to vocalize until, for 31 days straight, he blocked Cody's nose and placed him in a headlock until Cody made a guttural "gruh" sound he later associated with chocolate rewards. Cody eventually learned four additional sounds: "kuh," to refer to milk and drinks, "puh," to mean pick me up; "thuh," for brush; and "fuh," to refer first to chocolate fudge and then to all solid food. Laidler presumed that Cody's vocalizations were referential words.

Amid all their work in training apes to speak, all of these authors (as well as others – see Mitchell, 1999) recognized that their subjects understood speech quite well, and several tested their recognition with positive results. But the apes failed to develop spontaneous vocal babbling, speech, or other language-like abilities, and could only with great difficulty learn to utter a few sounds (Kellogg, 1968). This limited success, despite often grueling teaching sessions, was overgeneralized as researchers concluded that apes were incapable of productive language. Anatomical studies revealed that apes lacked a right-angle bend in their vocal-tract anatomy which permits the production of many human speech sounds, and possibly lacked strong neocortical control of their vocalizations (Lieberman, 1975; Goodall, 1986). However, recent evidence suggests that some ape vocalizations are referential, and that cross-fostered bonobos (at least) may imitate some aspects of human vocalizations (Miles, 1976; Hauser & Wrangham, 1990; Rumbaugh, Hopkins, Washburn, & Savage-Rumbaugh, 1993; Tomasello & Call, 1997).

SIGN-LANGUAGE STUDIES

Speech studies provided evidence not only of speech comprehension, but also of gestural communication. For example, Hayes and Hayes (1955) as well as Ladygina-Kots (1935/1982) noted that their cross-fostered chimpanzees made requests with manual gestures. Similarly, Laidler's orangutan, Cody, also imitated gestures and movements in response to a request for him to "do this" followed by an action, and made hand signs, for example, putting his hand on his head as a food request sign. Although there had been speculation about using gestures to communicate with apes (e.g., Yerkes, 1925), sign language was neither well known nor believed to be a full human language. By the 1960s, these notions had changed (see Klima & Bellugi, 1979).

In June 1966, Alan and Beatrix Gardner (1969; 1975) began to teach gestural signs to a chimpanzee, Washoe, whom they cross-fostered in a trailer in their yard with a small group of caregivers using only American Sign Language (ASL) gestures to communicate. The Gardners emphasized methods of sign acquisition, double-blind tests of signs, and vocabulary growth. Washoe learned signs through molding of her hands, shaping, and (less frequently) imitation (Fouts, 1973; Gardner et al., 1989). Washoe only infrequently heard speech because the Gardners' did not want her to be distracted from learning ASL. Washoe's signs and responses to signs were recorded, and a comparison of her early semantic relations and those of children showed that she could combine signs to encode meaning relationships, and was not merely stringing signs together. Washoe made her first sign in 3 months and developed a vocabulary of 132 signs by 51 months, and 150 signs by 8 years; she extended her vocabulary, and combined signs to create new meanings. The Gardners' criterion for Washoe's having learned a sign was that it occur spontaneously and appropriately on 15 consecutive days. Washoe's vocabulary, which includes signs for foods/drinks, objects, actions, persons, emphasis, locations, prepositions, animals, colors, locatives, attributes, emphasizers, places, pronouns, and time, is representative of the vocabularies of all the sign-language using great apes who followed her.

Washoe went to live with the Gardners' former student, Fouts, in Oklahoma in 1970, and stayed there until 1980. She went with Fouts to Washington in 1980, where she has remained. The Gardners' extended their studies to two additional chimpanzees, Dar and Moja (another chimpanzee, Pili, died during the study). Fouts expanded sign studies with other chimpanzees, including Ali, Booee, and Bruno, who developed smaller vocabularies, and focused on sign acquisition and intraspecific signing (Fouts, 1973; Miles, 1978a,b; Fouts with Mills, 1997). Washoe and several of the other signing chimpanzees, including Washoe's adopted son Loulis, actively use signs with each other at the Chimpanzee and Human Communication Institute at Central Washington University (Fouts, Fouts, & Van Cantfort, 1989).

Gestures based on ASL for the deaf became the most popular form of two-way communication with apes. ASL provides several important advantages over other symbolic systems: it is a fully complex human language, which allows rich and direct comparisons to be made between the use of symbols by apes and human children, and it is nonvocal, which avoids apes' problems with learning many human speech sounds (Klima & Bellugi, 1979). In addition, ASL is based on gestures, a modality for which apes are preadapted. Communicative gestures have been reported for captive gorillas, bonobos, and orangutans, as well as for captive and wild chimpanzees (e.g., Kellogg & Kellogg, 1933/1967; Miles, 1976; Tomasello et al., 1985; Goodall, 1986; De Waal, 1988; Bard, 1990; Gomez, 1990; Tanner & Byrne, 1993; this volume; see also Mitchell, 1999). Using gestures to communicate allows apes to incorporate their natural gestures into the interaction as well as to invent their own symbols. Most researchers used ASL gestures in pidgin Sign English (PSE), a variant produced by English speakers, and communicated in a "baby talk" style, appropriate for very young children, and often without many of the grammatical and other features of adult sign language. However, the ASL gestures used are readily interpreted by signers. One disadvantage of sign research is that recording and interpreting signs can be difficult, although claims that apes are not signing, but merely using natural gestures (Savage-Rumbaugh et al., 1983; Wallman, 1992) have not been supported (Ristau, 1996).

As in the speech studies of Hayes and Hayes, apes in most sign-language studies were cross-fostered in human settings, in which caregivers utilize a "pragmatic anthropomorphism" which presumes that apes can acquire language (Miles, 1997), and scaffold-simplified language forms much as human parents do with children (Vygotsky, 1978; Lock, 1980; Bruner, 1983; Vauclair, 1984). The human caregivers established empathy and understanding, and thereby enculturated the apes (Miles, 1997). I define enculturation as the deliberate process of raising an individual in a human setting, with the intention of transmitting cultural models and symbolic forms of communication, from how you brush your teeth to how you form a word, in which the individual identifies with the caregiver and comes to use the models and symbols themselves. (I suggest using the terms "traditioned," or "cultured," to refer to apes who learn their own natural behaviors from other apes in natural settings, "encultured" to refer to apes raised in human settings but not taught to communicate using symbols, and "dual-cultured" to refer to apes who have experienced both natural settings and human culture.)

The Gardners' approach was applied in teaching sign-language to a 1–year-old lowland gorilla named Koko starting in 1972, and later to another lowland gorilla named

Michael (Patterson, 1978; 1980; 1986; Patterson & Linden, 1981; Patterson & Cohn, 1994; Bonvillian & Patterson, this volume). Patterson made three major changes in the Gardners' methods: speech and signs were used simultaneously; a sign was considered to be "acquired" after spontaneous and appropriate usage on 15 days of the month; and all of the gorillas' communications were treated as meaningful. After early years living at a zoo, Koko was raised in a house trailer on the Stanford University campus, and then lived at the Gorilla Foundation in Woodside, California with Michael, another signing gorilla, plus an additional male gorilla. Koko's and Michael's environment was highly enriched; for example, Koko had a pet kitten named ALL-BALL, she set out cookies for Santa Claus, and played with toy tea-sets. Koko could comprehend both signed and spoken communications, and made two- and three-sign combinations of signs after just a few months (Patterson & Linden, 1981). Koko's vocabulary has been variously reported, but her core-acquired vocabulary approximates 250 signs (Patterson, 1978; Patterson & Cohn, 1994). (Using the Gardners' criteria, Koko has a vocabulary of about 185 signs [Ristau, 1996], which is comparable to Washoe's vocabulary of about 150 at 8 years of age.) Koko coined a number of sign combinations including BAREFOOT HEAD (for a bald man), GIRAFFE BIRD (for an ostrich), and ELEPHANT BABY (for a Pinocchio doll) (Patterson & Linden, 1981). Koko also responds to, and very occasionally uses, Wh-questions, such as WHO?

Koko showed some evidence of rule-following behavior in her signing, such as placing the MORE sign at the beginning of her combinations. Whether or not such ordering represents semantic-order preferences depends upon analysis of the other signs in the emphasizer class. Koko also placed verbs correctly either before or after a pronoun 76–84% of the time, which indicates a rule-following preference (Patterson & Linden, 1981). Koko used language for humor, metaphors, jokes, rhymes, and puns (Patterson & Linden, 1981). Interestingly, her rhymes are in English, not ASL, for example, HAIR/BEAR, and ALL/BALL.

After several years, Michael developed a vocabulary of about 110 signs which he combined meaningfully. He engaged in rudimentary exchanges with his caregivers (Patterson & Linden, 1981, p. 172), such as:

Barbara: WHAT THIS? (picture of a bird)
Michael: CAT EAT
Barbara: WHAT (did) YOU SAY ABOUT CATS? WHAT (do) CATS EAT?
Michael: BIRD

Like Koko and other signing apes, Michael understands both signing and speech, and has extended sign meanings and coined neologisms, for example, the "insult" TOILET DEVIL.

In 1973, Terrace began to teach signs to a week-old chimpanzee, Nim Chimpsky, who lived with a human family until he was 18 months, and then with a small group of student "teachers" who used ASL signs with him (Terrace, 1979; Terrace, Pettito, Sanders, & Bever, 1979). Nim learned to use individual signs and exhibited some semantic relations. The intensity of the teaching techniques apparently caused him to combine his signs in lengthy strings, and to interrupt and imitate his caregivers' signing, provoking Terrace to conclude that an ape could not create a sentence. Later experiments with Nim in a more naturalistic situation, and with sign-using orangutan Chantek in his usual relaxed circumstances (see below), indicated that these apes signed in shorter sequences, with little interruption or imitation (Miles, 1983; O'Sullivan & Yeager, 1989).

Shapiro (1982; this volume) also used the sign-language model with a dual-cultural rehabilitant orangutan, Princess, in a natural setting in Borneo. Like the Gardners, he employed reinforcement, but he explicitly avoided enculturating the orangutan. Princess acquired 32 signs, mostly requests or items desired, until she lost interest in the lessons.

I also applied the Gardners' methods to the orangutan, Chantek, who I raised starting in 1978 on the campus of the University of Tennessee at Chattanooga with a small group of caregivers (Miles, 1980; 1983; 1986; 1990; 1994; Miles & Harper, 1994; Miles, Mitchell, & Harper, 1996). I emphasized enculturation, contextual analysis of Chantek's discourse, semantic analysis, cognitive abilities such as deception and imitation, and the evolutionary implications of ape language abilities. Chantek developed a vocabulary of approximately 150 signs. His largest sign categories were for actions and objects; food and drink signs formed only 22% of his vocabulary. He invented signs himself, for example, VIEWMASTER, BALLOON, and DAVE-MISSING-FINGER. Chantek showed comprehension of speech, could break down a sign into its elements, and improved his sign articulation when asked to SIGN BETTER. His use of signs was referential, and occasionally deceptive (Miles, 1986). He modulated the meanings of his signs, and also displayed rule-ordered regularities. He learned the signs for CHANTEK, YOU, ME, and his caregivers. He learned a number of cultural scripts and games.

VISUAL SYMBOL STUDIES

Some researchers turned to nonvocal means to try to establish language-like communication with apes, using visual symbols presented as plastic tokens on a board, or geometric shapes called lexigrams displayed on a computer monitor and keyboard. Although tokens and lexigrams are artificial codes rather than natural human language, they serve as an effective semantic system that can be used to test language-related abilities, including cognitive processes and aspects of grammar. They have the advantage of simplicity because they offer a finite vocabulary and representation without rapid fading or much ambiguity. Computer lexigrams also leave a permanent record for easy recording and analysis, and direct comparisons can, and have been, made with human children (Savage-Rumbaugh et al., 1993). However, many elements of human language are missing with the visual-symbol systems, including invention of new symbols by apes, ease of use and initiation of communication, prosodic features, modulations of meaning, and contractions. Additionally, lexigram boards must be carried about, tokens can only be used in a restricted area, and pointing to lexigrams falls prey to the same interpretive problems that affect gestural modes. Studies using visual codes also utilized speech to communicate with the apes, and were generally conducted under strict experimental conditions, although some researchers who continued in these methods have subsequently recognized the need for more normative cultural conditions to produce language-like abilities (Savage-Rumbaugh et al., 1993).

Premack (1971) used visual symbols in the form of colored plastic tokens with a female chimpanzee, Sarah. He was interested in the nature of intelligence rather than language *per se*, but was able to teach Sarah an artificial code with the tokens. Sarah was not encouraged to initiate communications herself; instead she answered questions with her placement of plastic "names" for objects ("pail"), actions ("insert"), qualifiers/attributes ("yellow"), and conditionals ("same," "different"). For example, when Sarah was required to describe both an apple and the blue triangle token representing the concept of apple, she gave an identical description, from which Premack (1976) concluded that she understood the representation. In addition to responding to questions based on words and simple sentences, Premack taught metalinguistic distinctions, for example, class, color, shape, size, copula, quantifiers, and logical if/then relations.

Lexigrams were first taught to the chimpanzee Lana (Rumbaugh, Gill, & Von Glasersfeld, 1973). Lana was placed in a small research room and communicated through the keyboard and computer screen. To bring them to the apes' attention, lexigrams were highlighted, electronically activated on a display, or pointed to on the lexigram board each while presenting the referent person, object, or action. To be acceptable as sentence-like communications, lexigrams had to follow a rigid phrase-structure correlational grammar. Lana was trained in long sessions on individual problems which the researchers ultimately concluded led to the formation of learning sets rather than language-like abilities (Rumbaugh, 1977; Savage-Rumbaugh & Lewin, 1994). Still, lexigrams were used communicatively. Two chimpanzee subjects, Sherman and Austin, were raised together and taught to categorize tools and foods using lexigrams, thereby showing that chimpanzees can use symbols for intraspecific information transfer (Savage-Rumbaugh, 1979; 1986; Savage-Rumbaugh, Rumbaugh, & Smith, 1980). Matsuzawa (1990) replicated the computer lexigram studies of the Rumbaugh team with chimpanzee Ai.

Bonobos were also taught to communicate with lexigrams (Savage-Rumbaugh, 1984), but their skills were developed using the cross-fostering enculturation model pioneered by sign-language researchers, which provided a more enriched social and physical environment including more opportunity for agency by the ape. Based on a preliminary lexigram acquisition criterion of 9 out of 10 correct usages, one bonobo, Kanzi, showed a vocabulary of 125–150 lexigrams. Kanzi learned lexigrams through observation, rather than direct instruction, and also used his lexigrams to comment on the environment or his actions, although most of Kanzi's communications were holophrastic (single lexigram) requests (Savage-Rumbaugh & Lewin, 1994). Placing emphasis on comprehension of requests and comparisons with children, the research team concluded that Kanzi's performance was comparable to that of a 2-year-old child (Savage-Rumbaugh et al., 1993).

Lexigrams were also taught to juvenile and subadult orangutans at the National Zoo

(Robert Shumaker, personal communication, November 22, 1997, Washington, DC). Shumaker focused on comprehension of speech to foster two-way communication using lexigrams displayed on a computer and on flash cards. He conducted relaxed sessions with the orangutans, who chose to travel to the Think Tank exhibit and interact with Shumaker in front of zoo visitors. After two years of training, orangutans Azy and Indah acquired eight lexigrams: "apple," "banana," "grape," "carrot," "chow," "cup," "bag," and "open," which they generalized to new referents.

A COMPARISON OF GREAT APE LANGUAGE ABILITIES

Although apes have not acquired adult human language, they do have the intelligence for a rudimentary, referential, generalizable, imitative, displaceable, symbol system (see Ristau, 1996 for a review). The results of the studies so far show that apes are capable of mastering a simple nonvocal symbol system which can be referential at a level of 2–year-old children, while other cognitive skills are comparable to those of older children. Although they can acquire a few spoken words, it is with great difficulty, probably even if an artifical spoken system were developed. Systems that are based on signing result in greater production; both sign and computer systems result in comprehension of symbols.

Apes are motivated to communicate about the here and now, the immediate episodic present, although they do, to a limited degree, make displaced reference to things separated in time and space. Their communications are heavily supported by scaffolding from their caregivers, and they primarily make requests to meet their immediate needs (this is encouraged by the fact that they are under the control of caregivers who restrict foods and special events, even more so than human children). They make reference to the environment and themselves and their feelings, make declarative and evaluative statements, use symbols with each other, engage in rudimentary conversational exchanges, comprehend requests for actions involving relational terms, understand basic metacommunications about symbols and symbol performance, utilize some rule-following grammatical structures, describe basic properties of their experience, and engage in deceptive communications. However, they do so at a lesser rate than human children, and they require a great deal of patience, repetition, and caregiver support.

Initially, when most language studies used chimpanzees as subjects, differences among the chimpanzees in the different projects were considered "more reflective of ways in which different experimenters have interpreted the behaviors of the chimpanzees than of true differences in the capacities of the apes themselves" (Savage-Rumbaugh et al., 1983, p. 459). Later, when attention focused on bonobos, Savage-Rumbaugh et al. (1986, p. 231) concluded that bonobos have "a greater propensity for the acquisition of symbols" based on the performance of Kanzi. Now, with the Language Center's change to more naturalistic enculturation methods, and the equal performance of chimpanzees on many tasks, Savage-Rumbaugh (1991) has acknowledged the importance of enculturation rather than species distinctions, once again minimizing the differences among the apes.

Table 10.1 *A comparison of the linguistic abilities of the great apes*

Ability	Orangutan	Gorilla	Chimpanzee	Bonobo
Comprehension of natural referential vocalization			X	X
Production of natural referential vocalization			X	X
Comprehension of natural referential gestures	X	X	X	
Production of natural referential gesture	X	X	X	
Comprehension of speech	X	X	X	X
Production of a few spoken words	X		X	X
Comprehension of ASL signs	X	X	X	
Production of ASL signs	X	X	X	
Comprehension of plastic tokens			X	
Productive use of plastic tokens			X	
Comprehension of computer symbols/lexigrams	X		X	X
Productive use of computer symbols/lexigrams	X		X	X
Deixis (linguistic pointing)	X	X	X	X
Comprehension of 150+ symbols	X	X	X	X
Productive use of 150+ symbols	X	X	X	X
Symbols for names	X	X	X	X
Symbols for actions	X	X	X	X
Symbols for objects	X	X	X	X
symbols for food/drink items	X	X	X	X
Symbols for attributes (e.g., RED)	X	X	X	X
Symbols for emphasis (e.g., ALL, MORE)	X	X	X	X
Symbols for feelings/evaluations (e.g., GOOD, HURT, SAD, COLD)	X	X	X	X
Symbols for basic time concepts (e.g. TIME, FINISHED, NOW)	X	X	X	X
Combinations of symbols	X	X	X	X
Symbol contractions	X	X	X	X
Novel recombinations of symbols in coining metaphorical terms	X	X	X	X
Invented symbols	X	X	X	
Environmental reference (e.g., HEAR BELL, BREAK THAT)	X	X	X	X
Occasional use of a Wh-question (e.g., WHAT, WHO)	X	X	X	X

Table 10.1 *Continued*

Ability	Orangutan	Gorilla	Chimpanzee	Bonobo
Nonlinguistic means to ask Wh-questions (e.g., questioning facial expression)	X	X	X	X
Deceptive use of symbols	X	X	X	
Joking use of symbols		X	X	
Insults with symbols (e.g., DIRTY, BAD, DEVIL)	X	X	X	
Self-directed symbols (e.g. ME, name sign)	X	X	X	X
Symbols used with conspecifics		X	X	X
Symbols used with other animals	X	X		
Symbols used for planning	X	X	X	X
Spontaneous use of symbols	X	X	X	X
Initiation with symbols	X	X	X	X
Discourse with symbols	X	X	X	X
Mean length of utterance and upper bound comparable to children	X	X	X	X
Comprehension of in/on/under/out	X	X	X	X
Productive use of in/on/under/out	X	X	X	X
Comprehension of similar but grammatically different commands	X		X	X
Rhymes, puns, or poetry		X		X
Displaced reference	X	X	X	X
Generalization of symbols	X	X	X	X
Semantic over- & under-extension of symbols	X	X		
Referential use of symbols	X	X	X	X
Rule-following symbol use	X	X	X	X
Rudimentary grammar	X	X	X	X
Problem-solving with symbols	X	X	X	X
Symbolic play with symbols	X	X	X	
Generalizing hand gestures to other body parts (e.g., feet)	X			
Comprehension of request to improve communication form	X			

Given that all four great ape species have received some level of language training consisting of from two to over a dozen individual animals, it is now possible to make some preliminary comparisons. Table 10.1 lists some of the features of language related to studies of great apes. The Table shows that, on the majority of features, all four of the great apes are similar in having learned simple symbol systems. Areas of difference are minimal, and most have to do with differing emphases and interpretations of evidence; in some cases apes of a given species have not been tested using a given methodology. Overall, all great ape species show the rudiments of human languages, but they fail to show fully developed human language, lacking in (e.g.) complex social emotional vocabulary, extensive use of Wh-questions, and complex grammar.

Thus, it is likely that when members of all great ape species have been thoroughly enculturated, tested using similar criteria, not pressured or interrupted, and allowed a degree of conversational agency, they will demonstrate similar abilities, perhaps with minor species variation. The differences that have appeared over the years on which claims of uniqueness of one species have been based are likely false or an artifact of the method and particular emphasis of that research program.

ACKNOWLEDGMENTS

This research was supported by the National Institute of Child Health and Human Development grant NICHD 14918, National Science Foundation grant BNS 8022260 and grants from the UC Foundation. The loan of Chantek was provided by the Yerkes Regional Primate Research Center supported by National Institutes of Health grant RR 00165. The assistance of Warren Roberts, Sue Parker, and Robert Mitchell is gratefully acknowledged.

REFERENCES

Bard, K. A. (1990). "Social tool use" by free-ranging orangutans: A Piagetian and developmental perspective on the manipulation of an animal object. In S. T. Parker & K. R. Gibson (eds.), *"Language" and intelligence in monkeys and apes: Comparative developmental perspectives* (pp. 356–378). Cambridge University Press.

Bruner, J. (1983). *Child's talk: Learning to use language*. New York: Norton.

Desmond, A. (1979). *The ape's reflexion*. New York: The Dial Press.

De Waal, F. B. (1988). The communicative repertoire of captive bonobos (*Pan paniscus*) compared to that of chimpanzees. *Behaviour*, **106**, 183–251.

Fouts, R. S. (1973). Acquisition and testing of gestural signs in four young chimpanzees. *Science*, **180**, 978–980.

Fouts, R. & Mills, S. T. (1997). *Next of kin: What chimpanzees have taught me about who we are*. New York: Morrow.

Fouts, R., Fouts, D., & Van Cantfort, T. (1989). The infant Loulis learns signs from cross-fostered chimpanzes. In R. Gardner, B. Gardner, & T. Van Cantfort (eds.), *Teaching sign language to chimpanzees* (pp. 280–292). Albany, NY: State University New York Press.

Furness, W. H. (1916). Observations on the mentality of chimpanzees and orangutans. *Proceedings of the American Philosophical Society*, **55**, 281–290.

Gardner, R. A. & Gardner, B. T. (1969). Teaching sign language to chimpanzee. *Science*, **165**, 664–672.

Gardner, R. A. & Gardner, B. T. (1975). Early signs of language in child and chimpanzee. *Science*, **187**, 752–753.

Gardner, R. A., Gardner, B. T., & Van Cantfort, T. E. (eds.). (1989). *Teaching sign language to chimpanzees*. Albany, NY: State University of New York Press.

Garner, R. L. (1900). *Apes and monkeys: Their life and language*. Boston: Ginn and Co.

Gomez, J. C. (1990). The emergence of intentional communication as a problem-solving strategy in the gorilla. In S. T. Parker & K. R. Gibson (eds.), *"Language" and intelligence in monkeys and apes* (pp. 333–355). Cambridge University Press.

Goodall, J. (1986). *The chimpanzees of Gombe: Patterns of behavior*. Cambridge, MA: Harvard University Press.

Hauser, M. D. & Wrangham, R. W. (1990). Recognition of predator and competitor calls in nonhuman primates and birds: A preliminary report. *Ethology*, **86**, 116–130.

Hayes, C. (1951). *The ape in our house*. New York: Harper and Brothers.

Hayes, K. J. & Hayes, C. (1955). The cultural capacity of chimpanzee. In J. A. Gavan (ed.), *The non-human primates and human evolution* (pp. 110–125). Detroit: Wayne University Press.

Hayes, K. J. & Nissen, C. (1971). Higher mental functions of a home-raised chimpanzee. In A. M. Schrier & F. Stollnitz (eds.), *Behavior of nonhuman primates*, (vol. 4, pp. 60–115). New York: Academic Press.

Kellogg, W. N. (1968). Communication and language in the home-raised chimpanzee. *Science*, **162**, 423–427.

Kellogg, W. N. & Kellogg, L. A. (1933/1967). *The ape and the child*. New York: Hafner Publishing Co.

Klima, E. & Bellugi, U. (1979). *The signs of language*. Cambridge: Harvard University Press.

Ladygina-Kots, N. N. (1935/1982). Infant ape and human child: Their instincts, emotions, games, and expressive movements. *Storia e Critica della Psicologia*, **3**, 113–189.

Laidler, K. (1980). *The talking ape*. New York: Stein & Day.

Lieberman, P. (1975). *On the origins of language: An introduction to the evolution of human speech*. New York: Macmillan.

Lock, A. (1980). *The guided reinvention of language*. New York: Academic Press.

Matsuzawa, T. (1990). Spontaneous sorting in human and chimpanzee. In S. T. Parker & K. R. Gibson (eds.), *"Language" and intelligence in monkeys and apes: Comparative developmental perspectives* (pp. 451–468). New York: Cambridge University Press.

Miles, H. L. (1976). The communicative competence of child and chimpanzee. *Annals of the New York Academy of Sciences*, **260**, 592–597.

Miles, H. L. (1978a). Language acquisition in apes and children. In F. C. C. Peng (ed.), *Sign language and language acquisition in man and ape: New dimensions in comparative pedolinguistics* (pp. 103–120). Boulder, CO: Westview Press.

Miles, H. L. (1978b). The use of sign language by two chimpanzees. *Dissertation Abstracts International*, **39**, 11A.

Miles, H. L. (1980). Acquisition of gestural signs by an infant orangutan (*Pongo pygmaeus*). *American Journal of Physical Anthropology*, **52**, 256–257.

Miles, H. L. (1983). Apes and language: The search for communicative competence. In J. de Luce & H. T. Wilder (eds.), *Language in primates: Perspectives and implications* (pp. 43–61). New York: Springer-Verlag.

Miles, H. L. (1986). How can I tell a lie? Apes, language, and the problem of deception. In R. W. Mitchell & N. S. Thompson (eds.), *Deception: Perspectives on human and nonhuman deceit* (pp. 245–266). Albany, NY: State University of New York Press.

Miles, H. L. (1990). The cognitive foundations for reference in a signing orangutan. In S. Parker & K. Gibson (eds.), *"Language" and intelligence in monkeys and apes: Comparative developmental perspectives* (pp. 511–538). Cambridge University Press.

Miles, H. L. (1994). ME CHANTEK: The development of self-awareness in a signing orangutan. In S. T.

Parker, R. W. Mitchell, & M. L. Boccia (eds.), *Self-awareness in animals and humans* (pp. 254–272). Cambridge University Press.

Miles, H. L. (1997). Anthropomorphism, apes, and language. In R. W. Mitchell, N. S. Thompson, & H. L. Miles (eds.), *Anthropomorphism, anecdotes, and animals* (pp. 383–404). Albany, NY: State University of New York Press.

Miles, H. L. & Harper, S. (1994). "Ape language" studies and the study of human language origins. In D. Quiatt & J. Itani (eds.), *Hominid culture in primate perspective* (pp. 253–278). Denver: University Press of Colorado.

Miles, H. L., Mitchell, R. W., & Harper, S. (1996). Simon says: The development of imitation in an enculturated orangutan. In A. Russon, K. Bard, & S. T. Parker (eds.), *Reaching into thought: The minds of the great apes* (pp. 278–299). Cambridge University Press.

Mitchell, R. W. (1999). Scientific and popular conceptions of the psychology of great apes from the 1790s to the 1970s: Déjà vue all over again. *Primate Report*, **53**, 1–118.

Müller, F. M. (1887). *The science of thought*. London: Longmans.

O'Sullivan, C. & Yeager, C. P. (1989). Communicative competence and linguistic competence: The effects of social setting on a chimpanzee's conversational skill. In R. Gardner, B. Gardner, & T. Van Cantfort (eds.), *Teaching sign language to chimpanzees* (pp. 269–279). Albany, NY: State University of New York Press.

Patterson, F. G. P. (1978). The gestures of a gorilla: Language acquisition in another pongid. *Brain and Language*, **5**, 72–97.

Patterson, F. G. P. (1980). Innovative uses of language by a gorilla: A case study. In K. E. Nelson (ed.), *Children's language* (vol. 2, pp. 497–561). New York: Gardner Press.

Patterson, F. G. P. (1986). The mind of the gorilla: Conversation and conservation. In K. Benirschke (ed.), *Primates: The road to self-sustaining populations* (pp. 933–947). New York: Springer-Verlag.

Patterson, F. G. P. & Cohn, R. (1994). Self-recognition and self-awareness in lowland gorillas. In S. T. Parker, R. W. Mitchell, & M. L. Boccia (eds.), *Self-awareness in animals and humans* (pp. 273–290). New York: Cambridge University Press.

Patterson, F. & Linden, E. (1981). *The education of Koko*. New York: Holt, Rinehart & Winston.

Premack, D. (1971). Language in chimpanzee? *Science*, **172**, 808–822.

Premack, D. (1976). *Language in ape and man*. Hillsdale, NJ: Erlbaum.

Ristau, C. A. (1996). Animal language and cognition projects. In A. Lock & C. R. Peters (eds.), *Handbook of human symbolic evolution* (pp. 644–685). Oxford: Clarendon.

Rumbaugh, D. M. (ed.) (1977). *Language learning by a chimpanzee: The Lana project*. New York: Academic Press.

Rumbaugh, D. M., Gill, T. V., & Von Glasersfeld, E. C. (1973). Reading and sentence completion by a chimpanzee (*Pan*). *Science*, **182**, 731–733.

Rumbaugh, D. M., Hopkins, W., Washburn, D. A., & Savage-Rumbaugh, E. S. (1993). Chimpanzee competence for comprehension in a video-formatted task situation. In H. L. Roitblat, L. M. Herman, & P. E. Nachtigall (eds.), *Language and communication: Comparative perspectives* (pp. xx–xx). Hillsdale, NJ: Erlbaum.

Savage-Rumbaugh, E. S. (1979). Symbolic communication: Its origins and early development in the chimpanzee. *New Directions for Child Development*, **3**, 1–15.

Savage-Rumbaugh, E. S. (1984). *Pan paniscus* and *Pan troglodytes*: Contrasts in preverbal competence. In R. Susman (ed.), *The pygmy chimpanzee: Evolutionary biology and behavior* (pp. 395–414). New York: Plenum Press.

Savage-Rumbaugh, E. S. (1986). *Ape language: From conditioned response to symbol*. New York: Columbia University Press.

Savage-Rumbaugh, E. S. (1991). Language learning in the bonobo: How and why they learn. In N. Krasnegor, D. M. Rumbaugh, R. L. Schiefelbusch, & M. Studdert-Kennedy (eds.), *Biological and behavioral determinants of language development* (pp. 209–233). New York:

Savage-Rumbaugh, [E.] S. & Lewin, R. (1994). *Kanzi: The ape at the brink of the human mind*. New York: John Wiley & Sons.

Savage-Rumbaugh, E. S., Rumbaugh, D. M., & Smith, S. T. (1980). Reference: The linguistic essential. *Science*, **210**, 922–925.

Savage-Rumbaugh, [E.] S., McDonald, K., Sevick, R. A., Hopkins, W. D., & Rubert, E. (1986). Spontaneous symbol acquisition and communicative use by pygmy chimpanzees (*Pan paniscus*). *Journal of Experimental Psychology: General*, **115**, 211–235.

Savage-Rumbaugh, E. S., Pate, J., Lawson, J., Smith, S., & Rosenbaum, S. (1983). Can a chimpanzee make a statement? *Journal of Experimental Psychology: General*, **112**, 457–492.

Savage-Rumbaugh, E. S., Murphy, J., Sevcik, R. A., Brakke, K. E., Williams, S. L., & Rumbaugh, D. M. (1993). *Language comprehension in ape and child. Monographs of the Society for Research in Child Development*, **58**(3–4, Serial No. 233).

Shapiro, G. L. (1982). Sign acquisition in a home-reared/free-ranging orangutan: Comparisons with other signing apes. *American Journal of Primatology*, **3**, 121–129.

Tanner, J. E. & Byrne, R. W. (1993). Concealing facial evidence of mood: Perspective-taking in a captive gorilla? *Primates*, **34**, 451–56.

Terrace, H. S. (1979). *Nim: A chimpanzee who learned sign language*. New York: Knopf.

Terrace, H. S., Petitto, L. A., Sanders, R. J., & Bever, T. G. (1979). Can an ape create a sentence? *Science*, **206**, 891–902.

Tomasello, M. & Call, J. (1997). *Primate cognition*. New York: Oxford University Press.

Tomasello, M., George, B. L., Kruger, A. C., Farrar, M. J., & Evans, A. (1985). The development of gestural communication in young chimpanzees. *Journal of Human Evolution*, **14**, 175–186.

Vauclair, J. (1984). Phylogenetic approach to object manipulation in human and ape infants. *Human Development*, **27**, 321–328.

Vygotsky, L. S. (1978). *Mind in society: The development of higher psychological processes*. Cambridge, MA: Harvard University Press.

Wallman, J. (1992). *Aping language*. Cambridge University Press.

Witmer, L. (1909). A monkey with a mind. *Psychological Clinic*, 3, 179–205.

Yerkes, R. M. (1925). *Almost human*. New York: The Century Co.

11

The development of spontaneous gestural communication in a group of zoo-living lowland gorillas

JOANNE E. TANNER AND RICHARD W. BYRNE

When zoo-living gorillas perform communicative physical motions that seem to resemble those of signing gorillas, what does it mean? What are the processes involved in developing these gestures? What communicative behavior is universal for the species and what is individually learned? Though some gestures are shared by all gorillas and others are unique to individuals, such a simple dichotomy does not really tell the whole story. We have found that, in the gorilla group at the San Francisco Zoo, "species-typical" expressions such as slapping, clapping, pounding, and chestbeating develop quite differently in each individual. Individuals also create gestures, some of which have not been described for other gorillas. Other gestures are shared by a few, but not all, individuals in a group. Gestural repertoires of different individuals at the same ages vary both in type and quantity. There is variation in individual gorillas' usage of gestures over time and in accord with changing social conditions. The purpose of the present research is to describe this variation, explore the diverse physical and functional properties of these gestures, and learn why and how they have developed in this particular captive group of gorillas.

Before beginning the observations at the San Francisco Zoo that are described here, the first author (JT) had worked for 8 years with the signing gorillas Koko and Michael of the Gorilla Foundation, developing a particular interest in untaught signs and "natural" gestures of sign language tutored gorillas (Patterson, Tanner, & Mayer, 1988; Patterson & Tanner, 1988). When JT visited the zoo as a casual observer after having worked with signing gorillas, she was struck by the amount of gestural communication used by the zoo gorillas, some of which resembled untaught gestures or even taught signs used by signing gorillas. The iconic nature of some of the gestures used by the zoo gorillas as well as the sign-instructed gorillas seemed to be a feature held in common. Some zoo gestures had physical properties that appeared to depict activity desired or intended by the gesturing gorilla. For the sign-instructed gorillas, gestural inventions often seemed to outline physical features of objects or locations which were brought to their attention or to which they wished to draw attention. Exploration of this iconic aspect of gorilla gesture has been another of the aims of study.

SUBJECTS AND SETTING

The San Francisco Zoo's present gorilla enclosure has been the group's home since 1980. It has an outdoors area of 2300 square meters, or 38 × 50 m at maximum parameters. It is covered with grass and other vegetation and contains large, climbable live trees as well as several dead trees, large stumps, and two artificial rock 'hills' including arches and cave-like areas. The enclosure is below ground (viewer) level, except for one windowed viewing area where gorillas and humans can interact face to face.

The subjects, the gorillas at the San Francisco Zoo, are members of a stable social group and all have spent either most, or the entirety, of their lives at this zoo. The group at present includes the son, Kubie, and grandsons, Shango and Barney, of the wild-caught founder, Bwana, as well as unrelated individuals. A wild-born but human-reared female, Pogo, grew up at the zoo with Bwana; and two younger females whose early rearing was by humans in zoo nurseries, Bawang and Zura, joined the group in 1981 and 1982 respectively, after the death of two older females. Bawang is the mother of Kubie's offspring, Shango and Barney. Further information about each of the gorillas in the group during the study is given in Table 11.1 (and see Parker, chapter 18, this volume). Though they interact daily with keepers and observe zoo visitors, none of these gorillas has had any intentional human instruction in gestural communication.

The two senior gorillas, Bwana and Pogo, had important roles in the group though they gestured very little. Until his death Bwana was a strong leader, continually watchful and alert to every event. The coexistence of two silverback males is unusual in zoos, and the interaction between Bwana and his son Kubie influenced the dynamic of the whole group. Though frequently challenged by Kubie, Bwana was himself rarely the instigator of display or aggression. On a one-to-one basis, Bwana usually retreated from Kubie unless severely provoked, but the rest of the group would rally with Bwana and take aggressive action to subdue Kubie if he harassed a youngster or female.

Pogo, the oldest female, who has never mated, has consistently repelled male advances over the years, though she would often teasingly court male attentions. Recently, a medical examination ascertained that she had a constriction in her vagina. Until the birth of Bawang's babies, Pogo spent a lot of time in the trees or on the periphery of the social group, trying to avoid harassment from Kubie in particular. She was extremely interested when Bawang's infant, Shango, was born and tried to get near him for months, but Bawang did not allow contact until Shango was 6 months old. After that, Pogo gradually became a preferred playmate and frequent "baby-sitter" for Shango. She has also played with Barney, but not as much, because Barney has older brother Shango as play partner. Pogo, like Bwana, rarely gestured, though once in a while in the course of play she would engage in a surprising burst of communicative activity.

GENERAL METHODOLOGY

Beginning in October 1988, observations were made one morning each week for approximately 3 hours when the gorillas were outdoors (conditions permitting).

Table 11.1 *Gorillas of the San Francisco Zoo*

Name	Sex	Age in March 1989	Age in March 1996	Birthplace and rearing	Parents
Bwana	male	31	deceased 1994	wild born, Cameroon	unknown
Pogo	female	31	38	wild born, human reared in Cameroon	unknown
Kubie (Mkubwa)	male	13	20	captive born, mother reared in San Francisco	Bwana and Jackie
Bawang	female	8	15	captive born, nursery reared in Cincinnati	Ramses and Amani (Cincinnati)
Zura	female	7	14	captive born, nursery reared in Columbus	Oscar and Toni (Columbus)
Shango	male	born March 1989	7	captive born, mother reared in San Francisco	Kubie and Bawang
Barney (Ike-ozo)	male	born October 1993	2.5	captive born, mother reared in San Francisco	Kubie and Bawang

Regular observations have continued to the time of this writing, except for one 6–month hiatus in the study from September 1989 to March 1990. Videotaped records were chosen as the only adequate way to study gestural communication; tiny variations in physical form of motions and direction of gaze are important elements for study and cannot always be instantly perceived nor adequately described in the form of written notes; further, any aversion of an observer's gaze to write may mean loss of data.

The research team has consisted of both JT and the camera operator, her husband Charles Ernest (CE), throughout the study except for 9 months in 1992–1993 when in their absence a team of trained students followed the same videotaping procedure JT and CE had established: the principal researcher (JT) scans the activity taking place and suggests to the camera operator the most relevant area to film. The camera follows a given social interaction continuously wherever possible (sequence sampling). For the first few years of the study, the choice of subjects was easy because virtually all of the gesturing occurred during play sessions between a young silverback, Kubie, and a young adult female, Zura. Very little gesturing took place in agonistic contexts or between other gorillas, and none in feeding situations. Later in the study, when play sometimes occurred simultaneously between two of the older gorillas and between an infant and older gorilla, the camera continuously followed the interaction thought most likely to contain gestural communication. If no apparent interaction was occurring and the gorillas were all resting, eating, or spatially separated, the camera was turned off. This procedure has resulted in approximately 200 hours of videotape used for the various analyses described in this chapter. (On average, a 3–hour visit would yield an hour of usable videotape.) The videotaped records are time coded, date and time stamped, and often include verbal commentary about context or behavior of other members of the gorilla group while the camera was focused on a single ongoing interaction. Further, JT took diary-style written notes about any events affecting the gorillas' activity that the camera was unable to record (such as annoyance from zoo visitors, airplanes or construction noise causing distraction, or information from the zoo-keeper about events during the previous week).

Each videotaped instance of gesture was catalogued on the Filemaker Pro database computer program. Only gestures clearly visible on the videotape were included. If the same gesture was rapidly repeated consecutively several times it was counted only once. Most gorilla vocalizations were too quiet to analyze and are ignored here, but bodily gestures which produced sound are included.

GROUP HISTORY AND STUDY PERIODS

The data analyzed here is divided into six "study periods" of roughly equal duration. A "period" was delineated on the basis of the occurrence of important social events or changes in group composition. Such changes often seemed to affect social communication. A brief social history explains what events and changes demarcated study periods (see Table 11.2).

Table 11.2 *Study periods and social history of group*

Study Periods	Duration	Age of juveniles	Social conditions and activity
Period 1	October 31, 1988–September 20, 1989 (Then a 6–month break in data collection occurred.)	from beginning of observations until Shango's age is 6 months	Kubie plays exclusively with Zura; Bawang is pregnant, then stays continually with newborn Shango.
Period 2	March 19, 1990–June 17, 1991	Shango, 1 to 2.3 years	Noise stress from nearby construction causes some decrease in play (Gold & Ogden, 1992); Kubie plays with both Bawang and Zura; Shango plays with everyone.
Period 3	July 1, 1991–September 22, 1992	Shango, 2.3 to 3.5 years	Kubie begins to pursue and guard Bawang, presumably from his father Bwana. Kubie does not play at all. Shango plays with all the females; this is the only play in the group.
Period 4	October 2, 1992–June 24, 1993	Shango, age 3.5 to 4.3 years	Kubie's pursuing and guarding of Bawang continues until mating and into the first few months of her pregnancy (February 1993). He does not engage in play. Shango plays with the females.
Period 5	July 14, 1993–September 1, 1994	Shango, 4.3 to 5.6 years; Barney, birth to age 11 months	Reappearance of play on Kubie's part; this period extends from the latter part of Bawang's pregnancy, through Barney's birth in October 1993 to his age 11 months. Kubie's play is nearly all with Shango; Shango plays with everyone. This period ends with Bwana's death.
Period 6	September 6, 1994–January 26, 1996	Shango, 5.6 to 6.10 years; Barney, 11 months to 2.3 years	This period begins right after Bwana's death; Barney is becoming independent; Kubie plays with both youngsters; they play with everyone and particularly each other.

DEFINITION OF GESTURE

Gesture has been defined in many different ways. For animals, gesture frequently refers to bodily motion or facial expression that is "ritualized" to some degree. For human behavior, gesture can mean almost any intentional and voluntary kind of nonverbal communication. Because an aim of study has been to focus on the degree to which signs and signals distinct from ordinary whole body action are used by gorillas, here body postures and locomotory gaits have been excluded from categorization as gesture.

As a working definition, the term *gesture* is here used for *all discrete, non-locomotor limb and head movements, regardless of receptive sensory modality* (sight, sound, touch), *that occurred when gorillas were in proximity and were engaged in social interaction immediately before, after, or during the movements.* We presumed these gestures to be communicative, in the sense of attempting to influence other gorillas' behaviors, and then examined the data to see if such gestures in fact were consistently responded to with particular behaviors by others.

In *tactile* communication, gestures may be defined as different from ordinary motions in that they involve transformations of purposive behaviors so that they are no longer forceful enough to be mechanically effective (Bretherton & Bates, 1979; Goldin-Meadow & Mylander, 1984; Gómez, 1990). For example, lightly brushing a hand downward on another's body to indicate a desire for downward movement on another's part would be a *tactile gesture*, as opposed to the direct *action* of forcefully pushing the other down. A purely *visual gesture* with similar function would be performed by a downward motion of the hand and arm in the space in front of the signaler's body while having the visual attention of the other.

CLASSIFICATION OF GESTURES

All gestures by the gorillas, whether visual, tactile, or auditory, were catalogued. Table 11.3 provides a descriptive list of all the gesture types specifically referred to in this chapter, with comparisons to similar gestures in other ape species.[1] The gestures in Table 11.3 are only part of the San Francisco gorillas' actual repertoire; other gestures did not occur with enough frequency to provide sample sizes sufficiently large for analysis.

[1] These comparisons were difficult because in many sources a careful physical description of gestures was not given. For instance, terms such as patting, poking, tapping, hitting, and swiping may have been interchangeable in ethograms from different zoos. Therefore the comparisons in Table 11.3 are not exact, and in some cases guesses. Many gestures and other behaviors that have not been observed in San Francisco are listed for gorillas in other zoos, but likewise some San Francisco gestures are not reported elsewhere. Some gestures used socially in San Francisco are listed only as solitary or stereotyped behavior in other zoos. The number and type of behavioral elements in ethograms for the same species also vary tremendously from one to the other, perhaps because of the individual goals and observational focus of the studies they were for, but probably also because behaviors differ from zoo to zoo and in the wild.

Table 11.3 *Some social gestures of the gorillas at the San Francisco Zoo*

Gesture type and class; gorillas who used that gesture[a]	physical description	usual context at S. F. Zoo	similar gesture in other gorillas in zoos or wild?	similar gesture in other apes?
armshake silent visually received Kubie, Zura, Bawang	P: space in front of or at sides of the body C: one or both relaxed, open hands M: arms and hands shaken loosely; may vary from prolonged motion of entire upper body to minimal motion of hand(s) shaken from wrists	play, sometimes warning	several zoos, also called wave or hand shake; solitary or sterotyped only in some gorillas, 1	bonobo: nervous request or threat, De Waal, 1988 chimpanzee: wristshaking, threat, Goodall, 1986; warning, approach invitation, De Waal, 1988, Plooij, 1984
armswing under silent visually received Kubie, Zura	P: space in front of body, ends between legs at crotch C: open hand or both hands M: arm(s) swings from space in front of body back to body between legs	male–female play		chimpanzee: beckoning?, Goodall, 1968b; stretch over, Van Hooff, 1973, Plooij, 1984 bonobos: ? Savage-Rumbaugh et al., 1977
backhand pound audible Kubie, Zura, Shango	P: any environmental surface C: fist M: back of hand hits surface forcefully, usually audible	play, agonistic display	frequent at zoos, 1; in wild, 2	chimpanzee: "hitting away" (threat), Goodall, 1968a
body beats audible Kubie, Zura, Bawang, Shango, Barney	any location on body except chest beaten with alternating open hands or fists. Often audible but not as resonant as real chestbeating	play	frequent at zoos, 1; in wild, 2	

Table 11.3 *Continued*

Gesture type and class; gorillas who used that gesture[a]	physical description	usual context at S. F. Zoo	similar gesture in other gorillas in zoos or wild?	similar gesture in other apes?
chest beat audible Kubie, Zura, Bawang, Shango, Barney	chest is slapped with alternating open palms, audible effect	play, agonism, display	species typical: details on use in mountain gorillas, 2	
chest pat silent visually received Kubie	P. chest C: one cupped hand M: hand taps chest lightly, no audible effect	play	at several zoos, 1; in wild, 3	bonobo: De Waal, 1988
clap audible Kubie, Zura, Bawang, Shango, Barney	flat hands, palms contact in space in front of body. Audible effect	play	frequent in zoos, 1; in wild, 3, 4, and Fay, 1989	bonobo: Ingmanson, 1987, De Waal, 1988, Thompson 1993
head nod silent visually received Kubie	head moves abruptly downward and then returns to vertical position	play	in wild, 3	chimpanzee: inviting approach, De Waal, 1988; also Savage in Fouts unpublished and undated training material
hide playface silent visually received Zura	P: open mouth C: open, curved hand M: hand covers mouth	avoiding play		

knock/pound audible Kubie, Zura	P: any environmental surface C: fist M: knuckles or side of hand hits surface, sometimes audible	play	frequent at zoos, 1	
slap (surface) audible Zura, Bawang, Shango, Barney	P: any environmental surface C: open palm M: palm contacts surface forcefully, usually with audible effect	play	frequent at zoos, 1; in wild 2, Mori, 1983	chimpanzee: frequent in wild and captivity
tactile gestures tactile close Kubie, Zura, Bawang, Shango, Pogo	touching of the recipient's body with directional indication but short of force to actually move the body; includes hand moved down the back vertically, or across horizontally; patting, gentle pulling of a hand, pushing away, and others	positioning, mating play	probably at other zoos, described as push, brush, or nudge, 1; in wild 2	bonobo: Savage-Rumbaugh et al., 1977 chimpanzee: Kohler, 1925, Yerkes, 1943
tap other silent visually received Kubie, Zura, Shango	P: body of other gorilla, most often head or chest C: open hand, or fingers bent at knuckles M: fingertips or knuckles contact body of other gorilla then quickly move back	play	at several zoos, may be called poking, tagging or touch with hand, 1; in wild 3	bonobo: punching? Savage-Rumbaugh et al., 1977, De Waal, 1988 chimpanzee: poke? Van Hooff, 1973

References for comparisons to other gorillas: 1. Ogden & Schildkraut, 1991; 2. Schaller, 1963; 3. Robert Campbell films at National Geogographic, viewed by J. T., 1991; 4. Fossey, 1983.

[a] "used" means gesture was observed to be performed at least 5 times in at least one study period by the listed gorilla(s)

P = place (location on body, or location in space)
C = hand configuration or shape
M = motion (direction, force) of gesture
(PCM system as by Stokoe et al., 1965)

KUBIE'S GESTURES

The young silverback male Kubie, in his play with the female Zura, had the largest repertoire of gestures of all the gorillas during Period 1. The first classification project undertaken was an analysis of Kubie's gestures during this period. A corpus of over a thousand gestures performed during interaction with Zura, gleaned from 22 hours of videotape, was used for this analysis (Tanner & Byrne, 1996). Three main classes of gesture were discerned:

(1) *Tactile close* gestures trace or mime on the body of the receiver movements that the gesturing gorilla apparently desires the other gorilla to make. In the majority of cases these gestures do not require visual attention of the receiver.
(2) *Silent visually received* gestures are usually made when the signaler has the recipient's visual attention. They invite further attention, elicit activity, or promote contact (*armshake, armswing under, head nod, chest pat, tap other*).
(3) *Audible* gestures elicit the receiver's attention through sound as well as motion. They often are performed without the receiver's visual attention (*chestbeat, knock, pound*).

Each class of gestures had consistent, but different, communicative effects: *Tactile close* gestures usually resulted in movement of Zura's body in the direction indicated by Kubie's gesture; *silent visually received* gestures resulted in a high rate of bodily contact in play activity. These first two classes of gestures are not regularly seen in all gorillas and have properties that we consider iconic or deictic. (For further discussion of iconicity see Tanner & Byrne, 1996, and the general discussion at the end of this chapter.) *Audible* gestures, which are species-typical, resulted in a low rate of subsequent physical contact, but did frequently result in redirection of the attention, or alteration of the locomotion, of the recipient gorilla.

The frequency distributions of these different classes of gestures varied in different kinds of play sessions between the gorillas. Their frequencies corresponded to differing degrees of physical proximity and visual access of the gorillas to each other. For instance, *tactile close* gestures were most frequently used in sessions of play involving positioning for sexual inspection and mating play; *silent visually received* gestures were most frequently used for a game of tag around a tree and for a tug-of-war game where both players were within a few feet of each other; *audible* gestures were most frequently used in a long-distance "king of the mountain" game with one gorilla on top of a rock formation and the other below on the grass, repeatedly trading places.

A further finding in this study of Kubie's gesture (Tanner & Byrne, 1996) was that playfaces (open-mouthed faces), as used by Kubie, were not *signals* promoting play, but always responses to imminent contact by Zura. Visually received gestures, on the other hand, were useful in promoting play contact in the face of resistance or reluctance on the part of Zura. Recent work by Pellis & Pellis (1996) reanalyzing the "playface" argues that the open mouth of the playface is a preparation to fend off the "attack" of another animal by biting, not a metacommunicative "play signal"; Kubie's usage seems to fit this interpretation.[2]

ZURA'S *HIDE PLAYFACE*

Zura, Kubie's play partner, also had a large repertoire of gestures, some of which have yet to be analyzed. One gesture that was unique to Zura, however, seemed to call for closer study. This was Zura's behavior of hiding her playface with a hand or both hands during interactions with Kubie. (See Figure 11.1: Zura *hides playface* while Kubie performs *head nod* and *hands behind back*, a possibly deictic inviting gesture; and Figure 11.1e: Zura again *hides playface* as Kubie completes *armswing under*.) The *hide playface* was found to be significantly associated with postponement or nonoccurrence of play and to take place in several contexts: when the older male, Bwana, was in proximity or directly interfering with play, where play with the larger Kubie was getting particularly rough, and in situations where she may have hoped to deceive Kubie about her play motivation (Tanner & Byrne, 1993). Zura's behavior is similar to that of a chimpanzee who hid his fear grin with his hands (De Waal, 1982). This is the only other report of an pe manually concealing a facial expression.

A reinterpretation of the playface by Pellis and Pellis (1996) as a response rather than signal (on the face-maker's part) fits well with our conclusion that Zura was concealing her play motivation to prevent the occurrence of play in situations in which it would cause trouble. Zura restrained her presumably automatic and involuntary playful response, apparently aware that after her open-mouthed playface contact play was imminent. This playface-hiding took place in Kubie's view in 21 of 22 cases where line of sight was detectable on the video. Therefore, though perhaps the playface was not an intentional signal by Zura, it may have functioned as a signal to Kubie. Zura behaved as if she was aware of this by producing a canceling signal. She may have anticipated that the onset of contact play would cause interference from Bwana or cause her to get rough treatment from Kubie. Zura was, in essence, fashioning a simple negative by separating a signal from its referent, a first step toward some of the properties of human language (Bateson, 1968).

GESTURAL REPERTOIRE OVER THE LONG TERM

We now turn to the larger picture of change in gesturing over the longer term for the entire group. Some of the changes in number of gestures both within and between

[2] For comparison to chimpanzees, see Goodall (1968a) who observed that the playface was used to *initiate* play by only one individual among the entire troop at Gombe. For others, the playface only became evident during the play session: "the full play face was in fact usually displayed as soon as contact play (wrestling, tickling, etc.) became at all vigorous" (Goodall 1968a, p. 258). The "play walk," a ritualized rolling gait, was used by adolescent and older males to initiate play; for juveniles and females, often a simple approach and reach initiated play. In later writing about play initiation, Goodall (1986) does not mention the playface as a play initiation at all, but discusses additional play invitations such as "finger wrestling," "back present," or approach with a "play twig" in the mouth or brandished in the air for the invitee to attempt to snatch.

a 25.25
Zura *slaps* tree+*armshake*
Kubie *hands behind back*

b 25.28
Zura *hides playface*
Kubie *head nod,*
hands behind back

c 25.30
Kubie *tap other*

d 25.31
Kubie *armswing under* (come...)

e 25.32
Zura *hide playface*
Kubie *armswing under* (hand arrives between legs), *headnod*

f 25.33
Kubie and Zura make contact and wrestle

Figure 11.1 Series of gestures by Kubie and Zura, Study Period 1 (time on videotape in minutes and seconds).

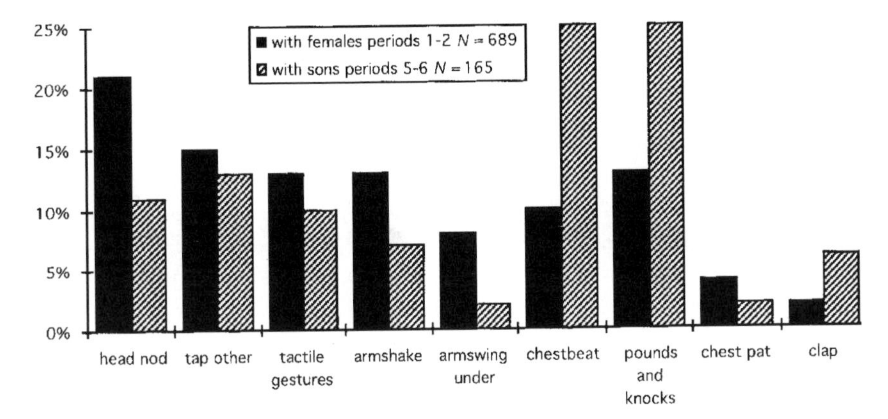

Figure 11.2 Kubie's most common social gestures during periods with different play partners, with percentage of each gesture type used per period (for this comparison, Periods 1 and 2 were combined and Periods 5 and 6 were combined; Periods 3 and 4, when Kubie did almost no gesturing, were not included).

individuals from period to period were so drastic that, even without statistical analysis or strictly equal observation times, they call for explanation. The most striking changes in quantity of gesture over time were clearly related to social circumstances (see Table 11.2). The number of gestures Kubie was observed to use during Periods 1 and 2 when he was playing regularly with Zura was 100 times greater than the number he was observed to use in Periods 3 and 4, when he stopped playing and spent a great deal of time pursuing Bawang.[3] Kubie's observed gestures increased again by about 14 fold during Periods 5 and 6 when he again began to play, this time with his sons. As gestures are found predominantly in play contexts, the cessation of play behavior by Kubie for Periods 3 and 4 meant virtual cessation of his communicative gesture.

Change in gestural repertoire with different social partners

Whether a gorilla's type of gestures might differ in interaction with different individuals seemed worth investigating, since we had already found that gestures used vary according to style and setting of play. There was, indeed, a difference in distribution of types of gestures observed to be used by Kubie at different times with different play partners (Figure 11.2). For analysis, Periods 1 and 2 were combined into an "early" period, and Periods 5 and 6 into a "late" period. During Periods 1 and 2, the early period, Kubie's play was with females; during Periods 5 and 6, the late period, his play was almost exclusively with his young sons. Proportions, within each period, of the nine most frequent gesture types observed for Kubie varied very significantly overall between the earlier and later periods ($\chi^2[8, N = 854] = 67.26, p < 0.0001$). (Here, and in

[3] An earlier lesser drop in Kubie's gesturing from Period 1 to Period 2, when he still played regularly, may have been related to the occurrence of noise stress from nearby construction; an overall decrease in play in the group at this time was documented (Gold & Ogden, 1992).

all succeeding analyses, N is the total number of Kubie's gestures rather than, as normally the case, the number of individuals contributing data. Note that for this reason statistically valid conclusions do not generalize to all gorillas.) In order to investigate whether this variation was particularly associated with certain gestures, we examined each gesture separately, in comparison to all other gestures pooled. Thus, for each gesture type, we constructed a 2×2 table in which the frequency of that gesture was compared with that of all others combined, for each of the two observation periods, early and late. Some gestures did not show significant variation relative to the whole set: *tap other* ($\chi^2[1, N = 854] = 0.67$, $p > 0.4$), *tactile close* gestures ($\chi^2[1, N = 854] = 1.10$, $p > 0.3$) and *chest pat* ($\chi^2[1, N = 854] = 2.11$, $p > 0.1$). All the other gesture types did show significant variation in proportion of use from earlier to later periods. There was relative decrease in the later period of the silent inviting gestures *head nod* ($\chi^2[1, N = 854] = 9.28$, $p < 0.01$), *armshake* ($\chi^2[1, N = 854] = 5.03$, $p < 0.03$), and *armswing under* ($\chi^2[1, N = 854] = 7.99$, $p < 0.005$), but increase in *chestbeat* ($\chi^2[1, N = 854] = 28.32$, $p < 0.0001$), *knock* and *pound* ($\chi^2[1, N = 854] = 12.97$, $p < 0.001$) and *clap* ($\chi^2[1, N = 854] = 8.85$, $p < 0.01$).

The gestures *head nod, armshake, armswing under* (see Figure 11.1 b, c, d and e) were used extensively by Kubie during Period 1 to get and direct visual attention and encourage approach during play with females (Tanner & Byrne, 1996). These gesture types were used less by Kubie when he played with his young sons in Periods 5 and 6, probably because, in play with Kubie, older son Shango was usually the initiator. The gestures that increased for Kubie in Periods 5 and 6 were typical audible male displays (*chestbeat, knock*, and *pound*), plus *clap*, which was a frequent gesture of Shango's. It might seem odd that *tactile close* gestures were observed in the same relative frequency in Kubie's play with his sons as in play with Zura and Bawang, because that type of gesture was associated with positioning for mating play. Kubie, however, sometimes engaged in mounting behavior and anal inspection with his sons, and *tactile close* gestures also functioned to adjust location in ordinary play when the gorillas were already in close contact.

Finally, the number of Kubie's gestures with his sons during Periods 5 and 6 was far less than during Periods 1 and 2 with the females, even though observation time was greater for Periods 5 and 6. Perhaps the females required greater persuasion to participate in interactions than the rambunctious youngster Shango. Shango often actively tried to initiate play, and used gestures to do so, even when Kubie initially showed no interest. (Number of gestures observed of Shango during Periods 5 and 6 was roughly double the number of observations for Kubie.) For Kubie's gestures in Period 1 where the sequels were contact, on the other hand, Zura initiated the actual contact 70 to 80% of the time; it seems that Kubie's gestures were important in influencing her choice to take part in contact interaction. These differences in Kubie's usage of gesture in differing social situations (interaction with females vs. young males) support the finding in our earlier work that gestures function as specialized communicative elements and vary according to context.

Individual differences in gesturing in brothers at the same age

In order to discover what kind of differences might appear in the development of gesture in gorillas of the same age and sex, a comparison was made of the gestures that brothers Shango and Barney each were observed to use socially when aged 12 to 27 months (Figure 11.3). Barney seemed to use a larger overall number of gestures than his older brother Shango did at the same age; though observation time was not strictly equal, Barney still was observed to use nearly 3 times as many social gestures. For both, the most frequent gestures were primarily *slap* (on surfaces in the environment such as the rocks, ground, and trees), *clap*, and *chestbeat*, all typical in gorillas as a species (see Fay 1989 regarding clapping as species-typical for lowland gorillas). Barney also performed a substantial number of *armshakes*. The four gestures seemed interchangeable in function; all were used the large majority of the time when in front of another gorilla in fairly close proximity and therefore can be considered to be communicative. Proportions of these four gesture types observed within the matching age periods varied significantly overall between the two brothers ($\chi^2[3, N=287]=28.13$, $p < 0.0001$). Further analysis did not show variation in the frequency of *chestbeat* by each brother for the matching age periods, relative to other gestures ($\chi^2([1, N=287]=2.09$, $p > 0.1$). However, Barney was seen to use *clap* ($\chi^2[1, N=287]=11.13$, $p < 0.001$) and *armshake* ($\chi^2[1, N=287]=6.57$, $p < 0.01$) relatively more than Shango, and *slap* ($\chi^2[1, N=287]=18.85$, $p < 0.0001$) relatively less. Barney also used *clap* in nonsocial ways not seen in Shango (these instances are not included in the numbers above for *clap*, which are for social, communicative gestures only).[4]

Two young hand-reared male gorillas studied by Redshaw and Locke (1976) also varied from our subjects in their gesture usage. Both strongly preferred slapping, like Shango, at a similar age (1.10 to 2.2 years) to our subjects. Clapping was rare, and at this age their subjects did not use chestbeating in a friendly social context. Slapping was always followed by the approach of the other gorilla for Redshaw's subjects; this was not the case for Shango or Barney, perhaps because their slaps were often performed in front of the older gorillas, who frequently did not choose to play. Redshaw's subjects had only each other as companions.

At times Barney also performed *claps* and *chestbeats* in immediate response to claps or chestbeats by other gorillas (or zoo visitors); this was never observed in Shango during the same age period. Parker (1993, and chapter 18, this volume) reports imitative response by Kubie to his own older brother's chestbeating and slapping between age 2 and 3 years. However, Barney developed an *armshake* gesture that was never a part of

[4] Barney, unlike Shango, seemed to use *clap* as a "marker"; he would frequently clap just *before* jumping off an object such as a tree trunk, rock, or ledge, or clap immediately *after* accomplishing some physical feat such as stripping the bark off a stick, jumping off something, or climbing up and balancing on top of an object like a tree trunk, tub, or large pile of branches. *Chestbeat* also was performed several times prior to jumping, but *slap* was not, perhaps for anatomical reasons. This aforementioned "marker" usage did not seem to be an attention-getting device and did not appear to be communicatively directed toward others; it might be similar to expressions used by very young children to encode success (called non-nominal autoprotodeclaratives by Gopnik [1982]), such as *"there"* upon placing a block or puzzle piece successfully.

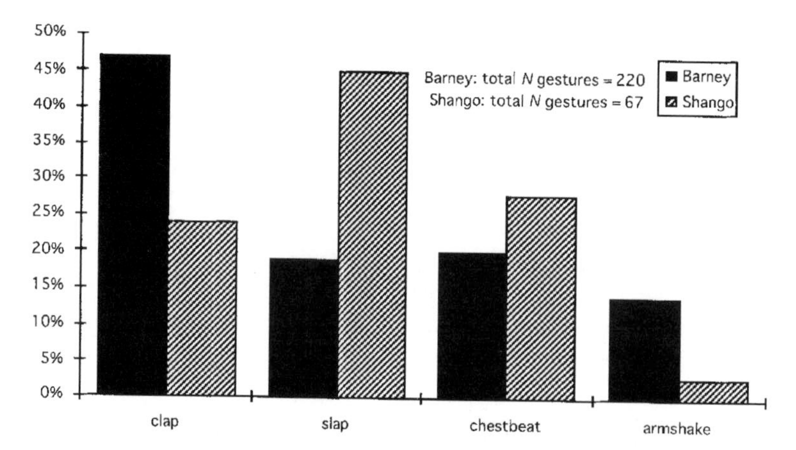

Figure 11.3 Most frequent gesture types (social) between ages 12 and 27 months in two young male gorillas, as percentage of total repertoire of each gorilla.

Shango's standard repertoire despite his frequent exposure as a youngster to its usage in play between Kubie and Zura. Barney seemed to discover *armshaking* in solitary situations, a few weeks later extending it to social usage. Parker reports similar progression of development of *armshake* in Kubie, and also lists twice as many types of play "enticements" used by Kubie as for his older half brother, Sunshine (now at another zoo). Perhaps the greater quantity and variety of gesture use seen for Barney and Kubie, the younger brothers, were some kind of "sibling effect" resulting from early exposure to an older brother's gestures, or were simply because the younger sibling tends to be less controlled by the mother and has more motivation to initiate play when an older sibling is available as playmate.

Developmental changes in gesturing in the individual

To learn about development in gesture use from juvenile period to maturity in an individual gorilla we followed the usage of gestures by the young male Shango across time from age 1 year to age 6 years 10 months (Figure 11.4). Proportions of observations of *slap, clap*, and *chestbeat* within five different Study Periods varied very significantly overall ($\chi^2[8, N=465]=134.53, p < 0.0001$). With increasing age, Shango not only was observed to use a larger number of gestures, but showed a change in preferred types of gestures. After age 5 years 6 months (Period 6) Shango played not only with his father, Kubie, but increasingly actively with his younger brother Barney. During this period there is a relative decrease in both *clap* and *slap* and a sharp increase in *chestbeat* compared to the earlier Period 5. Comparison of the frequency of observations of each of the three commonest gestures against the other gestures pooled, in 2×2 tables, showed significant variation in Shango's proportion of use of each gesture between Period 5 and Period 6: for *clap*, ($\chi^2[1, N=305]=22.96, p < 0.0001$); for *slap*, ($\chi^2[1, N=305]=9.68, p < 0.01$); for *chestbeat*, ($\chi^2[1, N=305]=61.04, p < 0.0001$). Period 6

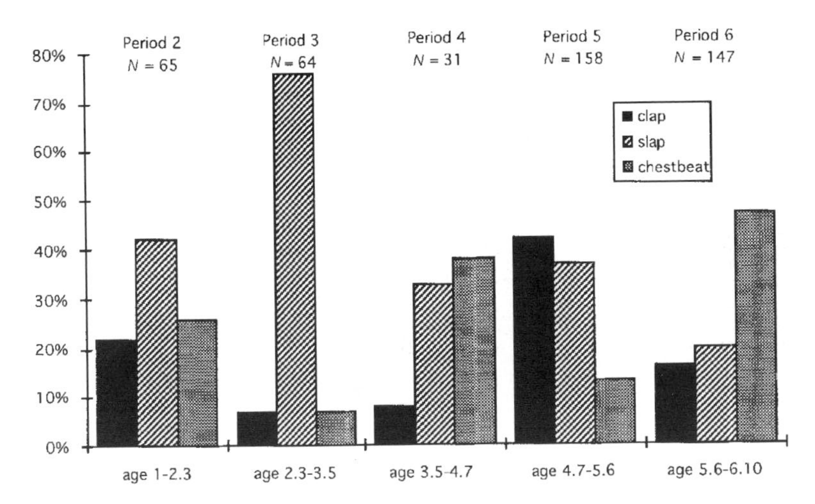

Figure 11.4 Shango's most common social gestures over a 5–year span (Study Periods 2–6) with percentage of most frequent gesture types out of total gestural repertoire for each period.

also saw the new appearance of *backhand* and an increase in *tactile close* gestures (not shown in Figure 11.4).

As Shango matured, his gesture use began to approximate that of his father, Kubie: he was seen to perform *chestbeat, backhand*, and *tactile close* gestures more frequently, and *clap* and *slap* were performed less frequently. Perhaps *clap* and *slap* are "baby" gestures, gradually dropped with increasing maturity; similar dropping of earlier gestures in maturity was found by Tomasello et al. (1989, 1994) for chimpanzees. However, this may be different for females; the preferred gestures of Bawang, Shango's mother, at the age of 9 (when Shango was between age 1 and 2.3 years) were *slap, clap*, and *body* (other than chest) *beating*; she rarely *chestbeat*. The same gestures remained prevalent for Bawang at age 14 (when her younger son Barney was between 1 and 2.3 years old), though Bawang was observed to gesture less overall at the later age. There may be functional reasons for females to use audible gestures other than chestbeating. Chestbeating is not an anatomically comfortable behavior for a lactating female, as Bawang has been during most of the duration of this study; clapping would serve far better as auditory communication during lactation. Zura, a female without young, chestbeat frequently at the same ages when Bawang did not. The fact that, in the wild, clapping has been observed only in females and young (Fay, 1989) seems to support this idea.

The two oldest individuals in the group, Bwana and Pogo, did very little gesturing at all. Until we can follow other individuals to the same age it will be impossible to know whether reduced gesturing is related to age, and perhaps available energy, or is a matter of individual differences. Much of the gesturing done by the other gorillas is performed bipedally, and the oldest gorillas rarely assumed a bipedal stance.

DEVELOPMENT OF GESTURES: GENERAL DISCUSSION

The questions posed at the beginning of this chapter are really one question. How does an animal born with a certain set of typical species behaviors move from these to creation and comprehension of unique ways of communicating through movement and sound? It is apparent from these zoo observations that this creative potential is present in the untutored ape, and we see its furthest development in the innovative signs of apes taught sign languages and other symbol systems. We discuss three aspects of the development of gestures: first, how and why iconic gestures might develop; next, the stages when different classes of gesture may appear (and sometimes disappear) in the development of individual lifespans; and, finally, the extent to which there seem to be differences in the kind of gestures developed by zoo-living and symbol-taught apes, and what this indicates we might find in wild populations.

The development of iconic gestures

The anatomical element that unifies the gestural abilities of all the great apes is the complete rotational movement of their joints. Though individual digit control evolved earlier, only the apes can produce movement in all directions from wrist, elbow, and shoulder joints. This locomotory adaptation that allows movement through trees also makes possible subtle gestures of the hands and limbs (Morbeck, 1994, and see Chevalier-Skolnikoff, Galdikas, & Skolnikoff, 1982; Povinelli & Cant, 1995, for diverse ideas regarding the effect of ape locomotion on cognitive ability). We propose that the ability to function in three dimensions underlies a shared emergent ability of the great apes to depict motion iconically with the limbs. With these fine rotational movements, a single gesture type can vary in configuration, direction, size, and forcefulness on a continuous scale according to the circumstances of its use. Though to produce mimetic gesture the necessary anatomy must be supplemented by an adequate level of cognition, feedback to the brain from the results of use of the brachiator's distal anatomy would certainly have facilitated development of iconic gesture like that which we observe in the great apes.

The importance for the history of humanity of the development of the ability to depict actions and objects iconically has often been noted; whether or to what degree apes share this ability with humans has been a topic of debate (Donald, 1991; Burling, 1993; Mitchell, 1994). Our zoo observations may help to show how an iconic faculty might have developed in its earliest manifestations in our hominoid ancestors; signing apes demonstrate this faculty a step or two further in advancement (Miles & Harper, 1994).

For sign language, Klima and Bellugi (1979) consider a sign to have an iconic relationship to its referent when 'elements of the form of a sign are related to the visual aspects of what is denoted" (Klima & Bellugi, 1979, p. 21). Because we have not observed our zoo subjects to depict objects, here we limit our definition to the depictions of action that these gorillas do produce. We also limit the concept "iconic" to gesture representing action inferred to be desired by the signaler of the recipient.

Though iconic gesture could certainly also depict a signaler's own intended action, in practice such gestures in animals cannot be distinguished from automatic "intention movements," or incipient action (Savage & Rumbaugh, 1977) unless employed deceptively (see Mitchell, 1994).

In our study, therefore, we call a gesture (by hand, arm or head) *iconic* when its motion path in space, or on another animal's body, follows a path of movement, or form of an action, that we infer to be desired by the gesturing animal of another (remembering that, as defined earlier, a gesture lacks the physical force to actually move or change the activity of the recipient, unlike a simple social action). We can infer that a gesture (for example, in open space, moving an outstretched arm down from above the head) may be depicting intentions regarding others' actions, rather than being an "intention movement," when the depicted act itself (in this example, moving the body down) does not come to be performed by the signaler; further, the gesture may be quite discrete and different from the action itself (for instance, an arm motion as opposed to a whole body motion).

The finding that Kubie's communications included gestures that are in our definition iconic (Tanner & Byrne, 1996) is similar to results of Savage-Rumbaugh, Wilkerson, & Bakeman (1977). They found that pygmy chimpanzees expressed intent for different types of sexual positioning with iconic gestures, both tactile and in space. Observations in the wild of pygmy chimpanzees dragging branches to indicate proposed direction of movement (Ingmanson, 1996), the possibility of directional trail-marking (Savage-Rumbaugh, Williams, Furvichi, & Kano, 1996), and, in common chimpanzees, the representation of direction and duration through sound placement in drumming (Boesch, 1991), provide further support for the presence of an ape faculty to plan future movement and represent proposed movement iconically. There has been no study of the gestures of uninstructed common chimpanzees or orangutans specifically in terms of iconicity, though observers such as Köhler (1925), Crawford (1937), Yerkes (1943) and Van Hooff (1973) discerned the mimetic character of the gestures of the chimpanzees they watched. In their study of young chimpanzees, Tomasello and colleagues have described some chimpanzee gestures that appear to be similar to those we have classified as iconic, but did not discuss that aspect of young chimpanzees' gesturing (Tomasello, Gust, & Frost, 1989, Tomasello et al., 1994) . For untutored gorillas, both wild and captive, any study of gesture other than the anecdotal has been of species-typical actions (Redshaw & Locke, 1976; Schaller, 1976 [1963]; Fossey, 1983).

For both Kubie and Zura, development of novel silent forms of visual communication, including the iconic, may have been promoted by social and physical conditions in their particular group. The presence of Bwana, another mature male, at times may have meant a need for silence. Gorillas in the wild have been observed to suppress normal vocalizations in mating situations when other more dominant males are nearby (Byrne & Whiten, 1990), and chimpanzees have been found to suppress sound in numerous situations (Goodall, 1986). Captive gorillas have been observed to use distractive manual activities to achieve interaction (Mitchell, 1989, 1991). At the San Francisco Zoo, an additional factor was a physical environment that permitted female choice as to proximity with males. During the first several periods of our study, the door to the gorillas' indoor quarters was usually kept open to a narrow width that allowed free entry and exit by the two younger females and infants but was too narrow for the males and the older female to be able to come back inside, once released outdoors in the morning (see Kerr, 1993). Another escape route was the large trees in the enclosure that the females readily climbed, but the mature males rarely did.

It might seem likely that Kubie and Zura developed their usage and comprehension of *tactile close* and other silent gestures through a process of "conventionalization" or "ontogenetic ritualization." This process was posited for chimpanzees by Tomasello (1990) and Tomasello et al. (1989, 1994). A similar progression was described by Savage-Rumbaugh et al. (1977) for pygmy chimpanzees, whose gestures showed two kinds of extensions from the ordinary social action they represented. These were (1) gestures gently touching the partner's body and (2) gestures in space, with both versions of a gesture meant to attain the same goal, such as ventro-ventral sexual positioning. In a study with a young gorilla, Muni, as subject, Gómez (1990) reports a related process. The gorilla first used a human companion as a physical object to climb upon or manipulate in reaching a goal, then later began communicating intent, through gaze and gesture, to the human in order to influence the human to collaborate in reaching the goal (Gómez, 1990). For all these apes, depictions of activity desired from partners may have developed from force or whole body motions into gentle directional touches, then into arm or head motion in space, "shaped" by the responses of the recipient.

For an example of how a conventionalization process might work, let us look at Kubie's frequent gesture *armswing under*, which was usually preceded by the gesture *tap other* (see Figure 11.1; c, d, and e). Kubie would tap Zura without force (Figure 11.1c), presumably gaining her attention as subject of activity (a deictic gesture), then, having her visual attention, would swing his arm toward himself (Figure 11.1d; an iconic depiction of the motion desired from Zura), moving his open palm to a final position between his legs (deictic, Figure 11.1e). This might have developed as follows: when Kubie wanted Zura to come to him he might first have pulled Zura to him; when co-operative, Zura might move toward him if his arm only brushed down her body then towards himself (tactile gesture), then eventually simply the motion of his arm swinging toward himself would carry the same message (visually received gesture in space).

On the other hand, longitudinal data on Kubie's development argue against this "conventionalization" or "ontogenetic ritualization" interpretation. In addition to the studies described earlier in this chapter illustrating the similar repertoire Kubie employed with different individuals in diverse time periods, we are fortunate to have a bit of information about Kubie at an even earlier period. A 20-minute piece of film (Sandra Keller, *Friends of the Zoo*, 1984) shows Kubie at age 8 with the older female Pogo, attempting to get her to raise her bottom for copulation and/or estrus checking. The film has 83 gestures by Kubie, and only 4 by Pogo. With Pogo, Kubie utilized many of the same gesture types (including many *tactile close* gestures) frequently used with the similarly uncooperative Zura five years later. This implies that if conventionalization was the process by which Kubie's gestures were established, he went through a similar process three different times with three different females and later again with his sons.

We have no evidence that this happened; on the contrary, there is evidence that it did not. It is quite possible that Kubie might generalize his gestures to use with other individuals, but, unless he repeated the whole conventionalization process with each new partner, *the other (receptive) half of the communicative equation would not be able to function*; the other females, and later his sons, *would not understand* his gestures. Though Kubie's earliest use of his gestures with each female occurred before the beginning of this study, we did observe Kubie's early usage of gestures with each of his young sons. Gesture types used earlier with females were performed in finished form with the sons. It is unlikely that with our weekly observations we would totally have missed observing some of a presumably extensive shaping process, had it been taking place.

Was observational learning involved on the part of the younger females, then? This also seems unlikely; Zura was not in the group when Kubie originally developed his gestures, though as a youngster she perhaps had opportunity to observe Kubie using gestures with the older female, Pogo. A simpler explanation might be that *comprehension* of the motion depicted in gestures is biologically encoded for primates, thus eliminating the need for conventionalization on the receiver's side of the interchange (Perrett et al., 1989). It is also possible that actions of the signaler other than the gestures help the receiver read his intentions; but then, why gestures?

None of the above explanations for development of iconic and deictic gesture completely satisfies. In Kubie's case as well as that of Savage-Rumbaugh's pygmy chimpanzee subjects and Gomez's young gorilla, it seems that something like the process of conventionalization led not simply to a learning of a gradually more finely shaped association of stimulus and response, but to an understanding of the partner in communication as an intentional and responding being. Whether the receiving partner was a human or another ape, the signaling ape made sure that visual contact was established (except for *tactile close* gestures), and seemed to understand both the other's potential actions and what the partner might, in turn, understand from his (the signaler's) performance of gestures. The *tactile close* gestures performed by Kubie support this explanation; these gestures varied so greatly in type and size and force (being performed in a three-dimensional space on the body of another gorilla) that an individual conventionalization of each gesture would seem an unwieldy process. The understanding of iconic motions on the body and in space may instead be a normal part of great ape development when conditions make such gestures useful. Even the far more complex symbols of human language are often learned, after all, not by a gradual process of shaping or careful teaching and demonstration, but by active integration into contexts of daily action where the experienced language-user and the child, or student, commonly interact. This "untaught" learning of language or symbols systems has been pertinently described for the symbol-adept pygmy chimpanzee, Kanzi (Savage-Rumbaugh, 1984).

Development of gestures over the life of an individual

Though we have no observations of the first emergence of gestures for Kubie because JT's regular observations began after he was a mature adult, we are fortunate to have observations by Parker (1993 and this volume) that tell us about Kubie's development during the first 3 years of his life. Parker's observations of Kubie document his increasing awareness between ages 2 and 2.9 years of the communicative significance of his behaviors, as evidenced by his watching of other animals' responses to his displays and provocations. Parker suggests that trial-and-error learning regarding the relative success of various sorts of play enticements was effective in the building of a repertoire of displays. Kubie's play repertoire as listed by Parker, including behaviors we define here as gestures, is very similar to that of Shango's and Barney's at the same ages. Parker does not report *tactile close* directional gesture for Kubie at this age, nor was it commonly observed in either Shango or Barney at age 1 to 2.3; those few instances tentatively recorded were directed to much larger gorillas where "force" applied by a small gorilla would have been ineffective. Nor were silent directional gestures in space, self-indicating gestures, or gestures indicating location observed in any youngsters during their first two years. After age 3.5 years, Shango began to use some *tactile close* gestures, but has not yet, at age 6, been observed to use *silent visually received* gestures in space nor silent self-indicating deictic gestures as Kubie does. This may imply that a certain level of cognitive development over the course of maturation is necessary for a gorilla to employ iconic and deictic gestures. The ages at which an ape reaches certain levels of cognitive development may, of course, be subject to individual variation (as for mirror use in apes; see Parker, Mitchell, & Boccia 1994), probably depending on a combination of genetics, environmental input, and degree of human influence.

We hope soon to follow Shango through the adolescent "blackback" stage. For now, information for adolescent male gorillas comes from different individuals, but can perhaps still help bridge this age gap in the observations. In two different cases, extensive gesturing by an adolescent male gorilla was directed toward an unco-operative female who was considerably larger and older; this age and size difference was also the case for the pygmy chimpanzees whose gesturing was described by Savage-Rumbaugh et al. (1977). Our gorilla evidence comes from the 1984 film, described above, of adolescent Kubie with the older female Pogo; and from video collected at the Rio Grande Zoo in Albuquerque, New Mexico by the first author (JT) in 1992 (transcribed in detail in Tanner, 1993). The Rio Grande video is a record of a "sneak" mating by a 7-year-old male and a 27-year-old female. Many silent gestures, both *tactile close* and *silent visually received*, were used by the young male and also the older female, and were seemingly of a similar iconic and deictic quality to those recorded in our San Francisco observations, though not identical in type. Gestures by the 7-year-old and the female seemed not only to promote an actual copulation, but also first to serve to elude the 27-year-old silverback male, and then to reach agreement upon a meeting out of sight in the enclosure's deep dry moat.

Only future observation can tell us whether such gestures will appear in the course of Shango's maturity, and whether they will appear only when Shango is confronted with uncooperative females. We have not yet observed extended attempts on Shango's part to interact with Zura or Pogo sexually. So far, much of Shango's play, including the sexual, is directed toward his mother Bawang, who quickly rebuffs most of his approaches or turns them into wrestling play.

Gorillas in the wild, zoo gorillas, signing apes, and humans: differences and similarities in gesture use

All human children introduce nonverbal symbolic activity into object play around age 13 months (Bates, 1979). Deaf children exposed to no sign language create iconic gestures for both objects and actions in order to communicate (Goldin-Meadow & Mylander, 1984). Such activities exceed what has been observed of apes untaught by humans. Symbolic play and creation of iconic gestures for both objects and actions is not, however, beyond apes who have intensive interaction with humans and/or have been taught symbol systems by humans. Why, then, is symbol use for objects, and symbolic play, not found in gorillas in zoos or in the wild? Iconic gestures as representations or simulations of motions and shapes seem closely related to mimetic ability. But the fact that repertoires of non-species-typical gestures were not shared among all members of our zoo study group implies that imitation of particular gestural forms is not a common method of transmission.[5] The "imitation" that takes place in the production of an iconic gesture is a reproduction of previously experienced, and imminently desired, functional body motion. Imitation of another's *gestural* motion is removed from "real" activity, and perhaps this second order of detachment is where untutored apes have difficulty. Imitation of gestures as well as more complex activities are, however, frequent in signing or intensely human reared apes (Hayes & Hayes, 1952; Gardner & Gardner, 1971; Patterson, Tanner, & Mayer, 1988; Tomasello, Savage-Rumbaugh, & Kruger, 1993; Custance & Bard, 1994; Miles, Mitchell, & Harper, 1996). Free-ranging rehabilitant orangutans exposed to human activities and interactions imitate complex sequences of people's activities, even without any encouragement from humans (Russon & Galdikas, 1993). Gorillas in the wild show evidence of the ability to imitate organization of action in the context of complex feeding techniques (Byrne & Byrne, 1993; Byrne & Russon, 1998). But it seems that, for apes to go beyond action-based simulation to imitation of gestural shapes made by others and to representation of object shapes, enculturation with human symbol systems and turn-taking interactions with humans are necessary, and may result in a different kind of learning process (Bard & Vauclair, 1984; Savage-Rumbaugh, 1984; Tomasello et al., 1993; Kranz, 1993; Call & Tomasello, 1996).

Still, symbol-taught apes, in spite of their expanded communicative vocabulary, do not lose, but rather elaborate upon "natural" or species-typical gestures. This fact has been seized upon by some as evidence that apes do not really learn human sign languages, but only perform gestures that they would "naturally" use anyway (Wallman, 1992; Pinker, 1994). This can, instead, be interpreted as an illustration of the universality of the ape and human capacity for iconic expression. The rapid shaking of

[5] One gesture that is not species-typical, though seen at low frequency in some other zoo gorillas (see ethograms in Ogden & Schildkraut 1991), *armshake*, is used by more than one individual in this group; Kubie and Zura both used *armshake* frequently and sometimes in synchrony. In earlier observations when Kubie was an infant, it was noted that *armshake* was performed by every member of the group (Parker 1993, and this volume). We have observed the emergence of armshaking by Barney (though it seems to have faded from his repertoire at the time of writing), but it was not observed in Shango, Bawang, Pogo, or Bwana during the present study.

an arm and hand can express urgent activity for an untutored ape and also can likewise express the concept of "hurry" in American Sign Language. Anyone familiar with signing apes knows that their gestural repertoire is much enlarged compared to untaught apes, not only with ASL signs that are unlike any "natural" ape gesture, but also with spontaneous inventions, many of which are iconic for some aspect of the actions or objects referred to (see Bonvillian & Patterson, 1993, and this volume, for discussion of prevalence of iconic signs in early sign acquisition by two gorillas). In addition to invented signs, taught signs are often modulated and performed consistently in ways that better suit ape hands and anatomy. Some innovations by the signing gorilla Koko even appear to be iconic for sound, involving cross-modal transfer of the sound of an English word for an already known sign, and using that sign for a different concept that has a similar sound in English, such as "knee" for "need," "witch" for "which," "brows" for "browse" (Patterson & Tanner, 1988; Patterson & Gordon, 1993). This is not a result of human reinforcement of chance novelty; these "acoustic errors" and other invented signs were generally not comprehended by Koko's human companions and were discouraged until repeated context made her meaning clear.

Apes trained in symbol systems other than sign language also employ many gestures to express themselves; it would be of great interest if the researchers working with these apes would describe the physical form of these gestures in more detail (Savage-Rumbaugh, 1986; Greenfield & Savage-Rumbaugh, 1990). The fact that symbol-taught (but non-signing) apes invent gestures and signing apes do not slavishly imitate human signs but rather adapt, change, and invent them, even in the face of human discouragement, would seem to be evidence that a process of real cognitive and behavioral assimilation is taking place (Patterson, 1980; Patterson & Linden, 1981; Patterson & Tanner, 1988; Gardner, Gardner, & Nichols, 1989; Patterson & Cohn, 1990; Miles, 1993; Miles et al., 1996).

With our present knowledge of the potential for iconic simulation in apes, we should not negate the possibility of some inventive use of gesture in the wild by gorillas and orangutans, though it might not necessarily be universal in the species or in every population (as in Van Schaik's report of orangutan tool use, this volume). There have been rare but telling observations of gorillas in their native habitats that hint at directional indications through sound and/or gesture. Beating the ground was associated with the starting or turning of group movement in eastern lowland gorillas in Zaire (Mori, 1983). Clapping by female and young lowland gorillas in the Central African Republic apparently indicated location to the silverback (Fay, 1989). In Gabon, a group of gorillas was observed crossing the savanna between two patches of forest. One gorilla, who appeared to be older and to have difficulty in walking, was closely accompanied by another younger gorilla, who sometimes walked backwards facing the older gorilla. When the older gorilla stopped, the younger gorilla faced him and waved one arm up in front him, appearing to urge the older gorilla on with a gesture that humans might interpret as 'come on' (personal communication 1993, C. Tutin & R. Parnell). An arm-lifting gesture by females, possibly related to female transfer, seemed to be directed at other groups in a clearing in the Mbeli Bai, Nouabale-Ndoki National

Park (Fay, 1994). In the Central African Republic, tree-slapping then chestbeating was performed by a female who was up a tree with her infant when she detected a human below, followed by the appearance a minute later of a silverback, who charged the human (personal communication, M. Goldsmith, 1996).

Great variability is now known to exist among gorilla groups in social composition as well as in ecological settings in their native habitats; it is evident that there are many different kinds of problems to be solved in group living, and perhaps as many different kinds of solutions. One of these may be the invention of gesture, as well as other forms of simulation, both in captivity and in the wild, to negotiate social relations and group cohesion.

SUMMARY

All gorillas share some species-typical communicative behaviors; but it seems that, with maturity and the appearance of certain social needs, further gestures may develop in some individuals. Some of these gestures have iconic or deictic properties. Individuals in the study group employed differing repertoires of gesture types which have remained stable over years, though the quantity of gestures used fluctuates over time relative to social situation and stresses in the environment. The fact that some zoo gorillas spontaneously develop iconic and deictic gestures helps explain the potential of gorillas, and all great apes, not only to acquire sign language and the use of other symbol systems with human instruction, but also to create further symbols of their own invention.

ACKNOWLEDGMENTS

We would like to thank Charles Ernest for assistance with camera work and video transcription throughout the span of these observations, and Mary Kerr, gorilla keeper at the San Francisco Zoo, for helpful contextual information throughout the project. For suggestions and comments that have greatly improved this chapter, we thank Bob Mitchell, Sue Parker, and Lyn Miles. And we thank Francine Patterson and the gorillas Koko and Michael for pointing us in the direction that has led to this research.

REFERENCES

Armstrong, D. F., Stokoe, W. C., & Wilcox, S. E. (1995). *Gesture and the nature of language*. Cambridge University Press.

Bates, E. (1979). The emergence of symbols: Cognition and communication in infancy. New York: Academic Press.

Bateson, G. (1968). Redundancy and coding. In T. Sebeok (ed.), *Animal communication*, (pp. 614–626). Bloomington, IN: Indiana University Press.

Bard, K. & Vauclair, J. (1984). The communication context of object manipulation in ape and human adult–infant pairs. *Journal of Human Evolution*, **13**, 181–190.

Boesch, C. (1991). Symbolic communication in wild chimpanzees? *Human Evolution*, **6**, 81–90.

Bonvillian, J. & Patterson, F. (1993). Early sign language acquisition in children and gorillas: vocabulary content and sign iconicity. *First Language*, **13**, 315–338.

Bretherton, I. & Bates, E. (1979). The emergence of intentional communication. In I. C. Užgiris (ed.), *Social interactions and communication during infancy* (pp. 81–100). San Francisco: Jossey-Bass.

Burling, R. (1993). Primate calls, human language, and nonverbal communication. *Current Anthropology*, **34**, 1, 25–53.

Byrne, R. W. & Byrne, J. M. (1993). Variability and standardization in the complex leaf-gathering tasks of mountain gorillas (*Gorilla g. beringei*). *American Journal of Primatology*, **31**, 241–261.

Byrne, R. W. & Russon, A. (1998). Learning by imitation: a hierarchical approach. *Behavioral and Brain Sciences*, **21**.

Byrne, R. W. & Whiten, A. (1990). Tactical deception in primates: the 1990 database. *Primate Report*, **27**, 1–101.

Call, J. & Tomasello, M. (1996). The effect of humans on the cognitive development of apes. In A. E. Russon, K. A. Bard, & S. T. Parker (eds.), *Reaching into thought: the minds of the great apes* (pp. 371–403). Cambridge University Press.

Chevalier-Skolnikoff, S., Galdikas, B., & Skolnikoff, A. Z. (1982). The adaptive significance of higher intelligence in wild orangutans: a preliminary report. *Journal of Human Evolution*, **11**, 639–652.

Crawford, M. P. (1937). The cooperative solving of problems by young chimpanzees. *Comparative Psychology Monograph*, **14**(2), 1–88.

Custance, D. & Bard, K. (1994). The comparative study of self-recognition and imitation: the importance of social factors. In S. T. Parker, R. W. Mitchell, & M. L. Boccia (eds.), *Self-awareness in animals and humans* (pp. 207–226). Cambridge University Press.

De Waal, F. (1982). *Chimpanzee politics*. New York: Harper & Row.

De Waal, F. (1988). The communicative repertoire of captive bonobos (*Pan paniscus*), compared to that of chimpanzees. *Behavior*, **106**(3–4), 183–251.

Donald, M. (1991). *The origins of the modern mind*. Cambridge, MA: Harvard University Press.

Fay, J. M. (1989). Hand-clapping in western lowland gorillas. *Mammalia*, **53**(3), 457–458.

Fay, J. M. (1994). The Nouabale–Ndoki Project, Northern Congo. *Gorilla Conservation News*.

Fossey, D. (1983). *Gorillas in the mist*. Boston: Houghton Mifflin.

Gardner, B. T. & Gardner, R. A. (1971). Two-way communication with an infant chimpanzee. In A. M. Schrier & F. Stollnitz (eds.), *Behavior of nonhuman primates* (pp. 117–184). New York: Academic Press.

Gardner, B. T., Gardner, R. A., & Nichols, S. G. (1989). The shapes and uses of signs in a cross-fostering laboratory. In R. A. Gardner, B. T. Gardner, & T. E. van Cantfort (eds.), *Teaching sign language to chimpanzees* (pp. 55–180). Albany, NY: State University of New York Press.

Gold, K. C. & Ogden, J.J. (1992). The effects of construction noise on captive lowland gorillas. Unpublished presentation, 1992 Gorilla Workshop, Milwaukee, Wisconsin.

Goldin-Meadow, S. & Mylander, C. (1984). Gestural communication in deaf children: the effects and noneffects of parental input on early language development. *Monographs of the Society for Research in Child Development*, **207**(49), Nos. 3–4.

Gomez, J. C. (1990). The emergence of intentional communication as a problem-solving strategy in the gorilla. In S. T. Parker & K. R. Gibson (eds.), *"Language" and intelligence in monkeys and apes* (pp. 333–355). Cambridge University Press.

Goodall, J. Van Lawick- (1968a). *The behaviour of free-living chimpanzees in the Gombe Stream Reserve. Animal Behavior Monographs*, **1**(3), 161–311.

Goodall, J. Van Lawick- (1968b). A preliminary report on expressive movements and communication in the Gombe Stream chimpanzees. In P. C. Jay (eds.), *Primates: Studies in adaptation and variability* (pp. 313–374). New York: Holt, Rinehart and Winston.

Goodall, J. Van Lawick- (1986). *The chimpanzees of Gombe*. Cambridge, MA: Harvard University Press.

Gopnik, A. (1982). Words and plans: Early language and the development of intelligent action. *Journal of Child Language*, **9**, 303–318.

Greenfield, P. M. & Savage-Rumbaugh, E. S. (1990). Grammatical combination in *Pan paniscus*: processes of learning and invention in the evolution and development of language. In S. T. Parker & K. R. Gibson (eds.), *"Language" and intelligence in monkeys and apes* (pp. 540–578). Cambridge University Press.

Hayes, K. J. & Hayes, C. (1952). Imitation in a home-raised chimpanzee. *Journal of Comparative and Physiological Psychology*, **45**, 450–459.

Ingmanson, E. J. (1987). Clapping behavior: non-verbal communication during grooming in a group of captive pygmy chimpanzees. *American Journal of Physical Anthropology*, **72**, 214.

Ingmanson, E. J. (1996). Tool-using behavior in wild Pan paniscus: Social and ecological considerations. In A. E. Russon, K. A. Bard, & S.T. Parker (eds.), *Reaching into thought: the minds of the great apes* (pp. 190–210). Cambridge University Press.

Kerr, M. (1993). The captive management and social dynamics of a two silverback gorilla group. Unpublished Masters' thesis, San Francisco State University.

Klima, E., & Bellugi, U. (1979). *The Signs of Language*. Cambridge, Massachusetts: Harvard University Press.

Köhler, W. (1925). *The mentality of apes*. London: Routledge and Kegan Paul.

Kranz, L. (1993). Communicating with Ndume. *Journal of the Gorilla Foundation*, **17**(1), 5.

Miles, H. L. (1993). Language and the orang-utan: the old 'person' of the forest. In P. Cavalieri and P. Singer (eds.), *The great ape project* (pp. 42–57). London: Fourth Estate.

Miles, H. L. & Harper, S. E. (1994). "Ape language" studies and the study of human language origins. In D. Quiatt and J. Itani (eds.), *Hominid culture in primate perspective* (pp.253–78). Niwot, CO: University Press of Colorado.

Miles, H. L., Mitchell, R. W., & Harper, S. E. (1996). Simon says: the development of imitation in an enculturated orangutan. In A. E. Russon, K. A. Bard, & S. T. Parker (eds.), *Reaching into thought: the minds of the great apes* (pp. 278–299). Cambridge University Press.

Mitchell, R. W. (1989). Functions and social consequences of infant–adult male interaction in a captive group of lowland gorillas. *Zoo Biology*, **8**, 125–137.

Mitchell, R. W. (1991). Deception and hiding in captive lowland gorillas. *Primates*, **32**, 23–527.

Mitchell, R. W. (1994). The evolution of primate cognition: simulation, self-knowledge, and knowledge of other minds. In D. Quiatt & J. Itani (eds.), *Hominid culture in primate perspective* (pp. 177–232). Niwot: University Press of Colorado.

Morbeck, M. E. (1994). Object manipulation, gestures, posture, and locomotion. In D. Quiatt & J. Itani (eds.), *Hominid culture in primate perspective* (pp. 117–135). Niwot: University Press of Colorado.

Mori, A. (1983). Comparison of the communicative vocalizations and behaviors of group ranging in Eastern gorillas, chimpanzees and pygmy chimpanzees. *Primates*, **24**(4), 486–500.

Ogden, J. & Schildkraut, D. (eds.) (1991). *Compilation of gorilla ethograms*. Published by Gorilla Behavior Advisory Group, affiliated with the Gorilla Species Survival Plan.

Parker, S. (1993). The ontogeny of social play in an infant gorilla: a comparative cognitive perspective. Unpublished presentation, Animal Behavior Society Meeting, University of California at Davis, July 1993.

Parker, S. T., Mitchell, R. W., & Boccia M. L. (eds.) (1994). *Self-awareness in animals and humans.* Cambridge University Press.

Patterson, F. (1980). Innovative uses of language by a gorilla: a case study. In K. E. Nelson (ed.), *Children's Language* (vol. 2, pp. 497–561). New York: Gardner Press.

Patterson, F. & Cohn, R. (1990). Language acquisition by a lowland gorilla: Koko's first ten years of vocabulary development. *Word*, **41**(3), 97–143.

Patterson, F. & Gordon, W. (1993). The case for the personhood of gorillas. In Cavalieri, P. & Singer, P. (eds.), *The great ape project* (pp. 58–7)7. London: The Fourth Estate.

Patterson, F. & Linden, E. (1981). *The education of Koko.* New York: Holt, Rinehart and Winston.

Patterson, F. & Tanner, J. (1988). Gorilla gestural communication. *The Journal of the Gorilla Foundation*, **12**(1), 2–5.

Patterson, F., Tanner, J., & Mayer, N. (1988). Pragmatic analysis of gorilla utterances: early communicative development in the gorilla Koko. *Journal of Pragmatics*, **12**, 35–54.

Pellis, S. & Pellis V. (1996). On knowing it's only play: the role of play signals in play fighting. *Aggression and Violent Behavior*, **1**(3), 249–268.

Perrett, D. I., Harries, M. H., Bevan, R., Thomas, S., Benson, P. J., Mistlin, A. J., Chitty, A. J., Hietanen, J. K., & Ortega, J. E. (1989). Frameworks of analysis for the neural representations of animate objects and actions. *Journal of Experimental Biology*, **146**, 87–113.

Pinker, S. (1994). *The language instinct.* New York: Morrow and Company.

Plooij, F. X. (1984). *The behavioral development of free-living chimpanzee babies and infants.* Norwood, NJ: Ablex Publishing.

Povinelli, D. J. & Cant (1995). Arboreal clambering and the evolution of self-conception. *Quarterly Review of Biology*, **70**(4), 393–421.

Redshaw, M. & Locke, K. (1976). The development of play and social behaviour in two lowland gorilla infants. *Dodo: The Journal of the Jersey Wildlife Preservation Trust, 13th Annual Report*, 71–86.

Russon, A. & Galdikas, B. (1993). Imitation in free-ranging rehabilitant orangutans (*Pongo pygmaeus*). *Journal of Comparative Psychology*, **107**, 147–161.

Savage-Rumbaugh, E. S. (1984). *Pan paniscus* and *Pan troglodytes*: contrasts in preverbal communicative competence. In R. Susman (ed.), *The pygmy chimpanzee: Evolutionary biology and behavior* (pp. 395–413). New York: Plenum Press.

Savage-Rumbaugh, E. S. (1986). *Ape language: From conditioned response to symbol.* Oxford University Press.

Savage-Rumbaugh, E. S., Wilkerson, B., & Bakeman, R. (1977). Spontaneous gestural communication among conspecifics in the pygmy chimpanzee (*Pan paniscus*). In G. Bourne (ed.), *Progress in ape research* (pp. 97–116). New York: Academic Press.

Savage-Rumbaugh, E. S., Williams, S., Furuichi, T., & Kano, T. (1996). Language perceived: paniscus branches out. In W. C. McGrew, L. Marchant, & T. Nishida (eds.), *Great ape societies* (pp. 173–184). Cambridge University Press.

Savage, E. S. & Rumbaugh, D. M. (1977) Communication, language, and Lana: a perspective. In D. M. Rumbaugh (ed.), *Language learning by a chimpanzee* (pp. 287–309). New York: Academic Press.

Schaller, G. (1976 [1963]). *The mountain gorilla: Ecology and behavior.* University of Chicago Press.

Stokoe, W., Casterline, D., & Croneberg, C. (1965). *A dictionary of American Sign Language on linguistic principles.* Washington DC: Gallaudet College Press.

Tanner, J. E. (1993). Unpublished first year Ph.D. report, University of St. Andrews, Scotland.

Tanner, J. E. & Byrne, R. W. (1993). Concealing facial evidence of mood: perspective-taking in a captive gorilla? *Primates*, **34**(4), 451–457.

Tanner, J. E. & Byrne, R. W. (1996). Representation of action through iconic gesture in a captive lowland gorilla. *Current Anthropology*, **37**(1), 162–173.

Thompson, J. M. (1993). Cultural diversity in the behavior of Pan. In D. Quiatt and J. Itani (eds.), *Hominid culture in primate perspective* (pp. 95–115). Niwot, CO: University Press of Colorado.

Tomasello, M. (1990). Cultural transmission in the tool use and communicatory signaling of chimpanzees?. In S. T. Parker & K. R. Gibson, *"Language" and intelligence in monkeys and apes* (pp. 274–311). Cambridge University Press.

Tomasello, M., Gust., D., & Frost, G. T. (1989). A longitudinal investigation of gestural communication in young chimpanzees. *Primates*, **30**(1), 35–50.

Tomasello, M., Savage-Rumbaugh, E. S., & Kruger, A. C. (1993). Imitative learning of actions on objects by children, chimpanzees, and enculturated chimpanzees. *Child Development*, **64**, 1688–1705.

Tomasello, M., Call, J., Nagell, K., Olguin, R. & Carpenter, M. (1994). The learning and use of gestural signals by young chimpanzees: a transgenerational study. *Primates*, **35**(2), 137–154.

Van Hooff, J. (1973). A structural analysis of the social behavior of a semi-captive group of chimpanzees. In M. von Cranach and I. Vine (eds.), *Expressive movement and non-verbal communication* (pp. 75–162). London: Academic Press.

Wallman, J. (1992). *Aping language*. New York: Cambridge University Press.

Yerkes, R. M. (1943). *Chimpanzees: A laboratory colony*. New Haven: Yale University Press.

12

Early sign-language acquisition: comparisons between children and gorillas

JOHN D. BONVILLIAN AND FRANCINE G. P. PATTERSON

In recent years, claims about the acquisition of sign language in chimpanzees, gorillas, and orangutans have been shrouded in controversy. On one side of this controversy have been researchers who have reported impressive language abilities in these species (Fouts, 1973; Gardner & Gardner, 1974; Patterson, 1978, 1980; Miles, 1983, 1990; Gardner, Gardner, & Van Cantfort, 1989; Patterson & Cohn, 1990). On the other side have been investigators who have dismissed these claims as largely unfounded (Seidenberg & Petitto, 1979; Sebeok & Umiker-Sebeok, 1980; Sebeok & Rosenthal, 1981; Terrace, 1985). Throughout this controversy, great effort often has been expended trying to answer the question "Do apes have language?" as if this question could be resolved unequivocally "yes" or "no." A much more fruitful approach would be to devote greater effort determining more precisely the particular language and communicative capabilities of these species (Greenfield & Savage-Rumbaugh, 1990; Miles, 1990, 1997). An important early step in this approach to the issue of language in nonhuman primates would be to focus on how sign-language learning apes and humans both resemble and differ from each other (Rimpau, Gardner, & Gardner, 1989).

In most of the cross-species comparisons of language acquisition that have been conducted, the gestural or sign-language productions of apes were contrasted only with the spoken-language productions of young children. A major shortcoming with this approach is that it is not clear what conclusions can be drawn from such comparisons. Any differences that were observed could be attributed to differences between the species, to differences between the languages, or to both. A more appropriate approach would be to include comparisons of the apes' early sign usage with that of children learning to sign. A critical limiting factor in conducting such comparisons, however, has been that until relatively recently little was known about the course of early sign-language acquisition in young children.

We believe that the present investigation represents an important step in rectifying this problem in cross-species comparisons. In this investigation, data from two longitudinal studies of early sign-language development in young children are compared systematically with findings from an ongoing investigation of sign acquisition in two lowland gorillas. By making these comparisons, we are able to determine how much gorillas and children resemble each other in their rates of sign learning, vocabulary content, nonsign gestural communication, and referential sign usage. Furthermore, we hope that the information we obtain on these important aspects of early language and communicative development will lead to a greater understanding both of apes' communicative abilities and of language acquisition more generally.

The young children who participated in the two longitudinal studies of sign-language acquisition had parents who reported that American Sign Language (ASL) was the principal language used in their homes. (ASL is the primary means of communication among most individuals who become deaf prelingually in the United States.) Children, deaf or hearing, of ASL-using deaf parents typically acquire ASL as their first or native language. Only in the past several decades has ASL been the focus of systematic investigation by linguists. These investigations revealed that ASL has an extensive vocabulary or lexicon, a set of rules that guide the formation of signs, a rich system of morphological processes, and a variety of sophisticated ways to convey meaning (Stokoe, 1960; Stokoe, Casterline, & Croneberg, 1965; Klima & Bellugi, 1979; Wilbur, 1987). Today, most linguists accord ASL "full membership in the family of languages" (Brown, 1977, p. 9). The linguistic structure of ASL, however, is quite different from that of English and most other auditory–vocal languages.

Most hearing persons in this country who learn to sign, unless they grow up amidst deaf friends or relatives, never fully master ASL. Rather these hearing individuals typically acquire a vocabulary of ASL signs and, when communicating manually, put these signs into English word order. This mixture of ASL and English structure often is referred to as pidgin sign English or PSE. In the present investigation, both the gorillas and the young children were exposed to ASL and to PSE, although there was undoubtedly wide variation among the participants in the amount of their exposure to each system. Before beginning our account of the sign-language and communicative development of the children and gorillas, however, we first need to set the stage by providing important background information about each of the research programs.

THE RESEARCH PROGRAMS

The present comparative investigation rests on the findings of two substantial programs of research on sign-language acquisition. Project Koko chronicles the development of two lowland gorillas over a period of more than two decades. In the second research program, a total of 22 young children of deaf parents were studied in a pair of longitudinal investigations of sign-language acquisition (Bonvillian, Orlansky, Novack, & Folven, 1983b; Folven & Bonvillian, 1991). The similarities in methodology across the two research programs provided a rare opportunity to conduct systematic cross-species comparisons.

The gorillas

Project Koko was begun in 1972 when Koko was one year old. A half year earlier Koko had been removed from the lowland gorilla group at the San Francisco Zoo.

This action was undertaken because Koko was seriously ill and malnourished; concerns had been expressed about her survival. After receiving medical treatment, Koko was placed with a human family. During the several months she spent with this family, Koko regained her health. In their interactions with her, the members of this family used spoken English. Koko has continued to be exposed to English speech on a daily basis ever since.

Francine Patterson (FGPP) began her work with Koko at the San Francisco Children's Zoo just after Koko's first birthday. For the next 11 months, FGPP typically spent 5 hours each day interacting with Koko. When Koko was 23 months old, the project was relocated to a 5-room, 10' × 50' house trailer on the Stanford University campus. This relocation enabled Koko's interaction with human caregivers and her sign-language input to be increased to between 8 and 12 hours each day. Moreover, individuals fluent in ASL (deaf and hearing sons and daughters of deaf parents) joined the project. For the next several years, Koko received between 10 and 20 hours of ASL input and teaching from these individuals each week.

In September 1976, the research program was expanded with the addition of a young male lowland gorilla named Michael. Michael was born in Africa, but spent most of his life prior to joining Project Koko in a zoo-like environment in Austria. There he received most of his care from a German-speaking woman. After joining Project Koko at the estimated age of 3.5 years, Michael was immersed in a language environment that consisted primarily of ASL signs and spoken English.

Koko and Michael were raised in an environment that was similar in many ways to that in which human children are reared. For most of their early upbringing, Koko and Michael lived in a furnished two-bedroom trailer. The gorillas had a host of toys to play with, frequent outside exercise, occasional travel excursions, various pets, and extensive daily human contact.

Written diary accounts of Koko's and Michael's development were kept on a daily basis from the time each entered the project. These records detailed the signs the gorillas produced as well as information on the context of their sign usage. These written accounts also included descriptions of other aspects of the gorillas' development, including observations of their play and social behavior, physical growth, cognitive skills, motor development, and health. These written accounts were supplemented by periodic film, audiotape, and videotape records. Reliability of sign records was ensured by having two independent observers maintain simultaneous accounts of the gorillas' sign productions. The records of the two observers were compared regularly to determine any disagreements. These disagreements were then discussed and decisions made about how to code future instances of such signs or gestures. Over the course of the project, new observers were trained until they achieved a high level of coding agreement with that of a current observer.

Human participants

The sign-language acquisition of 22 young children (8 boys, 14 girls) was examined in a pair of longitudinal investigations. Most of the children began their participation prior to their first birthday. Two of the children were congenitally deaf; there were no other discernible disabilities among the children. For 17 of the children, both their parents were deaf; the remaining 5 had 1 deaf and 1 hearing parent. With a single exception, the hearing parents were actively involved in both deaf education and the deaf community. Most of the parents had graduated from state schools for the deaf and had attended Gallaudet University, an institution of higher learning for deaf students.

All of the parents reported that the principal means of communication in their family was a manual one, typically ASL. The hearing parents and some of the deaf parents supplemented their sign-language input by speaking to their children. There was, however, considerable variation in the amount, quality, and intelligibility of the deaf parents' speech. The 20 hearing children also were exposed to the spoken English of hearing relatives, friends, babysitters, television shows, and, in some cases, older hearing siblings. Despite these wide individual differences in exposure to English speech, all the children received extensive ASL input.

The young children with deaf parents were visited in their homes about once every 5 to 6 weeks for periods ranging from 5 months to 2 years. Each visit lasted about 1 hour. During these visits, videotape records were made of each child's signs, parent–child interactive play, and parent–child communicative exchanges. Each set of parents also kept diary records of their child's sign-language development. In these accounts, the parents recorded the English gloss (or translation equivalent) of each newly acquired sign their child produced, the date each new sign was first produced, how their child formed or pronounced each new sign, and the situational context of each newly acquired sign. During each home visit, an adult experimenter reviewed the previous month's diary records with the parents. These reviews both helped to inform the parents about the type of detailed information the researchers were seeking and to clear-up any ambiguities in the diary accounts.

Research program differences

Although the upbringing of the gorillas and human children was similar in many ways, there were also important differences in their environments. One difference was in the stability of caregivers. All the children received their primary care from their parents; relatives, friends, and babysitters helped out periodically. Several of the children also participated in organized daycare programs. For the gorillas, one adult human female (FGPP) provided much of the primary care from the time Koko was one year old and Michael 3.5 years old. A large number of both voluntary and paid assistants, however, also were involved in the care and teaching of the gorillas. The length of participation and degree of involvement of these assistants varied widely. Some worked with Koko and Michael regularly for years; others were involved for only several hours each week over a period of several months. The most noteworthy difference in personal histories across the two research programs belonged to Michael. His movement from Africa to Europe to the United States and the accompanying interruption in his caretaking clearly made his the most disruptive personal background of any of the participants in either program. Furthermore, because Michael did not join Project Koko until he was about age 3.5, it was not possible to include Michael in a number of the cross-species comparisons of early development.

The linguistic environments of the two species also differed in various ways. Whereas the deaf parents typically were highly fluent ASL users, there was a greater range in the signing skills of those individuals who worked with Koko and Michael. While some were highly fluent ASL users, other assistants had much more limited skills. Moreover, many of those individuals who worked with Koko and Michael combined ASL signs and spoken English. This resulted in simultaneous bimodal (sign and speech) input. In this simultaneous input, English word order was followed. Although both Koko and Michael received considerable ASL input and interaction over the years from fluent signers, they probably received less ASL-only input than the children did during their early years.

Because the children of deaf parents received mostly ASL input, and Koko and Michael were exposed more to ASL signs in English word order (i.e., PSE), one might expect that these input differences would have an impact on early sign production. The likelihood of such an outcome, however, is lessened in light of previous observations about children's ASL acquisition (Newport & Meier, 1985). ASL-learning children begin their vocabulary development by producing lexical items in ASL citation form and do not add inflectional or derivational morphemes to their signs until much later. Thus, for the cross-species comparisons in early vocabulary development undertaken in the present study, differences between ASL and PSE input probably had only a minimal effect. Any effects of input differences would be more likely to emerge once signs were being combined into multisign utterances.

Finally, in comparing the early sign-language productions of Koko and Michael to those of the children, it is important to remember that the ages at which the gorillas were first introduced to signs differed considerably from that of the children. Koko's sign language input did not begin until she was slightly over 1 year old (53 weeks) and Michael's not until he was about 3.5 years old. In contrast, the deaf parents reported that they typically began to sign to their children in their early infancy. The parents also reported that they would occasionally mold their children's hands into signs beginning in their first year. So, at the age that Koko was initially being introduced to signs, the children already had been exposed to signing for a number of months. The reason we underline this point of age differences at sign introduction is that, although maturational age has been shown to be a very important factor in determining language onset and early development (De Hirsch, Jansky, & Langford, 1964; Minkowsi et al., 1966; Landau & Gleitman, cited in Meier & Newport, 1990), exposure time and experience in using language also are important factors in language development (Menyuk et al., 1991). Lastly, it is important to recognize that Michael's absence of any sign input and relatively little close contact with humans until he was about 3.5 years of age may have had an adverse effect on his sign-language acquisition. The level of sign-language mastery that one achieves may be dependent, in part, on the early interaction one has with other language-users (Mayberry, 1994).

EARLY VOCABULARY DEVELOPMENT

In this section, we examine the early sign-language vocabulary development of both the 2 lowland gorillas and the 22 children of deaf parents. In conducting this examination, we focus primarily on the rate that new signs were learned and on vocabulary content. In many of our analyses, the gorillas' and the children's sign-acquisition data also are compared with the findings on early vocabulary development reported by Nelson (1973). In Nelson's study, 18 children with normal hearing were studied longitudinally as they learned to speak English. Following Nelson's approach, we selected two measures of early expressive language development for particular emphasis: the 10–word (or sign) vocabulary and the 50–word (or sign) vocabulary. These two points in early lexical development were selected because they capture important aspects of initial vocabulary acquisition, can be determined reliably, and have been used frequently in other investigations of early language development.

Rate of early vocabulary acquisition

Almost from the beginning of the project, Koko responded to the sign input she received. Her responsiveness necessitated establishing specific acquisition criteria to help in the evaluation of her performance. Unfortunately, there are not widely accepted criteria as to what constitutes the acquisition of a word or a sign. In most of their reports on gorilla sign-language development, Patterson and her associates used two different sets of criteria to decide what constituted the acquisition of a lexical item. In one approach, the *Patterson* criteria, a sign was considered to be acquired only if it was observed and recorded by two different observers and used spontaneously and contextually appropriately on at least half the days for a period of a month. In the second index of gorilla vocabulary development, the *Emitted* criteria, each use of a different sign that was produced both spontaneously and in an appropriate context was accepted as a new vocabulary entry. This latter approach more closely resembles the approach used in studies of young children's vocabulary development. The more conservative Patterson criteria were employed as a partial check against overattribution of lexical skills to the gorillas.

It should be noted that both these measures of lexical acquisition count only those instances where a sign was used spontaneously in a contextually appropriate manner. Koko and Michael, however, frequently would use a sign in a contextually appropriate way at one time during a day and then use the same sign in a contextually inappropriate manner at another time that same day. Of course, young children also do this to some extent in their early word or sign usage.

As early as the second week of the project, Koko began to demonstrate that she had some notion of the role of signs. In these early instances, volunteers reported that Koko would make approximations to the signs FOOD/EAT and DRINK. (By convention, English glosses of ASL signs are printed in capital letters.) By one month into the project, and after only a few days of formal sign training, Koko consistently made a close approximation of the sign FOOD/EAT when she was offered bits of fruit. Within a couple of months of the project's commencement, Koko was making a number of gestures that closely resembled ASL signs. These gestures included recognizable versions of such ASL signs as FOOD/EAT, DRINK, MORE, TOOTHBRUSH, DOG, HAT, and OUT. Koko often would generate these items on a daily basis, but frequently not in contextually appropriate situations. For example, Koko, upon being shown a banana, might first reach for the fruit and, when that overture was not successful, then make gestures that resembled the signs DRINK and MORE. When those gestures did not result in her being given the desired banana, she might then produce her FOOD/EAT sign and be rewarded with a piece of banana.

From early on in the project, it seemed as if Koko had learned the routine that, if she generated certain gestures that were in her repertoire, then the desired outcome might be forthcoming. By her third month of sign-language input and instruction, Koko's sign interactional skills had progressed to where she would often reply with her "signs" when asked simple questions. Overall, during the first months of the project, Koko gradually increased the number of gestures she produced that formationally resembled signs. These initial vocabulary items, however, often were not used in a consistent, systematic manner. Rather, Koko would generate a series of recognizable signs or gestures until she successfully attained what she desired. To some extent, Koko may have learned to produce these "signs" or gestures in a tool-like manner to help her obtain desired objects and actions. (See Miles, 1986, for observations about the orangutan Chantek's sign learning. Chantek apparently made associations between signs and their referents or outcomes. In some cases, Chantek would consistently use a sign to obtain a goal while seemingly being unaware that his sign–referent association was quite different from that of his caregivers.)

Although Koko emitted her 10th different sign at 16 months, many of these "signs" were not used spontaneously in a consistent manner. In comparison, the sign-learning children emitted their 10th different sign at a younger mean age (13.3 months) and their usage appeared to be more consistent than Koko's. Only gradually over her first year of training did Koko become relatively consistent in her application of the appropriate signs to objects or actions. Moreover, some of her contextually inappropriate uses of signs may have reflected Koko's incomplete knowledge of the concept boundaries or defining features of certain signs. Much like young children learning to speak or to sign, Koko often overextended or overgeneralized signs. It should be pointed out, though, that some of Koko's early sign-like gestures (e.g., her gestural requests to be lifted up or to be tickled) appeared to be used in a contextually appropriate manner from the first weeks of the project onward.

By the time Koko was 25 months old – after 13 months of sign-language input and instruction – she had produced 50 different signs (Emitted criteria). In contrast, children learning to speak or to sign typically reach this milestone 6 months earlier (Rescorla, 1980; Bonvillian et al., 1983a). Thus, Koko's age at acquisition of early vocabulary milestones was considerably later than that of most signing or speaking children. Using the more conservative Patterson criteria, Koko progressed even more slowly: she produced her 10th different sign at 28 months and her 50th different sign at 35 months.

An additional concern about recording Koko's (and Michael's) sign productions was that some of her signs might be the product of cues or prompts from her human caregivers. In such instances, Koko's signs would not be spontaneously produced but be dependent on extrinsic factors. Because of these concerns, an additional check on Koko's lexical development was conducted: double-blind testing of many of her vocabulary items. In these tests Koko was able to see the test objects, but the person recording her responses was not able to see these objects. This was accomplished by having one examiner put a test object in a plywood box with a plexiglass door so that Koko could see it and by having the assistant recording Koko's response stand behind the box. This approach eliminated possible cueing or prompting of Koko by the recorder. This approach also reduced any bias to overinterpret any rough approximation of a sign as a correct response. Koko's responses to the double-blind test items were correct, on average, on about 60 percent of the trials. Koko thus produced the correct sign label for objects from among the many dozens of signs in her lexicon most of the time. Although this level of performance indicates considerable systematicity in her sign naming, it also raises concerns as to why she did not achieve a higher score. Presumably, most young children would earn a higher score, but this must remain conjecture as equivalent double-blind testing has not been carried out with children. It should also be pointed out that the double-blind testing often failed to elicit Koko's co-operation. When that happened, Koko would miss a number of items that she had previously identified correctly. Koko's frequent failure to co-operate with the testing probably resulted in her percentage correct score not reflecting her true vocabulary knowledge.

Koko's rate of acquisition of new vocabulary items accelerated when she was 2.5 years old. This period of rapid vocabulary growth, consisting of about 200 new signs per year (Emitted criteria), continued until she was about 4.5 years of age. In children learning to speak, a vocabulary growth spurt typically has its onset in the latter half of the second year (Bloom, Lifter, & Broughton, 1985). Koko's primary vocabulary growth spurt thus appears to have a later onset date than that reported for children. Children learning to speak also typically acquire many more than 200 different words per year during their preschool years (Owens, 1996).

As we noted earlier, Michael's introduction to the project did not occur until he was about 3.5 years old. When Koko was this age, she already had an extensive sign vocabulary. Koko also had spent most of her life in close contact with human caregivers in a more family-like environment; this experience may have facilitated her sign learning. Furthermore, Michael's age of 3.5 years at his introduction to signing coincided with the midpoint in Koko's primary sign-acquisition spurt. From this, one might surmise that Michael might be either maturationally primed to start learning signs rapidly or that he might have nearly missed a principal period for sign learning (or both). Regardless, after joining the project, Michael began acquiring signs. Many of these signs met both the Emitted and Patterson criteria.

Despite Michael's more advanced age at his introduction to signs, Koko's rate of vocabulary development generally was ahead of Michael's development. A single exception to this trend was that Michael learned more signs according to the Patterson criteria during his first year of being taught to sign than did Koko during the equivalent period. More specifically, Michael acquired 18 signs according to the Patterson criteria in his first year of sign input to Koko's 6 during her first 12 months of instruction. Koko's rate of acquisition of signs (according to the Patterson criteria), however, increased during her second year of sign instruction; she surpassed Michael on this dimension after each had received two years of sign input. Koko, moreover, from the beginning of her instruction, generated more signs that met the Emitted criteria than did Michael. After one year of sign input, Koko had emitted nearly twice as many signs as Michael had in his initial year of being taught to sign. Furthermore, it took Michael 18 months to achieve a vocabulary of 50 different emitted signs, a milestone achieved by Koko in 13 months. Koko has continued to maintain her lead over Michael in vocabulary development in the years that have followed.

Because of the gorillas' relatively late introduction to signs in comparison with that of the children of deaf parents, a focus solely on age of attainment probably is not a fair measure of the gorillas' rate of sign acquisition. For this reason, we elected to calculate vocabulary acquisition rates using a method devised by Nelson (1973). In this approach, a vocabulary acquisition rate per month was determined for each participant based on the number of months between that individual's acquisition of a 10–item and a 50–item lexicon. For the children, there was a mean acquisition rate of 7.8 signs per month, with the range extending from a child who acquired only 3.6 new signs per month to one who acquired 13.3 new signs per month (Bonvillian et al., 1983a). The gorillas, for the period between emitting their 10th and 50th signs, had a mean acquisition rate of 4.0 signs per month (Bonvillian & Patterson, 1993). (Koko acquired new signs at a mean rate of 4.4 per month, whereas Michael's rate was 3.6 new signs per month.) By using Nelson's approach to correct for the gorillas' later start, the two species' early sign-acquisition rates did not appear to differ nearly as greatly. The gorillas typically progressed more slowly than the children, but this mean difference in acquisition rates should not obscure the overlap between the species in individual participants' rates of vocabulary acquisition.

Cross-species comparisons of vocabulary content

The content of Koko's and Michael's early lexicons was quite similar to that of children learning to sign and children learning to speak. In arriving at this conclusion, the English glosses of the sign-vocabulary items of the young signing children and the gorillas were compared with each other and with the records provided by Nelson (1973) for young children learning to speak. Moreover, this conclusion held both for analyses conducted at the 10-item and 50-item vocabulary levels. This resemblance can be seen in Table 12.1. In this Table, the children's and the gorillas' sign vocabularies are classified into different grammatical categories according to the criteria established by Nelson.

The similarity between the early signing of the children and gorillas is particularly striking. Perusal of the Table reveals that the rank order of the frequencies of entries in

Table 12.1 *Early grammatical categorization: Comparisons of mean percentages of children's and gorillas' 10-item and 50-item vocabularies by category in three studies*

Categories used by Nelson	Examples	Speech data from Nelson		Children's sign-acquisition data		Koko's and Michael's signs: Emitted criteria		Koko's and Michael's signs: Patterson criteria	
		10-word	50-word	10-sign	50-sign	10-sign	50-sign	10-sign	50-sign
Nominals									
Specific	Mommy, Mike	24	14	14	7	5	5	0	4
General	apple, ball, girl	41	51	54	60	55	53	50	62
Actions	go, sleep	16	13	19	13	35	23	40	20
Modifiers	big, clean, mine	8	9	9	12	5	15	10	12
Personal–social	thank-you, no	5	8	3	6	0	4	0	2
Functions	for, what	6	4	0	1	0	0	0	0

the different grammatical categories for the gorillas and the signing children was identical at the 50-item level. Moreover, this finding held for rankings of the gorillas' signs regardless of whether the classifications were determined from signs that met the Emitted or Patterson criteria. Considerable similarity across species was also evident at the 10–sign vocabulary level.

The one grammatical category scored by Nelson that was not represented in either Koko's or Michael's sign lexicons was function items. Instances of this category also were quite rare in the vocabularies of the signing children, accounting for only one percent of all the entries in their 50-sign vocabularies. This very low incidence of function items (or functors) in both signing children and gorillas may reflect inherent differences between ASL and English. Functors typically are not specifically signed in ASL. Instead, the structural or relational information that function words convey in English utterances is often incorporated into the gestural production of individual content-type signs in ASL. (This is often accomplished by a signer through the placement and movement of signs on an imaginary visual stage.) Moreover, the few signs (e.g., WHERE, WHAT) generated by the children that were classified as function items according to Nelson's criteria were used by the children as interrogatives. These signs did not play a structural or relational role as is the case for most function words in English and other spoken languages.

The resemblance in vocabulary between the young signing children and the gorillas also is evident in the overlap of specific lexical entries. The majority of the items present in both Koko's and Michael's 50-sign lexicons appared as well in the 50-sign vocabularies of the children (Bonvillian & Patterson, 1993). Moreover, this pattern held when the comparisons were conducted on both the gorillas' signs that met the Patterson criteria and those that met the Emitted criteria. Of the signs that appeared in the gorillas' vocabularies and not in the children's, many consisted of specific nominals (e.g., PENNY), names of preferred foods (e.g., NUT), and favored activities of the gorillas (e.g., CHASE). The latter differences, moreover, are probably more indicative of species' differences in dietary and activity preferences than in sign-learning abilities. Even with these exceptions included, the overall pattern that emerges is one of considerable resemblance across species in specific vocabulary entries.

How might one account for this striking overlap in lexical entries between the gorillas and children? One explanation would be to underline the similarities in their physical environments. Both the children and the gorillas were reared in homes or home-like settings; typically they were surrounded by furniture, toys, pets, appliances, and people. Additional resemblance in their environments was ensured by participation in similar activities: walks, meals, bike and car rides, caring for pets, stories, and bed-time experiences. From this approach, the overlap in vocabulary may be largely the product of commonality in items and activities across environments. A second explanation would be to underscore the importance of language input in early vocabulary development. Because adult human caretakers were the source of language input and instruction for all participants, the overlap also might be accounted for by the caretakers' frequent labeling of the same objects, activities, and properties. A third explanation would be to emphasize similarities in the language acquisition and cognitive mechanisms across individual participants and across species. Clearly, many of the same items, activities, and properties appeared to stand out from others in their environments for both the young children and gorillas: toys were acted upon and disappeared, caretakers comforted and chastised them, foods were too cold or too hot, etc. And it is these aspects of the environment that might be deemed worthy of being named or requested by child and gorilla alike. Thus, some of the agreement in lexical entries might be attributable to similarities in internal processing mechanisms across the language learners. Unfortunately, the structure of the language-acquisition studies does not allow for the acceptance or rejection of any of these explanations. Moreover, all three factors – similarities in physical environments, language input, and internal processing mechanisms – probably contributed to the overlap in vocabulary development.

ICONICITY AND SIGN ACQUISITION

Contrary to popular belief, most signs in the sign languages used by deaf persons are not pantomimic. Rather, the meanings associated with most signs are primarily dependent on conventions of usage within the deaf community. In ASL, there is substantial evidence to indicate that the incidence of iconic signs is not particularly large. Observers unfamiliar with ASL typically are unable to guess the correct meanings of the large majority of signs shown to them (Hoemann, 1975; Klima & Bellugi, 1979). The opportunity for resemblance between a sign and its referent, however, may be inherently greater for visual languages than for spoken languages. In English, only a very small number of words are clearly onomatopoeic. In contrast, although only a minority, there are still quite a few ASL signs that are pantomimic or iconic in that they clearly resemble the objects, actions, or properties for which they stand.

The iconic nature of some signs has been seen by various investigators as making the meaning of these signs transparent, and hence contributing to making sign languages easier to learn than spoken languages. From this perspective, the clearly observable ties between certain signs and their referents would facilitate the early learning and retention of these signs. Indeed, scholars of language origins (e.g., Whitney, 1867; McBride, 1971, 1973; Armstrong, Stokoe, & Wilcox, 1995) have long speculated that pantomimic gestures and iconic signs probably played a critically important role in the emergence and evolution of human language.

The view that iconicity might play an important role in children's sign language acquisition has been advanced by Brown (1977) for early vocabulary development. There are at least several reasons to expect that sign iconicity might play such a major role in early vocabulary development. One reason is that most of the first signs to be acquired are nouns at what Brown calls the Basic Object Level (e.g., car, milk). The signs for many of these concrete objects are iconic. Another factor is that sign-using parents or caregivers might believe that iconic or pantomimic signs are easier for infants to learn, and thereby produce a disproportionate number of iconic signs in their early language input. Finally, investigators frequently have reported that iconic signs are acquired more readily than noniconic signs in certain severely speech-disordered populations, such as low-functioning autistic individuals (Konstantareas, Oxman, & Webster, 1978).

In light of the potential effect of iconicity on sign-language acquisition, we undertook an analysis of the incidence of iconic signs in the early sign vocabulary development of both the young children and the gorillas. We reasoned that if iconic signs predominated in the early lexicons of the children and the gorillas, then that would indicate a prominent role for iconicity in early sign learning and retention. Correspondingly, if there were relatively few iconic signs in the children's and gorillas' early sign vocabularies, then that would indicate a relatively minor impact for iconicity. Finally, if the incidence of iconic signs in their early lexicons differed widely between human infants and gorillas, then that would suggest that iconicity played somewhat different roles in the early learning of the two species.

In conducting our analyses, we followed the sign-classification system developed by Stokoe et al. (1965). In this approach, *iconic* signs are described as signs that clearly resemble their referent, have readily transparent meanings, and that are largely pantomimic or imitative in nature. Signs that represent a relatively minor or obscure feature of the referent are designated *metonymic*. In these signs, the tie between a sign and its meaning is not readily apparent – one would be unlikely to guess the meaning of a metonymic sign simply by seeing it produced. A third category of signs is designated as *arbitrary*. A sign in this category has no discernible tie to its referent. Although a more complex classification system than the present three groups might have captured a wider range of sign-to-referent relationships, the present approach was selected because it effectively captures the principal sign–referent relationships and is relatively easy to use.

The role of iconicity in vocabulary development was probed initially in the young children learning to sign. In the first longitudinal study of children's sign-language acquisition, the proportions of iconic, metonymic, and arbitrary signs were determined for the 13 children at the time they attained a 10-sign vocabulary (mean age = 13.2 months) and again when they were 18 months old (Orlansky & Bonvillian, 1984). (These classifications were conducted independently by three coders with an average percentage agreement of 86%; any disagreements were resolved by using the classification reached by the majority of the coders.) Analysis of the children's 10-sign vocabularies revealed that iconic signs comprised 30.8%, metonymic signs 33.8%, and arbitrary signs 35.4% of the vocabularies. By the time these children were 18 months old, the incidence of iconic signs in their vocabularies (mean vocabulary size = 48.2 signs) had changed relatively little. Iconic signs now accounted for 33.7%, metonymic signs 30.5%, and arbitrary signs 35.8% of the children's lexicons. There were, however, wide individual differences among the children in the proportion of iconic signs in their vocabularies at both assessment points. The proportion of iconic signs extended from a low of 20% to a high of 60% in individual children's 10-sign vocabularies. At 18 months, the range extended from a low of 13.5% iconic signs in one child to a high of 46.7% in another.

In the second longitudinal study of children's sign-language acquisition, the extent of

Table 12.2 *The number (and percentages) of the gorillas' iconic, metonymic, and arbitrary signs at two vocabulary levels*

Sign classification	Emitted criteria			Patterson criteria		
	Koko	Michael	Mean	Koko	Michael	Mean
10-sign vocabulary						
Iconic	6 (60%)	6 (60%)	6 (60%)	5 (50%)	7 (70%)	6 (60%)
Metonymic	3 (30%)	1 (10%)	2 (20%)	3 (30%)	1 (10%)	2 (20%)
Arbitrary	1 (10%)	3 (30%)	2 (20%)	2 (20%)	2 (20%)	2 (20%)
50-sign vocabulary						
Iconic	20 (40%)	22 (44%)	21 (42%)	23 (46%)	20 (40%)	21.5 (43%)
Metonymic	17 (34%)	15 (30%)	16 (32%)	18 (36%)	14 (28%)	16 (32%)
Arbitrary	13 (26%)	13 (26%)	13 (26%)	9 (18%)	16 (32%)	12.5 (25%)

sign iconicity in the children's early vocabularies was examined in a slightly different way. In this study, the relative proportions of iconic, metonymic, and arbitrary signs were determined for the different lexical entries in the children's early vocabularies – defined for this study as before 13 months of age. (Three independent coders performed the classifications, as in the first study; percentage agreement averaged 82%. These same three coders subsequently carried out the iconicity classifications for Koko's and Michael's signs.) Of the 44 different signs produced by the children before the age of 13 months, 36% were classified as iconic, 30% as metonymic, and 34% as arbitrary. These percentages are quite close to those found in the first study of children's sign acquisition.

Across the two studies of children's sign acquisition, the proportion of iconic or pantomimic signs in the children's vocabularies was about one-third of their lexical entries. Moreover, this proportion did not change appreciably regardless of whether the analyses were conducted early in the children's vocabulary development or months later. Although this proportion is considerably higher than estimates of the incidence of iconic signs in ASL overall, this finding that about one-third of the children's signs were iconic or pantomimic also can be interpreted as indicating that iconic or pantomimic aspects did not play a major role in the early sign-language acquisition of children of deaf parents. With about two-thirds of the children's signs from the beginning of both longitudinal studies classified as noniconic, factors other than sign iconicity clearly are playing important roles in the children's early sign learning.

In comparison with the young children's early signing, iconic signs were quite common in Koko's and Michael's initial 10-sign lexicons. For Koko's first 10 emitted signs, 6 were classified as iconic, 3 as metonymic, and 1 as arbitrary. Moreover, as can be seen in Table 12.2, relatively similar distributions by sign classification were evident for both gorillas across both sets of sign-acquisition criteria. Furthermore, the percentage of iconic signs in the 10-sign lexicons of the gorillas was significantly higher than that determined for the children (Bonvillian & Patterson, 1993). Nevertheless, although over half of Koko's and Michael's first ten signs were classified as primarily iconic in nature, it should be noted that signs coded as arbitrary also were acquired during this period. Thus, whereas sign iconicity may have facilitated each gorilla's initial sign acquisition, from early on both gorillas were able to learn signs that had no discernible ties to their referents. (For a more thorough discussion of the significance of iconicity in primate cognition, please see Mitchell, 1994.)

By the time Koko and Michael had achieved 50-sign vocabularies, the proportion of iconic signs in their lexicons had declined appreciably. For Koko's first 50 emitted signs, 20 (40%) were classified as iconic, 17 (34%) as metonymic, and 13 (26%) as arbitrary. Similar classification distributions were obtained for Michael and for signs meeting the more conservative Patterson acquisition criteria (see Table 12.2). Clearly, by the time each gorilla had attained a 50-sign lexicon, iconic or pantomimic factors were no longer playing a predominant role in sign acquisition.

Overall, the incidence of iconic signs in the 10- and 50-item vocabularies of Koko and Michael exceeded the average proportions of iconic signs in the children's vocabularies at these same points. Whereas iconic signs on average constituted slightly less than one-third of the signs present in the 10-sign vocabularies of the children, more than half of Koko's and Michael's signs at this stage were classified as iconic. When comparisons were conducted using the 50-sign lexicons, the differences between species were much smaller. (Indeed, the slightly larger proportion of iconic signs in the gorillas' 50-sign lexicons is attributable to the higher incidence of iconic signs in the gorillas' initial 10-sign lexicons.) Although iconic signs remained the modal sign-classification category for Koko and Michael, they now accounted for fewer than half the gorillas' signs. Moreover, several of the children had as high or higher proportions of iconic signs in their 50-sign lexicons as Koko and Michael.

Additional data will be needed before strong conclusions can be reached about the role of iconicity in gorilla sign-language learning. The findings based on Koko and Michael would appear to indicate that iconicity plays a somewhat greater role in the very early sign vocabulary development of the gorillas than it does in human children with normal intelligence. At the same time, the substantial number of metonymic and arbitrary signs in the 50–sign lexicons of both gorillas also indicates that iconicity is not a critical factor in determining gorillas' overall sign acquisition. It should also be noted that it is quite possible that neither the gorillas nor the human infants perceived the ties or resemblances between signs and their referents that led adult observers to classify these particular signs as iconic or metonymic. That is, for human and gorilla infants alike, most signs may be essentially arbitrary in nature. Finally, since every participant acquired at least some signs classified as arbitrary, then factors other than iconicity are needed to account for the children's and gorillas' early sign-language acquisition.

NONSIGN GESTURAL PRECURSORS OF LANGUAGE

Children learning to speak often make considerable use of gestures prior to producing recognizable words. This early gestural production has been described as playing an important role in children's language acquisition. These gestures have been characterized as providing the framework for subsequent language usage through their introduction of turn-taking and communicative exchange (Bruner, 1978). Language, according to this approach, is seen as developing largely out of functional communicative gestural routines. Moreover, the infant's acquisition of certain communicative gestures is seen as indicating a capacity for shared reference and conventionalized communication (Bates, Camaioni, & Volterra, 1975).

The characteristics and processes of early gestural communication in children have been examined in depth by Bates and her associates (Bates, Camaioni, & Volterra, 1975; Bates, et al. 1979). Working within the perspective of a functional communicative gestural basis of language, Bates and her associates (Bates et al., 1979) identified a sequence of communicative gestures that both preceded the onset of spoken language and that positively predicted vocabulary size. These gestures typically emerged in young children between 9 and 13 months of age. In this sequence, *ritual requests* (conventionalized requests such as touching an adult's hand and then waiting) often were the first to emerge, typically around 9 to 10 months of age. Children, at around 10 months of age, usually would then begin to *show* objects to others (e.g., holding an object out to another for viewing); this was apparently done in many cases to gain an adult's attention. Often it was not until 2 months after the onset of showing that children reliably demonstrated *giving*, the voluntary surrender of a proffered object to another. Finally, at around 13 months of age, children made use of *communicative pointing* by pointing to an object and looking to an adult, often apparently for confirmation. This onset of communicative pointing usually followed by several weeks the onset of *noncommunicative pointing* (e.g., using the index finger to examine an object), in which the child engages in pointing-for-self acts. Taken together, these gestures (the four members of the communicative sequence and noncommunicative pointing) afford the child a relatively effective early means of communicative interaction and referencing, although such communication would be largely constrained to the immediate context.

Until relatively recently, little was known about the course of sign-language acquisition and the emergence of these gestures. Analyses of the videotape and written records of the children in both sign-language acquisition studies revealed that there were a number of parallels between the emergence of these gestures in sign-learning children and the findings reported by Bates et al. (1979) for prespeech communication development in infants. The sign-learning children proceeded through the sequence of communicative gestural production at ages close to those reported for children learning to speak (Folven & Bonvillian, 1991). Another similarity was evident from analyses conducted on the children in the first sign-acquisition study (Folven, Bonvillian, & Orlansky, 1984/1985). In this investigation, it was found that the strongest predictors of the size of the children's sign lexicons between 9 and 26 months of age were their communicative pointing and giving gestures, just as they were the two strongest predictors for speech-vocabulary size in Bates et al. (1979). Thus, children's predominant language modality, sign or speech, does not affect the onset age, order of emergence, or vocabulary predictive capacity of these nonsign gestures.

Koko's nonsign gesture production

Would the same sequence of nonsign gestures emerge in Koko much as it did in the signing and speaking children? To answer this question, we turned primarily to the written diary records of Koko's early development. The information in the diary accounts was supplemented by reviewing selected film and videotape records that had been made periodically to document Koko's development. These different sets of records were examined and scored for the onset of the same five gestures delineated in the Bates et al. (1979) project as preceding language production in human infants. It should be noted, however, that differences between the projects existed in the participants' ages at the time the projects commenced and in the duration of the projects. Whereas Bates et al. (1979) examined infants between 9 and 13 months of age, Project Koko's observational records did not begin until she was one year old. Thus, there was only a single month overlap in the projects' time frames. It is possible, then, that Koko's records do not capture the true onset dates for some of the nonsign gestural items because of Koko's later starting date. This concern is particularly relevant for the two nonsign gestures observed soon after Project Koko commenced, specifically noncommunicative pointing and ritual requests. The much longer duration of Project Koko, however, enabled the full range of gestures to emerge.

The first of the five "prelinguistic" gestures to be evident in Koko's behavioral repertoire was noncommunicative pointing. In the very first days of the project, when Koko was 53 weeks old, she was observed exploring cracks and bubbles with her index finger. Soon after, she was seen using her fingers to help her examine the wheels on a toy truck. Ritual requests emerged when Koko was 13 months old. Here Koko would often extend her hand if she wanted to be given something or raise her arm when she wished to be picked up. When Koko was 14 months old, she began to demonstrate communicative pointing. An early instance of this behavior occurred when Koko put her finger first to FGPP's mouth, then close to the window, repeated this sequence, and finally looked at FGPP. (This sequence was interpreted as a request for FGPP to make 'fog' on the window; after she obliged by blowing on the window, Koko drew on it with her fingers.) Not until Koko was about 20 months old did clear instances of giving behaviors appear. When she was 21 months old, Koko repeatedly gave FGPP different objects (for example, a glass after Koko had emptied it). Some 6 months before Koko reliably demonstrated the giving of objects to others, Koko was observed voluntarily releasing objects. Finally, showing behaviors began to emerge in Koko around her second birthday. For example, at the age of 23 months, Koko presented a pole to one of her male caregivers where he could easily see it; this took place when Koko apparently wanted a hard-to-reach item. In another instance, when she was 25 months old, Koko moved her face close to that of her principal caregiver (FGPP) and opened her mouth to reveal a piece of bean lodged between her teeth. She then took her caregiver's hand in her own and guided it to the stuck bean. Overall, then, instances of each of the five gestures identified by Bates et al. (1979) were evident in Koko's behavioral records during her first 2 years.

Child and gorilla comparisons

There were several readily apparent differences between Koko and the signing and speaking children in their acquisition of the different nonsign gestures. One clearly noticeable difference was that the young children's rate of acquisition was quite rapid; they demonstrated each of the different gestures either by the end of their first year or by the beginning of their second. Koko, in contrast, progressed much more slowly, not acquiring all the different gestures until she was about 2 years old. A second difference was in the overall pattern of acquisition of the gestures. This was particularly evident in the relatively late acquisition by Koko of giving and showing behaviors. Whereas these behaviors typically emerged in the human infants before their first birthday and prior to the onset of communicative pointing, these gestures were not produced by Koko until late in her second year and months after the onset of communicative pointing.

Koko's gestural production deviated most from that of the young children in the domain of showing. Whereas the children often were observed showing objects to other individuals during their first 2 years, this was not the case for Koko. Unlike the children, there were very few instances of showing recorded for Koko during her first 30 months, and these instances often were not clear-cut examples. It is, of course, possible that other gestural behaviors that Koko frequently produced might play, in gorillas, a largely equivalent role to that of showing in humans. For example, beginning in the latter half of her second year, Koko was observed frequently putting food and other objects in her caregivers' mouths. It simply may be the case that visual showing of items may not be as important in gorillas as it is in humans. It may also be the case that the visual showing of items might serve a different purpose or be considered threatening in gorillas. Although Koko's showing of objects to others differed considerably from that of the children, this is not to say that she was not aware of others' visual capacities at one level: Koko typically would produce her ASL signs and her natural communicative gestures where they could be easily seen by another.

In summary, Koko demonstrated each of the five nonsign gestures described by Bates et al. (1979) as preceding language use in children. The onset ages and pattern of emergence of these gestures in Koko, however, differed considerably from those reported for children learning to sign or to speak.

REFERENTIAL LANGUAGE

Various investigators (e.g., Bonvillian et al., 1983a) have reached the conclusion that young children often acquire certain early language milestones more rapidly in the manual mode. Critics of these studies, however, have responded by contending that the investigators frequently ignored the context of sign usage and attributed linguistic status to a sign entirely on the basis of its form (Volterra & Caselli, 1985; Petitto, 1988). A focus on context, these critics claimed, would show that much early sign usage is either imitation of a caregiver's modeling or part of a restricted interactive routine. Because of these concerns, an analysis of the context of early sign production was conducted both for a number of the young sign-learning children and for Koko. In particular, the contexts of the early sign productions were examined to determine the onset of *referential* language. Referential signing was identified as the use of vocabulary items to name new instances of objects or events (Bates et al., 1975; Bates et al., 1979).

The acquisition of referential language ability represents an important advance for the young language-learner. An individual who employs a sign or a word to name a new exemplar demonstrates an understanding that language can be used to categorize objects, people, and events, that things have names or labels, and that language is not tied to a particular item or context. Such use of referential language constitutes an important step for the language learner beyond earlier usage that was either imitative or that elicited desired responses from others. (It should be noted that the meaning of the word "referential" as in "referential signing" or "referential language ability" in this chapter differs from how this word is used in Miles, 1990. Whereas "referential" is employed in this chapter to indicate the use of signs to name or to label new instances of things, signs become referential to Miles when they are "intentionally meaningful indicators related to the real world" p. 511.)

To determine the onset of referential language, it is necessary to have relatively complete contextual information on early word or sign usage. Such information (written and videotape records) was available on the early sign productions of eight of the children (Bonvillian & Folven, 1990; Folven & Bonvillian, 1991) and for the gorilla, Koko. For both the children and Koko, their initial recognizable signs were all used nonreferentially. These early productions often were imitations of others' signs, requests for familiar items, or parts of gestural routines. Many of the initial sign productions, moreover, appeared to be used relatively indiscriminately. With time, these early signs gradually were employed more systematically for objects, events, and persons. For the children, their first referential sign usage occurred at the mean age of 12.6 months, with a range extending from 10.1 to 16.2 months (Folven & Bonvillian, 1991). This age of onset of referential usage in sign does not differ from the age of 13 months reported as typical for first referential spoken-word usage in children learning to speak (Bates et al., 1975; Bates et al., 1979).

Koko's initial referential sign usage followed in age that of all of the young signing children examined. Koko was credited with her first referential sign usage when she was 20 months old. At this time, Koko initially produced the sign BABY to a specific toy, then to a new doll shown to her, and, shortly after, to a human baby that she encountered. Over the next month, there were other examples of such spontaneous referential usage. This ability to use signs to name or label new instances of a concept was thus relatively well established in Koko by the end of her 21st month.

Not only was Koko chronologically older than any of the children at the time she began to use signs referentially, but she also differed from the children in that the onset of her referential usage occurred long after she had attained a 10–sign lexicon (Emitted criteria). In contrast, for 6 of the 8 children whose referential signing had been systematically probed (Folven & Bonvillian, 1991), their onset of referential usage preceded their attainment of a 10–sign vocabulary. In the 2 children who were exceptions to this pattern, the 2 events nearly coincided: their onset of referential usage followed their acquisition of a 10–sign lexicon by an average of only 3 days. In Koko's case, her onset of referential usage followed her attainment of a 10–sign vocabulary (Emitted criteria) by 4 months. Furthermore, with a vocabulary of 29 signs (Emitted criteria) at the time she began to use signs referentially, Koko's vocabulary was more than twice as large as any of the children's at the onset of referential signing. Koko may have needed more experience using different signs before she recognized that signs could be utilized to name or label things. (If one were to use the more conservative Patterson acquisition criteria, then yet another pattern emerges. Koko's onset of referential usage preceded her attainment of a 10–sign lexicon by 8 months, a period much longer in the other direction from that of the children.)

In addition, there were similarities and differences between Koko and the children in how the onset of referential usage interrelated with the onset of production of the nonsign communicative gestures. As with both signing (Folven & Bonvillian, 1991) and speaking children (Bates et al., 1979), Koko's referential usage did not emerge until after she had demonstrated communicative pointing. Unlike the children, Koko's referential usage preceded her production of object showing and nearly coincided with the onset of her production of giving gestures. Koko thus acquired referential language skills both at an older age than any of the children and after following a pattern of development that differed somewhat from that of the children.

The fact that the onset of giving and showing gestures in Koko did not precede her referential signing calls into question the view that these gestures are precursors to referential language use. Rather, these gestures may be behaviors that typically appear early in the course of human development and that are positively, but not critically, related to referential-language onset. In the future, it will be of interest to determine whether other nonhuman primates learn to use language referentially (in the sense of naming or labeling) and, if so, how their onset of referential language interrelates with their production of other nonsign communicative gestures. It is quite possible that these species might demonstrate additional patterns of development.

Finally, we should note that there is evidence that the ability to use signs referentially, similar to that of Koko's, may also have emerged in the chimpanzee Washoe. A likely instance of such signing occurred in Washoe when she also was 20 months old. (Washoe had been taught by the Gardners to communicate using ASL signs beginning when she was about 10 months old.) During a visit to the Gardners' home, Washoe climbed up on the bathroom counter, looked at the collection of toothbrushes, and signed TOOTHBRUSH. The Gardners wrote:

> At the time, we believed that Washoe understood the sign TOOTHBRUSH, but we had never seen her use it. She had no reason to ask for the toothbrushes in the Gardner bathroom, because they were well within her reach; and it is very unlikely that she was asking to have her teeth brushed. This was one of the earliest examples of a situation in which Washoe named an object with no obvious motive other than communication. (Gardner & Gardner, 1989, p. 18)

CONCLUDING REMARKS

In this chapter we have brought together diverse observations on early sign-language acquisition in children and gorillas. In our analyses, we have noted that there are many developmental aspects on which the two species are similar and a number on which they differ. Similarities across species are evident in vocabulary content, the acquisition of referential signing, and in the generation of various nonsign gestures. Differences are evident in the two species' rates and patterns of early sign language and gestural development. In general, Koko and Michael attained many of the same early language and gestural milestones, but at a considerably slower rate than the children. Overall, we believe that the similarities in early development across species outweigh the differences.

Of particular significance in this investigation is Koko's acquisition of referential language skills. This ability to name or to label new instances of concepts reveals both an ability to identify the critical features common to objects or actions and that such objects or actions have names. This ability to use language referentially was long believed not to be a characteristic of animals, including nonhuman primates (Locke, 1690; Müller, 1865, 1889; Terrace, 1985).

In light of these findings on early development, it is clear that gorillas show important language-like abilities. At least, as seen in Koko's and Michael's sign usage and gestural production during their first couple of years of sign input, there are many parallels in development with that of young children. Moreover, these similarities are more apparent when the gorillas' development is compared with that of young sign-learning children than with young children learning to speak. Indeed, the differences in early language and communicative development between the gorillas and children in their first years appear to be more ones of rate of acquisition than of qualitatively different mechanisms. Regardless, one should recognize that the gorillas' performance is not identical to that of the children. Gorillas and humans constitute different, albeit closely related, species, and differences in capabilities are to be expected.

Although Koko's and Michael's attainment of many early language and gestural milestones is quite impressive, it is important to keep in mind the long-term outcome differences between the children and the gorillas. Briefly, whereas the children's sign-language skills continued to develop rapidly until they became highly fluent signers, the same level of success has not yet been achieved by the gorillas. Over the years, Koko and Michael continued to add new signs to their vocabularies, to create new sign-like gestures, to initiate conversations with human signers on a frequent basis, and to spontaneously generate contextually appropriate multisign utterances (Patterson & Cohn, 1990). Yet the gorillas' vocabulary size, their acquisition and use of diverse morphological processes, and the complexity of their utterances do not approach those of fluent signers. In comparison with the mature signing skills achieved by children of deaf parents, Koko's and Michael's signing skills are at a much lower level.

A different evaluative outcome will be reached if the gorillas' sign-language skills are compared with those of various language-impaired human population groups. The gorillas' sign-language achievements often exceeded those reported for persons with profound disabilities who received manual communication training. For example, although sign training has proved to be an effective communicative approach for many mute, low-functioning children with autistic disorder, a large number of these individuals acquire only a small core sign vocabulary and make only the most rudimentary sign combinations (Bonvillian & Blackburn, 1991). Similarly, many aphasic adults, multi-handicapped hearing-impaired individuals, and severely and profoundly mentally retarded persons make only relatively minimal progress in acquiring sign-language skills (Orlansky & Bonvillian, 1983; Christopoulou & Bonvillian, 1985). If one were to argue that Koko and Michael did not demonstrate a wide range of language-like skills, then one would need to conclude that most members of these other groups also fail to do so.

In the present investigation, we uncovered a number of similarities in the early sign-language acquisition and communicative development of children and gorillas. Yet our distinct impression is that the two species' levels of signing skill tend to diverge more and more with increasing age. In the future, it will be important to determine whether this impression is indeed correct, and, if so, when this divergence occurs. The basis for this apparent divergence, and whether or not it is maintained as the gorillas develop, should be the focus of a future investigation.

ACKNOWLEDGMENT

During the preparation of this chapter, John Bonvillian was the recipient of a Sesquicentennial Associateship from the University of Virginia.

REFERENCES

Armstrong, D. F., Stokoe, W. C., & Wilcox, S. E. (1995). *Gesture and the nature of language*. Cambridge University Press.

Bates, E., Camaioni, L., & Volterra, V. (1975). The acquisition of performatives prior to speech. *Merrill-Palmer Quarterly*, 21, 205–226.

Bates, E., Benigni, L., Bretherton, I., Camaioni, L., & Volterra, V. (1979). *The emergence of symbols: Cognition and communication in infancy*. New York: Academic Press.

Bloom, L., Lifter, K., & Broughton, J. (1985). The convergence of early cognition and language in the second year of life: Problems in conceptualization and measurement. In M. Barrett (ed.), *Children's single-word speech* (pp. 149–180). Chichester, UK: John Wiley and Sons.

Bonvillian, J. D. & Blackburn, D. W. (1991). Manual communication and autism: Factors relating to sign language acquisition. In P. Siple & S. D. Fischer (eds.), *Theoretical issues in sign language research: Vol. 2. Psychology* (pp. 255–277). University of Chicago Press.

Bonvillian, J. D. & Folven, R. J. (1990). The onset of signing in young children. In W. Edmondson & F. Karlsson (eds.), *SLR '87: Papers from the Fourth International Symposium on Sign Language Research* (pp. 183–189). Hamburg: Signum Press.

Bonvillian, J. D. & Patterson, F. G. P. (1993). Early sign language acquisition in children and gorillas: Vocabulary content and sign iconicity. *First Language*, 13, 315–338.

Bonvillian, J. D., Orlansky, M. D., & Novack, L. L. (1983a). Developmental milestones: Sign language acquisition and motor development. *Child Development*, **54**, 1435–1445.

Bonvillian, J. D., Orlansky, M. D., Novack, L. L., & Folven, R. J. (1983b). Early sign language acquisition and cognitive development. In D. Rogers & J. A. Sloboda (eds.), *The acquisition of symbolic skills* (pp. 207–214). New York: Plenum Press.

Brown, R. (1977). Why are signed languages easier to learn than spoken languages? In W. C. Stokoe (ed.), *Proceedings of the National Symposium on Sign Language Research and Training* (pp. 9–24). Silver Spring, MD: National Association of the Deaf.

Bruner, J. S. (1978, September). Learning the mother tongue. *Human Nature*, 42–49.

Christopoulou, C. & Bonvillian, J. D. (1985). Sign language, pantomime, and gestural processing in aphasic persons: A review. *Journal of Communication Disorders*, **18**, 1– 20.

De Hirsch, K., Jansky, J., & Langford, W. (1964). Oral language performance of premature children and controls. *Journal of Speech and Hearing Disorders*, **29**, 60–69.

Folven, R. J., & Bonvillian, J. D. (1991). The transition from nonreferential to referential language in children acquiring American Sign Language. *Developmental Psychology*, **27**, 806–816.

Folven, R. J., Bonvillian, J. D., & Orlansky, M. D. (1984/1985). Communicative gestures and early sign language acquisition. *First Language*, **5**, 129–144.

Fouts, R. S. (1973). Acquisition and testing of gestural signs in four young chimpanzees. *Science*, **180**, 978–980.

Gardner, B. T. & Gardner, R. A. (1974). Comparing the early utterances of child and chimpanzees. In A. Pick (ed.), *Minnesota Symposium on Child Psychology* (vol. 8, pp. 3–23). Minneapolis: University of Minnesota Press.

Gardner, R. A. & Gardner, B. T. (1989). A cross-fostering laboratory. In R. A. Gardner, B. T. Gardner, & T. E. van Cantfort (eds.), *Teaching sign language to chimpanzees* (pp. 1–28). Albany, NY: State University of New York Press.

Gardner, R. A., Gardner, B. T., & Van Cantfort, T. E. (eds.). (1989). *Teaching sign language to chimpanzees*. Albany, NY: State University of New York Press.

Greenfield, P. M. & Savage-Rumbaugh, E. S. (1990). Grammatical combinations in *Pan paniscus*: Processes of learning and invention in the evolution and development of language. In S. T. Parker & K. R. Gibson (eds.), *"Language" and intelligence in monkeys and apes* (pp. 540–578). Cambridge University Press.

Hoemann, H. W. (1975). The transparency of meaning of sign language gestures. *Sign Language Studies*, **7**, 151–161.

Klima, E. S. & Bellugi, U. (1979). *The signs of language*. Cambridge, MA: Harvard University Press.

Konstantareas, M. M., Oxman, J., & Webster, C. D. (1978). Iconicity: Effects on the acquisition of sign language by autistic and other severely dysfunctional children. In P. Siple (ed.), *Understanding language through sign language research* (pp. 213–237). New York: Academic Press.

Locke, J. (1690). *An essay concerning human understanding*. London: Thomas Bassett.

Mayberry, R. I. (1994). The importance of childhood to language acquisition: Evidence from American Sign Language. In J. C. Goodman & H. C. Nusbaum (eds.), *The development of speech perception* (pp. 57–90). Cambridge, MA: MIT Press.

McBride, G. (1971). On the evolution of human langauge. In J. Kristeva, J. Rey-Debove, & D. J. Umiker (eds.), *Essays in semiotics/Essais de sémiotique* (pp. 560–566). (*Approaches to semiotics*, vol. 4). The Hague: Mouton.

McBride, G. (1973). Comments. *Current Anthropology*, **14**, 15.

Meier, R. P. & Newport, E. L. (1990). Out of the hands of babes: On a possible sign advantage in language acquisition. *Language*, **66**, 1–23.

Menyuk, P., Liebergott, J., Schultz, M., Chesnick, M., & Ferrier, L. (1991). Patterns of early lexical and cognitive development in premature and full-term infants. *Journal of Speech and Hearing Research*, **34**, 88–94.

Miles, H. L. (1983). Apes and language: The search for communicative competence. In. J. de Luce & H. T. Wilder (eds.), *Language in primates: Implications for linguistics, anthropology, psychology and philosophy* (pp. 43–61). New York: Springer-Verlag.

Miles, H. L. (1986). How can I tell a lie? Apes, language, and the problem of deception. In R. W. Mithcell & N. S. Thompson (eds.), *Deception: Perspectives on human and nonhuman deceit* (pp. 245–266). Albany, NY: State University of New York Press.

Miles, H. L. (1990). The cognitive foundations for reference in a signing orangutan. In S. T. Parker & K. R. Gibson (eds.), *"Language" and intelligence in monkeys and apes* (pp. 511–538). Cambridge University Press.

Miles, H. L. (1997). Anthropomorphism, apes, and language. In R. W. Mitchell, N. S. Thompson, & H. L. Miles (eds.), *Anthropomorphism, anecdotes, and animals* (pp. 383–404). Albany, NY: State University of New York Press.

Minkowski, A., Larroche, J. C., Vignaud, J., Dreyfus-Brisac, C., & Dargassies, S. S. (1966). Development of the nervous system in early life. In F. T. Falkner (ed.), *Human development* (pp. 254–325). London: Saunders.

Mitchell, R. W. (1994). The evolution of primate cognition: Simulation, self-knowledge, and knowledge of other minds. In D. Quiatt & J. Itani (eds.), *Hominid culture in primate perspective* (pp. 177–232). Boulder, CO: University Press of Colorado.

Müller, F. M. (1865). *Lectures on the science of language*. New York: Charles Scribner.

Müller, F. M. (1889). *Three lectures on the science of language and its place in general education*. London: Longmans, Green.

Nelson, K. (1973). Structure and strategy in learning to talk. *Monographs of the Society for Research in Child Development*, **38**(1–2, Serial No. 149).

Newport, E. L. & Meier, R. P. (1985). The acquisition of American Sign Language. In D. I. Slobin (ed.), *The crosslinguistic study of language acquisition: Vol. 1. The data* (pp. 881–938). Hillsdale, NJ: Erlbaum.

Orlansky, M. D. & Bonvillian, J. D. (1983). Recent research on sign language acquisition: Implications for multihandicapped hearing-impaired children. *Journal of the National Student Speech Language Hearing Association*, **11**, 72–87.

Orlansky, M. D. & Bonvillian, J. D. (1984). The role of iconicity in early sign language acquisition. *Journal of Speech and Hearing Disorders*, **49**, 287–292.

Owens, R. E., Jr. (1996). *Language development: An introduction* (4th edn). Columbus. OH: Merrill.

Patterson, F. G. (1978). The gestures of a gorilla: Language acquisition in another pongid. *Brain and Language*, **5**, 72–97.

Patterson, F. G. (1980). Innovative uses of language by a gorilla: A case study. In K. E. Nelson (ed.), *Children's language* (vol. 2, pp. 497–561). New York: Gardner Press, John Wiley.

Patterson, F. G. P. & Cohn, R. H. (1990). Language acquisition by a lowland gorilla: Koko's first ten years of vocabulary development. *Word*, **41**, 97–143.

Petitto, L. A. (1988). 'Language' in the pre-linguistic child. In F. Kessel (ed.), *The development of language and language researchers* (pp. 187–221). Hillsdale, NJ: Erlbaum.

Rescorla, L.A. (1980). Overextension in early language development. *Journal of Child Language*, **7**, 321–335.

Rimpau, J. B., Gardner, R. A., & Gardner, B. T. (1989). Expression of person, place, and instrument in ASL utterances of children and chimpanzees. In R. A. Gardner, B. T. Gardner, & T. E. van Cantfort (eds.), *Teaching sign language to chimpanzees* (pp. 240–268). Albany, NY: State University of New York Press.

Sebeok, T. A. & Rosenthal, R. (eds.). (1981). *The Clever Hans phenomenon: Communication with horses, whales, apes, and people.* Annals of the New York Academy of Sciences, vol. 364. New York: New York Academy of Sciences.

Sebeok, T. A. & Umiker-Sebeok, J. (eds.). (1980). *Speaking of apes: A critical anthology of two-way communication with man.* New York: Plenum Press.

Seidenberg, M. S. & Petitto, L. A. (1979). Signing behavior in apes: A critical review. *Cognition*, 7, 177–215.

Stokoe, W. C., Jr. (1960). Sign language structure: An outline of the visual communication systems of the American deaf. *Studies in Linguistics, Occasional Papers 8*. Buffalo: University of Buffalo.

Stokoe, W. C., Jr., Casterline, D., & Croneberg, C. (1965). *A dictionary of American Sign Language on linguistic principles*. Washington, DC: Gallaudet College Press.

Terrace, H. S. (1985). In the beginning was the "name." *American Psychologist*, **40**, 1011–1028.

Volterra, V. & Caselli, M. C. C. (1985). From gestures and vocalizations to signs and words. In W. Stokoe & V. Volterra (eds.), *SLR '83: Proceedings of the Third International Symposium on Sign Language Research* (pp. 1–9). Silver Spring, MD: Linstok Press.

Wilbur, R. B. (1987). *American Sign Language: Linguistic and applied dimensions* (2nd edn). Boston: College Hill.

Whitney, W. D. (1867). *Language and the study of language: Twelve lectures on the principles of linguistic science*. New York: Scribner's.

13

Early sign performance in a free-ranging, adult orangutan

GARY L. SHAPIRO AND BIRUTÉ M. F. GALDIKAS

INTRODUCTION

Young apes have been preferred subjects in language studies for a number of reasons. As dependent creatures, young apes engage in social behaviors which provide the interactive context in which many signs are learned. Young apes appear to learn certain behaviors more rapidly than older primates (Kawai, 1965). Young apes are also easier to handle and manage than older apes, who test their strength and dominance against caretakers as they enter adolescence (personal experience). Older apes are also stronger and potentially more dangerous than younger apes. Programs requiring close interaction, such as those carried out by ape language researchers, are more readily implemented when the subjects are young apes.

For these reasons sign-learning abilities in adult apes have not been as actively examined. Nevertheless, sign-learning abilities of adult apes should be of interest to those who study great ape cognitive or linguistic abilities. Likewise, although sign learning has not been studied in free-ranging apes, their motivation and ability to learn signs is of great interest. This chapter reports Project Rinnie, a study of the early sign-learning performance of an adult orangutan free-ranging within Tanjung Puting National Park (then a nature reserve).

METHODS

Subject

Rinnie, a captive-born, free-ranging female orangutan was approximately 10 to 12 years old at the onset of the Project. She was released as a rehabilitant at the Orangutan Research and Conservation Program (Galdikas, 1979) where she established a home range in a freshwater swamp forest across the river from the Camp Leakey Research Station. Rinnie was the oldest and the most dominant among the several ex-captive orangutans.

Training sessions

The study was conducted from July 1978 through May 1980. Most of the training sessions occurred at the river's edge or at a feeding platform across the river; however,

Rinnie occasionally crossed the river on her own during the dry season. If she was present, training was conducted in the camp area. Prior to the training session, the trainer (carrying referents, edibles, and recording equipment) swam or paddled across the river from the camp landing to a dry region on the opposite bank which then became the site for that day's training session. If Rinnie was not in view, the trainer would call her name repeatedly until she appeared from the surrounding forest. On most occasions, Rinnie would choose to join the trainer. She would either sit next to or across from the trainer. During the training session, the trainer and Rinnie would conduct a series of transactions described below. The sessions ranged in length from several minutes (initially) to over 2 hours. Most often they lasted for approximately 1 hour. Frequently during the session, other ex-captive orangutans would observe the sign-learning transactions. Only one session was administered on a given day.

Method of training:

Rinnie was trained using the molding technique following the methods of Gardner and Gardner (1969) and Fouts (1972). Rather than time to a criterion of performance, the trial (consisting of a series of specific trainer/subject actions) was chosen as the training unit. The trial can be described by the following events: (1) *referent presentation* – the referent is held in front of the subject (if an object) or an activity is demonstrated (e.g., scratching). The trainer then asks the subject in Ameslan (American sign language) and vocal Indonesian, WHAT IS THIS? Outside of object referent training, the subject also participates in situations which prompt signing to acquire an object or activity of interest (e.g., WHAT DO YOU WANT? or HOW?). (2) *response* – the subject produces one of several gestural responses including the target signed response. (3) *molding* – the subject's hands are physically manipulated or prompted through the complete range of motions defining the specific sign. (4) *reinforcement* – a small amount of food or drink is given following molding; a slightly larger amount of food or drink is given following a correct response. Trials are then repeated for either the same sign or a different sign. Normally five successive trials are given for a specific sign prior to changing to a different sign.

Types of responses

Following presentation of the referent and questioning, Rinnie's response was assigned to one of the following classes: (1) *no response* – a noncommunicative response such as turning the head, scratching the body or not responding with the hands at all; (2) *incorrect response* – a signed response is made unprompted (without guidance by the trainer) but is judged to be definitely incorrect due to context and referent (e.g., calling a nut, PINEAPPLE; (3) *poor response* – the response is made unprompted, however, it is either a partial sign (cheremic) or a poorly formed version of an acceptable sign, Poor responses may or may not be recognizable as to their meaning; (4) *good response* – the sign is made unprompted and is judged by the trainer as being within the range of acceptability.

Table 13.1 *Signs trained to orangutan: Rinnie*

Contact actions	Names	Modifiers	Consumable referents	Natural referents	Other referents
brush	Rinnie	more	food	tree	hat
tickle	Gary	sweet	drink	leaf	comb
hug	Benny		banana	flower	pipe
play			nut	bug	watch
		Personal	sweet (candy)	twig	glasses
			rice	fungus	pillow
		please	pineapple		tuning fork
		hurt	fruit		dirt
		sorry	biscuit		
			berry		
			soap		

Recording responses

Following the presentation of a training trial, data were recorded and a subsequent trial begun. The following types of elements were recorded: (1) *the referent* and/or signs presented during the trial, (2) *the response* made by the subject, (3) *any prompting* that occurred during the trial, and (4) *other events* of importance and interest (e.g., time, weather, other orangutans in area). The descriptions were made as they occurred. The recording of spontaneous utterances for food and contact were interspersed between regular training trials.

Trainers and observers

The first author was the primary trainer throughout the study. Observers (Indonesian university students, visitors, researchers, and staff at Camp Leakey) aided in data collection during training and testing sessions. None of the observers were deaf signers. Each was taught signs being taught to Rinnie, and asked to look for them during the session. Some observers assisted as secondary trainers on occasion.

Signs trained

The signs that were intentionally taught to Rinnie over the course of the 22-month study are presented in Table 13.1. Most of the referents for these signs were objects which permitted relatively easy vocabulary training. Of the object referents, many were edibles as it became increasingly clear that Rinnie was most interested in edibles and that her attendance to the sessions was based primarily on the fact that she would receive edibles. She was taught other signs chosen to examine the relative ease or difficulty of learning when the referents for the signs were nonedibles (e.g., hat, Gary, Rinnie). Because the subject was free-ranging in the nearby forest, it was assumed she might learn signs for natural referents (e.g., tree, leaf, flower, fungus, bug) easily inasmuch as she had presumably developed concepts concerning these referents prior to training. She also was taught other object signs whose referents she had (presumably) not observed prior to training (e.g., toy pillow, tuning fork). In addition, Rinnie was taught other nonobject signs (e.g., more, open, hurt) that had been taught early to the chimpanzee, Washoe, the gorilla, Koko, and the orangutan, Chantek. The presumption was that she could learn to use them correctly as well.

Criteria of sign acquisition

The criteria for sign acquisition varies. Gardner and Gardner (1969, 1971) accepted a sign by Washoe as acquired (a reliable part of her vocabulary) when she signed it spontaneously and appropriately for fourteen consecutive days. Additionally, the sign in question had to have been observed by two independent observers on each of those days. Patterson (1978), less stringent in her criterion, required that two independent observers record a sign's spontaneous and appropriate occurrence on at least half the days in a given month. Other criteria have been used by various researchers (Terrace, 1979; Miles, 1983).

Because of the special nature of the study, a different criterion of sign proficiency was used. When a sign was observed for at least fourteen consecutive sessions unprompted and in good form and context by the trainer, it was considered a potentially acquired sign (PAS). Because only one trainer worked with Rinnie regularly, later, an observer familiar with Rinnie's responses was brought to the training site to observe the session. Rinnie's responses during the session were independently recorded as to type and quality. These responses were later compared to the PAS. If verified by the observer, the PAS was classified as an acquired sign following the Gardner criteria. A similar modified assessment was made with the Patterson criteria.

Testing

At 6-month intervals, Rinnie was given a set of tests designed to measure her vocabulary without the use of a positive reinforcement. The purpose of the tests was to demonstrate her signing competence in a more controlled environment than training permitted. Both single (observer) and double (tester and observer) blind testing were used over the course of the project when feasible. A testing stand was fabricated in order that testing referents could be presented to Rinnie without the observer's knowledge of the item. The observer could only report on what the subject actually signed.

The principal trainer and two observers assisted in recording the test results as well as administering the test itself. It was presumed that Rinnie would be able to transfer information acquired during training to the testing environment and communicate that information even to someone other than the trainer. Testing was given over the course of 1 month, during which time each person presented 16 training referents to Rinnie in a varied but predetermined order. No molding or reward was given following the response. Each referent was presented 8 times over the course of 4 testing sessions by each tester.

Table 13.2 *Signs invented by orangutan: Rinnie and integrated into her signed combinations*

Contact actions	Reference by deixis (pointing)	Requesting actions
groom	you	give
scratch	this/that/there	
grab-hair		

Table 13.3 *Signs performed by orangutan: Rinnie within & outside trainining and testing sessions*

Contact actions	Names/ references	Modifiers	Consumable referents	Natural referents	Other referents
brush	you[a]	more	food[a]	tree[a]	hat[a]
tickle	this/that	sweet[a]	drink[a]	leaf	pillow
hug	Rinnie[a]		banana	flower[a]	tuning fork
scratch	Gary[a]		nut[a]	bug[a]	
groom	hurt[a]	*Other*	sweet (candy)[a]	twig	
grab-hair		*Actions*	rice[a]	fungus	
			pineapple[a]		
		come/give	fruit[a]		
		open	biscuit[a]		
			soap		

[a] Acquired according to Gardner & Gardner (1969) and Patterson (1978)

RESULTS

Within the first month of sign training, Rinnie began to produce signs singly and in combination. While the trainer had to prompt and mold the gestural responses, Rinnie was using them in the correct context. Rinnie also invented signs not intentionally taught (Table 13.2) and integrated them into her combinations. These signs were typically contact activities (groom, scratch, or grab-hair) that were ritualized variations of the activities themselves or referents specified by pointing or deixis (you, this/that/there). Table 13.3 shows the complete listing of signs performed by Rinnie both within and outside training and testing sessions. Of these 32 signs, 13 were acquired vocabulary according to the criteria set by Gardner and Gardner (1969) and 17 were performed according to criteria set by Patterson (1978) during the first 13 months of training (Figure 13.1 and Table 13.4).

Outside of vocabulary training, Rinnie spontaneously created signs both singly and increasingly in combination. Figure 13.2 shows Rinnie's first signed combinations in

Table 13.4 *Sign acquisition by Rinnie based on criteria set by Gardner & Gardner (1969) and Patterson (1979)*

	Months following the initiation of training				
Sign	G	P	Sign	G	P
Drink	5	5	Hat	8	8
Nut	5	5	Gary	8	8
Pineapple	5	5	Tree		9
Biscuit	5	5	Fruit		10
You	5	5	Hurt	10	10
Food		5	Tree	10	10
Flower		8	Bug	11	10
Sweet	9	8	Rini	11	10
			Rice	13	13

G – gardner & Gardner: Sign observed in correct form and context over 14 consecutive days.
P – Patterson: Sign observed in correct form and context during more than half the days of a given month.

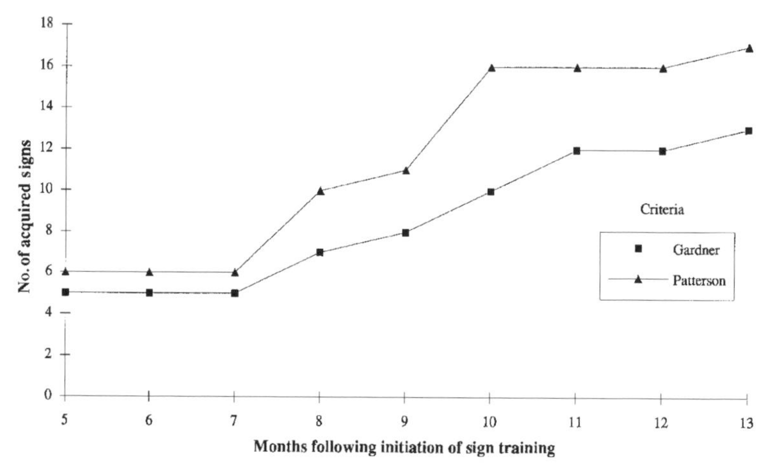

Figure 13.1 Accumulated sign acquisition by orangutan Rinnie during the first 13 months of sign training based on criteria set by Gardner & Gardner and Patterson.

diagrammatic form after 1 month training. Her signs were typically 2- to 3-sign combinations. Her combinations were usually requests for food or contact activities. She would request these items or activities either by first referring to the trainer (YOU) followed by the signs for the items or activities (Figure 13.2a) or she would begin by signing, FOOD-EAT (Figure 13.2b), GIVE (Figure 13.2c), or the repeating sign, MORE (Figure 13.2d). Occasionally Rinnie would simultaneously sign with both hands, "YOU" "THAT-THERE" or "YOU" "FOOD-EAT". Interestingly, when FOOD-EAT was

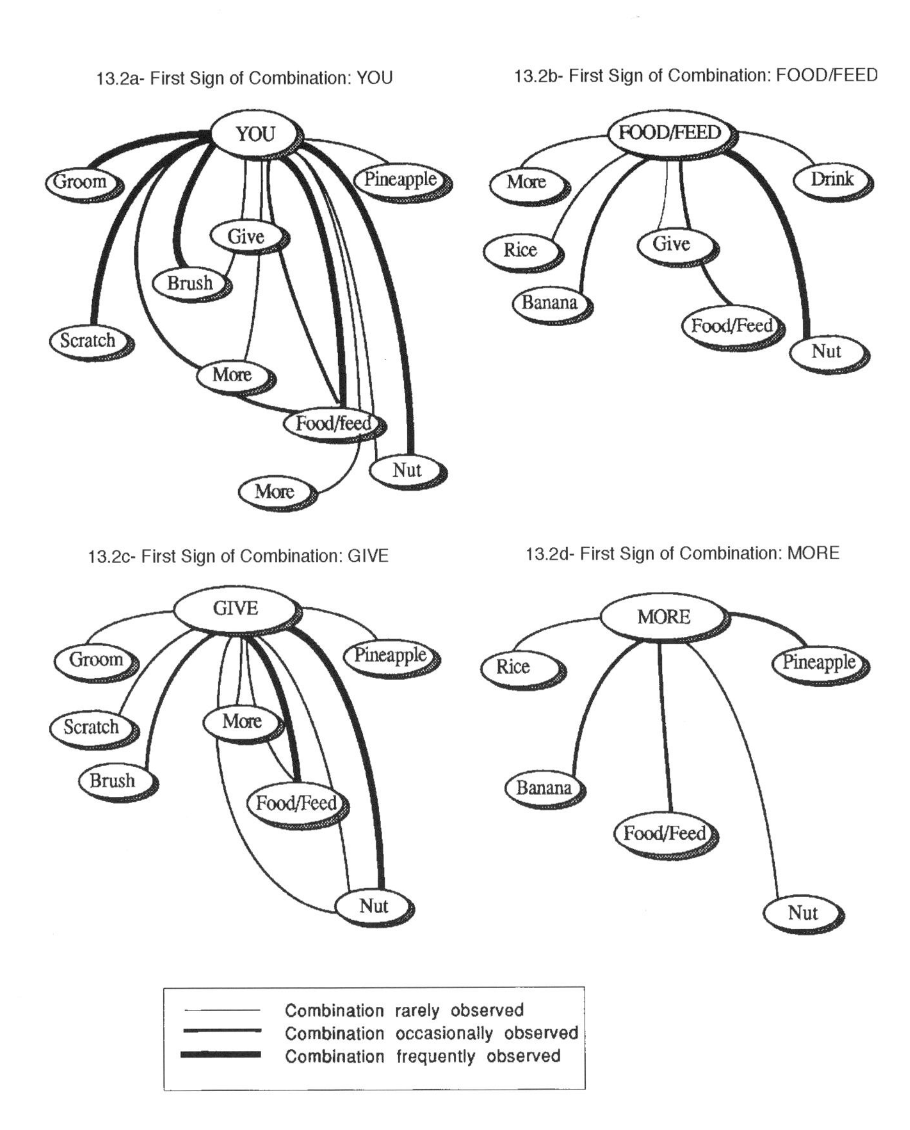

Figure 13.2 Diagrammatic representation of Rinnie's first combinations – 1 month after training begins.

the lead sign, only signs representing food objects were appended (Figure 13.2b). Rinnie's signed requests for contact were made usually once the food items had been consumed or when she seemed apprehensive during the session.

After 4 months of training, Rinnie's spontaneous combinations began to increase in number and in length. An increased number of combinations showed duplication of a sign, for example, "RICE YOU RICE" and combinations that were not as logically

Table 13.5 *Rinnie's combinations – 4 months after training begins – sign combination by rank order*

Combination	Frequency	days observed	Combination	Frequency	days observed
You Rice	105	20	Give Rice	5	3
More Rice	63	14	More Nut	4	3
You Biscuit	32	6	You Grab-hair	3	3
You Scratch	27	9	You Groom	3	3
You Nut	17	5	Rice You Rice	3	3
More You Rice	8	5	You Leaf	3	2
Open Rice	6	4	Rice More	2	2
You More Rice	5	5	Food/feed Rice	2	2
More You Scratch	5	4	More You	2	2
You Rice More	5	4	Rice More You	2	2
			You Give	2	2

Combinations observed in at least 2 days during the month – 39 different combinations.

ordered as earlier observed, for example, "RICE YOU MORE." When her signing combinations were examined (Table 13.5), it seems clear she was very interested in requesting rice during this period. While 39 different signed combinations were observed during the month, only 21 different combinations were observed in at least 2 days during the month. From these 21 different combinations, 304 signed utterances were observed, and of these, 206 (67%) were requests for rice. "YOU RICE" and "MORE RICE" were the two combinations accounting for the greatest number of spontaneous utterances and number of days observed. Requests for contact activities (scratching, grooming, grabbing hair) accounted for only 38 or 12.5% of the 304 qualified utterances.

The results of single and double-blind tests for each referent are displayed in Figure 13.3. Only seven signs were performed correctly on at least one occasion, while other signs were performed poorly, incorrectly, or not at all. Interestingly, the sign RICE was not performed correctly during the testing sessions even though it was a highly preferred referent. The two best performed signs were those with edible referents (nut and drink) and acquired by both the modified Gardner's and Patterson's criteria just prior to testing. However, other recently acquired signs (PINEAPPLE and BISCUIT) were performed incorrectly or not at all (with the exception of one correct response for PINEAPPLE).

DISCUSSION

Methodological concerns

The methods employed in Project Rinnie intentionally focused on quantifying training effort to elicit an acceptable response by Rinnie who later received a food reward. Such

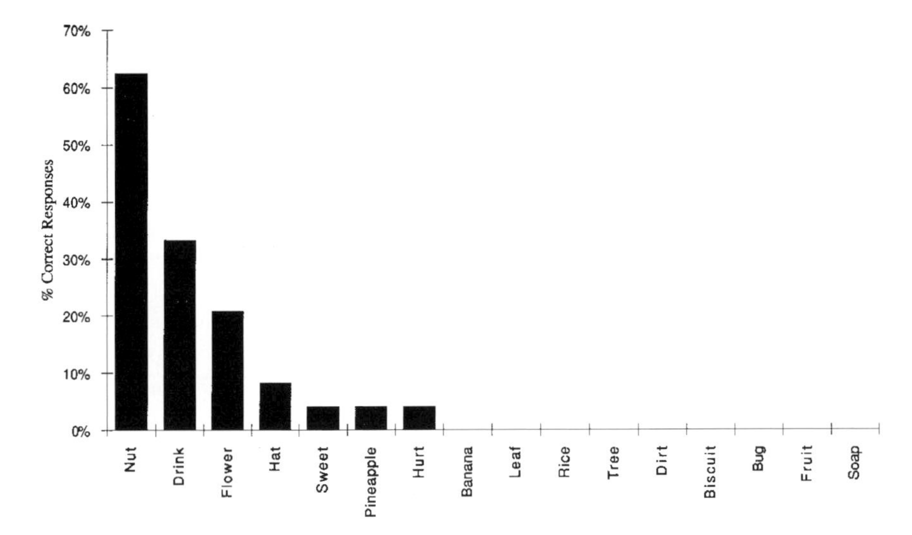

Figure 13.3 Test results by sign trained to Rinnie (double-blind test given 6 months after the start of training.

approaches have been criticized for producing a form of conditioned behavior that is considered less language-like than those observed in more open, enculturated, and less-structured environments. However, because Rinnie was free-ranging and available for a limited period of time, a decision was made to maximize experimenter control within an environment that was difficult to control. It would also be impractical in such an environment to enculturate the adult, free-ranging orangutan into the type of language-enriched environments observed with young, social apes. Indeed, it would have been unethical to modify the trajectory of Rinnie's rehabilitation progress by doing so. The techniques used were effective and appropriate in eliciting and documenting early sign-learning behavior of an adult, free-ranging orangutan.

Vocabulary

Rinnie's vocabulary competence was inferred in a number of ways. According to the criteria set out by Gardner and Gardner (1969) and Patterson (1978), She learned to produce 13 and 17 signs, respectively, during her first 6 months. These signs were observed both within and outside of her vocabulary drill sessions. For example, before and after drill sessions, Rinnie would spontaneously sign requests for food, drink, and contact using some of the signs learned in the sessions. Many signs were observed in cheremic form (Stokoe, Casterline, & Croneberg, 1965), that is with 2 of the 3 motoric elements that define the sign, and were not counted as part of her vocabulary. As observed in other ape language studies, some of Rinnie's signs were used in combinations much earlier and more frequently than when they might have been acquired according to the above criteria (e.g., RICE). Finally, test results suggest that the vocabulary competence of Rinnie was far less than that observed during training and spontaneous signing. Part of her poor performance during testing may have been due to confusion or loss of motivation within an altered transaction environment.

Signed combinations

Signed combinations became a part of Rinnie's spontaneous signing behavior a month after beginning sign training. They were typically requests for food such as, YOU FOOD and GIVE FOOD. After nearly 2 months she changed her strategy of requesting food and replaced the sign for food with the sign for the specific item, e.g., YOU PINEAPPLE or YOU MORE RICE. Her ability to invent a sign and then integrate it into a sequence of signs suggests she had learned part of the "game," but was not as good as other animals taught to create novel sequences of behavior (e.g., dolphins). She was praised and/or given an edible reward for producing combinations which showed logical sign order as in English or Indonesian (e.g., subject–action–object) and for occasionally molding logical sign order (e.g., YOU FEED PINEAPPLE following her production of YOU PINEAPPLE).

Most of Rinnie's combinations of signs employed the use of one hand, typically the hand with which the sign was shaped; however, she was observed to use both hands to sign some combinations. For example, when Rinnie signed, YOU GIVE FOOD, she occasionally used two hands to simultaneously sign, YOU, and GIVE, followed by FOOD.

While most early signed combinations were produced to gain the attention of the trainer (YOU), or specify the repetition of an activity (MORE) prior to producing the actual item or activity, a few appeared to exhibit elements of productivity. For example, she signed, PINEAPPLE DRINK when asked to name the container holding fresh pineapple juice after drinking it for the first time. On another occasion, when asked to name a large ant she had consumed previously, she signed, PINEAPPLE NUT. It could be argued that the signed response was a conceptual error rather than productivity, however, she may have been combining attributes of the ant in making her response (a crunchy head [nut-like] and a sour abdomen [pineapple-like]). She produced the combination more than once, suggesting that the response was not a random combination of signs. Yet the number of such suggestive signed combinations was limited, and the claim for productivity inconclusive.

Signed exchanges

It appears that, for Rinnie, most of the "conversations" or signed exchanges were one-way. Signs were learned as a means to acquire edibles and contact from a provider. This should not be surprising for a species that spends over 60% of its time in search for food or actual foraging (Galdikas, 1979). Her focus on edibles could also have been a by-product of the method of training where positive signing behavior was rewarded with food. Intraspecific grooming is fairly uncommon among older orangutans, though receiving such offered contact, including scratching, appears to be welcomed. There may also be a psychological value for such contact, as Rinnie would spontaneously sign for scratching or grooming when she appeared apprehensive if the trainer raised his voice.

Signed exchanges were begun by either the trainer asking Rinnie, "What want?" or by Rinnie spontaneously signing to the trainer. It was found that spontaneous signing could be elicited when the trainer did not begin the exchange or when the trainer ignored the initial signed response of Rinnie. In these latter cases, the conversation might proceed as the following: Trainer: "What want?" Rinnie: FOOD. Trainer waits. Rinnie: YOU FOOD. Trainer waits. Rinnie: YOU THAT FOOD. The sign, OPEN enabled another set of transactions which could increase the length of exchange. The sign was taught by putting items of interest (usually food) into containers which could be temporarily sealed. Rinnie's hands were placed upon the container and then molded to form the sign, OPEN. The container would be opened by the trainer and the item would be taken out. At this point, she might spontaneously name the item or sign a combination. Alternatively, the trainer would ask her to name the item before giving it to her. Typically, exchanges were a series of signed and nonsigned events leading to the acquisition of an item or activity of interest. It is difficult to discount the possibility that Rinnie learned to spontaneously sign as a means to an end, that is her signing was associative and pragmatic rather than symbolic and productive.

It is also difficult to assess whether Rinnie's signed conversations showed interrogatory features, i.e., were used to pose a question rather than a statement. The Gardners (1969) suggested that Washoe posed implied questions (e.g., TIME EAT?) when she maintained the position of her hand at the terminal location for a longer period of time than normal while at the same time looking at the caretaker. This integration of a sign and eye contact was rarely observed with Rinnie. Normally Rinnie avoided eye contact during the training portion of the sessions (Shapiro & Galdikas, 1995) and engaged in prolonged eye contact when she appeared apprehensive towards the trainer and sought reassurance following a verbal dispute. Such periods were also contexts for requests for grooming and scratching. Occasionally she would maintain eye contact with the trainer while signing, YOU GROOM, but more commonly she would look away.

Signing errors during training

Rinnie's signing errors were primarily motoric in nature. Though some of the signs might be considered difficult to produce, many of her errors were due to sloppy signing in which an incomplete version of the sign was produced. During sign training, the sign would typically have 2 of the 3 cheremic elements (hand configuration, hand movement, and sign location) produced (hence poor) prior to it being correctly produced. Rinnie would also incorrectly name a referent that had motoric elements in common with a correctly produced sign. For example, she might sign, DRINK when asked to name a peanut (nut) as the hand configuration and the location of the sign were the same for both the signs DRINK and NUT. Rinnie would also produce conceptual errors (e.g., signing LEAF when asked to name tree.

Additionally, she would sign erroneously, when asked to name a new referent or one in training, she would use names of established foods. For example, when she was asked to name biscuit she signed PINEAPPLE, NUT, and DRINK. These may have been conceptual errors (all are edibles) or an attempted strategy to acquire food, as there is less risk in signing something that is known and desired, but incorrect, than in not producing anything at all. Either way, the correct sign was molded following an incorrect response and a small piece of food was given. Other errors were recorded when Rinnie merely strung sets of food and nonfood signs together. These errors occasionally occurred when she was overly excited to obtain food from the trainer early in the signing session.

Signing errors during testing

Rinnie's relatively poor performance during testing as opposed to training may have been due to key differences in the transaction procedures. Frequently Rinnie seemed apprehensive during testing, despite efforts to keep her interested and attentive. Certain changes in the normal transaction procedure were unavoidable. For example, classic testing requires that Rinnie does not receive a reward following a correct response. Additionally, in this setting, poor, incorrect, and no responses were not followed by the typical molding and reinforcement transactions. Finally, Rinnie had to interact with individuals who were not usually present during training. Rinnie's apprehension regarding other adult humans or apes in the area may have caused her to be less interested in signing to less familiar individuals.

This explanation seems to account for some of her poor performance. Figure 13.4 illustrates the differential performance by Rinnie as a response to the tester. When good and poor responses are combined, it seemed that Rinnie performed increasingly better with testers with whom she was most familiar and who provided edibles. Conversely, she showed an increased number of nonsigned responses to individuals who were less familiar and who provided fewer edibles.

Comparison with a younger orangutan

Juvenile orangutans seem to interact in a wider range of affiliative behaviors with their caretakers than do adults. Princess, a juvenile, ex-captive, female orangutan (Shapiro, 1982), learned to produce signs in a "home-reared" environment during the same period as Rinnie. Princess learned to produce a larger set of signs as requests for contact activities, including HUG and TICKLE – signs Rinnie did not learn or which could not be taught due to motivation problems. Rinnie did not like to be tickled nor did she show any interest in being hugged. Princess, on the other hand, did not show the type of inventiveness displayed by Rinnie for the contact signs, groom, scratch, and grab-hair.

Spontaneous sign combinations also occurred earlier with Rinnie than with Princess. This may have been due to differences in neuromotor maturation or improved problem-solving abilities by the adult. Rinnie may have learned that producing combinations of signs was something that could lead to a desirable goal. Certainly, the trainer appeared pleased as Rinnie's combinations increased in number and sign length.

While Princess's signing performance has not been analyzed within a communicative context, it seems that she posed implied questions more clearly than Rinnie. Eye contact and the maintained terminal hand was observed usually in feeding contexts, especially when the caretaker did not immediately provide high-quality food following initial

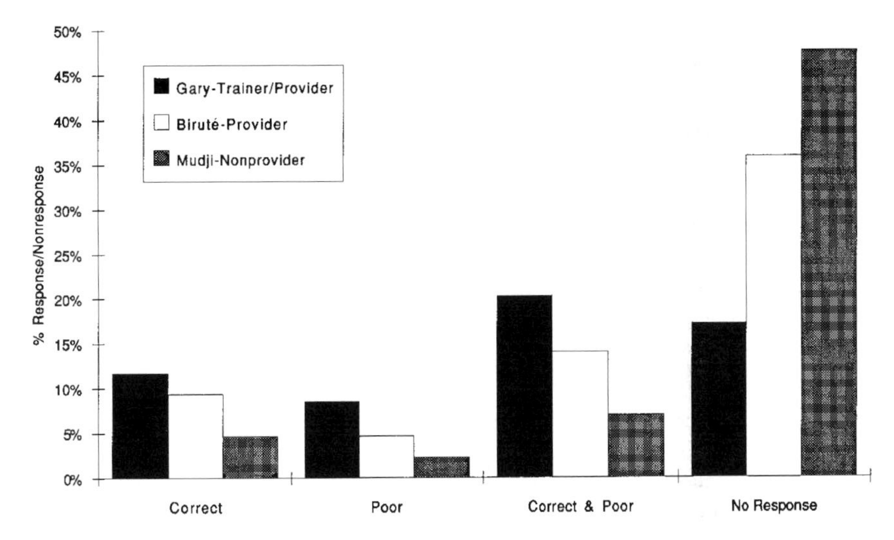

Figure 13.4 Test results for 16 signs trained to Rinnie based on tester (double-blind test given 6 months after training).

signing but waited for Rinnie to sign again. In these cases, hand gestures were of better form and eye contact was more frequently maintained.

Subject motivation and attitude

Rinnie's apparent motivation at the sessions varied during the course of the project. During the first several months of training, Rinnie appeared to show extreme interest in attending the sessions, attending to the training procedure, and interacting with the trainer. She would frequently wait for the trainer and show signs of frustration through redirected aggression toward ground foliage. As discussed earlier, Rinnie's main interest for attending the sessions was to acquire food and drink. Once she was satiated, her motivation to participate in the vocabulary training waned. At these times she would be content to sit back and regurgitate. Signing during these periods was typically for contact, though requests for contact also took place when food was available.

Rinnie occasionally showed signs of frustration when she could not, or was not allowed to, take the highly preferred bottle of sweetened milk from the trainer. She attempted to take the milk forcibly from the trainer during several of the sessions. When this occurred she was verbally reprimanded. Nevertheless, Rinnie's behavior was primarily positive during the project, and she was very manageable within the training context.

Rinnie displayed a notable species-specific tendency during the course of the project. This was the noticeable aversion of eye contact from important target locations in the sign-learning environment. During the sign response and molding portion of the training trial, Rinnie typically looked away towards the nearby trees or the food referents. The trainer would have to physically move her head to look toward the site of training. It is assumed that Rinnie would learn to make the referent/gestural association more rapidly if she closely observed this key transaction. Gaze aversion has been observed in other orangutans (Shapiro & Galdikas, 1995; Miles, personal communication) and may represent a species-specific tendency to redirect attention toward items of interest (other orangutans, future paths of travel). It may also be a limiting feature to orangutan sign training.

Ethical issues and the future of orangutan research

The Rinnie Project represents one of the first in which a great ape was given a choice to participate in an experiment where she was physically and psychologically manipulated over a period of many months. At the end of the Project, Rinnie continued on her self-directed program of rehabilitation and eventually returned to the forest for good (Galdikas & Shapiro, 1996).

Rinnie was never chained, leashed, or caged, and, while she showed occasional irritation toward the trainers for withholding a drink, she never posed a significant threat to them. Rinnie elected to attend the sessions on her own terms. Occasionally she did not attend. When she was sated with food, drink, or attention, she either left the training area to return to the surrounding forest or she contentedly stayed on until after the trainers left. During the course of the hour or so, she was given a set of tasks which, from the trainer's perspective, provided information about the learning and language capabilities of the orangutan.

There is still much to learn about orangutan cognition and mentality. Shapiro (1991) has argued that rehabilitation stations can provide other scientists with potential subjects for noninvasive research in the areas of cognition, learning, and experimental behavior, especially as it applies to subjects' adaptations to native habitats. This approach is contingent upon the experimenter's willingness to lose some control in manipulating subjects and environmental conditions. The benefit, however, would be not only acquiring new information about this marvelous red ape, but treating the individual orangutan with the dignity and respect it deserves.

Animal welfare has taken on an increased dimension within academia during the past decade, especially as it relates to the primates. This has been part of a growing trend toward the social responsibility of scientists; it is an extension of similar trends in the human and environmental arena (e.g., nuclear proliferation, bio-engineering, indigenous cultures). While great apes have recently been accorded attention from bioethicists, philosophers, and scientists regarding their inclusion into the "moral community of equals" (Cavalieri & Singer, 1994), they have suffered and continue to suffer from the hands and greed of humankind.

SUMMARY

Rinnie's signing performance during the first 6 months of training indicated that a free-ranging adult orangutan could, on her own free will, enter into a long-term relationship with an experimenter and learn a body of signs. The study marks the first time any great ape, albeit an ex-captive, free-ranging one, has been given sign learning training in the species own natural habitat.

Project Rinnie shows that adult orangutans will use signs as symbolic instruments to acquire edibles and contact activity. Additionally, it appears that older orangutans quickly begin to produce combinations of signs and will learn that certain actions can be ritualized to form signs not intentionally taught. These signs are naturally integrated into the spontaneous sign combinations of the ape.

REFERENCES

Cavalieri, P. & Singer, P. (eds.) (1994). *The great ape project*. New York: St. Martin's Press.

Fouts, R. (1972). The use of guidance in teaching sign language to a chimpanzee. *Journal of Comparative and Physiological Psychology*, **80**, 515–22.

Galdikas, B. M. F. (1979). Orangutan adaptation at Tanjung Puting Reserve: mating and ecology. In D. Hamburg and E. McCown (eds.), *The behavior of the great apes*. Menlo Park: Staples Press.

Galdikas, B. M. F. & Shapiro, G. L. (1996). Orangutan ethics. In P. Cavalieri (ed.), *Etica & Animali* (vol. 8/96, pp. 50–67. Milan: Animus.

Gardner, B. T. & Gardner, R. A. (1971). Two-way communication. With an infant chimpanzee. In A. Schrier & F. Stollnitz (eds.), *Behaviour of nonhuman primates* (pp. 117–184). New York: Academic Press.

Gardner, R. A. & Gardner, B. T. (1969). Teaching sign language to a chimpanzee. *Science*, **165**, 664–672.

Kawai, M. (1965). New acquired precultural behavior of the natural troop of Japanese macques. *Primates*, **6**, 1–30.

Miles, H. L. (1983). Apes and language: the search for communicative competence. In J. de Luce and H. T. Wilder (eds.), *Language in primates – perspectives and implications*. New York: Springer-Verlag.

Patterson, F. (1978). The gestures of a gorilla: Language acquisition in another pongid. *Brain and Language*, **5**, 72–97.

Shapiro, G. L. (1982). Sign acquisition in a home-reared / free-ranging orangutan: Comparisons with other signing apes. *American Journal of Primatology*, **3**, 121–129.

Shapiro, G. L. (1991). The use of rehabilitation stations as adjuncts to orangutan research. Presented at the *Second International Conference on Great Apes of the World*, Jakarta/Pangkalan Bun/Tanjung Puting, Kalimantan Tengah, Indonesia.

Shapiro, G. & Galdikas, B. M. F. (1995). Attentiveness of orangutans within the sign learning context. In R. D. Nadler, B. M. F. Galdikas, L. Sheeran, & R. Rosen (eds.), *The neglected ape* (pp. 199–212). New York: Plenum Press.

Stokeo, W. C., Casterline, D., & Croneberg, C. (1965). *A dictionary of American Sign Language*. Washington, DC: Gallaudet College Press.

Terrace, H. (1979). *Nim*. New York: Knopf.

PART IV

Social cognition in gorillas and orangutans

14

Comparative aspects of mirror self-recognition in great apes

KARYL B. SWARTZ, DENA SARAUW, AND SIÂN EVANS

Mirror self-recognition (MSR) is a phenomenon with a more than 20-year empirical history that continues to stir controversy and debate (see Parker, Mitchell, & Boccia, 1994b; Swartz & Evans, 1997). Since Gallup's (1970) first description of MSR, the challenge to comparative psychology has been to interpret the phenomenon (see Gallup, 1970, 1977, 1982, 1994; Anderson, 1984; Parker, 1991; Mitchell, 1993a,b, 1994, 1997a,b; Parker & Mitchell, 1994; Swartz, 1997; Swartz & Evans, 1997). Gallup's subjects were four common chimpanzees (*Pan troglodytes*), aged approximately 3.5 to 6 years (Swartz & Evans, 1994, 1997). The phenomenon, as described by Gallup (1970), is as follows.

Chimpanzees were provided with visual access to a mirror 8 hours a day for 10 days. The subjects were caged singly for the duration of mirror exposure. The initial responses to the mirror-image were social in nature, as though the mirror image were an unfamiliar conspecific. Over the course of 2 to 3 days of mirror exposure, social behaviors waned, and behaviors that Gallup called "self-directed" began to occur. These self-directed behaviors included the use of the mirror to investigate and manipulate body areas (such as the face) that were otherwise invisible to the animal, as well as such behaviors as blowing bubbles or making faces while looking at the mirror-image. Gallup interpreted the demonstration of self-directed behaviors as indicating that the chimpanzees recognized themselves in the mirror; however, because of the potentially subjective nature of this interpretation, he devised a more objective test of self-recognition, often called the "mark test."

In the mark test, the chimpanzees were anesthetized and a red mark was placed on the brow and one ear, locations that were invisible to the animal without the use of the mirror. Following a no-mirror control period, the marked subjects were provided with the mirror. Compared to the control period in which only one mark touch occurred, all four subjects demonstrated a high level of mark-touching, using the mirror to guide the hand to the mark. One subject smelled her hand following a touch to the mark. Gallup (1970) proposed that the results of the mark test with chimpanzees demonstrated that they recognized themselves in the mirror. Chimpanzees with no mirror exposure prior to a mark test did not touch the mark, and macaque monkeys (*Macaca* spp.) showed no self-directed behavior during mirror exposure nor did they pass the mark test.

Since Gallup's (1970) first paper on MSR, the common chimpanzee (*Pan troglodytes*) has been the model for the phenomenon. The progression from social behavior to self-directed behavior, and the manner of responding to the mirror-image, as well as the structure of the mark test, have been the standard against which other species were measured. The phenomenon has been investigated with a large number of individuals from several nonhuman primate species (see Anderson, 1984), but the capacity for MSR appears to remain limited to great apes. The phenomenon has been replicated with chimpanzees (e.g., Swartz & Evans, 1991; Povinelli, Rulf, Landau, & Bierschwale, 1993) and orangutans (*Pongo pygmaeus*; Lethmate & Dücker, 1973; Suarez & Gallup, 1981), but it is only recently that gorillas (*Gorilla gorilla*; Patterson & Cohn, 1994; Parker, 1994; Swartz & Evans, 1994) have been added to the list of nonhuman primates who have passed the mark test. Preliminary studies with bonobos (*Pan paniscus*; Hyatt & Hopkins, 1994; Westergaard & Hyatt, 1994; Walraven, Elsacker, & Verheyen, 1995) suggest that they show self-directed behavior when presented with a mirror. In general, monkeys do not show MSR (see Anderson, 1984); however, with special training, some monkeys have touched marks on their faces while looking in the mirror (Itakura, 1987; Thompson & Boatright-Horowitz, 1994; Howell, Kinsey, & Novak, 1994; but see also Swartz, 1997). It should be noted here that pigeons (*Columba livia*), with careful and prodigious training, have demonstrated a phenomenon that has been called MSR (Epstein, Lanza, & Skinner, 1981); but that result has not been replicable (Thompson & Contie, 1994). As Gallup (1982) has argued so thoroughly, the demonstration of mark-touching that arises following careful training of component responses is not indicative of the same psychological process as that demonstrated by chimpanzees and other great apes who spontaneously demonstrate self-directed behaviors and who require no additional training beyond mirror exposure to pass the mark test.

Because the demonstration of self-directed behavior and passing the mark test have become the benchmarks of MSR, it is important to consider each of these indicators from a methodological standpoint before specifically comparing species. There are differences across empirical studies with respect to measurement or determination of MSR. First, Gallup's initial study as well as several that followed (Gallup, McClure, Hill, & Bundy, 1971; Hill, Bundy, Gallup, & McClure, 1970; Swartz & Evans, 1991), included a wide range of behavior in the self-directed category. Specifically, a class of behaviors that other researchers (Parker, 1991; Lewis & Brooks-Gunn, 1979; Lin, Bard, & Anderson, 1992) labeled contingent behaviors were included in the self-directed category. Swartz and Evans (1991), differentiated these behaviors from self-directed behavior, referring to them as "self-referred" behaviors, i. e., behaviors with the self (or body) as a referent but not necessarily directed at the body. This category included such behaviors as making faces, looking in the mirror while manipulating food wads with the lips, watching in the mirror while directing an action toward the environment such as catching water in the mouth from an overhanging spout or manipulating the cage lock with the hand. These behaviors were called self-referred by Swartz and Evans because the animal watched an action or its effect in the mirror rather than looking at it directly, but, in keeping with Gallup (1970), they were reported as self-directed behaviors.

The idea here, suggested by Lewis and Brooks-Gunn (1979) in their studies of MSR in human infants, is that contingent behavior involves testing the contingent relationship between the mirror-image and the animal's body. This contingent relationship can be observed by moving the body while watching the mirror-image. These behaviors may, in fact, provide the organism with important information regarding the nature of the mirror-image. Contingent behaviors may precede the onset of actual self-directed behaviors. Early studies, then, may have inflated estimates of the onset and frequency of self-directed behaviors. Because of this practice of collapsing behaviors into three major functional categories in studies of MSR, it has been unclear exactly what specific behaviors fall into each category. Further, because the category of contingent behaviors was introduced into comparative studies later than other categories, it is unclear what relationship exists between contingent and self-directed behaviors in nonhuman primates. From a theoretical perspective, it would appear that understanding the contingent relationship between the animal's own behaviors and those of the mirror-image should lead to understanding of the mirror-image, and more importantly, knowledge of the source of the mirror-image (the animal's own body) (Parker, 1991). Such understanding should lead to self-directed behaviors. As will be discussed later, this analysis of contingent behaviors is not the only theoretical perspective on contingent and self-directed behaviors (see Mitchell, 1993a).

The use of functional categories such as self-directed, contingent, and social responses requires interpretations about the motivation or purpose underlying each individual behavior. It presupposes an understanding of the function of each individual behavior, an assumption that may not always hold. In the category of social behavior, for example, some behaviors are difficult to classify (e.g., lip flip in chimpanzees) and others that may appear to be social may occur in nonsocial situations as well (see Thompson & Boatright-Horowitz, 1994, for discussion). Further, the use of functional categories assigns equal importance to each behavior within a category. Although for purposes of communication it is necessary to reduce the number of individual behaviors reported, caution must be used in developing and communicating these categories. It is helpful that more recent reports have provided details concerning which specific behaviors are included in each functional category (Lin et al., 1992; Povinelli et al., 1993).

Also at issue are certain methodological concerns with the mark test. Gallup (1994) suggests that such variables as color and salience of the mark as well as lack of obtrusiveness in applying the mark are important considerations. In the initial study, Gallup (1970) used an odorless red dye (Rhodamine B). The color red was chosen because of its possible biological significance (i.e., the color of blood or a wound). In addition to Rhodamine B, researchers have used clown paint (Suarez & Gallup, 1981; Swartz & Evans, 1991), children's make-up (Lin et al., 1992), or white typewriter correction fluid (Calhoun & Thompson, 1988; Thompson & Boatright-Horowitz, 1994). In order to apply the mark unobtrusively, animals that were not anesthetized were trained to allow their faces to be touched on a regular basis with a cloth or implement that was used on the day of the mark test to apply color. The sham marking procedure appears to be effective in allowing application of the mark without detection by the animal, and it eliminates possible problems in interpretation of behavior following anesthesia. For example, Heyes (1994) has suggested that an increase in mark touches during the mirror-present condition compared to the preceding control period could be an artifact of head-rubbing motions that become more extensive and frequent as the animal recovers from anesthesia. However, Heyes' suggestion loses all credibility in light of repeated demonstrations of mark-directed behavior in animals who were not subjected to anesthesia prior to the mark test (Lin et al., 1992; Parker, 1994).

In order to determine salience of the mark to the subject, Gallup (1994) suggested two control procedures. The first involves observing the reactions of cagemates to a marked individual. If the other animals investigate the mark on the marked animal, that suggests that the mark is sufficiently salient to induce a mark- directed response if the target animal understands the nature of the mirror-image. Second, a subject can serve as its own control if a mark is placed on the body in a location visible without the use of the mirror, such as the wrist. If the animal investigates the mark on the wrist but does not touch the mark on the face or ear, that procedure eliminates one possible artifactual explanation for failure to pass the mark test. In discussing the role of the mark test in determining MSR, Gallup (1994) suggests that the presence of self-directed behavior during mirror exposure should be a prerequisite to conducting the mark test, although that procedure has not always been followed (Suarez & Gallup; 1981; Swartz & Evans, 1991; Povinelli et al., 1993; see discussion in Mitchell, 1993b, p. 355).

But the major question is "What is passing?" with respect to the mark test. The easy answer is "touching the mark on the head while using the mirror to guide the hand to the mark." However, behavior is rarely as simple as that. Questions arise about the supposed relationship between looking and touching the mark. What if the animal touched the mark after looking at its marked image in the mirror, but did not look at the mirror-image while touching? What about touches that do not coincide with looking, but occur immediately after? If we agree that immediate touches to the mark following inspection of the marked mirror-image are consistent with MSR, how long after mirror-image inspection can the touch occur? Does the latency of touching vary with species in our criterion for passing the mark test? What is the interpretation of mark touches if other behaviors intervene between seeing the marked image in the mirror and touching the mark? Although some of these issues were discussed by Patterson & Cohn (1994) in their description of an inconclusive mark test with the gorilla Michael, such issues have not been directly assessed, either theoretically or empirically.

In addition to the above definitional issues, the question of appropriate controls during the mark test must be addressed. There are at least two types of control measures. The first addresses the possibility that the animals might detect the presence and location of the mark by some means other than mirror-based visual inspection, for example, by feeling or smelling the mark and locating it tactually before mirror presentation allows its discovery by visual means. Most studies have controlled for this by using a no-mirror control period prior to the mirror period during the mark test (Gallup, 1970). Still others have presented the nonreflective side of the mirror during the control period (e.g., Swartz & Evans, 1991). Others have counted mark touches while looking in the mirror and compared them to mark touches while not looking in the mirror (Lin et al., 1992). In this case if mirror-guided touches to the mark exceed non-mirror-guided touches, the animal is said to pass the mark test. A second concern relates to the baseline rate for touching marked areas of the face. A control measure for this problem is to count the number of touches, mirror-guided or not, that occur to the other (unmarked) brow. Swartz & Evans (1991) included this measure in the observations, although we did not report it. Looking at the data for nonmarked brow touches, we found that, of the 8 animals who showed some interest in the mirror and some self-directed behaviors (Amélie, Berthe, Henri, JoJo, Koula, Makata, Masuku, Nestor, and N'tébé), only 1 (Amélie) animal touched the other brow once during any mark test. That touch was not mirror-guided.

Species differences in response to the mirror are also a source of interest and controversy. Individuals from all four great ape species have shown evidence of self-directed behavior. Table 14.1 presents a summary beginning with Gallup's (1970) first report of research addressing MSR with chimpanzees, bonobos, orangutans, and gorillas. In some studies, only mirror exposure was given with no mark test. In the Table, the total number of animals studied is presented, with a break-down of the number who showed self-directed behavior during mirror exposure and the number who passed the mark test. At the end of the list for each species, a total is presented, along with the percentage of those tested who showed evidence of MSR through self-directed behavior or touching the mark. An attempt was made to present data only once from each individual animal, so that when data from the same individuals were reported in two separate papers, the results from only one study were presented. In some cases, the judgment about the presence of self-directed behavior is based on descriptive passages in the text of the published article. Although these few instances may inflate the estimate of how many animals have been studied for MSR, the decision was made to be *in*clusive rather than *ex*clusive.

As the Table shows, chimpanzees are the most highly represented species. However, contrary to what was once believed about chimpanzees (Swartz & Evans, 1997), fewer than 50% of animals studied showed evidence of MSR using either criterion (self-directed behavior and passing the mark test). Similarly, of the few bonobos studied, only half demonstrated self-directed behavior.

Orangutans showed the highest proportion of MSR (85% showed self-directed behavior and 50% passed the mark test), and gorillas, the lowest (29% and 31%, respectively, on the two measures). However, chi-square analyses show that there is no significant difference across species in the proportion of those who do or do not show self-directed behavior [χ^2 (3) = 6.02, $p = 0.11$] or pass the mark test [χ^2 (2) = 0.89, $p = 0.64$]. Note that bonobos were not included in the latter analysis as to date none have been given a mark test.

The above finding is striking in light of previous assumptions about the robustness of the phenomenon in chimpanzees and its absence in gorillas; however the small number of gorillas studied leaves the situation with gorillas still somewhat unclear (see Swartz &

Table 14.1 *Results of* MSR *studies across great ape species*

Species	Reference	*N*	Self-directed behavior?	Mark test passed?
Pan troglodytes	Gallup, 1970	4	4 of 4	4 of 4
	Hill et al., 1970	3	2 of 3	2 of 3
	Gallup et al., 1971	6	3 of 6	3 of 6
	Lethmate & Dücker, 1973	6	4 of 6	2 of 2
	Suarez & Gallup, 1981	4	4 of 4	3 of 4
	Robert, 1986	1	0 of 1	0 of 1
	Calhoun & Thompson, 1988	2	2 of 2	2 of 2
	Swartz & Evans, 1991	11	4 of 11	1 of 11
	Lin, et al., 1992	12	8 of 12	7 of 12
	Povinelli et al., 1993	92	30 of 92	13 of 42
	Hyatt & Hopkins, 1994	8	7 of 8	Not given
	Parker, 1994	4	0 of 4	Not given
	Thompson & Boatright-Horowitz, 1994	10	5 of 10[a]	5 of 10
Summary for Pan troglodytes		163	73 of 163 (45%)	42 of 97 (43%)
Pan paniscus	Westergaard & Hyatt, 1994[a]	9	4 of 9	Not given
	Walraven et al., 1995	7	4 of 7	Not given
Summary for Pan paniscus		16	8 of 16 (50%)	—
Pongo pygmaeus	Lethmate & Dücker, 1973	2	2 of 2	1 of 2
	Suarez & Gallup, 1981	2	1 of 2	1 of 2
	Robert, 1986	1	1 of 1	0 of 1
	Miles, 1994	1	1 of 1	1 of 1
Summary for Pongo pygmaeus		6	5 of 6 (85%)	3 of 6 (50%)
Gorilla gorilla	Riopelle, Nos, & Jonch, 1971	2	1 of 2	Not given
	Suarez & Gallup, 1981	4	0 of 4	0 of 4
	Ledbetter & Basen, 1982	2	0 of 2	0 of 2
	Law & Lock, 1994	4	3 of 4	Not given
	Parker, 1994	4	1 of 4	2 of 2
	Patterson & Cohn, 1994	2	1 of 2	1 of 2
	Swartz & Evans, 1994	3	0 of 3	1 of 3
Summary for Gorilla gorilla		21	6 of 21 (29%)	4 of 13 (31%)

[a] Two of the animals included in Thompson & Boatright-Horowitz (1994) were those tested a second time in Calhoun & Thompson (1988). These animals are included only once in the count. There is an additional reference that deals with MSR in *Pan paniscus* (Hyatt & Hopkins, 1994) but the data from those animals are reported in Westergaard & Hyatt (1994).

Evans, 1997). Although it is clear that some gorillas have demonstrated evidence of MSR using the presence of self-directed behavior and/or passing the mark test as criteria, it is not yet clear why MSR has appeared to be difficult to find in gorillas. Although the above analyses provide no statistically significant differences among species, in 3 of 7 studies conducted with gorillas, no evidence of self-directed behaviors was reported. That contrasts with 2 of 13 studies using chimpanzees failing to find evidence of self-directed behavior, and no studies with bonobos or orangutans failing to find such evidence. Lack of interest in the mirror, low baseline levels of autogrooming, and gaze aversion have all been suggested as reasons for the low level of responsiveness of gorillas to their mirror-image (see Gallup, 1994; Patterson & Cohn, 1994; Swartz & Evans, 1994, for discussions). However, it is still unclear whether any or all of these factors is a likely explanation for obtained species differences. In unpublished data from King, a gorilla housed at Monkey Jungle in Miami, Florida, we (Evans and Swartz) observed that even though King showed a great deal of interest in the mirror, his behavior in front of the mirror was very subtle. Contingent behaviors included nostril flaring, eyebrow raising, and lip puckering. If the chimpanzee is used as the model for mirror behavior, these subtle responses might not be observed, or if observed, might be interpreted as rudimentary social behaviors and of little consequence. Clearly, more work remains to be done with gorillas as subjects.

Further, much more attention must be directed toward specifying the nature of species differences in mirror behavior. It is unclear what differences, if any, are shown among chimpanzees, bonobos, and orangutans. Studies of these three species have typically presented functional categories of self-directed, contingent, and social behavior. As above, there may be species differences in the onset and frequency of particular behaviors that are lost by using these functional categories. For example, in unpublished data collected with one chimpanzee and one orangutan, we (Swartz, Sarauw, & Evans) found that the orangutan showed more contingent and self-directed behavior during initial mirror exposure than did the chimpanzee; however, the chimpanzee passed the mark test but the orangutan did not. We are currently analyzing those data in detail to determine what, if any, species differences are shown as these animals explore their mirror-images.

The importance of determining species differences becomes more evident in light of individual differences reported within a particular species. Although chimpanzees have provided the model for MSR, there are individual differences among chimpanzees. Not all chimpanzees show MSR using the mark-test criterion (Swartz & Evans, 1991). Using the criterion of showing self-directed behavior, Povinelli et al. (1993) found significant individual differences in the demonstration of MSR among a large population of captive chimpanzees. In human infants, there are clear developmental differences, with MSR emerging at about 15–18 months in some infants but not until 24 months in others (Lewis & Brooks-Gunn, 1979). The developmental progression of MSR in chimpanzees is unclear. Lin et al. (1992) obtained evidence for MSR using a modified mark test in chimpanzees as young as 2.5 years; younger animals did not show evidence of MSR. Hill et al. (1970) reported evidence for MSR in chimps under 3 years of age. Robert (1986) failed to obtain evidence of MSR in an 11–month-old chimpanzee provided with limited mirror exposure. Calhoun & Thompson (1988) found evidence for retention of MSR across one year in a 3–year-old and a 4–year-old chimpanzee, which meant that these animals were 2 and 3, respectively, when they first passed the mark test. Povinelli et al. (1993) report that, with one exception, their chimpanzees did not show MSR until 5 years or older. The subjects in Gallup's (1970) original study were approximately 3.5 to 6 years (Swartz & Evans, 1994, 1997). It would appear the capacity to show MSR develops in the chimpanzee during the third year of life, and is shown robustly by 5 to 6 years of age.

In the only developmental study of MSR in great apes, Miles found that Chantek, a language-trained orangutan, first passed the mark test at the age of 2 years, 1 month, during the developmental stage in which he was beginning to show evidence of subjective representation, including linguistic self-reference and referential pointing (Miles, 1994). Chantek failed mark tests given when he was 2.5 to 3.5 years old, but passed them consistently following that. Miles suggested that the size of the mirror may have played a role in these failures. Patterson and Cohn (1994) describe mirror responses of Koko, a language-trained gorilla, during development. They reported that Koko showed self-directed behaviors using the mirror to guide her responses at 3.5 years of age. As Miles (1994) reported with Chantek, Patterson & Cohn observed that the onset of Koko's mirror-guided self-directed behaviors coincided with the development of related cognitive capacities such as the use of personal pronouns, symbolic play, and expressions of intentionality. Clearly, additional longitudinal developmental studies of MSR in great apes are in order, to determine not only the developmental time course of MSR, but also what early experiences, if any, affect its demonstration. The use of the comparative developmental approach demonstrated by Miles' (1994) analysis of Chantek's mirror responses (see also Parker, 1991, 1997; Parker, Mitchell, & Boccia, 1994b) is a strong approach to delineating and understanding species differences in MSR, and, as Parker (1991; 1997) has demonstrated, the cognitive abilities that underlie this capacity.

It may be the case that gorillas, orangutans, chimpanzees, and bonobos show MSR in similar but species-typical ways. Self-directed behaviors in other great apes often differ from those of the Gallup (1970) chimpanzee prototype that has guided our thinking. Species differences may reflect different ways of engaging the environment and must be taken into account.

But we still do not know exactly what the implications of showing MSR are. There remain various interpretations of the phenomenon. Gallup's (1970, 1975, 1977, 1982) interpretation is a very rich interpretation that imputes mind to self-recognizing organisms through increasing levels of awareness. Mirror self-recognition, suggests Gallup (1982) implies not only awareness, but also self-awareness. The presence of self-awareness provides the ability to monitor one's own state (mind) and to impute knowledge or emotional states to other like organisms (Theory of Mind). However, despite attempts to investigate the chimpanzee's ability to impute knowledge states to other organisms (Premack & Woodruff, 1978; Povinelli, Nelson, & Boysen, 1990), the results of such studies remain unclear, and the question about Theory of Mind in chimpanzees unresolved (Savage-Rumbaugh, Rumbaugh, & Boysen, 1978; Povinelli & Eddy, 1996; see Swartz, 1998).

Alternative explanations of MSR have been proposed that do not rely on or impute higher levels of awareness to organisms who show MSR (Parker, 1991; Mitchell, 1993a,b). Mitchell (1993a,b; 1997a,b) has proposed that the demonstration of MSR can be explained as a function of the organism's ability to show kinesthetic-visual matching, and is related to such cognitive capacities as planning (delayed imitation), recognition that one is being imitated, and pretense. Parker (1991), as does Mitchell, discusses the role of imitation and of understanding contingency in demonstrating MSR, in the framework of determining the Piagetian capacities demonstrated across species that may provide insight into the cognitive capacities that underlie MSR. Swartz (1997) has also argued that the demonstration of MSR does not necessitate attributing such cognitively complex functions as self-awareness or mind. That is, the animal may come to understand the nature of the mirror-image, i.e., that the source of the image is its own body, without having cognitive understanding of the self, and what it means to have a self.

At this point, we are left with the final conclusion that MSR is an interesting, challenging phenomenon, and our theoretical understanding of the phenomenon will grow considerably once we build species and individual differences into the theoretical model and address them specifically in our empirical investigations. Methodological issues may interact with species differences, leading to the need for more precise definitions of theoretical concepts. Finally, the relationships among MSR, Theory of Mind, and Piagetian capabilities will provide a rich area for theory-building and empirical investigation. It is expected that MSR, a phenomenon that has maintained the interest of comparative psychologists for over 20 years, will continue to provide a rich source of theory and empirical investigations.

REFERENCES

Anderson, J. R. (1984). Monkeys with mirrors: Some questions for primate psychology. *International Journal of Primatology*, **5**, 81–98.

Calhoun, S. & Thompson, R. L. (1988). Long-term retention of self-recognition by chimpanzees. *American Journal of Primatology*, **15**, 361–365.

Epstein, R., Lanza, R. P., and Skinner, B. F. (1981). "Self-awareness" in the pigeon. *Science*, **212**, 695–696.

Gallup, G. G., Jr. (1970). Chimpanzees: Self-recognition. *Science*, **167**, 86–87.

Gallup, G. G., Jr. (1975). Towards an operational definition of self-awareness. In R. H. Tuttle (ed.), *Socioecology and psychology of primates* (pp. 309–342). The Hague, The Netherlands: Mouton.

Gallup, G. G., Jr. (1977). Self-recognition in primates: A comparative approach to the bidirectional properties of consciousness. *American Psychologist*, **32**, 329–338.

Gallup, G. G., Jr. (1982). Self-awareness and the emergence of mind in primates. *American Journal of Primatology*, **2**, 237–248.

Gallup, G. G., Jr. (1994). Self-recognition: Research strategies and experimental design. In S. T.

Parker, R. W. Mitchell, & M. L. Boccia (eds.), *Self- awareness in animals and humans: Developmental perspectives* (pp. 35–50). Cambridge University Press.

Gallup, G. G., Jr., McClure, M. K., Hill, S. D., & Bundy, R. A. (1971). Capacity for self-recognition in differentially reared chimpanzees. *The Psychological Record*, **21**, 69–74.

Heyes, C. M. (1994). Reflections on self-recognition in primates. *Animal Behaviour*, **47**, 909–919.

Hill, S. D., Bundy, R. A., Gallup, G. G., Jr., & McClure, M. K. (1970). Responsiveness of young nursery reared chimpanzees to mirrors. Proceedings of the Louisiana Academy of Sciences, **33**, 77–82.

Howell, M., Kinsey, J., & Novak, M. (1994). *Mark-directed behavior in a rhesus monkey after controlled, reinforced exposure to mirrors*. Paper presented at the annual meeting of the American Society of Primatologists, Seattle, WA, July, 1994.

Hyatt, C. W. & Hopkins, W. D. (1994). Self-awareness in bonobos and chimpanzees: A comparative perspective. In S. T. Parker, R. W. Mitchell, & M. L. Boccia (eds.), *Self-awareness in animals and humans: Developmental perspectives* (pp. 248–253). Cambridge University Press.

Itakura, S. (1987). Use of a mirror to direct their responses in Japanese monkeys (*Macaca fuscata fuscata*). Primates, **28**, 343–352.

Law, L. E. & Lock, A. J. (1994). Do gorillas recognize themselves on television? In S. T. Parker, R. W. Mitchell, & M. L. Boccia (eds.), *Self-awareness in animals and humans: Developmental perspectives* (pp. 308–312). Cambridge University Press.

Ledbetter, D. & Basen, J. (1982). Failure to demonstrate self-recognition in gorillas. *American Journal of Primatology*, **2**, 307–310.

Lethmate, J., & Dücker, G. (1973). Untersuchungen zum Selbsterkennen im Spiegel bei Orang-utans und einigen anderen Affenarten. *Zeitschrift für Tierpsychology*, **33**, 248–269.

Lewis, M. & Brooks-Gunn, J. (1979). *Social cognition and the acquisition of the self*. New York: Plenum Press.

Lin, A. C., Bard, K. A., & Anderson, J. R. (1992). Development of self-recognition in chimpanzees (*Pan troglodytes*). *Journal of Comparative Psychology*, **106**, 120–127.

Miles, H. L. W. (1994). Me Chantek: The development of self-awareness in a signing orangutan. In S. T. Parker, R. W. Mitchell, & M. L. Boccia (eds.), *Self- awareness in animals and humans: Developmental perspectives* (pp. 254–272). Cambridge University Press.

Mitchell, R. W. (1993a). Mental models of mirror self-recognition: Two theories. *New Ideas in Psychology*, **11**, 295–325.

Mitchell, R. W. (1993b). Recognizing one's self in a mirror? A reply to Gallup and Povinelli, De Lannoy, Anderson, and Byrne. *New ideas in psychology*, **11**, 351–377.

Mitchell, R. W. (1994). Multiplicities of self. In S. T. Parker, R. W. Mitchell, & M. L. Boccia (eds.), *Self-awareness in animals and humans: Developmental perspectives* (pp. 81–107). Cambridge University Press.

Mitchell, R. W. (1997a) A comparison of the self-awareness and kinesthetic-visual matching theories of self-recognition: Autistic children and others. *Annals of the New York Academy of Sciences*, **818**, 39–62.

Mitchell, R. W. (1997b). Kinesthetic-visual matching and the self-concept as explanations of mirror-self-recognition. *Journal for the theory of social behavior*, **27**, 101–123.

Parker, S. T. (1991). A developmental approach to the origins of self-recognition in great apes. *Human Evolution*, **6**, 435–449.

Parker, S. T. (1994). Incipient mirror self-recognition in zoo gorillas and chimpanzees. In S. T. Parker, R. W. Mitchell, & M. L. Boccia (eds.), *Self-awareness in animals and humans: Developmental perspectives* (pp. 301–307). Cambridge University Press.

Parker, S. T. (1997). A general model for the adaptive function of self-knowledge in animals and humans. *Consciousness and Cognition*, **6**, 75–86.

Parker, S. T., & Mitchell, R. W. (1994). Evolving self-awareness. In S. T. Parker, R. W. Mitchell, & M. L. Boccia (eds.), *Self-awareness in animals and humans: Developmental perspectives* (pp. 413–427). Cambridge University Press.

Parker, S. T., Mitchell, R. W., & Boccia, M. L. (1994a). Expanding dimensions of the self: Through the looking glass and beyond. In S. T. Parker, R. W. Mitchell, & M. L. Boccia, *Self-awareness in animals and humans: Developmental perspectives* (pp. 3– 20). Cambridge University Press.

Parker, S. T., Mitchell, R. W., & Boccia, M. L. (eds.), (1994b). *Self-awareness in animals and humans: Developmental perspectives*. Cambridge University Press.

Patterson, F. G., & Cohn, R. H. (1994). Self-recognition and self-awareness in lowland gorillas. In S. T. Parker, R. W. Mitchell, & M. L. Boccia (eds.), *Self-awareness in animals and humans: Developmental perspectives* (pp. 273–290). Cambridge University Press.

Povinelli, D. J. & Eddy, T. J. (1996). What young chimpanzees know about seeing. *Monographs of the Society for Research in Child Development*, **61**(3).

Povinelli, D. J., Nelson, K. E., & Boysen, S. T. (1990). Inferences about guessing and knowing by chimpanzees (*Pan troglodytes*). *Journal of Comparative Psychology*, **104**, 203–210.

Povinelli, D. J., Rulf, A. B., Landau, K., & Bierschwale, D. (1993). Self-recognition in chimpanzees (*Pan troglodytes*): Distribution, ontogeny, and patterns of emergence. *Journal of Comparative Psychology*, **107**, 347–372.

Premack, D. & Woodruff, G. (1978). Does the chimpanzee have a theory of mind? *Behavioral and Brain Sciences*, **1**, 515–526.

Riopelle, A. J., Nos, R., & Jonch, A. (1971). Situational determinants of dominance in captive young gorillas. *Proceedings of the 3rd International Congress of Primatology*, **3**, 86–91.

Robert, S. (1986). Ontogeny of mirror behavior in two species of great apes. *American Journal of Primatology*, **10**, 109–117.

Savage-Rumbaugh, E. S., Rumbaugh, D. M., & Boysen, S. T. (1978). Sarah's problems of comprehension. *Behavioral and Brain Sciences*, **1**, 555–557.

Suarez, S. D. & Gallup, G. G., Jr. (1981). Self-recognition in chimpanzees and orangutans, but not gorillas. *Journal of Human Evolution*, **10**, 175–188.

Swartz, K. B. (1997). What is mirror self-recognition, and what is it not? *Annals of the New York Academy of Sciences*, **818**, 65–71.

Swartz, K. B. (1998). Review of "What young chimpanzees know about seeing" by D. J. Povinelli and T. J. Eddy. *International Journal of Primatology*, **19**, 379–382.

Swartz, K. B. & Evans, S. (1991). Not all chimpanzees (*Pan troglodytes*) show self-recognition. *Primates*, **32**, 583–496.

Swartz, K. B. & Evans, S. (1994). Social and cognitive factors in chimpanzee and gorilla mirror behavior and self-recognition. In S. T. Parker, R. W. Mitchell, & M. L. Boccia (eds.), *Self-awareness in animals and humans: Developmental perspectives* (pp. 189–206). Cambridge University Press.

Swartz, K. B. & Evans, S. (1997). Anthropomorphism, anecdotes, and mirrors. In R. W. Mitchell, H. L. Miles, & N. Thompson (eds.), *Anthropomorphism, anecdotes, and animals* (pp. 296–306). Albany, NY: State University of New York Press.

Thompson, R. K. R. & Contie, C. L. (1994). Further reflections on mirror usage by pigeons: Lessons from Winnie-the-Pooh and Pinocchio too. In S. T. Parker, R. W. Mitchell, & M. L. Boccia (eds.), *Self-awareness in animals and humans: Developmental perspectives* (pp. 392–409). Cambridge University Press.

Thompson, R. L., & Boatright-Horowitz, S. L. (1994). The question of mirror-mediated self-recognition in apes and monkeys: Some new results and reservations. In S. T. Parker, R. W. Mitchell, & M. L. Boccia (eds.), *Self-awareness in animals and humans: Developmental perspectives* (pp. 330–349). Cambridge University Press.

Walraven, V., Elsacker, L van, Verheyen, R. (1995). Reactions of a group of pygmy chimpanzees (*Pan paniscus*) to their mirror-image: Evidence of self-recognition. *Primates*, **36**, 145–150.

Westergaard, G. C. & Hyatt, C. W. (1994). The responses of bonobos (*Pan paniscus*) to their mirror images: Evidence of self-recognition. *Human Evolution*, **9**, 273–279.

15

Deception and concealment as strategic script violation in great apes and humans

ROBERT W. MITCHELL

In his books *Animal Intelligence* and *Mental Evolution in Animals*, Romanes (1882/1906; 1884/1900) argued that apes, being phylogenetically close to humans, should be capable of intentional deception. Until recently, scientists have had little more than Romanes' faith and a few observations upon which to base an expectation of skill in intentional deception in apes (Mitchell, 1999). But in the past two decades, descriptions of deceptive behavior in apes and other nonhuman primates, as well as discussions of its psychological requisites, have become commonplace, lending greater support to the expectations of intentional deception by apes and of commonalities in the underlying mental states in humans and apes (see, e.g., Menzel, 1974; De Waal, 1982, 1986; Miles, 1986; Mitchell, 1986, 1993, 1997; Mitchell & Thompson, 1986a,c; Smith, 1987; Whiten & Byrne, 1988; Byrne & Whiten, 1990, 1992; Byrne, 1997). As might be expected on phylogenetic grounds, many deceptions used by humans in sports, play, and teasing (e.g., Mawby & Mitchell, 1986; Mitchell & Thompson, 1986b, 1993; LaFrenière, 1988; Leekam, 1991; Reddy, 1991; Mitchell, 1996) are strikingly similar to those used by nonhuman primates. Just as the signing skills of great apes are comparable to those of 2- to 3-year-old children (Brown, 1970; Savage-Rumbaugh et al., 1993; Miles, 1997; this volume; Parker & McKinney, in press), their deceptions are also sometimes similarly comparable (e.g., Reddy, 1991), but often great apes' deceptions seem much more skillful than those of even 4-year-old children (e.g., LaFrenière, 1988). As humans age, their deceptions become far more complex in their psychology and preparation, involving extensive use of props and considerable linguistic and story-telling skills (see, e.g., Blum, 1972; Klein & Montague, 1977; Sexton, 1986; Werth & Flaherty, 1986; Mitchell, 1996).

Obviously, there is a need for a comparative framework within which to interpret similarities and differences in deceptions across species. In fact, there are multiple frameworks (Dennett, 1978, 1983; Miles, 1986; Mitchell, 1986, 1993; Whiten & Byrne, 1988; Byrne & Whiten, 1990) and, although there are discrepancies (e.g., compare Byrne, 1997; Mitchell, 1997), all share the idea that deceptive behaviors, whether intentional or not, depend upon recognizing regularities in others' responses to the deceiver's behavior. I build on these frameworks using script theory, initially developed to understand human event knowledge, to understand how deceptions arise in diverse species (see Mitchell, 1996).

SCRIPTS IN HUMANS

Scripts are schemas about the sequence of elements that constitute an event. They are "specific knowledge [we use] to interpret and participate in events we have been through many times" (Schank & Abelson, 1977, p. 37), such as eating in a restaurant or going to a movie. Described developmentally, a script is a "general event representation derived from and applied to social contexts . . . which specifies roles and props [for] an ordered sequence of actions appropriate to a particular spatial–temporal context, organized around a goal" (Nelson, 1981, p. 101). People can recognize scripts and apply them to "fill . . . in the causal chain between two seemingly unrelated events by referring to the script" (Schank & Abelson, 1977, p. 38). Scripts can also derive from vicarious experiences, such as watching an interaction. Children develop some scripts through discussions about event structure with other children and through repetitive interactions with parents and others (Nelson, 1981; Hudson, Sosa, & Shapiro, 1997). Because people easily extrapolate from unique events, scripts probably do not require extensive experience to develop. Although scripts can be quite elaborate, they do "not represent a very sophisticated category structure" in that they are "general but concrete" (Nelson, 1981, p. 116). Some deceptions used by criminals are literally scripted, in that there is a protocol to follow which specifies each agent's actions and reactions (Blum, 1972), and these trade on scripts used in normal interaction (Klein & Montague, 1977).

Among humans, deceptions are sometimes inserted into an ongoing script, such that adherence to the script allows the deceiver to benefit from the deception. An illustrative example is a practical joke I experienced after ordering in a pizza parlor:

> after I handed my payment for lunch over the clear countertop which sloped smoothly down in front of the pizza-cutting table below it, the cashier handed me my change below the countertop, with the result that I smashed my hand into the clear partition to obtain the change; the man laughed when he handed me change over the countertop. He deceived (with as much apparent pleasure) the next patrons as well. (Mitchell, 1996, p. 825)

Such practical jokes, common among humans (Anderson, 1986; Leekam, 1991; Reddy, 1991), are "expressive" deceptions (Mitchell, 1996), i.e., deceptions done for fun which, following their enactment and the victim's response, are usually immediately revealed. They are distinguished from unintentionally revealed deceptions by the indications of enjoyment, such as smiles or laughs, on the part of the deceiver. The payoff seems to be the enjoyment of the other's surprise, frustration, or embarrassment, otherwise they appear relatively harmless.

In the "fast-food restaurant" script utilized by the cashier, the sequence of ordering food, paying for it, and receiving change is fairly stereotypical, as is the part of the script which indicates that you receive change by moving your hand toward the cashier's hand. The deceptive cashier needed to know that the victim would move a hand toward him (such that it would hit the clear partition) when he moved his hand containing change toward the victim. He also had to sequence his actions within the enactment of the script. He understood that the money-giving gesture had to occur (1) after the victim had given him money, (2) after he had opened the cash-register, (3) after the victim had ordered the food, (4) while the victim was looking at him, and so on. Simply handing some change at any time – for example, before the victim gave him money, or when given exact change – would not work.

Like most uses of scripts, the cashier's use of the change-giving script (itself embedded in a fast-food restaurant script) does not require much psychological insight: "children and adults alike often act competently in a structured routine situation without either tacit or articulated knowledge of the script as viewed from another's point of view and without full knowledge of the rules and goals involved" (Nelson, 1981, p. 113). Although at some level the cashier had to be aware of the sequential components of his actions and those of the victim, he may not have thought about them consciously before handing the change under the counter. Even if the man initially thought up the deception, he need not have thought consciously about the sequential aspect of the deception. Scripts, including deceptive scripts like that of the cashier, can become so commonplace in structured routines that they are performed habitually, without thought (Mawby & Mitchell, 1986).

Deceptions can be conceived as minor variations on *behaviors* normally used while enacting a script, or as enactments of scripts in inappropriate *circumstances*. For example, in a basketball game, players may deceive by using part of a behavior (aiming the ball toward the basket as if to throw it in), and thus by varying the behavior; in this instance players also deceived by performing the behavior in inappropriate circumstances (when they were not going to throw it in the basket). In the cashier example, it was the placement of the cashier's hand (a behavior) below the transparent countertop (a circumstance), rather than above it, which was deceptive. Either way, the script is violated strategically, to obtain a given purpose.

A deceiver must have a script for the typical and sequential context within which to enact a behavior for deceptive purposes. A victim must also have scripts for interpreting the context and the deceiver's behavior in order for the deception to be enacted. Recognizing that he or she has been deceived before in the same context may make a victim suspicious (as well as create scripts to detect deception). This suspiciousness need not dissuade the deceiver from repeated deceptions if these cannot be detected until after their commission. For example, a poker player may use the same deceptions over and over again, with variation (Hayano, 1980), a professional fisherman may on multiple occasions indicate that he is having no luck when in fact his catch is quite large, to dissuade others from fishing nearby (Andersen, 1980), and a philandering spouse may repeatedly use the same sorts of excuses to account for his or her absence (Werth & Flaherty, 1986).

SCRIPTS IN GREAT APES

The script model allows great ape deception to be examined in a new light. Like humans (Piaget, 1947/1972; Nelson, 1981), great apes (as well as other mammals) develop schemas through play, and naturally observe and experience multiple regularities in actions and roles, all of which can be elaborated into scripts (Parker & Milbrath, 1993, 1994; Mitchell, 1994; Parker, chapter 18, this volume). Scripts depend upon knowing what another does when you perform an action. This knowledge can be used in deception. For example, if a chimpanzee knows that holding her hand out in supplication is likely to lead another chimpanzee to approach, she can use this behavior to entice this chimpanzee close enough to attack (De Waal, 1986). (A more complex script might employ knowledge about the other's reactions to your behavior, and your reactions to the other's reactions.) Such a script might be used to inhibit behavior, as when a chimpanzee knows that others will take (currently hidden) food away if the chimpanzee looks at and directs its attention toward the hidden food, so the animal waits until it can retrieve the food when others are not around (De Waal, 1986; see Thompson, 1986).

Through scripts, apes come to have expectations about what will happen when they do (or do not do) something. These expectations are based on what they know has happened in the past, either through personal experience or observation. Apes can then perform actions (or not perform them) to effect change in desired directions. Deceitful behaviors are likely to involve (1) actions apes use frequently enough to become aware of their consequences, or (2) inhibitions of these actions. Once the behaviors used for deceit are specified, researchers can look for both deceitful and nondeceitful uses of these behaviors to determine whether apes' nondeceitful uses imply scripts (Miles, 1986; Mitchell, 1986, 1988; Byrne & Whiten, 1988).

The relevance of scripts for understanding great ape deception in large part derives from the nature of deception. Deception occurs when one organism (the victim) perceives the action of another organism (the deceiver) to mean that something is true (when it is not) and responds appropriately to it, which benefits the deceiver (Mitchell, 1986). The only way an action could cause a victim to act with an appropriate response is if that action has *usually* resulted in a particular response. Repeated exposure to action–reaction sequences results in a script. Young apes are adept at testing the consequences of actions toward others (Yerkes, 1943, p. 30; Adang, 1984; Parker, chapter 18, this volume), which likely provides for development of scripts about many action–reaction sequences.

Scripts are also useful for understanding concealments that depend upon repeated experiences. Concealment occurs when one organism makes information inaccessible to another organism and benefits as a result (Whiten & Byrne, 1988). Some concealments are obviously learned from repeated experience. For example, young chimpanzees learn to conceal their excitement screams when silence is needed after other chimpanzees hit them, hold them, or cover their mouths until they quiet down (De Waal, 1982; Goodall, 1986). Similarly, young great apes develop scripts for hiding through experience, with the consequence of moving behind obstacles when playing hide-and-seek (Parker & Milbrath, 1994; Parker, chapter 18, this volume).

Some concealments appear to derive from inhibiting actions which are known to initiate a script's behavioral sequence. For example, a gorilla mother's typical intervention whenever a male showed interest in her infant, even by just looking at the infant, inspired a script in which males inhibited their interest in the infant until the mother could not intervene (Mitchell, 1991). Similarly, one chimpanzee slowly moved away from an aggressive chimpanzee, peering and grimacing at the aggressor and stopping movement when the aggressor moved (Goodall, 1968), and similar slow and halted movements are used in chimpanzees' deceptions. Given that Harrisson (1960) found it easy to predict what young orangutans were going to do by following their gaze, it seems likely that they would learn quickly to inhibit their gaze after being frustrated.

Frequent performance or observation of a sequence of events leads to the development of a script, so that knowledge can be effortlessly gained. Sometimes only one or a few exposures are necessary for script development. For example, some Gombe chimpanzees rapidly recognized regularities in responses to their own behaviors, and learned to inhibit these behaviors to thwart enactment of the script (Goodall, 1986). One learned after one trial not to make food calls upon receiving food, and another chimpanzee learned after several trials not to open the food box in the presence of other chimpanzees. (Another chimpanzee apparently never learned this simple rule, even though failure to learn resulted in loss of food to other chimpanzees. Not all chimpanzees are equal in intelligence [Köhler, 1925/1973].)

Script theory implies that event knowledge derives from observation and/or performance of repeated sequences of behavior and consequences. It suggests that deceits are based upon behaviors and consequences which have been repeatedly performed or observed by the deceiver. Consequently, behaviors used to deceive and conceal should be behaviors which are commonly enacted by apes, and whose consequences are likely to be known to them. Script theory should direct researchers to attend to such behaviors in young great apes *before* these behaviors are used deceitfully, to determine if their deceitful use is based upon previously encountered or observed uses incorporated into scripts. (Attention to such pre-deceptive uses of behaviors subsequently used to deceive was explicitly part of Miles' [1986] study of her orangutan's sign development.)

METHOD

To compare deceit across species, I examined literature on deception and concealment in apes and humans as well as looked for deceptions and concealments in literature about particular great ape species. This literature is extensive (Cuvier, 1811; Abel, 1819; Warwick, 1832; Brooke, 1841; Du Chaillu, 1868; Wallace, 1869/1962; Hornaday, 1879; Romanes, 1900, 1906; Christy, 1915; Sheak, 1917, 1924; Cunningham, 1921; Stern, 1924; Köhler, 1925/1973; Burbridge, 1926; Barns, 1928; Maxwell, 1928; Yerkes, 1929, 1943; Yerkes & Yerkes, 1929; Johnson, 1932; Raven, 1932; Ladygina-Kots, 1935/1982;

Pitman, 1935; Benchley, 1940; Hoyt, 1941; Hebb, 1946; O'Reilly, 1950; Hayes, 1951; Hayes & Hayes, 1952; Reuter, 1954; Hediger, 1955/1968; Gray, 1955; Martini, 1955; Oberjohann, 1957; Harrisson, 1960; Emlen, 1962; Riopelle, 1963; Reynolds & Reynolds, 1965; Goodall, 1968, 1986; Horr, 1969; MacKinnon, 1971; Blum, 1972; Menzel, 1974; Galdikas, 1975; Riopelle et al., 1976; Klein & Montague, 1977; Terrace, 1979; Woodruff & Premack, 1979; Andersen, 1980; Hayano, 1980; Sacks, 1980; Selman, 1980; De Waal, 1982; 1986; Adang, 1984, 1987; Quiatt, 1984; Anderson, 1986; Chevalier-Skolnikoff, 1986; Mawby & Mitchell, 1986; Miles, 1986, 1990; Mitchell, 1986, 1991, 1993, 1996; Mitchell & Thompson, 1986b, 1991, 1993; Sexton, 1986; Vasek, 1986; Werth & Flaherty, 1986; Smith, 1987; Dunn, 1988; LaFrenière, 1988; Savage-Rumbaugh & McDonald, 1988; Whiten & Byrne, 1988; McCornack & Levine, 1990; McCornack & Parks, 1990; Nishida, 1990; Byrne & Whiten, 1990; 1992; Buller, Strzyzewski, & Comstock, 1991; Reddy, 1991; Sodian, 1991; Van Lawick, 1991; Tanner & Byrne, 1993; Savage-Rumbaugh & Lewin, 1996; Miles, Mitchell, & Harper, 1996). In a few cases of human deception I used my own deceptions (sitting on an object to hide it from another) or observations of others' deceptions (acting as if to hit someone to make them flinch). Because human observation is not omniscient, comparisons between species are dependent upon available reports of ape and human deceit, and any comparison among species is a comparison among available reports. Compared to chimpanzees, very few orangutans or bonobos have been studied, making it less likely that particular deceptions will be observed in these species. New data will likely fill in some of the blanks. In this respect, it is unfortunate that the corpus of deceptions of the sign-using orangutan Chantek reported in his caregiver's daily record, which I analyzed with Lyn Miles during a post-doc at her laboratory, could not be included, as they have not yet been published. (Note that categories for which only humans showed evidence were ignored.)

Most of the categories used are borrowed from Byrne and Whiten (1990), who have catalogued most instances of deception in nonhuman primates; their catalogue has been an invaluable reference. Following these authors, I present instances of both concealment and deception, which together comprise what I call deceit. Although I have largely maintained the coding scheme of Byrne and Whiten (1990), I created some new categories (or modified some old ones) to make salient some species differences and similarities in deception:

(1) instances of deception involving objects, some of which had been categorized by Byrne and Whiten (1990) as "distraction (and attraction) using close-range behavior," I categorized as "distraction (and attraction) by showing object" and "distraction (and attraction) by throwing object," where "showing object" includes poking with an object;
(2) acts that thwart deception (as defined above) are "counterdeceptions" and
(3) acts that thwart concealment (as defined above) are "counterconcealments" (these last two acts may or may not themselves be deceptions or concealments); and
(4) instances of "expressive deceptions," which are deceptions (*not* concealments) done

in such a way that the victim discovers (or immediately recognizes) that he or she has been deceived (to the enjoyment of the deceiver).

In addition, I specified whether the observation occurred prior to 1960, or from 1960 on, and whether the deception was found in the wild or in captivity, and, if the latter, I further specified if the ape comprehended speech or used some form of language (e.g., sign language, signs on a keyboard) to communicate. (1960 is the approximate time of a renaissance in great ape studies – see Mitchell, 1999.)

In a few cases, particular instances of deception were coded under more than one category; for example, in a few instances in which a male gorilla acted interested in an infant but then play-attacked her mother when she came near, he attracted the mother by looking at the infant and by leading ("indirectly" – Byrne & Whiten, 1990, p. 9) the mother to the infant, and also used a social tool (the infant) to deceive the target (the mother). So his action was coded as "distracting by looking," "distracting by leading," and "using a social tool to deceive."

After completing this global description, I specified the particular behaviors the animals used to deceive and conceal. For example, a chimpanzee's turning his back to hide his erect penis from a dominant male was categorized into several behaviors: monitor gaze of other, turn body front away from other, and wait for detumescence of penis. This specification allowed me to see if deceptions were based on actions the apes were likely to have experienced repeatedly, such that they would have developed scripts for action–reaction sequences. (Again I specified whether the observation occurred before 1960, or from 1960 on.) Note that, given small sample sizes, I focus on types of behavior, asking whether or not at least one species member ever performed a behavior of a particular type for deceit; I am not concerned with the frequency of reports of use of a deceit in a given species.

RESULTS

A taxonomy of the general forms of deceit in orangutans, gorillas, chimpanzees, bonobos, and humans (borrowed largely from Byrne & Whiten, 1990, 1992) is presented in Table 15.1. All five species showed some form of concealment as well as deception by distraction, attraction, and creating an image, although not all subtypes of each were reported (Byrne & Whiten, 1992). Chimpanzees show the greatest diversity of deceits of all ape species in nature and in captivity, followed by gorillas and orangutans, and then bonobos.

One striking point of contrast is in the duration of deceptions: most deceptions by apes (as well as those by children) were remarkably short-lived, typically lasting from a few seconds to a few minutes, and provide no instances of probing by a victim long after a deception; in more mature humans both deception and probing may last for years (Werth & Flaherty, 1986; Sexton, 1986; McCornack & Levine, 1990; McCornack & Parks, 1990; Buller et al., 1991). Another point was that overall captive apes produced more types of deceit than their wild counterparts. Whereas gorillas and chimpanzees

Table 15.1 *Evidence of particular forms of behavioral deceit in great apes and humans*

	Orangutan	Gorilla	Chimpanzee	Bonobo	Human
Concealment by					
silence	n,h(n)	n,c,h(n,l)	n,c	s	s
hiding self	n,h(c)	n,c,h(l)	n,c,s	s,h(c)	s
hiding object/bodypart	s	c,s,h(l)	c,s,h(c)	s	s
inhibiting interest	c,s,h(c)	n,c,h(c)	n,c	s	s
ignoring	n	c,h(n)	n,c	—	s
Distraction by					
calling	—	—	c,s	s	s
looking	s	h(c)	n,c,s,h(c,l)	s	s
leading	—	h(c)	n,c	—	s
close-range behavior	s,h(c)	c	n,c	c,s	s
showing object	—	—	h(c)	—	s
throwing object	—	—	c	—	s
acquired signing	s	s	—	—	s
Attraction by					
calling	—	—	n	—	s
looking	—	c,h(c)	c	—	s
leading	s	c,h(c)	—	s	s
close-range behavior	s	c	c	—	s
showing object	s	—	c,h(c)	s	s
throwing object	s	—	—	—	s
acquired signing	s	—	—	s	s
Creating an image that is					
neutral	c,h(c)	n,c,s	n,c,h(c)	—	s
affiliative	—	c	n,c,h(c)	—	s
threatening	n,c,h(c)	n,h(n,c)	c,h(c)	—	s
Deflection to third party	—	—	s	—	s
Using social tool to deceive target	—	*c*	*n*	*s*	*s*
Counterdeception by nondeception	—	—	n,c	—	s
Counterconcealment by revealing					
hiding of others	—	—	c	—	s
by concealment	—	h(l)	n	—	s
Expressive deception	s	s,c,h(l)	c	s	s

Most instances of great ape deceit are taken from Byrne and Whiten's (1990) catalogue. In their catalogue, record no. 249 mislabels a *Pan paniscus* as a *Pan troglodytes*, and the animal is categorized as attempting to deceive the tool, when she is actually attempting to deceive the target. Also, although not marked as such, record no. 180 shows concealment by silence, and no. 216 shows counterdeception. (Note that some cases of distraction by close-range behavior, e.g., nos. 228–230 and no. 242, do not satisfy the definition of deception used in this chapter.) Finally, in this Table, "h" is used for great apes only.

n = in nature; c = in captivity or ex-captive in nature; l = language-trained [comprehension]; s = language-user; h = evidence prior to 1960; — = no evidence; parentheses surround codes for evidence prior to 1960.

who comprehended speech or used some form of language to communicate typically did not produce additional types of deceit compared to their captive counterparts, the language-using bonobo, Kanzi, and orangutan, Chantek, did. However, given the few observations of bonobos and orangutans in captivity and the wild, it seems likely that the difference here is not due to language, but rather to sampling bias. In addition, the observations of deceit prior to 1960 (marked with an "h" for "historical") stand up to more modern observations: of the 25 instances in which a particular type of deceit was observed in particular great ape species prior to 1960, only 3 failed to be described in recent accounts.

TYPES OF BEHAVIOR USED FOR DECEIT

All deceits in the sample are based on (combinations of) the 66 types of behavior in Table 15.2, most of which are shared across several species. (Note that the deceptive aspect of many of the behaviors used in Table 15.2 is not apparent; the point of Table 15.2 is to specify the actual behaviors used to deceive and conceal, not to describe their deceitful use.) According to the reports I found, chimpanzees employed 54 types of behavior for deceit, gorillas 41, bonobos 35, and orangutans 34 (see diagonal in Table 15.3). All great ape species shared many of their deceitful behaviors (from 32–46) with humans, and each species shared more than half of their behaviors with the other species (see Table 3). Indeed, comparison of the frequencies in types of behaviors used for deceit among the four great ape species indicates no statistically significant differences among species (Cochran Q (3, $N = 66$) = 5.98, ns). Those behaviors shared by all or most species have to do with when to act (act when others are not looking, act when others leave, act when others approach, act when other's back is turned), how to act (quietly, slowly, quickly), movement of self or other toward or away from something, gaze (look away, look at object/infant, look peripherally, monitor others and their gaze, look at nonexistent things), failure to look, and moving behind an obstacle. Note also that, of the 42 types of behaviors used for deceit by great apes described in accounts prior to 1960 (again designated with an "h" for "historical" in Table 15.2), 32 were also observed in more modern accounts. (In a few cases in which chimpanzees were described as hiding behind their nests [e.g., Christy, 1915; Reynolds & Reynolds, 1965; Goodall, 1968], they made noise, suggesting that they were not hiding at all; orangutans, by contrast, seemed to be silent when hiding [Wallace, 1869/1962; Harrison, 1960; MacKinnon, 1971].)

Objects were integral to some great ape deceits. Great apes commonly used objects to deceive and conceal, and used deceits to obtain objects (including food). Objects were thrown, offered, dropped, reached for, replaced after being removed, concealed by being sat on and held behind oneself, used to poke, covered with other objects, looked at, and not looked at. Infants of cautious, protective mothers were at times treated like objects by gorillas and chimpanzees, being looked at, not looked at, looked at peripherally, approached, sat on, and reached for surreptitiously (behind the back, with the foot). Fixed objects and enclosures were also used to deceive. One gorilla acted as if to

Table 15.2 *Behaviors used to conceal or deceive by apes and humans (x = behavior in reports from 1960 on, h = behavior in reports prior to 1960)*

	Orangutan	Gorilla	Chimpanzee	Bonobo	Human
fail to look at another		x,h	x	x	x
fail to look at infant/food/object	x	x	x	x	x
fail to look at one's own action	x	x	x		x
fail to make response/poker face	x,h	x	x,h	x	x
fail to exhibit sexual interest		x	x		x
fail to make food calls			x		
act when other(s) not looking	x	x,h	x	x	x,h
act after other(s) leave(s)	x	x	x	x	x
act when other approaches	x	x	x	x	x
act when other's back is turned	x	x	x		x,h
act quietly in noisy surroundings	x,h	x,h	x,h	x	x
act slowly	x	x,h	x		x
act quickly	x,h	x,h	x,h	x	x
reach for object	x	x	x		x
reach for infant/object with foot		x,h	x		
take/grab unallowed object/person	x			x	x
hold object/liquid in hand/mouth	x	h	x,h	x	x
hold lid of box containing objects			x		x
throw object at victim	x		x		x
show/offer object to other	x		x,h	x	x
drop object during object-keepaway				x	x
move objects back to former place				x	x
move infant via merry-go-round		x			
sit on desired object/infant		x,h	x		x
cover object with objects	x			x	x
cover action/object with body part	x	x,h	x		x
cover all or part of self with object	h		h	x	x
move behind obstacle	x,h	x,h	x	x	x
move away from others	x,h	x	x	x	x
build nest		x		x	
almost hit something		x			x
turn/keep body front away from other			x		h
hold out hand to other		x	x		x
direct body gesture toward other	h	h	x,h	x	x
point to inaccurate location	x		x		x
lead or push others to location		x	x	x	x
send or push other away				x	x
walk normally with food in hand			x	x	
make noises with objects	x		x		x
make hunting call			x		
make alarm call/act afraid	x		x	x	x

Table 15.2 *Continued*

	Orangutan	Gorilla	Chimpanzee	Bonobo	Human
make friendly face		x	x,h		x
make sexual invitation		x			x
act hurt/sick			x,h	x	x
keep arm stuck in wire mesh		x			
wait for detumescence of penis			x		
throw tantrum	h		x,h		x
groom/pet other			x		x
groom self/play with toes		x	x		
forage/search for object		x	x	x	
act playfully toward other			x	x	x
approach other with body/body part		x,h	x,h	x	x
approach infant/food/object		x	x	x	x
look at object/infant	x	x	x	x	x
look peripherally		x,h	x	x	x
look away	x,h	x,h	x	x	x
look at uninteresting/nonexistent things	x	h	x,h	x	x,h
monitor other(s)	h	x,h	x,h	x	x
monitor gaze of other(s)	x,h	x	x	x	x
move hand in front of face/penis		x	x		x
push grinning lips down with fingers			x		
sign/say presence of something	x				x
sign/say interest in getting object	x			x	x
sign/say interest in doing activity	x		x		x
sign/say inaccurate words/names	x	x	x		x
do "opposite" of what is requested		x	h		x

Table 15.3 *Number of types of behavior reportedly used to deceive and conceal by species (along the diagonal), as well as the number of those in common between species*

	Orangutan	Gorilla	Chimpanzee	Bonobo	Human
Orangutan	**34**	22	30	22	34
Gorilla		**41**	36	23	35
Chimpanzee			**54**	28	46
Bonobo				**35**	32
Human					**55**

Only those human deceitful behaviors also shown by great apes are presented.

hit the window of her cage. Another turned the merry-go-round an infant crawled on to make her easier to collect without calling attention to the act. Among sign-using chimpanzees, Austin made seemingly inexplicable noises on the outside of his enclosure to scare Sherman.

Sign-using apes used signs for deceit (summarized in Miles, 1986). The orangutan, Chantek, falsely indicated the presence of something, and interest in getting an object, engaging in an activity, or going somewhere. Chantek, the gorillas, Koko and Michael, and the chimpanzee, Lucy, used signs that were inappropriate in other ways: they either incorrectly named objects, incorrectly signed that others (rather than they themselves) did an (inappropriate) action, signed that they did an action which they had not, or did the opposite of what they were requested to do (e.g., moved an object up toward the ceiling when asked to put it under something, stated that a white blanket was red, signed SAD-FROWN when asked to smile). (A non-sign-using chimpanzee who was repeatedly tested also did the opposite of what was requested – see Gray, 1955.) In some of these cases, scripts were employed at two levels: (1) signs themselves resulted in scripted responses, when used by either caregiver or ape; and (2) the context of testing (during which many of these incorrect and inappropriate signs occurred) was highly scripted, and signs were directed toward disrupting this script.

Bodily control was present in many deceits. Chimpanzees and bonobos showed control over vocal signals when they made false alarm calls, and chimpanzees made false hunting calls and failed to make food calls in the presence of food. Gorillas and chimpanzees showed some control over their facial musculature, as when they appeared friendly but were not, or when chimpanzees kept their cheeks normal-looking while holding a liquid or objects in their mouth. Sometimes parts of the face or body were controlled by other means. One chimp used his hands to control his grimacing lips after turning his back to his opponent, and a gorilla covered her playface with her hand. Chimpanzees covered their erect penis with their hand or by turning their back to a dominant male (and waited for detumescence before turning around). Covering the self with a hand and by turning away seem likely to derive from attempts to thwart another's contacts in play.

Apes also showed voluntary control over other bodily actions, attempting to influence others by performing actions with one apparent motive or causal basis to achieve another motive. These performances included limping, coughing, throwing temper-tantrums, grooming or petting another, self-grooming, playing, nest building, foraging, searching for an object (which one already knows the location of), making sexual invitations, walking normally while holding food in hand, and holding one's arm to another in supplication or enticement. The actual motives for these deceitful actions are myriad, but are unimportant for understanding their usefulness in deceitful scripts. All of these actions have an obvious appearance which was then belied by action inconsistent with that appearance – for example, if a chimpanzee is really suffering from limping, why does he only limp when other chimpanzees can see him limp (De Waal, 1986)? Gorillas and chimpanzees hid expression of sexual interest until their desire could be satisfied without an audience. Chimpanzees, bonobos, and gorillas also falsely indicated that another had harmed them by either gesturing toward the other, expressing upsetment while looking at the other, and/or showing their wounds.

EXPRESSIVE DECEPTIONS

Expressive deceptions by apes were infrequent. Koko sometimes produced obviously incorrect signs for objects, or did something different from what the caregiver requested, largely as a way of toying with the caregiver. For example, Koko once made the well-known sign *drink* to her ear rather than to her mouth when requested to make the sign, or moved a toy toward the ceiling when asked to put it under a bag (Patterson & Linden, 1981, pp. 77). (Similar acts also allowed Koko, at times, to avoid testing.) Expressive deceptions by non-sign-using apes were rare, and all were reported among captive animals. A captive female gorilla teased a male by repeatedly sexually presenting but then moving away and laughing when the male showed interest, and another acted friendly toward a human but then pressed the human's finger against her cage and laughed (Chevalier-Skolnikoff, 1986). One other captive gorilla enjoyed watching humans start when she punched the glass wall of her cage, but kept them engaged by acting as if she were going to hit the wall and then stopping just before contact (Quiatt, 1984), an action which is reminiscent of older brothers who act as if about to hit a younger brother just to see him flinch. One captive chimpanzee appeared to be engaged by playing with his toes when, as soon as a human got close enough, the chimp sprayed him with a great deal of water shot from the mouth (Hediger, 1955/1968).

Expressive deceptions sometimes occurred in play in the context of games like object-keepaway. In object-keepaway, a player tries to maintain control over an object while enticing another to get it by making it appear available; it is a common mammalian game, and one of the earliest deceptions of children (7.5 months of age – see Reddy, 1991). The sign-using bonobo, Kanzi, frequently dropped an object near to himself and the play partner and appeared disinterested but, when his play partner moved to get it, quickly tried to get the object himself (Savage-Rumbaugh & McDonald, 1988), and Chantek once tossed a whistle toward the caregiver so that it came back toward Chantek, enticing the caregiver to get the object so that Chantek could grab him (Miles et al., 1996). Both of these share similarities to elaborated deceptive games by humans playing with dogs, with Chantek's game called "fakeout" (Mitchell & Thompson, 1991). Strangely, for some children games enticing others to attempt to gain an offered object which is then retracted apparently occur prior to "honest" offering of an object (Reddy, 1991), suggesting that experience with the consequences of offering is not necessary for its production. (Perhaps infants can learn from observing the consequences of *others'* offerings.)

DISCUSSION

As predicted by script theory, almost all of the deceitful actions are based on commonplace behaviors used by apes and humans in nondeceitful circumstances. Repeated experiences should have made the consequences of these behaviors salient. Most of the behaviors were repeatedly experienced early in life by apes and humans. An organism's failures to act also seem likely to have resulted from experiences with consequences of the organism's action (such that the resulting inhibitions became scripted). Apes and humans appear quite adept at developing scripts, sometimes quickly, from experience.

A few deceitful behaviors may seem unlikely to result from direct experience. Looking at nonexistent objects seems an aberrant activity, but apes have lots of experience of the consequences of looking at anything, and with looking away from another to avoid eyegaze, and so can simply perform the action of looking away without having any object. Reaching behind your back for an infant, and failing to look at your own actions while performing them, both display an unusual reliance upon nonvisual perception. But both also seem likely to derive from experiencing the consequences of more obvious actions like reaching for the infant (i.e., the mother takes the infant away) or watching what one is doing (i.e., other apes interfere with what one is doing) coupled with a continuing impulse to watch, interact, or obtain the infant or object. In addition, these behaviors sometimes rely on contextual support: if an infant appears behind a gorilla, he can take advantage of this by reaching behind.

This examination of deceit indicates that orangutans, gorillas, chimpanzees, and bonobos employ a limited number of behaviors, many of these identical across species (including humans). Combinations of these behaviors result in a variety of deceptive encounters for humans as much as for apes (see Mitchell, 1986, p. 16). Apes and humans are remarkably good at detecting regularities in social behavior and using these for their own ends. Scripts appear to be readily developed and utilized for deceit. They are important sources for deceptions based on foiled expectations. Even deceits based on inhibiting those actions which would allow the sequence of actions encoded in a script to unfold appear to be based on scripts. Deception and concealment are easily enacted in great ape and human society precisely because responses are so well scripted.

Human deceits can be more elaborate and complex than those of great apes for several reasons, mostly having to do with the fact that the means for deceit (and the victim's probing about the deceit) are so much more complex. Whereas both apes and humans use their body and objects for deceit, in their deceptions humans use a far more extensive array of props, and often use multiple props to deceive. Elaborate deceptions can require that a deceiver dress a part, open a bank account, rent a hall, write a diary or a check in the handwriting of another person, or produce artwork in the style of a particular artist (see Mitchell, 1996). Obviously, human deceptions are often dependent upon cultural institutions.

In addition, humans use language in deceit. There are at least three reasons why language complicates deception and concealment. First, human deceptions can be maintained for extensive periods largely because of planning, and planning and orchestrating complex deceptions are accomplished through language (Mitchell, 1996). Indeed, language may be an adaptation for elaborate planning (Parker & Milbrath, 1993). Plans become necessary in deception when scripts have not been developed or are extremely complex; plans are "the mechanisms that underlie scripts" (Schank &

Abelson, 1977, p. 70) and as such allow creation of scripts for *novel* sequences of action to achieve a deceptive goal. Second, language allows for falsehoods to be presented with no means of verification. For example, a therapist who was dying of cancer repeatedly claimed to her patient over a 3-month period that she was well, and a man provided excuses to his girlfriend for 3 years to conceal his sexual liaisons (Werth & Flaherty, 1986). Although both victims had an inkling that something was up, they both allowed the deceptions to go on because they were in part convinced by language (see Millikan, 1984, concerning the belief-inducing nature of language). Finally, many human deceptions require story-telling; in order to convince someone that something out of the ordinary is true, deceivers provide a (somewhat) convincing explanation in the form of a story, act consistently with the story, and probe to see if the victim continues to believe the story (see Sexton, 1986; Mitchell, 1996).

Whereas sign-using apes certainly have a linguistic skill comparable to that of a 2- to 3-year-old human (Brown, 1970; Savage-Rumbaugh et al., 1993; Miles, 1997, this volume; Parker & McKinney, in press), they do not sign stories, which children do at 3 years of age (Dunn, 1988). Although apes develop plans and use props, these are typically not very extensive unless raised by humans, in which case their plans and uses of props are to some degree comparable to, though not as extensive as, those of very young children (see discussion in Patterson & Linden, 1981; Miles, 1991, p. 16; Mitchell, 1993, 1994; Parker & Milbrath, 1993; Parker & McKinney, in press). Both language and our extensive array of props allow humans to have myriad signs with which to deceive, and language allows for declarative planning and co-ordination skills as well (Miller, Galanter, & Pribram, 1960; De Lisi, 1987). Human children develop their skill at planning through instruction in planning (Hudson, et al., 1997; De Lisi, 1987), which so far has not been described even for sign-using apes. It seems likely that, like young children, great apes have plans which are anticipations of familiar routines (Kreitler & Kreitler, 1987), a type of what De Lisi (1987, p. 90) calls the lowest form of planning, "plans in action." But apes' actions are also suggestive of "plans of action" (De Lisi, 1987, p. 90), which are deliberate sequences of behavior that are mentally represented (see discussions in Miles, 1986; Mitchell, 1986, 1993; Whiten & Byrne, 1988; Byrne & Whiten, 1990, 1992; Parker & Milbrath, 1993). Although great apes are remarkably good at devising deceptions based on the limited means available, without other means their deceptions are likely to remain limited, at least in comparison with those of more mature humans. (Unfortunately, little formal study has been devoted to children's naturally occurring deceptions, such that comparison of apes' and children's deceptions is difficult. If the lack of deception and concealment by 4-year-old and younger children in experimental situations can be reasonably compared to the naturalistic observations of deceit in great apes, the latter appear much more sophisticated.)

What remains to be done is careful developmental analysis of when and in what contexts behaviors used in deception first appear "honestly" in great apes and humans, and when these behaviors are first used to deceive or conceal. At present, there are published accounts of when deceits first appear in a few human children (e.g., LaFrenière, 1988; Reddy, 1991), but only one published account of when and in what contexts the "honest" behaviors used for the deceits first appeared (Reddy, 1991). In her study of the sign-using orangutan, Chantek, Miles specifically noted the appearance and context of all honest and dishonest uses of signs (as well as dishonest uses of non-sign behaviors), such that when these data are published they will provide insights into how deceit develops (for a preliminary account, see Miles, 1986). Only with knowledge like this can we understand if deceit develops through scripts, and usefully compare the developmental trajectory of skills in deceit across species (Morgan, 1894; Miles, 1986; Mitchell, 1986).

ACKNOWLEDGMENTS

I greatly appreciate the careful and insightful comments of my co-editors, whose discussions on the concept of scripts and thorough readings of the chapter were extremely helpful.

REFERENCES

Abel, C. (1819). *Narrative of a journey in the interior of China*. London: Longman, Hurst, Rees, Orme, and Brown.

Adang, O. M. J. (1984). Teasing in young chimpanzees. *Behaviour*, **88**, 98–122.

Andersen, R. (1980). Hunt and conceal: Information management in Newfoundland deep-sea trawler fishing. In S. K. Tefft (ed.), *Secrecy: A cross-cultural perspective*. New York: Human Sciences Press.

Anderson, M. (1986). Cultural concatenation of deceit and secrecy. In R. W. Mitchell & N. S. Thompson (eds.), *Deception: Perspectives on human and nonhuman deceit* (pp. 323–348). Albany: State University of New York Press.

Barns, T. A. (1928). Hunting the morose gorilla. *Asia and the Americas*, **28**, 116–123, 154–156.

Benchley, B. J. (1940). *My life in a man-made jungle*. Boston: Little, Brown and Co.

Blum, R. H. (1972). *Deceivers and deceived: Observations on confidence men and their victims, informants and their quarry, political and industrial spies and ordinary citizens*. Springfield, IL: Charles C. Thomas, Publisher.

Brooke, J. (1841). Letter (on orangutans). *Zoological Society of London*, **No. 52**, 55–60.

Brown, R. (1970). The first sentences of child and chimpanzee. In *Psycholinguistics: Selected papers by Roger Brown* (pp. 208–231). New York: Free Press.

Buller, D. B., Strzyzewski, K. D., & Comstock, J. (1991). Interpersonal deception: I. Deceivers' reactions to receivers' suspicions and probing. *Communication Monographs*, **58**, 1–24.

Burbridge, J. C. (1926). Miss Congo. *Nature Magazine*, **8**, 113–115.

Byrne, R. W. (1997). What's the use of anecdotes? Distinguishing psychological mechanisms in primate tactical deception. In R. W. Mitchell, N. S. Thompson, & H. L. Miles (eds.), *Anthropomorphism, anecdotes, and animals* (pp. 134–150). Albany: State University of New York Press.

Byrne, R. W. & Whiten, A. (1988). Toward the next generation in data quality: A new survey of primate tactical deception. *Behavioral and Brain Sciences*, **11**, 267–273.

Byrne, R. W., & Whiten, A. (1990). Tactical deception in primates: The 1990 database. *Primate Report*, **27**, 1–101.

Byrne, R. W. & Whiten, A. (1992). Cognitive evolution in primates: Evidence from tactical deception. *Man*, (N.S.), **27**, 609–627.

Chevalier-Skolnikoff, S. (1986). An exploration of the ontogeny of deception in human beings and nonhuman primates. In R. W. Mitchell & N. S. Thompson (eds.), *Deception: Perspectives on human and nonhuman deceit* (pp. 205–220). Albany, NY: State University of New York Press.

Christy, C. C. (1915). The habits of chimpanzees in African forests. *Proceedings of the Zoological Society of London, 1915*, 536.

Cunningham, A. (1921). A gorilla's life in civilization. *Zoological Society Bulletin*, **24**, 118–124.

Cuvier, F. (1811). Description of an orangutan. *Philosophical Magazine*, **38**, 188–199.

De Lisi, R. (1987). A cognitive-developmental model of planning. In S. L. Friedman, E. K. Scholnick, & R. R. Cocking (eds.), *Blueprints for thinking: The role of planning in cognitive development* (pp. 79–109). New York: Cambridge University Press.

De Waal, F. (1982). *Chimpanzee politics*. New York: Harper and Row.

De Waal, F. (1986). Deception in the natural communication of chimpanzees. In R. W. Mitchell & N. S. Thompson (eds.), *Deception: Perspectives on human and nonhuman deceit* (pp. 221–244). Albany, NY: State University of New York Press.

Dennett, D. C. (1978). *Brainstorms*. Montgomery, VT: Bradford Books.

Dennett, D. C. (1983). Intentional systems in cognitive ethology: The "Panglossian paradigm" defended. *Behavioral and Brain Sciences*, **6**, 343–390.

Du Chaillu, P. (1868). *Stories of the gorilla country, narrated for young people*. New York: Harper & Brothers, Publ.

Dunn, J. (1988). *The beginnings of social understanding*. Cambridge, MA: Harvard University Press.

Emlen, J. T., Jr. (1962). The display of the gorilla. *Proceedings of the American Philosophical Society*, **106**, 516–519.

Galdikas, B. Brindamour- (1975). Orangutans, Indonesia's "people of the forest." *National Geographic*, **148**, 444–473.

Goodall, J. van Lawick- (1968). A preliminary report on expressive movements and communication in the Gombe Stream chimpanzees. In P. C. Jay (ed.), *Primates: Studies in adaptation and variability* (pp. 313–374). New York: Holt, Rinehart and Winston.

Goodall, J. (1986). *The chimpanzees of Gombe: Patterns of behavior*. Cambridge, MA: Belknap Press.

Gray, G. W. (1955). The Yerkes laboratories. *Scientific American*, **192**(2), 67–77.

Harrisson, B. (1960). A study of orang-utan behaviour in semi-wild state, 1956–1960. *Sarawak Museum Journal*, **9**, 422–447.

Hayano, D. M. (1980). Communicative competency among poker players. *Journal of Communication*, **30**, 113–120.

Hayes, C. (1951). *The ape in our house*. New York: Harper and Brothers.

Hayes, K. J. & Hayes, C. (1952). Imitation in a home-raised chimpanzee. *Journal of Comparative and Physiological Psychology*, **45**, 450–459.

Hebb, D. O. (1946). Emotion in man and animal: An analysis of the intuitive processes of recognition. *Psychological Review*, **53**, 88–106.

Hediger, H. (1955/1968). *The psychology and behaviour of animals in zoos and circuses*. New York: Dover.

Hornaday, W. T. (1879). On the species of Bornean orangs, with notes on their habits. *Proceedings of the American Association for the Advancement of Science*, **28**, 438–455.

Horr, D. A. (1969). Our red-haired kin of the rain forest. *Life*, **66**(12), cover, 82–89, 91.

Hoyt, A. M. (1941). *Toto and I: A gorilla in the family*. Philadelphia: J. B. Lippincott Co.

Hudson, J. A., Sosa, B. B., & Shapiro, L. R. (1997). Scripts and plans: The development of preschool children's event knowledge and event planning. In S. L. Friedman & E. K. Scholnick (eds.), *The developmental psychology of planning: Why, how, and when do we plan?* (pp. 77–102). Hillsdale, NJ: Erlbaum.

[Johnson, M.] (1932). The mightiest of apes and tiniest of men. *Literary Digest*, **112**(9), 36, 40.

Klein, J. F. & Montague, A. (1977). *Check forgers*. Lexington, MA: Lexington Books.

Köhler, W. (1925/1973). *The mentality of apes*. New York: Liveright.

Kreitler, S. & Kreitler, H. (1987). Conceptions and processes of planning: The developmental perspective. In S. L. Friedman, E. K. Scholnick, & R. R. Cocking (eds.), *Blueprints for thinking: The role of planning in cognitive development* (pp. 205–272). Cambridge University Press.

Ladygina-Kots, N. N. (1935/1982). Infant ape and human child: Their instincts, emotions, games, and expressive movements. *Storia e Critica della Psicologia*, **3**, 113–189.

LaFrenière, P. J. (1988). The ontogeny of tactical deception in humans. In R. W. Byrne & A. Whiten (eds.), *Machiavellian intelligence* (pp. 238–252). Oxford University Press.

Leekam, S. R. (1991). Jokes and lies: Children's understanding of intentional falsehood. In A. Whiten (ed.), *Natural theories of mind* (pp. 159–174). Oxford: Blackwell.

MacKinnon, J (1971). The orangutan in Sabah today. *Oryk*, **11**, 141–191.

Martini, H. (1955). *My zoo family*. New York, Harper and Brothers.

Mawby, R. & Mitchell, R. W. (1986). Feints and ruses: An analysis of deception in sports. In R. W. Mitchell & N. S. Thompson (eds.), *Deception: Perspectives on human and nonhuman deceit* (pp. 313–322). Albany, NY: State University of New York Press.

Maxwell, M. (1928). The home of the eastern gorilla. *Journal of the Bombay Natural History Society*, **32**, 436–439.

McCornack, S. A. & Levine, T. R. (1990). When lies are uncovered: Emotional and relational outcomes of discovered deception. *Communication Monographs*, **57**, 119–138.

McCornack, S. A. & Parks, M. R. (1990). What women know that men don't: Sex differences in determining the truth behind deceptive messages. *Journal of Social and Personal Relationships*, **7**, 107–118.

Menzel, E. (1974). A group of young chimpanzees in a one-acre field. In A. M. Schrier & F. Stollnitz (eds.), *Behavior of nonhuman primates* (vol. 5, pp. 83–153). New York: Academic Press.

Miles, H. L. (1986). How can I tell a lie? Apes, language, and the problem of deception. In R. W. Mitchell & N. S. Thompson (eds.), *Deception: Perspectives on human and nonhuman deceit* (pp. 245–266). Albany, NY: State University of New York Press.

Miles, H. L. (1990). The cognitive foundations for reference in a signing orangutan. In S. T. Parker & K. Gibson (eds.), *"Language" and intelligence in monkeys and apes* (pp. 511–538). Cambridge University Press.

Miles, H. L. (1991). The development of symbolic communication in apes and early hominids. In W. von Raffler-Engel, J. Wind, & A. Jonker (eds.), *Studies in language origins, vol. 2* (pp. 9–20). Amsterdam: John Benjamins Publishing Co.

Miles, H. L. (1997). Anthropomorphism, apes, and language. In R. W. Mitchell, N. S. Thompson, & H. L. Miles (eds.), *Anthropomorphism, anecdotes, and animals* (pp. 383–404). Albany, NY: State University of New York Press.

Miles, H. L., Mitchell, R. W., & Harper, S. (1996). Simon says: The development of imitation in an enculturated orangutan. In A. E. Russon, K. A. Bard, & S. T. Parker (eds.), *Reaching into thought: The minds of the great apes* (pp. 278–299). Cambridge University Press.

Miller, G. A., Galanter, E., & Pribram, K. H. (1960). *Plans and the structure of behavior*. New York: Holt, Rinehart and Winston.

Millikan, R. G. (1984). *Language, thought, and other biological categories*. Cambridge, MA: Bradford Books/MIT Press.

Mitchell, R. W. (1986). A framework for discussing deception. In R. W. Mitchell & N. S. Thompson (eds.), *Deception: Perspectives on human and nonhuman deceit* (pp. 3–40). Albany, NY: State University of New York Press.

Mitchell, R. W. (1988). Ontogeny, biography, and evidence for tactical deception. *Behavioral and Brain Sciences*, **11**, 259–260.

Mitchell, R. W. (1991). Deception and hiding in captive lowland gorillas (*Gorilla gorilla gorilla*). *Primates*, **32**, 523–527.

Mitchell, R. W. (1993). Animals as liars: The human face of nonhuman deception. In M. Lewis & C. Saarni (eds.), *Lying and deception in everyday life* (pp. 59–89). New York: Guilford Press.

Mitchell, R. W. (1994). The evolution of primate cognition: Simulation, self-knowledge, and knowledge of other minds. In D. Quiatt & J. Itani (eds.), *Hominid culture in primate perspective* (pp. 177–232). Boulder, CO: University Press of Colorado.

Mitchell, R. W. (1996). The psychology of human deception. *Social Research*, **63**, 819–861.

Mitchell, R. W. (1997). Anthropomorphic anecdotalism as method. In R. W. Mitchell, N. S. Thompson, & H. L. Miles (eds.), *Anthropomorphism, anecdotes, and animals* (pp. 151–169). Albany, NY: State University of New York Press.

Mitchell, R. W. (1999). Scientific and popular conceptions of the psychology of great apes from the 1790s to the 1970s: Déjà vu all over again. *Primate Report*. 53, 1–118.

Mitchell, R. W. & Thompson, N. S. (eds.) (1986a). *Deception: Perspectives on human and nonhuman deceit*. Albany, NY: State University of New York Press.

Mitchell, R. W. & Thompson, N. S. (1986b). Deception in play between dogs and people. In R. W. Mitchell & N. S. Thompson (eds.), *Deception: Perspectives on human and nonhuman deceit* (pp. 193–204). Albany, NY: State University of New York Press.

Mitchell, R. W. & Thompson, N. S. (1986c). Epilogue. In R. W. Mitchell & N. S. Thompson (eds.), *Deception: Perspectives on human and nonhuman deceit* (pp. 357–361). Albany, NY: State University of New York Press.

Mitchell, R. W. & Thompson, N. S. (1991). Projects, routines, and enticements in dog–human play. In P. P. G. Bateson & P. H. Klopfer (eds.), *Perspectives in ethology, vol. 9: Human understanding and animal awareness* (pp. 189–216). New York: Plenum Press.

Mitchell, R. W. & Thompson, N. S. (1993). Familiarity and the rarity of deception: Two theories and their relevance to play between dogs (*Canis familiaris*) and humans (*Homo sapiens*). *Journal of Comparative Psychology*, **107**, 291–300.

Morgan, C. L. (1894). *An introduction to comparative psychology*. London: Walter Scott.

Nelson, K. (1981). Social cognition in a script framework. In J. H. Flavell & L. Ross (eds.), *Social cognitive development* (pp. 97–118). New York: Cambridge University Press.

Nishida, T. (1990). Deceptive behavior in young chimpanzees: An essay. In T. Nishida (ed.), *The chimpanzees of the Mahale Mountains* (pp. 285–290). Tokyo: University of Tokyo Press.

Oberhhohann, H. (1957). *My friend the chimpanzee*. London: Robert Hale Limited.

O'Reilly, J. (1950). Gorillas love Mom. *Collier's*, **125**(3), 33, 51–52.

Parker, S. T. & McKinney, M. (in press). *Origins of intelligence in apes and humans*. Baltimore: Johns Hopkins University Press.

Parker, S. T. & Milbrath, C. (1993). Higher intelligence, propositional language, and culture as adaptations for planning. In *Tools, language and cognition in human evolution* (pp. 314–333). New York: Cambridge University Press.

Parker, S. T. & Milbrath, C. (1994). Contributions of imitation and role-playing games to the construction of self in primates. In S. T. Parker, R. W. Mitchell, & M. L. Boccia (eds.), *Self-awareness in animals and humans* (pp. 108–128). Cambridge University Press.

Patterson, F. & Linden, E. (1981). *The education of Koko*. New York: Holt, Rinehart and Winston.

Piaget, J. (1947/1972). *The psychology of intelligence*. Totowa, NJ: Littlefield, Adams and Co.

Pitman, C. R. S. (1935). The gorillas of the Hayonsa Region, Western Kigezi, S. W. Uganda. *Proceedings of the General Meetings for Scientific Business of the Zoological Society of London*, **105**, 477–494.

Quiatt, D. (1984). Devious intentions of monkeys and apes? In R. Harré & V. Reynolds (eds.), *The meaning of primate signals* (pp. 9–40). Cambridge University Press.

Quiatt, D. (1987). Looking for meaning in sham behavior. *American Journal of Primatology*, **12**, 511–514.

Raven, H. C. (1932). Gorilla: Greatest of all apes. *Scientific American*, **146**(1), 18–21.

Reddy, V. (1991). Playing with others' expectations: Teasing and mucking about in the first year. In A. Whiten (ed.), *Natural theories of mind* (pp. 143–158). Oxford: Blackwell.

Reuter, H. (1954). Gorillas are funny people. *Nature Magazine*, **68**, 64–68.

Reynolds, V. & Reynolds, F. (1965). Chimpanzees of the Budongo Forest. In I. DeVore (ed.), *Primate behavior: Field studies of monkeys and apes* (pp. 368–424). New York: Holt, Rinehart and Winston.

Riopelle, A. J. (1963). Growth and behavioral changes in chimpanzees. *Zeitschrift für Morphologie und Anthropologie*, **53**, 53–61.

Riopelle, A. J., Jonch Cuspinera, A., Nos De Nicolau, R., Carbó, R. L., & Sabater Pi, J. (1976). Development and behavior of the white gorilla. In P. H. Oehser (ed.), *National Geographic Society Research Reports 1968* (pp. 355–367). Washington, DC: National Geographic Society.

Romanes, G. J. (1882/1906). *Animal intelligence*. New York: D. Appleton & Co.

Romanes, G. J. (1884/1900). *Mental evolution in animals*. New York: D. Appleton & Co.

Sacks, H. (1980). Button button who's got the button? *Sociological Inquiry*, **50**(3/4), 318–327.

Savage-Rumbaugh, [E.] S. & Lewin, R. (1996). *Kanzi: The ape at the brink of the human mind*. New York: John Wiley & Sons.

Savage-Rumbaugh, E. S. & McDonald, K. (1988). Deception and social manipulation in symbol-using apes. In R. W. Byrne & A. Whiten (eds.), *Machiavellian intelligence* (pp. 224–237). Oxford University Press.

Savage-Rumbaugh, E. S., Murphy, J., Sevcik, R. A., Brakke, K. E., Williams, S. L., & Rumbaugh, D. M. (1993). *Language comprehension in ape and child. Monographs of the Society for Research in Child Development*, **58**(3–4, Serial No. 233).

Schank, R. C. & Abelson, R. P. (1977). *Scripts, plans, goals and understanding*. Hillsdale, NJ: Erlbaum.

Selman, R. L. (1980). *The growth of interpersonal understanding*. New York: Academic Press.

Sexton, D. J. (1986). The theory and psychology of military deception. In R. W. Mitchell & N. S. Thompson (eds.), *Deception: Perspectives on human and nonhuman deceit* (pp. 349–356). Albany, NY: State University of New York Press.

Sheak, W. H. (1917). Disposition and intelligence of the chimpanzee. *Indiana Academy of Science*, **27**, 301–310.

Sheak, W. H. (1924). Some further observations on the chimpanzee. *Journal of Mammalogy*, **5**, 122–129.

Smith, E. O. (1987). Deception and evolutionary biology. *Cultural Anthropology*, **2**, 50–64.

Sodian, B. (1991). The development of deception in young children. *British Journal of Developmental Psychology*, **9**, 173–188.

Stern, W. (1924). *Psychology of early childhood up to the sixth year of age*. New York: Henry Holt and Co.

Tanner, J. E. & Byrne, R. W. (1993). Concealing facial evidence of mood: Perspective-taking in a captive gorilla? *Primates*, **34**, 451–456.

Terrace, H. S. (1979). *Nim: A chimpanzee who learned sign language*. New York: Alfred A. Knopf.

Thompson, N. S. (1986). Deception and the concept of behavioral design. In R. W. Mitchell & N. S. Thompson (eds.), *Deception: Perspectives on human and nonhuman deceit* (pp. 53–65). Albany, NY: State University of New York Press.

Van Lawick, H. (1991). *The people of the forest*. Bethesda, MD: Discovery Program Enterprises.

Vasek, M. (1986). Lying as a skill: The development of deception in children. In R. W. Mitchell & N. S. Thompson (eds.), *Deception: Perspectives on human and nonhuman deceit* (pp. 271–292). Albany: State University of New York Press.

Wallace, A. R. (1869/1962). *The Malay Archipelago, the land of the orang-utan and the bird of paradise: A narrative of travel with studies of man and nature*. New York: Dover Publications.

Warwick, J. E. (1832). The habits and manners of the female Borneo orang-utan (*Símia sátyrus*), and the male chimpanzee (*Símia troglódytes*), as observed during their exhibition at the Egyptian Hall, in 1831. *Magazine of Natural History*, **5**, 305–309.

Werth, L. F., & Flaherty, J. (1986). A phenomenological approach to human deception. In R. W. Mitchell & N. S. Thompson (eds.), *Deception: Perspectives on human and nonhuman deceit* (pp. 293–311). Albany, NY: State University of New York Press.

Whiten, A. & Byrne, R. W. (1988). Tactical deception in primates. *Behavioral and Brain Sciences*, **11**, 233–244.

Woodruff, G., & Premack, D. (1979). Intentional communication in the chimpanzee: The development of deception. *Cognition*, 7, 333–362.

Yerkes, R. M. (1929). The mind of a gorilla. Part III. Memory. *Comparative Psychology Monographs*, 8(24), 1–92.

Yerkes, R. M. (1943). *Chimpanzees*. New Haven: Yale University Press.

Yerkes, R. M. & Yerkes, A. W. (1929). *The great apes: A study of anthropoid life*. New Haven: Yale University Press.

16

Levels of imitation and cognitive mechanisms in orangutans

JOSEPH CALL

INTRODUCTION

> I have in mind not so-called instinctive and circular types of imitation . . . nor, at the other extreme, indubitably rational and purposive imitation with full consciousness of the objective and of the relation of act to result. The latter we know in ourselves, but only a naive or uncritical student of behavior could assert that chimpanzees are capable of it. I refer instead to varieties of imitative behavior which lie between these extremes. (R. M. Yerkes, 1943, p. 142)

The scientific study of imitation in primates dates back to the beginning of the century. Several scholars investigated the question of imitation in monkeys and apes using various experimental approaches such as puzzle boxes. While some of these early researchers claimed to have observed genuine cases of imitation (Hobhouse, 1901; Haggerty, 1909; Witmer, 1909; Furness, 1916), others maintained that the evidence was weak or nonexistent (Thorndike, 1901; Watson, 1908). As a consequence, whether or not primates were capable of learning by means of imitation remained a controversial and largely unresolved issue.

After a period of approximately 80 years in which little empirical research was conducted (see Tomasello & Call, 1997, for a review), the question of imitation has arisen with new force. In recent years, a number of studies of imitation have been conducted on capuchin monkeys, macaques, baboons, chimpanzees, and orangutans. While several studies have indicated that some species of monkeys do not seem to imitate (Beck, 1972, 1973; Chamove, 1974; Visalberghi & Fragaszy, 1990), whether or not chimpanzees and orangutans imitate is still an open question. Whereas some studies have reported cases of imitation in orangutans and chimpanzees (Russon & Galdikas, 1993, 1995; Custance, Whiten, & Bard, 1995; Miles, Mitchell, & Harper 1996) others have not confirmed those results (Tomasello, Davis-Dasilva, Camak, & Bard, 1987; Paquette, 1992; Nagell, Olguin, & Tomasello, 1993; Call & Tomasello, 1994). In addition, other studies have reported that individuals' imitative abilities are related to such variables as rearing or training history (Tomasello, Savage-Rumbaugh, & Kruger, 1993; Call & Tomasello, 1995; Miles et al., 1996) or the type of task employed (Call & Tomasello, 1995; Whiten et al., 1996).

We thus appear not to have advanced much after almost a century of research. However, we are in a better position now than we were a century ago in a number of regards. First of all, we have a better differentiation of social learning mechanisms. In this chapter, social learning refers to a group of learning mechanisms in which the observation of other individuals facilitates or enables the acquisition of novel behavior. The various kinds of social learning differ in the type of information (e.g., stimulus, motor patterns) that observers extract from the behavior of others. Yerkes' quote at the beginning of the chapter clearly illustrates the point that chimpanzees (and presumably other primates) may be capable of some forms of social learning (or imitation) but not others. In his classic book, Thorpe (1956) made a more formal distinction between three social learning processes: social facilitation, stimulus enhancement, and true imitation. In recent years, these three distinctions have been expanded to include a number of mechanisms all under the label of social learning (Galef, 1988; Whiten & Ham, 1992). Of special interest for our purposes are the distinctions among stimulus enhancement (i.e., local enhancement), emulation, and imitative learning.

Stimulus enhancement is defined as "an apparent imitation resulting from directing the animal's attention to a particular object or to a particular part of the environment" (Thorpe, 1956, p. 134). Note that, in this case, although there is an increased probability for action, it is not for any action in particular. Emulation can be defined as learning about the features and affordances of a situation or object as a result of a model's behavior. For instance, an animal may learn that a tool can be used to obtain food. Note that this use of the term "emulation" as originally used by Tomasello et al. (1987) differs from the term "goal emulation" as used by Whiten and Ham (1992), who focused on the model's goal, not necessarily on the changes in the environment that the model's behavior produced. Finally, true imitation consists of "copying of a novel or otherwise improbable act or utterance, or some act for which there is clearly no instinctive tendency" (Thorpe, 1956, p. 135). Galef (1985) has pointed out that Thorpe's definition presents a number of shortcomings. For instance, the phrase "some act for which there is clearly no instinctive tendency" is rather vague and difficult to interpret. Another important shortcoming is that Thorpe's (1956) definition of true imitation does not capture the important distinction between reproducing novel behaviors in service of a specific goal (i.e., imitative learning) and reproducing of behaviors without a goal beyond reproduction itself (i.e., mimicking). In the case of imitative learning, the copied action is a *means to an end* (e.g., a human punching a sequence of numbers to open an electronic door), while in mimicking the copied action is the goal itself (e.g., parrots reproducing speech). In other words, there is a clear means–ends dissociation (Piaget, 1952, 1962) in the case of imitative learning, but not in the case of mimicking. Tomasello (1996) has used this means–ends dissociation to argue that in imitative learning the observer perceives the model's intention to achieve a certain goal. The distinction between imitative learning and mimicking is important and will become especially relevant for evaluating the empirical information available on orangutan imitation presented in this chapter.

Another reason why we are in a better position to understand imitation in primates now is that currently there are more systematic data available on social learning in chimpanzees and orangutans than there were a few decades ago. Although some early studies briefly touched on the question of ape imitation (e.g., Witmer, 1909), the initial debate about primate imitation revolved mostly around monkeys' performance on problem solving tasks. Currently, there is a sizeable amount of information on orangutan and chimpanzee imitation in two main areas: (1) problem-solving and object manipulation, and (2) gestures and body movements (see Tomasello, 1996; Tomasello & Call, 1997, for a review). In problem-solving situations, individuals are typically faced with a directly unreachable goal (e.g., food) which can be obtained with the use of tools or some specific object manipulations. The area of gestural communication and body movements, includes the so-called "Do this" game (or "do-the-same-thing"), in which subjects reproduce a model's arbitrary gestures and body movements.

In this chapter, I organize the data available on imitation in orangutans and discuss their cognitive implications. First, I review and evaluate the evidence available on orangutan imitation for object manipulation and problem-solving, and then for gestures and body movements. Second, I discuss the question of what mechanisms are most likely to be responsible for orangutan behavior in these situations. Third, I analyze the cognitive components of orangutan imitation in an attempt to resolve some contradictory results. Finally, I attempt to relate imitation to other forms of learning.

THE EVIDENCE

Studies on problem-solving and object manipulation

Observational studies

A number of authors have cited casual observations to illustrate the notion that orangutans are especially good imitators. For example, Flourens (1845, cited by Yerkes & Yerkes, 1929) indicated that an orangutan imitated the posture, movements, and the use of a cane by an old man. More recently, Galdikas (1982) observed a number of tool-using behaviors among rehabilitant orangutans that she claimed had been likely learned by imitating humans. One example involved an orangutan using leaves for wiping his body (that according to Galdikas had been learned from a human using leaves for purposes of body care) and another involved the use of some man-made tools (i.e., hoe and shovel) to carry out similar activities to those of humans handling the same tools.

Galdikas (1982) cited a third example involving learning to use bridging materials (e.g., vines) to cross a river after observing a human model use them himself. Although the author noted that prior to the human demonstration orangutans had not used vines to bridge the river, it is true that some orangutans had already used boats or naturally placed logs to cross the river. Under those circumstances, it is conceivable that the use of vines by humans attracted the orangutans' attention toward them (a case of stimulus enhancement) and the orangutans solved the problem fortuitously. Alternatively, after one orangutan discovered the use of vines to bridge the river, the rest of the group benefited from the skillful subject leaving the vines in a propitious situation for learning. Anecdotes such as these are difficult to interpret because it is not clear what learning mechanisms are responsible for the acquisition of the behavior (Morgan, 1894; Galef, 1985).

Russon and Galdikas (1993, 1995) embarked on a more systematic observational study in search of actions that may qualify as cases of imitation. In these studies, the authors distinguished among various types of social learning and specifically focused their attention on imitative learning, that is, the reproduction of novel actions. They observed approximately 30 orangutans of mixed age and sex, concentrating the bulk of their observations on 6 adult females and 4 juveniles. These data were collected during focal observations and were supplemented with ad libitum observations and reports from a number of observers. Given the difficulty in detecting a behavior for the first time, a key issue in deciding what mechanism is responsible for the acquisition of a particular behavior, the authors used a number of criteria to decide whether a particular behavior qualified as imitative learning. In particular, they inferred imitative learning when the actions observed were incongruent (1) with the environmental conditions, (2) with species-specific orangutan behavior, (3) with a particular orangutan or (4) with a gradual (as opposed to sudden) acquisition as it is usually the case during associative learning. Most of the behaviors that Russon and Galdikas (1993, 1995) identified as imitation were human behaviors and, most of them involved tool using. Indeed, these authors indicated that much orangutan imitation (from orangutan models) may be very difficult to identify because it lacks the criteria outlined above. Perhaps the most striking cases reported by the authors are those in which orangutans reproduced complex behavior sequences of certain human activities (e.g., fire making, gas pumping). For instance, in the case of fire making, the orangutan removed the lid from the fuel container, poured fuel into a cup, dipped a burned stick in the fuel, fanned the burned stick, and finally, blew at the stick. Other examples included bridge making, painting, hanging hammocks, and boating, all of which the authors presented as evidence of imitative learning in orangutans.

It is clear that these later observational studies represent a significant improvement over past studies, and provide valuable information regarding orangutans' behavioral flexibility and creativity. Unfortunately, they still are not sufficient to decide with certainty what specific forms of social learning (if any) are responsible for the acquisition of those behaviors (Morgan, 1894; Galef, 1985). In addition, Russon and Galdikas point out that much of the imitation only partially matched the model. If that is the case, that opens the possibility for other forms of social learning such as emulation learning because individuals may have acquired the behavior through successive individual approximations.

Furthermore, in interpreting their results, Russon and Galdikas did not consider the potential effect that human enculturation (prior to the orangutans' arrival to the rehabilitation camp) may have had on the orangutans' ability to use imitative learning. It is thus unclear whether rehabilitant orangutans' social learning skills are comparable to those of wild orangutans without a history of enculturation. Several authors have noted the effect that human enculturation has on the cognitive development of apes (Hayes,

1951; Kellogg & Kellogg, 1933; Miles, 1994; Miles & Harper, 1994; Call & Tomasello, 1996; Call & Rochat, 1997). Recently, Call & Tomasello (1996) have articulated this view by systematically exploring the likely impact of human enculturation on a number of cognitive domains in apes. These authors distinguished various levels of human influence, from mere passive exposure to certain features of the human cultural environment to active modification via teaching. One example of the influence of passive exposure is illustrated by the tool-using skills that orangutans develop in captivity. Although it has been recently discovered that one group of wild orangutans use tools (Van Schaik, Fox, & Sitompul, 1996) it is still the case that tool use in wild orangutans is rare, whereas captive and rehabilitant orangutans use them regularly (Galdikas, 1982). The effect of passive exposure can be even more specific in the sense that human-made artifacts predispose orangutans to manipulate them in certain ways because these tools were designed with a specific function in mind. For instance, a shovel affords specific actions such as grasping by the handle to manipulate it, and that behavior in turn makes some other behaviors more likely to occur than others. Although this possibility may seem unlikely, since Russon and Galdikas' data set included some orangutan behavioral sequences with multiple steps that would be too complex to be learned by trial and error (e.g., fire-making), the authors did not have precise data on how much prior experience (including during their captive lives) subjects had had with those materials. Consequently, it is not possible to decide whether individual learning, or some sort of social learning (including imitative learning) was responsible for the acquisition of those behaviors.

In summary, Russon and Galdikas' studies are useful in pointing out the flexibility of orangutan behavior, as demonstrated by the range of actions that orangutans display in human environments. However, it is not clear how much information these studies provide about orangutan imitation, since individual learning or the various types of social learning apart from imitation may account for these results. It is almost certain that humans influence orangutans, and that orangutans socially learn a number of behaviors. Unfortunately, what we do not know yet is the specific mechanism of influence. Recent experimental studies are more helpful in this regard.

Experimental studies

The focus of early studies was problem-solving and tool use rather than imitation. Haggerty (1913) studied two female orangutans in two problems. In the platform problem, subjects were presented with a platform situated outside of their cage with a reward placed on top of it. Subjects were given a stick with a hooked end to retrieve the reward. Both subjects solved the problem without any need for demonstration. In the second problem, a metal pipe was attached to the cage in a horizontal orientation with a reward placed inside. Subjects needed to introduce a stick inside the pipe to extract the reward. Only one orangutan spontaneously solved the problem in less than five trials. The other orangutan was given five 15–minute trials on successive days without success. At that point, the unsuccessful orangutan was permitted to watch the skillful conspecific solving the problem five times. After the observer was released, she approached the apparatus, positioning herself on the same side where the model had been working, and solved the problem five times within the next few minutes. Haggerty (1913) concluded that the subject had learned from the other orangutan to perceive the relations between the different elements that constituted the problem and, consequently, had solved the problem by imitation.

Yerkes (1916) was also mainly interested in the problem-solving abilities of primates. In this study he tested two macaques and one orangutan on several problems. Yerkes presented a juvenile male orangutan (4–5 years of age) named Julius with four problems in which he had to use a tool to obtain a food reward. One of these problems consisted of obtaining an out-of-reach reward that was lying on a board by pulling in the board; another was analogous to the pipe problem previously described. Julius solved these two problems spontaneously without any model. The third problem consisted of piling up 2 or 3 boxes to climb and reach for an out-of-reach, suspended incentive. The orangutan was unsuccessful after several trials and Yerkes then facilitated the task by showing its solution to the subject. First, Yerkes modelled for the orangutan a complete solution. That is, he placed the boxes in the appropriate location, climbed them, and reached for the reward. After the boxes had been returned to their initial position and the reward had been replaced, Julius was given the opportunity to solve the problem. When the orangutan failed to obtain the reward, Yerkes showed the solution required to solve the problem (stacking the boxes) 5 more times in different sessions that lasted between 5 and 30 minutes. After these 5 sessions, Julius succeeded in stacking 2 boxes and reaching the reward. Yerkes interpreted these results as evidence of imitation but downplayed its role in problem-solving.

The fourth problem that Yerkes presented to the young orangutan consisted of a box that was locked with a padlock; a key was needed to unlock the padlock and open the box. Yerkes decided to try this experiment because he had observed Julius introducing splinters into the keyhole of the padlock. Yerkes reasoned that if Julius knew about keys and padlocks he must have learned its use from observing humans use them. In Julius' full view, Yerkes baited and locked the box with the padlock and offered the key to him. When the orangutan was released, he dropped the key and attacked the problem by biting and rolling the box, and hitting and pulling at the padlock. The orangutan also pushed the key through the hinges or into cracks under the lid of the box. Yerkes repeated his demonstration and the orangutan continued to use a number of behaviors, including inserting the key in the edge of the box, but never attempted to introduce it into the keyhole. As a consequence, Yerkes' idea that Julius' original behavior with the splinters might have resulted from imitating humans was not supported. Instead, it is likely that Julius' behavior simply reflected a species-specific behavioral propensity to introduce materials in holes.

Shepherd (1923) conducted the third experimental study on imitation in orangutans. He compared the ability of a 4–year-old male orangutan and a 7-year-old male chimpanzee in solving the platform problem used by Haggerty (1913). A human modelled the use of a stick to obtain an out-of-reach reward on 3 separate occasions. Subjects were given the opportunity to obtain the reward with the stick in four

2–minute trials, with 2 trials per day. Neither of the subjects secured the reward in the time allocated.

Nearly 50 years after Shepherd's study, Wright (1972) conducted another study on orangutan problem-solving and imitation. This was the first study on orangutans in which imitative learning rather than problem-solving in the form of tool-using and tool-making was the main focus of the study. Wright (1972) investigated whether Abang (a 5–year-old male orangutan) would be able to learn (1) to use a sharp stone flake to cut a string to obtain food that was locked inside a box, and (2) to manufacture stone flakes for that purpose. Accordingly, the study was divided into two phases. During phase 1 (tool-using phase), the orangutan was shown by a human model how to cut the cord with a flake on 9 occasions over a period of 4 days. After 13 trials without success, Abang's keeper guided the orangutan's hand to cut the rope. In the next trial, Abang succeeded in cutting the rope. Prior to success (and his keeper's guidance) the subject had engaged in trial-and-error behaviors of various sorts that included biting and hitting the rope. In the second phase of the experiment (tool-making phase), Abang was shown how to manufacture stone flakes to cut the rope and obtain the food. During a period that lasted 7 sessions, Abang witnessed the action modelled a total of 7 times. As in the previous phase, he engaged in a number of trial-and-error behaviors that included making noise with the stones, and using a hammer stone to hit the rope. Finally, after 12 trials, he succeeded in producing a stone sharp enough to cut the rope.

The picture emerging from these early studies on orangutan imitation is a mixed one. While some studies seem to support the idea of imitation (e.g., Haggerty, 1913; Wright, 1972), others provide only partially positive or negative evidence (e.g., Yerkes, 1916; Shepherd, 1923). Before jumping into a discussion of what mechanisms of social learning are responsible for the behavior of orangutans in these studies, it is important to note a number of limitations that make their interpretation problematic. None of these studies included a control group or any counterbalancing procedures in which individuals were presented with experimental and control conditions in different orders to control for the effect of experience. This is especially important since we have seen that orangutans were capable of solving some problems spontaneously. Also, some studies (e.g., Wright, 1972) used molding by physically guiding the actions of the subject as opposed to only modelling by visual example. Likewise, some studies also used an insufficient number of trials or subjects that were too young to succeed on the tasks presented.

Two recent experimental studies aimed at investigating imitative learning as a problem-solving strategy have taken into consideration the limitations of previous studies and have designed more controlled tests. These studies represent an improvement over past experimental studies, since they include multiple subjects, control groups, counterbalancing techniques, conspecifics as models (in some cases cagemates, and in one case in particular a mother–offspring pair), subadult and adult subjects, and a sufficient number of trials.

In the first of these studies, Call and Tomasello (1994) tested 16 subadult and adult orangutans using the procedure developed by Nagell et al. (1993) to test juvenile chimpanzees and human children. Call and Tomasello (1994) presented subjects with

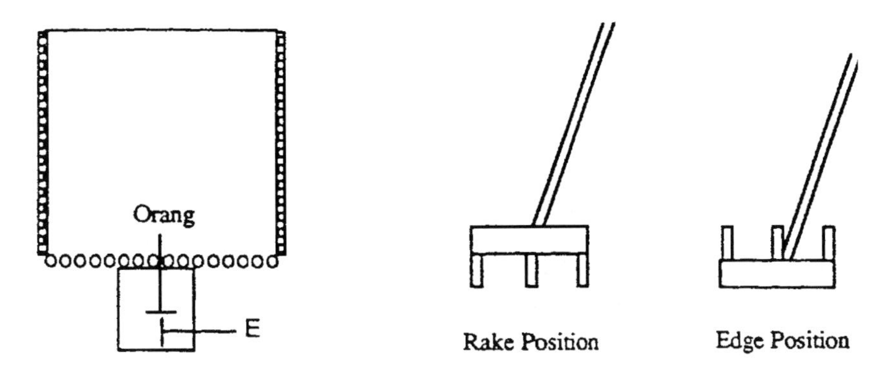

Figure 16.1 Experimental situation and tool employed in the Call and Tomasello (1994) study. The tool is shown in two different positions. Rake position is less efficient in retrieving the food that the edge position. (Reprinted with permission from the American Psychological Association).

the platform problem in which subjects had to get an out-of-reach food reward placed on a platform. In order to obtain the reward, subjects could use a T-shaped tool that could be used in two different orientations that resulted in different efficiencies (see Figure 16.1). Although previous studies had used the platform problem with orangutans, the tool efficiency was the same regardless of the tool orientation. In contrast, in this study when the tool was used in an edge orientation (prongs up), the reward was easily obtainable because the reward was easily moved with the edge of the head of the tool. In contrast, when the tool was in the rake orientation (prongs down), subjects had more difficulty in obtaining the reward because it slipped through the rake's prongs. A human demonstrator modelled the use of the tool with an edge orientation (edge condition) to one group of orangutans and with the rake orientation (rake condition) to another group of orangutans. In the edge condition, the experimenter placed the tool in the rake orientation, flipped it into the edge orientation, and pulled the tool, thus obtaining the reward. In the rake condition, the experimenter placed the tool in the edge orientation and simply pulled the tool to obtain the reward. Thus, no flipping was modelled in this last condition. In both experimental conditions, the subject's tool was always placed in the rake orientation at the beginning of each trial since we wanted to investigate the orangutans' tendency to flip the tool (from a rake to an edge orientation) in both experimental groups. Each subject received 50 models and was permitted to work on the problem for 1 minute after each model. If after a minute the subject had not obtained the reward but had moved it closer to him/herself, the experimenter did not interrupt the trial and simply performed another demonstration, permitting the subject to continue attempting to obtain the reward. If subjects were using imitative learning (i.e., copying the technique used by the model to obtain the food), it was expected that each group of orangutans would copy the technique that they had witnessed. Otherwise, the techniques adopted by subjects would be evenly distributed regardless of the model they had witnessed. Results indicated that model condition (edge vs. rake) had no effect on the techniques preferentially used by the two groups of subjects. Thus, this study failed to provide evidence of imitative learning in orangutans. Furthermore, analyses of the acquisition curves of the four most skilled subjects indicated that their preference for one or the other technique developed gradually, supporting the idea that individual learning in the form of trial and error was responsible for their performances.

A potential criticism of this study is that the authors used human as opposed to orangutan models (see Mitchell, 1993a). Orangutans may not have been motivated to copy a human model. For that reason, Call and Tomasello (1994) exposed a juvenile male orangutan to his mother, who had become a proficient edge user. The orangutan model demonstrated the edge strategy 10 times in each of three sessions. After every 10 demonstrations, the juvenile was given a chance to work on the problem for 10 trials at the end of each session. After the third session, the subject was given 70 additional trials without a model. At the end of these additional 70 trials the juvenile was preferentially using the edge technique as his mother had shown. However, a detailed analysis of the acquisition of the technique revealed a gradual shift from the rake to the edge technique analogous to the curves described for other animals in the previous testing phase. This gradual shift was characterized by a preferential use of the rake orientation in the four initial 10-trial blocks (including the three model sessions) followed by the alternation in each trial between the rake and the edge orientations in the next four 10-trial blocks. Finally, the subject preferentially used the edge orientation in the last two 10-trial blocks. Furthermore, when another male subadult orangutan was presented with the problem for the same number of trials as the juvenile orangutan but this time without a model, he also showed a gradual shift from the rake to the edge technique comparable to the juvenile's performance in the same number of trials. Consequently, the results of this additional study also failed to provide convincing evidence for the idea that orangutans use imitative learning as a problem-solving strategy.

Another possible criticism is that, given that all elements necessary for solving the problem were available to the subject (out-of-reach food, tool, platform, and their respective spatial relationships), there was no need for the animals to pay attention to the model. In other words, in a situation where the critical elements are available, subjects may not engage in imitative learning. Another potential pitfall is that subjects may have had difficulty in manipulating the rake or in understanding that the edge orientation was more likely to produce success than the rake orientation. Again, if that were the case, this experimental situation may not have been appropriate to elicit imitative learning.

With those criticisms in mind, Call and Tomasello (1995) designed a task that could only be systematically solved through imitative learning and that involved actions whose components were already in the subjects' repertoire. This represented quite a challenge, since the target actions could not be too complex because imitative learning may only work in a zone of proximal development (Bruner, 1972; Vygotsky, 1978). We presented twelve juvenile and adult orangutans and twelve 3- and 4-year-old children with an opaque box with a metal rod sticking out of it and a hole beneath the metal rod where a food reward was delivered (see Figure 16.2). The rod could be manipulated in three basic ways: pull, push,

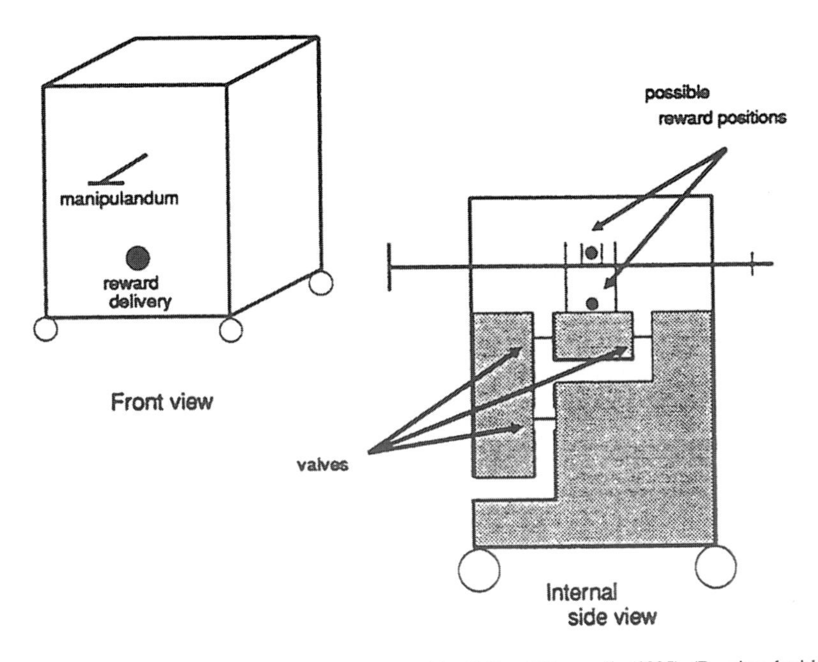

Figure 16.2 Box and rod problem apparatus used by Call and Tomasello (1995). (Reprinted with permission from the American Psychological Assocation).

or rotate, or any combination of these. When the subject performed the correct action, s/he received a food reward through the hole. The basic procedure consisted of the experimenter modelling the correct action or sequence of actions (e.g., pulling and rotating), which varied with the experimental phase, in order to obtain a reward. After his demonstration, he placed the apparatus against the fence and permitted subjects to manipulate the rod to obtain the reward. Call and Tomasello (1995) conducted more than 108 trials in which a single action or a combination of actions was used as target actions. Results indicated that none of the 12 orangutans consistently obtained the reward throughout testing and, consequently, none of them seemed to be copying the model's behavior to solve the problem. Instead, a number of animals developed preferences for certain actions and persisted in using them throughout testing. These preferences were not completely fixed, but evolved through trial-and-error to match to some degree the actions that were required to obtain the reward. We also repeated the same experiment (with 8 orangutans), this time using conspecifics (in most cases their own cagemates) as models. The results were analogous to those obtained with human models: none of the subjects reproduced the actions of their conspecifics.

One of the orangutans we tested was Chantek, who had been raised in a human environment from an early age as part of a project on cognitive and linguistic development (see Miles, 1983, 1990, for further details). Chantek had the distinction that he was able to reproduce bodily actions from a model as part of a "Do this" game (Miles et al., 1996). Given his ability to copy actions on command (see next section), we decided to put our task into the context of the "Do this" game and asked him to reproduce the actions that were needed to obtain the reward in the box problem (see Call & Tomasello, 1995, Experiment 3). Surprisingly, Chantek failed to copy the actions on the apparatus to obtain the reward even though he readily reproduced in the same session some actions like touching his head when asked to do so. Thus, Chantek showed an interesting dissociation between imitating actions within the well-established framework of the "Do this" gesture imitation game and a more open problem-solving situation.

In summary, in experimental studies, orangutans did not show any evidence of imitative learning in a problem-solving situation. Instead, they engaged in individual learning, disregarding both human and orangutan models. In contrast, 3- and 4-year-old children presented with the same situations quickly copied the actions modelled by the human demonstrator and, hence, provided evidence of their ability to use imitative learning in problem-solving situations (Call & Tomasello, 1995). It is possible that the difference between orangutans and humans was a product of the ontogenetic history of each species. Human adults encourage children from an early age to play imitation games with them, whereas orangutan mothers do not seem to encourage similar behaviors. Therefore, this ontogenetic difference may explain at least part of these results. However, this interpretation is undermined by the fact that Chantek was treated like a human child since early in life and yet he failed to use imitation in problem-solving situations.

Studies of imitation of gestures and body movements

A second avenue in the study of orangutan imitation has to do with the reproduction of gestures and body movements. Leaving aside the question of whether orangutans imitate or not in problem-solving situations, we may ask whether orangutans are capable of copying actions from a model at all. Most prominent in this research is the "Do this" game in which subjects are asked to reproduce actions after a model.

Nearly 80 years ago, Furness (1916) studied some cognitive abilities of two young wild-born orangutans. While attempting to teach one of them to speak (an enterprise that met with very limited success), he noted that the orangutan responded appropriately to the command "Do this." Unfortunately, Furness (1916) did not report what range of actions the orangutan would reproduce and, therefore, it is difficult to evaluate the possible significance of this study.

In the late seventies, Lyn Miles initiated Project Chantek, an 8–year longitudinal study aimed at investigating the cognitive and linguistic development of an infant male orangutan (see Miles, 1983, 1990, 1994; Miles & Harper, 1994; Miles et al., 1996 for further details). Over the course of Chantek's linguistic training, Miles noted that although molding (physically shaping hands) was the best method for teaching Chantek sign language in his first 4 years of life, he also acquired some gestures by imitation. For example, Miles et al. (1996) indicated that Chantek acquired the "wiper" sign (i.e., wiping a hand across the mouth) by observing others. Note, however, that the wiping motion may have already been in his repertoire and. consequently, Chantek may not have learned that or other gestures by imitation alone.

Before 2 years of age, Chantek was trained to play an imitation game consisting of touching some of his body parts (e.g., nose) corresponding to the parts that his caregivers were touching on their own bodies. By around 5 years of age, Chantek's caregivers taught him to play another imitation game, the "Do this" game in which he had to copy the same actions that his caregivers were producing. These actions included auditory (e.g., raspberry sounds), visual (e.g., eye-blinking), and kinesthetic modalities (e.g., touching nose). Miles et al. (1996) videotaped a total of 18 episodes of several demonstration–reproduction bouts, for a total of 24 actions. They classified Chantek's reproductions in three different ways depending on the level of accuracy: full imitations (salient aspects of the model were reproduced), partial imitations (less salient aspects of the model were reproduced), and non-imitation (failed to reproduce or produced an inaccurate action). Fifty-six percent of the reproductions were classified as full imitations, 34% as partial imitations, and 9% as nonimitation. Although these results suggest that Chantek was imitating on command, it would have been desirable if the authors had used blind scoring and had assessed inter-observer reliability.

In an attempt to replicate this study and to find out whether Chantek was still capable of playing this imitation game several years after Miles et al. (1996) collected their data, Call & Tomasello (1995) modelled 24 actions for Chantek while verbally asking him to "Do this" (Figure 16.3). These actions were all in the visual and kinesthetic modalities and some of them involved objects. Table 16.1 presents a complete list of the actions. After Chantek reproduced the correct action the experimenter gave him a food reward. This practice differed from the procedure of Miles et al. (1996), who did not use food rewards after Chantek's correct reproductions. Data were videotaped and scored by a coder after the action performed by the experimenter had been removed from the tape. Using this procedure, the coder was able to correctly identify what action had been modelled on 80% of the trials. This clearly indicated that Chantek was accurately reproducing the actions modelled by the experimenter in at least 80% of the trials. Chantek's accuracy was impressive and it was higher than the performance of two nursery-reared chimpanzees tested recently in a comparable setting (Custance, Whiten, & Bard, 1995). Still, there were some actions that Chantek did not reproduce. In explaining these results, the possibility of distraction seems unlikely since the actions were modelled twice, and only when Chantek was attending to the experimenter. Lack of motivation also seems unlikely since Chantek attempted actions in all trials. In some trials, when he failed to reproduce the appropriate action, he even produced "strings" of previously rewarded actions, suggesting that he had some special difficulty with particular actions (see also Hayes & Hayes, 1952; Custance, et al., 1995, for a comparison with chimpanzees). Although we used a set of arbitrary actions, it is possible that some of these actions closely resembled ASL signs in Chantek's repertoire which might have interfered with the imitation game (Miles, personal communication). If this were the case, the reproduction of these actions would not represent genuine cases of imitative learning but simply the production of previously learned gestures.

Figure 16.3 The orangutan Chantek mimicking the actions of a human model.

Table 16.1 *Repertoire of actions used in the "Do this" study*

1	Touch elbow	13	Open and close hand
2	Touch head	14	Sound with lips
3	Hand under chin	15	Rub arm
4	Touch both ears	16	Rotate head
5	Open mouth	17	Chest beating
6	Touch nose with finger	18	Block eyes with hand
7	Embrace	19	Blow air
8	Stick tongue	20	Ask palm up
9	Arm to sky	21	Rub hands
10	Arm in cross	22	Shake hand
11	Point	23	Clap hands
12	Touch tips of fingers	24	Pop (open mouth and touch open mouth)

In any case, Chantek's high degree of success points to one important fact. Orangutans under certain rearing conditions and training regimes can learn to play this imitation game, that is, they are able to systematically engage in kinesthetic–visual matching (Mitchell, 1993b). This ability has not been, to the best of our knowledge, observed outside of highly structured contexts in which other mechanisms of behavior acquisition such as emulation or trial-and-error learning have been ruled out.

DISSECTING IMITATION

After 100 years of research, the question of what the precise types of social learning orangutans use, and more specifically whether or not orangutans use imitative learning, is still a largely unresolved issue due to the existence of multiple alternative explanations. Table 16.2 lists all the studies reviewed above, indicating alternative mechanism besides imitative learning that might be responsible for the learning observed in those studies. None of the studies that have experimentally tested imitative learning applied to problem-solving has produced positive results, although the same procedures used with children of various ages have. Other observational studies that have been used to support imitative learning are problematic since they are without appropriate controls; even the most striking examples cited in support of imitative learning may be based on other mechanisms such as emulation, stimulus enhancement, or even individual learning. As a consequence, currently the evidence supporting imitative learning in problem-solving situations is still very fragile and at best what we have is evidence of emulation learning.[1] However, at least one orangutan is able to reproduce some presumably novel actions from a model with a substantial degree of accuracy in the

[1] It is important to emphasize, however, that currently emulation learning has not been compared with stimulus enhancement and therefore, strictly speaking, emulation learning has not yet been conclusively proven. However, since other apes such as chimpanzees are capable of emulation (e.g., Nagell et al., 1993), it is conceivable that orangutans are capable of it too.

Table 16.2 *Studies on imitation in orangutans specifying subjects, type of task, results, type of evidence, and alternative mechanisms other than true imitation*

Source	Subjects	Type of task	Results	Type of evidence	Alternative explanation(s)
Haggerty, 1913	2 females	platform pipe	not tested positive	experimental	individual learning stimulus enhancement emulation
Yerkes, 1916	1 juvenile male	support pipe box stacking padlock & box	not tested not tested mixed negative	experimental	individual learning stimulus enhancement emulation
Furness, 1916	1 juvenile female	Do this	positive	anecdotal	individual learning
Shepherd, 1923	1 juvenile male	platform	negative	experimental	individual learning
Wright, 1972	1 juvenile male	cut-string make stone flake	positive positive	experimental	individual learning stimulus enhancement emulation
Galdikas, 1982	various	object manipulation and problem solving	positive	anecdotal	individual learning stimulus enhancement emulation
Russon & Galdikas, 1993, 1995	6 adult females 4 juveniles	object manipulation	positive	observational	individual learning stimulus enhancement emulation
Call & Tomasello, 1994	4 adult females 12 juvenile and adult males	platform & rake	negative	experimental	individual learning stimulus enhancement emulation
Call & Tomasello, 1995	12	box & rod	negative	experimental	individual learning emulation
Call & Tomasello, 1995	1 adult male	Do this	positive	experimental	
Miles, Mitchell, & Harper, 1996	1 juvenile male	Do this	positive	experimental	

framework of the "Do this" imitation game. Interestingly, even though Chantek readily reproduced bodily actions in the "Do this" game, he failed to reproduce actions of the same human model in a problem-solving situation, even when he was explicitly asked to "Do this" in that situation (Call & Tomasello, 1995).

There are several explanations for this contradiction between the findings in the "Do this" game and those in the problem-solving situations. It is conceivable that the conditions in the problem solving situations presented were not conducive to eliciting imitation. In other words, Chantek and other orangutans might have been capable of imitating if we had presented other more appropriate settings (Russon & Galdikas, 1995). Alternatively, it is possible that these contradictory results arise due to the fact that imitation in the "Do this" task and problem-solving situations require different cognitive abilities. In other words, being able to imitate in one context may not necessarily translate into an ability to imitate in the other. This would explain why Chantek was able to succeed in the "Do this" task but not in problem-solving situations even when he was explicitly asked (the experimenter said "Do this") to reproduce the action needed to solve the problem (Call & Tomasello, 1995, Experiment 3). The possibility that the "Do this" task represents a different level of cognitive complexity than the problem-solving situations is explored in the next section.

Contrasting imitation in the "Do this" task and in problem-solving situations

Call and Tomasello (1995) have already indicated that the "Do this" and the problem-solving situations may differ in their cognitive requirements. Here I explore what those cognitive requirements might be. These two situations differ in at least two potentially important cognitive dimensions: means–ends dissociation and what I am calling the self–other distinction. Let me begin by analyzing the question of means–ends dissociation (Piaget, 1952).

In previous sections we have hinted at the relevance of the means–ends dissociation applied to imitation when a distinction was made between imitative learning and mimicking. Recall that in imitative learning an observer copies the model's actions *as a means* to obtaining a goal, whereas in mimicking the observer copies the model's action *as an end in itself.* Consequently, imitative learning has an intermediate step that may increase the complexity of the task. In regard to the two contexts that we are analyzing, the reproduction of the model's actions in the "Do this" game would correspond to mimicking, whereas the reproduction of the model's actions in problem-solving would correspond to imitative learning. However, equating mimicking with the "Do this" game may not be appropriate since it can be argued that the "Do this" game may have external rewards in the form of social rewards such as praise.[2] Furthermore, recall that

[2] Although we have argued that mimicking does not possess goals other than the reproduction of the action in itself, it could be argued that mimicking has goals, but of a different nature than those of imitative learning. For instance, individuals that mimic others may be attempting to engage in social interaction rather than to obtain some kind of material reward (e.g., food). If that were the case, the distinction between imitative learning and mimicking would become blurred, and perhaps it would be more appropriate to distinguish imitative learning from mimicking based on the *match/mismatch* between the model's and the observer's respective goals. Imitative learning would result when the

Chantek also reproduced actions in the "Do this" game as a means to obtaining food (Call & Tomasello, 1995). As a consequence, the differences between the "Do this" and the problem-solving situations may not be produced by a means–ends dissociation, or more specifically by the role of external rewards that increase the task's complexity. Having ruled out means–ends dissociation as a plausible explanation for the differences observed between tasks, let us turn our attention to the second cognitive dimension that we have identified: the self–other distinction.

The self–other distinction refers to the role that the observer /learner plays in the interaction with the model. Tomasello et al. (1994) have used this distinction to explore the acquisition of communicative gestures in young chimpanzees, but it can be extended into other domains of social learning. Figure 16.4 presents the two basic roles that a naive individual may play in a gestural acquisition situation (or other situations) in relation to a model.

In second-person imitation (dyadic interaction), a model directs an action toward a recipient and observer (second-person imitation, step 1, Figure 16.4), and later the recipient directs this novel action toward the model or any other individual (step 2, Figure 16.4). Thus, in second-person imitation the recipient acquires a novel behavior by *receiving* it from the model. In third-person imitation (triadic interaction), a model directs an action to a recipient while a third subject observes, without directly interacting with them (third-person imitation, step 1, Figure 16.4), and later the observer directs this novel gesture toward some other individual (step 2, Figure 16.4). Thus, in third-person imitation an individual acquires a novel behavior by *observing* (but not receiving) it from another individual. In sum, individuals can observe a novel behavior directed to the *self* (second-person imitation) or to *others* (third-person imitation) and acquire it either by receiving or observing it, respectively. The critical feature of this proposal is the individual's degree of involvement with the situation. An example will help clarify this point. Suppose that a stumptail macaque named Berryl is resting on top of her favorite branch. A youngster named Keplero wants her to move away from the branch so he jumps on top of the branch with enough force that Berryl loses her balance and falls to the ground. On a different occasion Berryl is resting on the ground near her favorite branch, which is occupied by another macaque. Again Keplero jumps on top of the branch in order to displace the other monkey which falls to the ground. In the first example, Berryl was directly involved in the situation, since she received the effect of Keplero's actions, whereas in the second example Berryl was not directly involved since she simply observed the effect of Keplero's behavior on a third party. If Berryl later jumped on the branch in order to knock someone off, she would have imitatively learned through second-person imitation in the first example, and third-person imitation in the second example.

Now let us analyze our two situations using the schema proposed in Figure 16.4. In the case of the "Do this" situation the model (i.e., experimenter) produces an action

observer copies an action from the model while pursuing the same goal as the model, whereas mimicking would result when the observer copies the model's behavior for a different goal than the model.

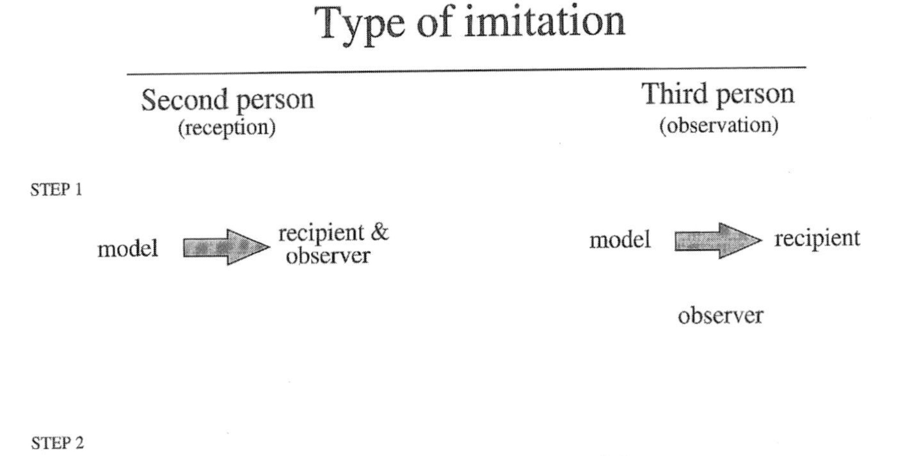

Figure 16.4 Diagram of the self–other distinction showing the various types of imitation based on the role that an invididual plays in each of them.

toward the recipient (i.e., orangutan) that he has to reproduce to obtain the reward. This represents a typical case of kinesthetic–visual matching (Mitchell, 1993b), and corresponds to the second-person imitation, in which the experimenter performs an action and the orangutan acts both as recipient and observer of that action. In contrast, in the case of the problem-solving situation, the model (i.e., experimenter) performs an action toward a recipient (i.e., object) while the observer (i.e., orangutan) witnesses it. In order to obtain the reward, the orangutan should copy the model's action on the object. This situation corresponds to third-person imitation in which the model performs an action, the object acts as the recipient, and the orangutan acts as the observer.

Recall that an orangutan failed to copy the model's actions in problem-solving situations but not in the "Do this" task. It is conceivable that in problem-solving situations orangutans perceive the model's action as unrelated to their own problem. This may constitute a critical difference between the "Do this" and the problem-solving situations. In the "Do this" case, the subject performs the action in a closed, well-established routine. This well-established routine is equivalent to using a matching-to-sample paradigm in which the subject has to match the action produced by the model. In a way, the orangutan incorporates the model's action into a routine that has been acquired by means of extensive training. Conversely, in the problem-solving situation, the orangutan has to transpose the model's action into his own sequence to solve the problem. In order to make the "Do this" situation comparable to the problem-solving situation at hand, the "Do this" game should be played in such a way that one orangutan reproduces the action of a model while Chantek watches, and Chantek, in turn, has to copy that same action.

In sum, we have explored the contribution of two factors (i.e., means–ends dissociation vs. self–other distinction) in explaining Chantek's differential performance in the "Do this" game and in problem-solving situations. The self–other distinction, and the complexity of triadic interactions, in particular, rather than a means–ends dissociation, appear as a more plausible explanation for his differential performance. In the next section I explore this possibility further by identifying what the components and levels of imitation are that subjects may use in problem-solving situations.

Components and levels of imitation in triadic situations

Assuming that the degree of subject involvement is responsible for the discrepancy in results observed between the "Do this" and the problem-solving situations, we can explore the specific problems that make the application of imitation more difficult in a triadic situation (i.e., third-person imitation). Observing a model performing an action to achieve a goal (a typical triadic situation) produces various sources of information, the understanding of which is crucial in investigating what type of social learning subjects engage in. Let us take the T-shaped tool study (Call & Tomasello, 1994) to illustrate this point. In this task, subjects had to flip and pull the tool to obtain the food. Figure 16.5 presents a number of possible sources of information available to the subject after a model has used the tool to obtain food. Each of these sources can be used by the subject to solve the problem.

Starting from the goal, subjects may simply observe that when food moves closer to the fence, the food becomes reachable. In this case, subjects may try to come up with ways to produce that result. For instance, they may throw objects at the food, spit water, move the platform, or use materials such as blankets to try to bring the food closer. Provided that subjects did not attempt those behaviors without the benefit of a model, we may assume that some sort of social learning occurred. In this case, it may qualify as stimulus enhancement (their attention has been drawn to the food) or even perhaps emulation of the simplest type – they saw the food dragged within reach and they attempted to reproduce that result with their own means.

Alternatively, subjects may concentrate their attention on the tool making contact with and dragging the food within reach, that is, on the relationship between the tool and the food and the effect it produces. Subjects may, for instance, hit the tool to obtain the food or simply push the tool, attempting to create the same effect. This would also constitute a more specific case of emulation since it was more directed to the critical features of the task. Another possibility is that subjects may concentrate on the orientation of the tool (edge vs. rake) and reproduce it without reference to the behavior of the model and then produce the other appropriate responses. This also would be a case of emulation. Finally, subjects may concentrate on the behavior of the model. They may realize that flipping and pulling the tool (even without reference to the effect it produces on the environment) produces a desirable goal. Unlike the other cases examined previously, this is the first time that the model's behavior *per se* plays a role.

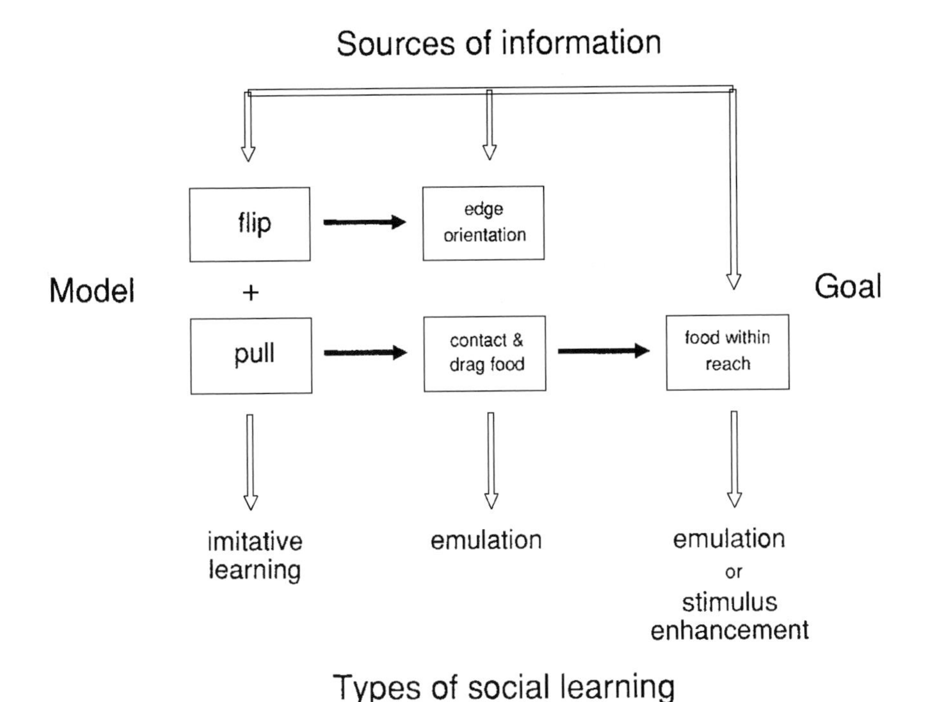

Figure 16.5 Types of social learning as a function of the sources of information picked up by the subject in the platform and rake problem (Call & Tomasello, 1994).

Imitative learning, however, should not be confused with mimicking in which subjects merely seek to reproduce behavior but not the model's goal. Interestingly, imitative learning may be understood as the combination between mimicking (copying the motor patterns) and emulation (copying the results). Results indicate that for problem-solving studies some sort of emulation is a more likely explanation than imitative learning.

In explaining the lack of imitative learning in problem-solving situations, there are at least two possibilities. One possibility is perceptual in nature. It is conceivable that orangutans do not perceive or chunk information in useful ways, or that they may have problems with the order or sequence of the actions. The second possibility is related to the understanding of the problem. According to Köhler (1925), imitation is not likely to occur unless subjects have some understanding of the problem (i.e., understanding of causal relations between its various elements). However, several authors (Piaget, 1952; Visalberghi & Limongelli, 1996; Tomasello & Call, 1997) have indicated that understanding of causality should not be treated as an all-or-none phenomenon. As a consequence, it is conceivable that the different sources of information stemming from a model are used depending on the level of causal understanding that subjects possess.

For instance, subjects could fail in the T-shaped tool problem because they may not realize that the orientation of the tool (edge vs. rake) is critical in solving the task efficiently. Alternatively, subjects may not understand that the tool has to contact the food in order to be effective. Finally, subjects also could fail because they overlook the importance of certain actions of the model. Of all the actions described in the T-shaped tool example (i.e., flipping, contacting, and dragging), only contacting and dragging are directly involved in getting the reward closer. As a consequence, subjects are likely to view them as directly responsible in food procurement. In contrast, flipping is different and it constitutes the crux of the problem. Flipping the tool is a behavior that may not be seen as clearly directed to the food (it produces no direct effect on the food's proximity to the model). In this sense, flipping is an *enabling* action in the same way that walking away from a food reward in a detour around a fence is enabling when obtaining the reward. It is true that subjects may learn to use these actions (through trial and error or insight) but they may not be capable of understanding their significance when another subject uses them and they are thus external to their own actions.

In sum, certain problem-solving situations present a special challenge for subjects trying to learn from a model. These situations commonly offer a number of sources of information resulting both from the model's behavior and the effects that this behavior produces on the environment. Various social learning mechanisms are grounded on the different sources of information available in these situations.

IMITATION IN RELATION TO OTHER FORMS OF LEARNING

The emphasis placed on imitation as a mechanism responsible for the behavior of orangutans has changed substantially over the years. Both Yerkes (1916) and Köhler (1925) considered imitation as a less interesting form of problem-solving than insightful problem-solving. Both authors viewed imitation, even though they did not study it in detail, as a "nuisance" that may compromise their experiments on individual problem-solving. Similarly, in the domain of sign language, Miles (1983) indicated that Chantek mostly used *spontaneous* signs as opposed to *imitated* signs when conversing with his human partners. Miles interpreted this result as indicating a greater flexibility in sign use than that demonstrated by one sign-using chimpanzee (Nim), whose conversations were mostly based on the reproduction of his caregivers' signs.

More recently, the interest has shifted and numerous studies emphasize the importance of imitation at the expense of other forms of social and individual learning. This emphasis on imitative learning is due to the fact that it is considered a more cognitively complex form of learning (see Byrne, 1995; Whiten & Ham, 1992). However, there are a number of potential problems with this position. First of all, it is not clear that imitative learning is more cognitively complex than some individual forms of problem-solving such as insight. It can be argued that imitative learning is cognitively complex since it implies some cross-modal skills such as perceiving others' behavior and self-producing it (Meltzoff, 1996). However, mimicking also involves this cross-modal quality (or kinesthetic–visual matching, Mitchell, 1993b) and it is not considered as complex as imitative learning. Furthermore, if imitative learning is cognitively complex because of its cross-modal quality, insight may be considered complex because it involves the creation and mental manipulation of a number of solutions to a problem.

One of the reasons for considering imitative learning as a particularly interesting and complex form of learning is based on the idea that imitative learning may be based on the ability to understand others' intentions. Tomasello and colleagues (Tomasello et al., 1993; Call & Tomasello, 1995) have argued that imitative learning of actions in problem-solving situations – in which an individual chooses from among several novel strategies the one it sees another demonstrating – requires an ability to understand the actions of others as intentional. In contrast, emulation learning would not require the understanding of intentions in others, since subjects may merely gather information about the changes that the model's behavior produces in the environment (Call & Tomasello, 1994, 1995; Tomasello, 1996). Choosing the same action performed by the model is the critical feature that Tomasello and colleagues use to infer understanding of intention in others. However, Meltzoff (1995) has shown that 18–month-old children are able to produce actions an adult meant to, but never succeeded in performing. For instance, in this study infants witnessed a model attempting without success to pull a toy apart. In other words, subjects were not shown the end result of the action (i.e., opening the toy), but simply what the model was attempting (i.e., pulling apart). Despite the fact that subjects had not seen the complete model (i.e., pulling and separating the object), infants produced the results the model was attempting significantly more often than two other control groups that were not shown the model's attempts. Meltzoff (1995) suggested that children were able to grasp the goals of others and reproduce the actions without the need to see a complete model. It can be argued that children in this experiment were emulating, or "goal emulating" to use Whiten and Ham's (1992) terminology. That is, they perceived the model's goal and attempted to reproduce it. As a consequence, even emulation learning in apes may be at least as complex as imitative learning in regard to the understanding of others' intentions.

The potential problems in ascribing a more cognitively complex organization to some social learning mechanisms over others are further illustrated by Russon and Galdikas' (1993) detailed observations. These authors pointed out that in some cases when orangutans reproduced actions without a clear goal, they went beyond emulation since they were impersonating (i.e., copying the model's behavior). However, these behaviors could be cases of mimicking. Russon and Galdikas (1993) also indicated that orangutans replicated some action components but not others. These authors interpreted these results as if the orangutans were engaging in something that went even beyond impersonation because they modified the model's sequence in a flexible manner. Interestingly, what this would demonstrate is precisely emulation which had previously been superseded by impersonation.

The problems of considering imitative learning as more important than other forms of learning are also evident when considering how adaptive imitative learning may be as opposed to other forms of learning. One of the main advantages of imitative learning is that it permits subjects to quickly and effortlessly copy a technique that may have been acquired by others through a long process of trial and error. To be effective in bypassing a long trial-and-error phase, copies have to be as similar as possible to the behavior of the model. In the absence of environmental changes, a population of imitators may be more successful than a population of nonimitators in exploiting resources since the former can quickly take advantage of the knowledge of others. However, under changing conditions a population of nonimitators who are likely to introduce changes in their behavior when needed would be in a better position than a population of imitators. Thus, it is best to understand the various learning mechanisms as complementary and take an ecological approach, keeping in mind that the usefulness of each social learning mechanism is going to be determined by the particular problems that subjects face (see Piaget, 1962). It is also important to remember that, when subjects confront a problem in their natural habitats, some forms of learning do not occur in isolation from others. For instance, when an orangutan observes another orangutan eating bark for the first time, and the observer tries to eat bark herself, her subsequent behavior will be modified by her current experience. Therefore, her future behavior will not only be modified by what she observes from her conspecific but also by her own experience and the state of the environment after her intervention (Russon, 1996). In this way, a feedback loop is created in which it is difficult to determine the social, individual, and environmental influences without the help of experiments which allow us to isolate the influence of the various factors. Orangutans have at their disposal all this information and, currently, it is not clear that copying the behavior of conspecifics as opposed to the consequences of their behavior plays a fundamental role in their problem-solving abilities.

CONCLUSION

Currently, there is no compelling evidence that orangutans are capable of reproducing a model's novel actions to obtain a goal in problem-solving situations (i.e., imitative learning). As a consequence, other forms of social learning such as emulation, or even stimulus enhancement, are as likely as imitative learning to explain the acquisition of novel behaviors by orangutans. Nevertheless, an orangutan has been trained to reproduce actions from others on command ("Do this" game). We have argued, however, that the "Do this" game is not comparable to imitative learning in problem-solving situations because the "Do this" game differs from problem solving situations in what we have called the self–other distinction. The self–other distinction focuses on the role played by the individual in the social learning context, and distinguishes between dyadic (second-person imitation) and triadic (third-person imitation) interactions. It was argued that the "Do this" situation would correspond to a dyadic interaction in which the individual acts both as recipient and observer in relation to the model's behavior. In contrast, a problem-solving situation would correspond to a triadic interaction in which the individual acts as a mere observer and does not directly receive the model's behavior, which is instead directed toward a third party.

Although at this point our conclusions must be taken as tentative due to the lack of strong empirical support, the self–other distinction might be a valid explanation of why the orangutan Chantek reproduced actions in the "Do this" game but failed to apply his kinesthetic–visual matching skills to problem-solving situations. Clearly, data derived from a single individual is informative but in no case definitive, and future studies should be directed at testing the importance of the self–other distinction in the context of social learning. Future studies should also investigate the role of processing demands and enabling actions, which are currently poorly understood. What level of understanding subjects possess of the causal relations among the various elements in these problems (e.g., tool, platform, food) may be critical in understanding what social learning mechanisms subjects are capable of using.

ACKNOWLEDGMENTS

I thank M. Carpenter and M. Tomasello for providing helpful comments and suggestions on an earlier version of this chapter.

REFERENCES

Beck, B. B. (1972). Tool use in captive hamadryas baboons. *Primates*, **13**, 276–296.

Beck, B. B. (1973). Observation learning of tool use by captive Guinea baboons (*Papio papio*). *American Journal of Physical Anthropology*, **38**, 579–582.

Bruner, J. (1972). The nature and uses of immaturity. *American Psychologist*, **27**, 687–708.

Byrne, R. W. (1995). *The thinking ape*. Oxford University Press.

Call, J. & Rochat, P. (1997). Perceptual strategies in the estimation of physical quantities by orangutans (*Pongo pygmaeus*). *Journal of Comparative Psychology*, **111**, 315–329.

Call, J. & Tomasello, M. (1994). The social learning of tool use by orangutans (*Pongo pygmaeus*). *Human Evolution*, **9**, 297–313.

Call, J. & Tomasello, M. (1995) The use of social information in the problem-solving of orangutans (*Pongo pygmaeus*) and human children (*Homo sapiens*). *Journal of Comparative Psychology*, **109**, 308–320.

Call, J. & Tomasello, M. (1996). The effect of humans on the cognitive development of apes. In A. E. Russon, K. A. Bard, & S. T. Parker (eds). *Reaching into thought: The minds of the great apes* (pp. 371–403). Cambridge University Press.

Chamove, A. (1974). Failure to find observational learning in rhesus macaques. *Journal of Behavioral Science*, **2**, 39–41.

Custance, D. M., Whiten, A., & Bard, K. A. (1995). Can young chimpanzees imitate arbitrary actions? Hayes and Hayes (1952) revisited. *Behaviour*, **132**, 839–858.

Flourens, M. J. P. (1845). *De l'instinct et de l'intelligence des animaux* (2nd edn) (pp. 40–45). Paris.

Furness, W. H. (1916). Observations on the mentality of chimpanzees and orang-utans. *Proceedings of the American Philosophical Society*, **5**, 281–290.

Galdikas, B. M. F. (1982). Orang-utan tool-use at Tanjung Puting Reserve, Central Indonesian Borneo (Kalimantan Tengah). *Journal of Human Evolution*, **10**, 19–33.

Galef, B. G., Jr. (1985). Social learning in wild norway rats. In T. D. Johnston & A. T. Pietrewicz (eds.), Issues in the ecological study of learning (pp. 143–166). Hillsdale, NJ: Erlbaum.

Galef, B. G., Jr. (1988). Imitation in animals: History, definition, and interpretation of data from the psychological laboratory. In T. R. Zentall & B. G. Galef Jr. (eds.), *Social learning: Psychological and biological perspectives* (pp. 3–28). Hillsdale, NJ: Erlbaum.

Haggerty, M. E. (1909). Imitation in monkeys. *The Journal of Comparative Neurology and Psychology*, **19**, 337–455.

Haggerty, M. E. (1913). Plumbing the minds of apes. *McClure's Magazine*, **41**, 151–154.

Hayes, C. (1951). *The ape in our house*. New York: Harper.

Hayes, K. J. & Hayes, C. (1952). Imitation in a home-raised chimpanzee. *Journal of Comparative Psychology*, **45**, 450–459.

Hobhouse, L. T. (1901). *Mind in evolution*. London: Macmillan.

Kellogg, W. N. & Kellogg, L. A. (1933). *The ape and the child*. New York: McGraw-Hill.

Köhler, W. (1925). *The mentality of apes*. London: Routledge and Kegan Paul.

Meltzoff, A. (1995). Understanding the intentions of others: Re-enactment of intended acts by 18–month-old children. *Developmental Psychology*, **31**, 838–850.

Meltzoff, A. (1996). The human infant as imitative generalist: A 20-year progress report on infant imitation with implications for comparative psychology. In B. G. Galef, Jr. & C. M. Heyes (eds.), *Social learning in animals: The roots of culture* (pp. 347–370). New York: Academic Press.

Miles, H. L. (1983). Apes and language: The search for communicative competence. In J. de Luce & H. T. Wilde (eds.), *Language in primates: Perspectives and implications* (pp. 43–61). New York: Springer-Verlag.

Miles, H. L. W. (1990). The cognitive foundations for reference in a signing orangutan. In S. T. Parker & K. R. Gibson (eds.), *"Language" and intelligence in monkeys and apes* (pp. 511–539). Cambridge University Press.

Miles, H. L. W. (1994). ME CHANTEK: The development of self-awareness in a signing orangutan. In S. T. Parker, R. W. Mitchell, & M. L. Boccia (eds.), *Self-awareness in animals and humans. Developmental perspectives* (pp. 254–272). Cambridge University Press.

Miles, H. L. W. & Harper, S. E. (1994). "Ape language" studies and the study of human language origins. In D. Quiatt & J. Itani (eds.). *Hominid culture in primate perspective* (pp. 253–278). Niwot, CO: University Press of Colorado.

Miles, H.L., Mitchell, R. W., & Harper, S. E. (1996). Simon says: The development of imitation in an enculturated orangutan. In A. E. Russon, K. A. Bard, & S. T. Parker (eds.). *Reaching into thought. The minds of the great apes* (pp. 278–299). Cambridge University Press.

Mitchell, R. W. (1993a). Kinesthetic-visual matching, perspective taking and reflective self-awareness in cultural learning. *Behavioral and Brain Sciences*, **16**, 530–531.

Mitchell, R. W. (1993b). Recognizing one's self in a mirror? A reply to Gallup and Povinelli, Byrne, Anderson and de Lannoy. *New Ideas in Psychology*, **11**, 351–377.

Morgan, C. L. (1894). An introduction to comparative psychology. London: Scott.

Nagell, K., Olguin, K., & Tomasello, M. (1993). Processes of social learning in the tool use of chimpanzees (*Pan troglodytes*) and human children (*Homo sapiens*). *Journal of Comparative Psychology*, **107**, 174–186.

Paquette, D. (1992). Discovering and learning tool-use for fishing honey by captive chimpanzees. *Human Evolution*, **7**, 17–30.

Piaget, J. (1952). *The origins of intelligence in children*. New York: Norton.

Piaget, J. (1962). *Play, dreams, and imitation*. New York: Norton.

Russon, A. E. (1996). Imitation in everyday use: Matching and rehersal in the spontaneous imitation of rehabilitant orangutans (*pongo pygmaeus*). In A. E. Russon, K. A. Bard & S. T. Parker (eds.). *Reaching into thought. The minds of the great apes* (pp. 152–176). Cambridge University Press.

Russon, A. E. & Galdikas, B. M. F. (1993). Imitation in ex-captive orangutans. *Journal of Comparative Psychology*, **107**, 147–161.

Russon, A. E. & Galdikas, B. M. F. (1995). Constraints on Great Apes' imitation: Model and action selectivity in rehabilitant orangutan (*Pongo pygmaeus*) imitation. *Journal of Comparative Psychology*, **109**, 5–17.

Shepherd, W. T. (1923). Some observations and experiments of the intelligence of the chimpanzee and ourang. *American Journal of Psychology*, **34**, 590–591.

Thorndike, E. L. (1901). Mental life of monkeys. Psychological Review Monograph Supplement, **15**, 442–444.

Thorpe, W. H. (1956). *Learning and instinct in animals*. London: Methuen and Co.

Tomasello, M. (1996). Do apes ape? In B. G. Galef, Jr. & C. M. Heyes (eds.), *Social learning in animals: The roots of culture* (pp. 319–346). New York: Academic Press.

Tomasello, M. & Call, J. (1997). *Primate cognition*. New York: Oxford University Press.

Tomasello, M., Savage-Rumbaugh, E. S., & Kruger, A. C. (1993). Imitative learning of actions on objects by children, chimpanzees, and enculturated chimpanzees. *Child Development*, **64**, 1688–1705.

Tomasello, M., Davis-Dasilva, M., Camak, L., & Bard, K. (1987). Observational learning of tool-use by young chimpanzees. *Human Evolution*, **2**, 175–183.

Tomasello, M., Call, J., Nagell, K., Olguin, R., & Carpenter, M. (1994). The learning and use of gestural signals by young chimpanzees: A trans-generational study. *Primates*, **35**, 137–154.

Van Schaik, C., Fox, E., & Sitompul, A. (1996). Manufacture and use of tools in wild Sumatran oranutans. *Naturwissenschaften*, **83**, 186–188.

Visalberghi, E. & Fragaszy, D. M. (1990). Do monkeys ape? In S. T. Parker & K. R. Gibson (eds.), *"Language" and intelligence in monkeys and apes*. (pp. 247–273). Cambridge University Press.

Visalberghi, E. & Limongelli, L. (1996). Acting and understanding: Tool use revisited through the minds of capuchin monkeys. In A. E. Russon, K. A. Bard, & S. T. Parker (eds.), *Reaching into thought.* The minds of the great apes (pp. 57–79). Cambridge University Press.

Vygotsky, L. (1978). *Mind in society*. Cambridge, MA: Harvard University Press.

Watson, J. B. (1908). Imitation in monkeys. *Psychological Bulletin*, **5**, 169–178.

Whiten, A. & Ham, R. (1992). On the nature and evolution of imitation in the animal kingdom: Reappraisal of a century of research. In P. J. B. Slater, J. S. Rosenblatt, C. Beer, & M. Milinsky (eds.), *Advances in the study of behavior* (pp. 239–283). New York: Academic press.

Whiten, A., Custance, D. M., Gómez, J. C., Teixidor, P., & Bard, K. A. (1996). Imitative learning of artificial fruit processing in children (*Homo sapiens*) and chimpanzees (*Pan troglodytes*). *Journal of Comparative Psychology*, **110**, 3–14.

Witmer, L. (1909). A monkey with a mind. *The Psychological Clinic*, **3**, 179–205.

Wright, R. V. S. (1972). Imitative learning of a flacked stone technology – The case of an orangutan. *Mankind*, **8**, 296–306.

Yerkes, R. M. (1916). *The mental life of monkeys and apes*. Delmar, New York: Scholars' Facsimiles and Reprints.

Yerkes, R. M. (1943). *Chimpanzees. A laboratory colony*. New Haven, CT: Yale University Press.

Yerkes, R. M. & Yerkes, A. W. (1929). *The great apes. A study of anthropoid life*. New Haven, CT: Yale University Press.

17

Parental encouragement in *Gorilla* in comparative perspective: implications for social cognition and the evolution of teaching

ANDREW WHITEN

INTRODUCTION

Ten years ago, studies of animal behavior had little to say about the existence of "teaching," despite the years which had elapsed since the innovative attempts of Barnett (1968) and Ewer (1969) to put it on the agenda. In recent years, we have seen a new flowering of interest in the phenomenon in primates – the principal subjects of this chapter – and also in other animals. Descriptions of quite elaborate forms of behavior which have been argued to fit concepts like teaching, encouragement, and discouragement have accumulated in the course of long-term primate field studies (e.g., Van de Rijt-Plooij and Plooij, 1987; Hiraiwa-Hasegawa, 1990; Boesch, 1991) as well as in captive studies (e.g., Fouts, Fouts & Cantfort, 1989). More recently, largely qualititative observations like these have been supplemented by the first substantial quantitative analyses designed not only to describe the phenomena but to test hypotheses about the power of these types of parental behavior to influence infant development (Caro & Hauser, 1992; Maestripieri, 1995a,b, 1996). Caro and Hauser showed that, in conjunction with their own data on felids and primates, by 1992 the accumulation of empirical reports was sufficient to merit the first major review of "teaching in nonhuman animals," teaching here being defined by its *functional consequence* of facilitating a juvenile's developmental progress, rather than by any deliberate *intention to educate*.

The great apes contribute some of the most complex and interesting components of this body of evidence and these form the basis of this chapter. However, we are also immediately presented with an enigma: why has it taken so long for the observations to accumulate? Decades of field research have generated a corpus of observations numerous enough in absolute terms to demand some explaining, yet, relative to those decades of study, episodes of putative teaching appear to occur infrequently. They do not give the appearance of being a regular part of the everyday lives of the apes concerned, but instead are noted with excitement by observers as rather rare and treasured signs that teaching *can* occur in somewhat human-like forms in nonhumans. But, if such behavior is so rare, what can its significance be? It seems difficult to believe that it has evolved because it fulfills a function akin to that of human teaching, enhancing infants' developmental progress, if only a minority of juvenile apes ever experience it, and even then quite rarely. In this chapter I shall explore the hypothesis that the findings may have a more important message for us about the performers' minds – their social cognition – than about the role of education *per se* in the everyday lives of apes. However, what it has to say about ape social cognition has implications for humans' mental ancestry, including that underlying the evolution of the teaching that occurs in our species.

THE INEVITABLE ANTHROPOCENTRIC PERSPECTIVE

For many, the reason for interest in the topic of teaching in animals, and particularly in apes, is the evolutionary reconstruction of a capacity that is thought to be one of the most important components of human culture (Bruner, 1972; King, 1991, 1994; Galef, 1992; Parker & Russon, 1996). However, even when the reasons for interest are not explicitly human-centered, reference to human teaching as the comparative baseline for any study of nonhumans seems not only sensible but inevitable. "Teaching" is a term with well-known, everyday scope when applied to humans, and to stray too far from this would simply lead to talking of something quite different. The solution adopted by Caro and Hauser was to broaden the comparative definition of teaching so as to focus on its functional consequences rather than on any intent which motivates it, as follows:

> [1] An individual actor **A** can be said to teach if it modifies its behavior only in the presence of a naive observer, **B**, at some cost or at least without obtaining an immediate benefit for itself. [2] **A**'s behavior thereby encourages or punishes **B**'s behavior, or provides **B** with experience, or sets an example for **B**. [3] As a result, **B** acquires knowledge or learns a skill earlier in life or more rapidly or efficiently than it might otherwise do, or that it would not learn at all. (P. 153)

Behavior meeting these criteria has certain benefits for B: whether or not A intends these benefits for B is another issue, beyond the scope of this definition. Citing Pearson (1989), Caro and Hauser note that, to those who study human teaching, this definition will be unsatisfactory because it omits what such authorities often consider crucial – the *intention* of bringing about learning.

However, if we turn for our anthropcentric starting-point to empirical studies of human parent–infant and other expert–novice interactions, we find phenomena and classifications of these phenomena which are actually closer to those of Caro and Hauser than they may have imagined. One now-classic study with 3– to 5–year-old children dissected the ways in which an adult behaved while supporting a child solving a constructional problem-solving task (Wood, Bruner, & Ross, 1976). The categories which Wood et al. discriminated during these informal social interactions were described as "scaffolding functions" of the tutor. "Scaffolding" was a term used seemingly to avoid the formal and intentional connotations of the conventional term, "teaching": it was described as a process that enabled a child or novice to achieve a goal which would otherwise be beyond their unaided efforts at that stage of development. Scaffolding "consists essentially of the adult 'controlling' those elements of the task that are initially beyond the learner's capacity, thus permitting him to concentrate upon and complete only those elements that are within his range of competence," and, the authors note, this "may result, eventually, in development of task competence by the learner at a pace that would far outstrip his unassisted efforts" (Wood et al., 1976).

Wood et al. identified six major scaffolding functions in their tutors: *recruitment* to the task, *simplifying* the task demands, *encouraging persistence*, *marking* critical features, *avoiding frustration*, and *demonstrating* certain techniques. These will be discussed in greater detail further below. For the moment it is important to note that these functional categories were based on observational analyses of the actions of a tutor who had been asked to gear her behavior to the needs of the child, while leaving the child to perform as much of the task as they could for themselves. Thus the tutoring had an overall pedagogic intent imposed by the experiment, but was informal and relatively unconstrained in its specific contents (the functional categories discerned by the authors). Whiten (1975), Boesch (1991), and Parker (1996) therefore used the functional categories from the Wood et al. study as a basis for comparisons with ape behavior and we shall similarly revisit them below.

However, the Wood et al. study may have overemphasised the relatively formal aspects of teaching which are so readily assumed to be the human norm: their study was based on an experiment in which the child was set up with a problem to solve and a tutor was asked to help them in this. A study by Whiten and Milner (1984) offers a potentially more relevant comparison with ape behavior, in so far as it was based on naturalistic observations of spontaneous behavior, and also concerned an earlier period of development (6–15 months) in which several peaks of human infants' cognitive achievement correspond with those achieved by apes (Parker & Gibson, 1977, 1979; Parker, 1996). In the Whiten and Milner study, the behavior and experiences of human infants were observed in their homes and other everyday environments, both in the UK and in Nigeria. The "scaffolding" behavior of those individuals (mostly adults, but sometimes children) with whom the infants interacted in the context of object play was classified into two broad types which included seven principal categories (Figure 17.1).

Several findings from this study are of interest as a basis for comparative studies. Much of what was observed to occur under these categories of action does not appear to be formal or intentional teaching: if parents are asked what they are doing when acting in these ways, a typical response will be that they are "just playing" with the infant. In this study we reserved the term "teaching" for a subcategory of the general *functional* category of scaffolding, in which there appears to be an *intent* to educate the infant, at least in the short term. Our operational criterion for this category, appropriate to this preverbal stage of infancy, was demonstration of some action, followed by support of the infant's attempt to achieve it. Quantitative records on 24 British and 24 Nigerian infants, each observed for a total of 12 hours over a 4-month period, showed that teaching of this character is actually quite rare both in absolute terms and relative to the frequency of actions in the other categories of scaffolding (Figure 17.2). Of over 100 incidents of scaffolding occurring per hour in the eldest British sample, less than 1%

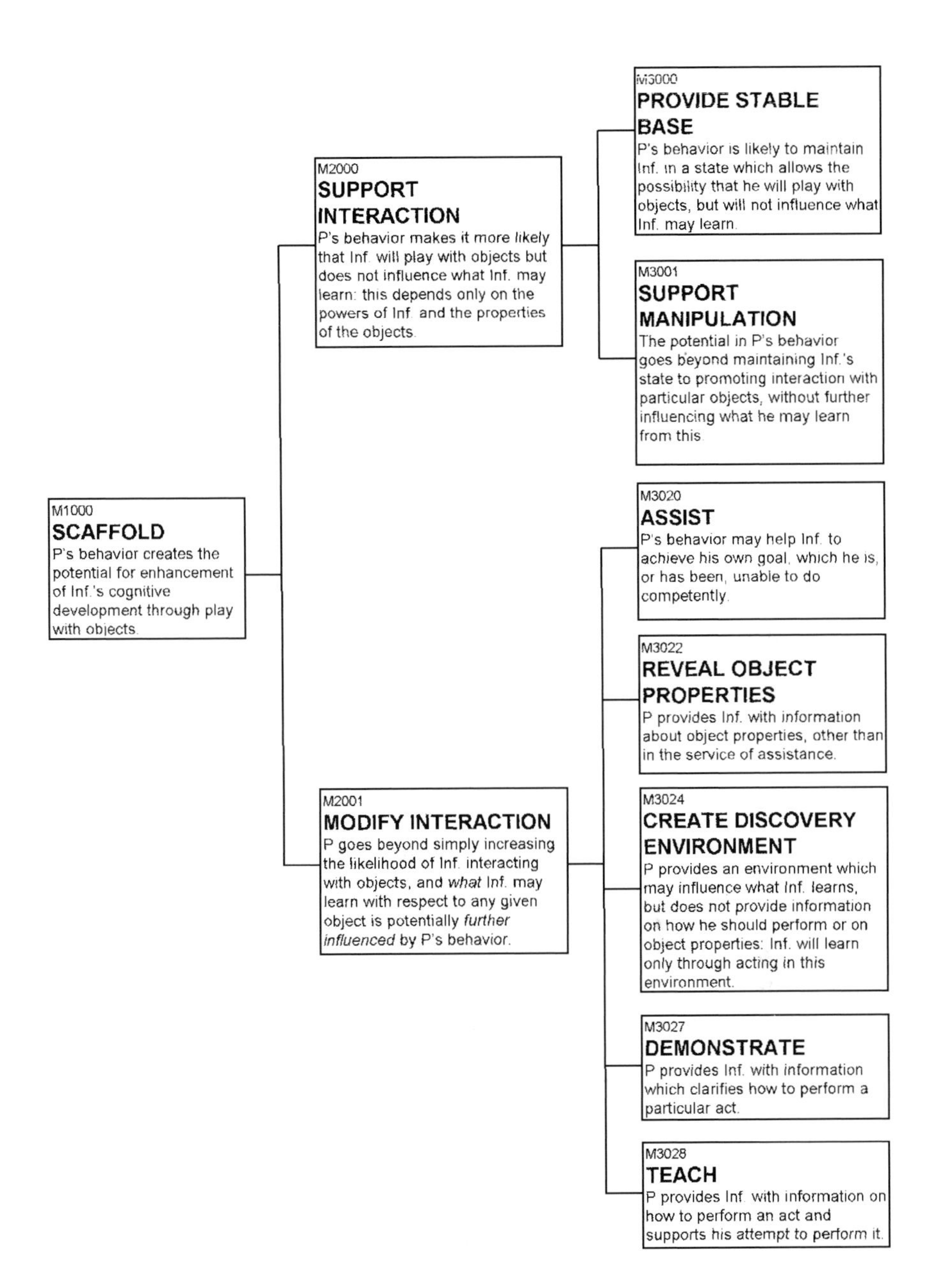

Figure 17.1 A taxonomy of categories of human "scaffolding" observed in adults and others interacting with human infants in everyday contexts at infant ages 6–15 months. These are the top-level categories from a hierarchical classification that additionally explicates their content at several lower levels not shown here (Whiten & Milner, 1984, to which code numbers refer).

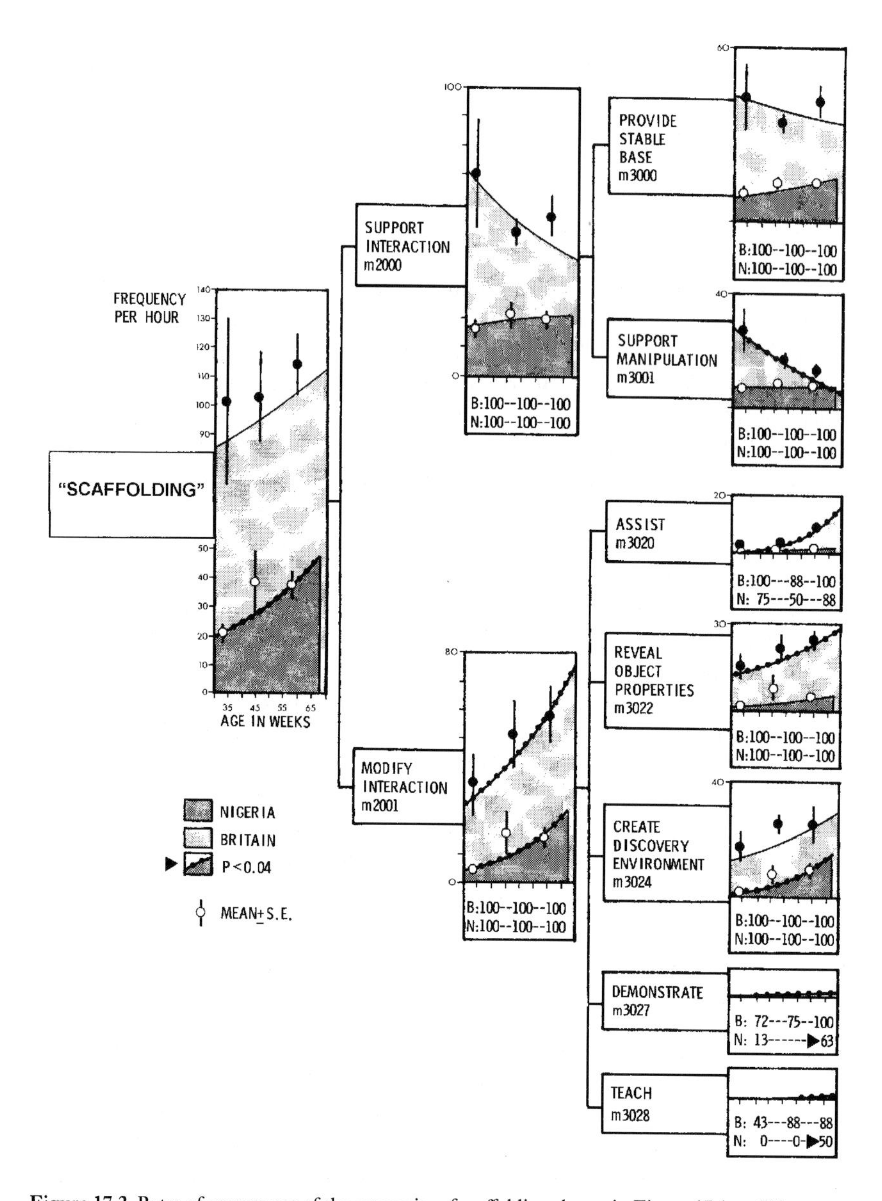

Figure 17.2 Rates of occurrence of the categories of scaffolding shown in Figure 17.1 at different infant ages, in Britain and Nigeria. Mean and S.E. are shown for samples of three different ages, together with simple curves fitted to the data for all 24 infants at different ages across the span. Beneath the curves, the percentage of the sample that showed the category of behavior is given (always 100% except for *assist*, *demonstrate*, and *teach*). Categories of *support interaction* tended to remain stable or (in the case of *support manipulation* for the British sample) decrease with developmental age. Categories of the more complex *modify interaction*, by contrast, rose in frequency. Relative to other categories, *demonstrate* and *teach* were very rare.

were "teaching": it was not even observed in all British samples, and only in half the Nigerian ones. Even the specific category of demonstrating how to perform some act (i.e., without the further encouragement to copy which we labeled "teaching") is only a little more common: it was never seen in a third of the Nigerian sample. By contrast, common categories of modifying the infant's interactions with objects include channeling its attention to particular properties or features, and "creating a discovery environment," in the context of which the infant is more likely to discover something new than it would through its own unmodified efforts: one or other of these occurred with a frequency rising to about 1 per minute for the British infants, 1 per 3 minutes for the Nigerian infants (Figure 17.2). "Create discovery environment" is similar to what Caro and Hauser later described as "opportunity teaching," in which an adult puts a juvenile (typically its offspring) "in a situation conducive to learning a new skill or acquiring new knowledge." This was the category to which they ascribed most of their nonhuman examples. Our studies indicate that much of parental scaffolding for human infants has this nondirective character: the parent intervenes, but the way in which the infant benefits depends very much on how it responds to what may be a very open-ended set of potentials the parent has created, even in interaction with the world of tools and other objects, which it is so important for the human infant to master. Nevertheless, if the potential for the infant learning something new *is* increased, then an important scaffolding function is being served.

How do the putative teaching behaviors of apes compare with the categories of scaffolding identified in these studies of human development ? I shall first describe a series of observations of maternal gorilla behavior which is relevant to this question. For each category of human scaffolding in turn, I shall examine the extent to which the functions concerned are served in the gorilla case and in the observations of others who have studied great apes and other animals. Finally, I shall discuss the cognitive implications of the scaffolding acts observed.

ENCOURAGEMENT BEHAVIORS IN GORILLA MOTHER–INFANT INTERACTION

The following observations were made on a western lowland gorilla (*Gorilla gorilla gorilla*) mother and infant at the Bristol Zoo, UK. Delilah, the mother, was wild-born but had lived in a small social group at the zoo from the age of 3 years. She was estimated to be about 8 years old when she gave birth to her first infant, Daniel. She reared the infant herself, in tactual but not visual isolation from another female and a male gorilla for the duration of the observations reported below. Alongside quantitative records of mother–infant interactions, extensive *ad-libitum* notes were kept and it is from these that the following descriptions are extracted (Whiten, 1975).

1. Physical support of crawling

This was first seen during the sixth week after birth. For 2 weeks Delilah had been putting the infant down prone for increasingly longer periods before he whimpered and

Figure 17.3 Mother gorilla supporting infant's head with her left hand while it made early crawling movements (Figures 17.3 to 17.6 redrawn from photographs by R. J. Spijkerman).

was picked up. Now, in this prone position, he started to make clumsy limb movements which infrequently and briefly either raised parts of his body off the ground or moved him laterally. He appeared to be trying to crawl. Late in the sixth week, Delilah started to support Daniel's head in her hand (Figure 17.3) and he would continue the movements. During week 7 he managed to raise his trunk by pushing with his legs when he was supported in this way, and a few days later was seen to crawl forwards a few centimetres when on his mother's ventrum (she lying on her back) or on the ground unsupported. Such crawling is rarely seen in the wild (Fletcher, personal communication) or even in mother-reared captive gorillas, by contrast with those reared by humans (Meder, 1989).

2. Tugging

During the seventh week also, Delilah was first seen to move out of arm's reach. She moved to about 2 metres away and remained there for 20 seconds. Daniel did not protest, but turned his head to watch her. She returned and picked him up, held him ventro-ventrally for a few seconds, then put him down prone and "tugged" him, slowly pulling him along by one arm over a distance of several feet. Daniel made no movements and merely slid over the smooth tiled floor. This was observed several times over the next few weeks, disappearing as Daniel became more competent at independent crawling.

3. Luring

During week 8, when Daniel was able to crawl about 30 centimetres unaided, Delilah was first seen to bend down quadrupedally with her head bowed down onto the floor and her face about 30 centimetres in front of the infant. As Daniel crawled about 15 centimetres towards her, she moved away to 30 centimetres again and this time when he approached her she picked him up. Periods of this "luring" behavior became increasingly frequent and prolonged – up to 5 minutes in week 9 – but were always terminated by ventral contact with or without suckling.

4. Physical support of climbing

During week 14, Daniel was first seen to lean bipedally against the wall and bars of the cage. During week 17, Delilah took Daniel from this position and held him against the bars of the cage about 70 centimetres off the ground. She did not let go. Daniel had at this point made no attempt to climb, and even when held in this fashion he did not grip the bars. He was put down after a few minutes, but Delilah often repeated this action in the following days. During week 18, Daniel was first seen to grip when held in this position, and later this week started to climb unaided. Delilah was then seen to hold him over some horizontal logs suspended from the ceiling of the cage (Figure 17.4), as she had held him against the vertical bars. When by week 25 Daniel had some competence in climbing both up and down vertical bars, Delilah hung him by his arms from the ceiling bars. She then moved away to a position where it seemed she would be unlikely to catch him if he fell (and he did not fall) before returning to take him down.

5. Dangling

A type of maternal behavior which seemed qualitatively different to those already described was seen during week 10. In one corner of the cage was a raised platform about 80 centimetres high, covered with smooth tiles like the floor. Delilah and Daniel often sat on this platform. Daniel had once fallen off it during week 8, but during week 10 his more active crawling led him to fall off again several times, particularly from one of the two edges. Later this week, Delilah often took the infant by his arm and suspended him over the side of the platform so that he would swing gently above the area where he had fallen before. She also later held him in similar fashion from the logs near the top of the cage several times (Figure 17.5). Several further falls from the platform were followed by this type of maternal behavior only a few minutes later. Over the next few weeks, falls became less frequent and then stopped.

These episodes appear to meet the criteria for teaching advocated by Caro and Hauser in so far as Delilah modified her behavior in specific ways in Daniel's presence (indeed apparently in response to the development of *specific behavior* – see below) with no immediate benefit to herself (Caro and Hauser's first criterion) and thereby encour-

Figure 17.4 Mother supporting infant so that he can climb on suspended logs.

aged his efforts (behavior types 1–4 described above) or offered him special experiences (behavior type 5) (Caro and Hauser's second criterion).

Whether these cases meet Caro and Hauser's third criterion, that the maternal acts enhance the development of the infant's behavior, is more difficult to determine. Daniel's rate of locomotor development was within the quite wide variance reported by Meder (1989) for human and mother-reared gorillas. That advancement of the infant's progress is at least a plausible outcome is suggested by the immediate, positive responses Daniel made in relation to several of his mother's acts, as described in the observations above. However, rigorous demonstration of even medium-term developmental effects is not available for virtually any of the purported cases of teaching or encouragement in animals, including those reviewed by Caro and Hauser. The reason is that this is an issue of cause and effect, which can only be satisfactorily resolved through experimentation. Maestripieri (1995b, 1996) has provided extensive quantitative data on behavior similar to "luring" in macaque mothers and shown that correlations exist between the extent of such behavior and the advancement of infant locomotory progress such as one would predict if the encouragement affects this progress: however, since correlations do not prove cause and effect, Maestripieri notes that developmental effects

Figure 17.5 Mother "dangling" the infant from suspended logs (compare Figure 17.8).

of encouragement have not been established. The correlations might reflect maturational processes and/or individual differences. Experimental manipulation of juveniles' experiences of maternal support appear to be restricted to Caro's (1980a,b) work with domestic cats. While the results of this are again consistent with a developmental effect of maternal encouragment on infant abilities (in this case predatory behavior), the difficulty of properly controlling for simpler effects of maternal presence alone means that a specific effect of teaching remains unclear, as Caro and Hauser acknowledge.

Accordingly, we are at a stage where evaluation of animal teaching by Caro and Hauser's third criterion remains largely a question of provisionally inferring that, given all we know of infant development, the experiences the juvenile manifestly receives have obvious potential to support the faster or better development of particular abilities or skills. Recognizing this caveat, we can attempt to compare the functional varieties of human and nonhuman primate teaching behavior: what we can examine at this stage then, is really *apparent functional potential.*

COMPARISON WITH SCAFFOLDING FUNCTIONS AS DESCRIBED BY WOOD ET AL., 1976

The first of Wood et al.'s functions is *recruitment*, in which the tutor takes the initiative in attracting the pupil's attention to a task. Delilah took the initiative in the sense that, for example, she helped Daniel's crawling by supporting his head and luring him *before* he had achieved any significant crawling: she put him on the horizontal logs *before* he had come anywhere near them. However, the very concept of recruitment in Wood et al.'s scheme appears to derive very much from these authors having set up a specific identifiable task, with respect to which a simple dichotomy of tutor or pupil initiative can be relatively easily identified. What this dichotomy may hide is that the maternal "initiative" may be *elicited* by cues from the infant, such as, in the case of Delilah, fumbling attempts at crawling or climbing on the infant's part. This comment may also apply to the luring behavior of macaque mothers described by Maestripieri.

At one step further removed, a mother's "initiative" may be elicited not by the immediate efforts of the infant but by more general cues reflecting its level of developmental competence. This may be true, for example, with respect to the behavior of cat mothers who begin to bring live prey to kittens at a particular stage of infant maturation: in one sense the mother "recruits" the kittens to predatory opportunities, but in another sense she may be simply responding in a quite reflexive way to maturational cues from them. The question of "initiative" is thus not a simple one. Even in the human case, Whiten and Milner's (1984) observations of the natural range of scaffolding behaviors which occur during human infancy indicated that in the vast majority of cases parents are responding and elaborating on something the infant is already engaged on. The same is true of the most elaborate teaching described in wild apes, in which mother chimpanzees appeared to show their infants how to perform nut-cracking with hammers (discussed more below): this appeared in direct response to infants' incompetent attempts.

Wood et al.'s second function is *reduction in degrees of freedom,* which involved "simplifying the task by reducing the number of constituent acts required to reach a solution." Delilah appeared to achieve this by offering support in Daniel's early attempts to crawl and climb: because he was supported in these ways, he was able to achieve at higher levels than he could have done unaided. What the mother does here is to provide some element which the infant cannot (yet) provide itself: working as a team, as it were, mother and infant provide several elements of an achievement such as crawling or climbing, which the infant will later manage all the elements of itself.

This function has been observed in other apes. For example, Van Rijt-Plooij and Plooij (1987) have described the way in which a chimpanzee mother repeatedly backed up to a rock on which she had placed her infant, who had not yet made the transition from riding ventrally on her to the dorsal position. This mother so "reduced the degrees of freedom" that the infant was both *able* to crawl onto her back from the rock, and also *could do little else*. Thus, she both facilitated his achievement of an act he was not yet doing unaided, and funnelled his behavior towards this.

The function of reducing degrees of freedom is not restricted to primates, however. An obvious example is domestic cat and cheetah mothers bringing live prey to their kittens, and retrieving it when the partially competent efforts of their offspring allow it temporarily to escape (Ewer, 1969; Caro, 1980a, 1994). Such scaffolding of object-directed skills appears rarer in primates. Boesch (1991) reported a mother chimpanzee removing a nut clumsily placed on an anvil by her 6–year-old son, then cleaning the anvil and replacing the nut in the correct position, allowing the youngster succesfully to hammer it. It seems likely she reduced the degrees of freedom in a helpful way. However, Boesch reports only one case of this.

Wood et al.'s third function is *direction maintenance*, "the role of keeping the child in pursuit of a particular objective." The behavior of Delilah and other primates in "luring" seems quite directly to fulfill this function. In a more general sense, much of her supporting behavior probably did the same however: for example, by supporting him on the vertical bars, Delilah probably encouraged him to climb more.

The fourth function is *marking critical features*: "a tutor by a variety of means marks or accentuates certain features of the task that are relevant" (other than by demonstrating an act – see below). Human parents can direct their infants' attention rather accurately to task-relevant features of the environment, by both verbal and nonverbal means, like pointing. In luring, Delilah attracted attention to a specific locus – herself – but this does not amount to marking critical features of the task of crawling. "Pointing out" of critical features may be one feature differentiating human and nonhuman scaffolding.

However at another extreme, parents may merely position the infant in relation to its environment, such that its attention is likely to fall on signficant features (Whiten & Milner, 1984). Delilah's holding Daniel over the edge of the platform appeared to be of this class, and, if we see the task as the avoidance of falling off, Daniel's attention would have fallen on task-relevant features of the environment. Instead of abruptly approaching the edge of the platform and going over, Daniel was suspended and swung such that extensive binocular and parallax "visual cliff" depth information would have been available.

The fifth function is *frustration control*: "problem solving should be less dangerous or stressful with a tutor than without." It is plausible that Delilah's scaffolding behavior made Daniels' progress (e.g. crawling when lured) more efficient, which is what seems to lie behind the function of frustration control. "Dangling" matches this category for a different reason: it was apparently a relatively safe way to learn about dangerous "cliffs." The vagueness of this functional category makes it difficult to evaluate its wider distribution in animals: it may be more applicable to scaffolding for predatory behavior in carnivores than to most primates.

Wood et al.'s sixth and final function is *demonstration*. This was the one category which Whiten (1975) noted as absent in Delilah's behavior and indeed in apes generally. However, Boesch (1991) has since described an episode in which a mother chimpanzee appeared to demonstrate. Her 5–year-old daughter had been unsuccesfully attempting to open nuts using an irregularly shaped hammer stone for 8 minutes. The mother approached and, when the juvenile gave her the hammer, she "in a very deliberate manner, slowly rotated the hammer into the best position . . . it took her a full minute to perform this simple rotation" (p. 532). The daughter then took up the hammer and, by adopting the same grip as her mother, succesfully cracked nuts. Boesch's description fits the concept of demonstration, in so far as the slow and deliberate acts of the mother are exactly the kind of change which turns "doing X" into "demonstrating X." However, this was a single incident, so the interpretation must remain speculative.

COMPARISON WITH SCAFFOLDING FUNCTIONS AS DESCRIBED BY WHITEN AND MILNER, 1984

Whiten and Milner's taxonomy of scaffolding was in one sense more wide-ranging than that of Wood et al., since it was based on events observed throughout infants' everyday experiences. However, it was also restricted in one particular way, focusing only on the infant's interactions with the world of manipulable objects – for the human infant, a major activity and an important arena of cognitive development. For comparative purposes, we will set aside this focus and allow the definition of the categories in Figure 17.1 to extend not just to manipulable objects, but to all aspects of the infant's environment.

The first of the two most primitive forms of scaffolding, *providing a stable base*, is essentially concerned with managing the motivational state of the infant so that it is more likely to do things from which it learns. Much could be said about variance in this function in nonhuman primates, but this is not appropriate here: we need only note that this is a common function of maternal behavior in wild primates. Indeed, the concept as we applied it to human parenting derived from attachment research (Ainsworth, Blehar, Waters, & Wall, 1978), which in turn learned from the early work of Harlow, particularly the demonstration that even an artificial, yet familiar, mother-figure could provide infant monkeys with a secure base from which they were prepared to explore novel environments (Harlow, 1959).

Next is *support manipulation*, where the scaffolding influences contact between the infants and certain environmental features, without further affecting what happens as a result of the infant's response (the subject of the more "interventive" functions described below). Examples of this in nonobject-focused contexts include Delilah's holding of her infant up to the horizontal logs (otherwise not part of its locomotory world) as described earlier, and, in object-related contexts, mother chimpanzees' tendency to leave hammers and good nuts available for their offspring to use (Boesch, 1991).

There is a negative or reverse form of this category which was not part of our child study but is exemplified in some observations of primate behavior. In the clearest cases, mother chimpanzees have been observed to remove and discard items which their infants are exploring but which are not eaten by the mother herself (Hiraiwa-Hasegawa, 1990). It is possible that a similar function may be achieved, but in a less overt fashion, when adult howler monkeys avoid exploring new foods themselves in the presence of an infant (who might follow their example, and suffer more from toxic components) (Whitehead, 1986).

Whiten and Milner discriminated five subcategories which all shared the characteristic of *further affecting* the way in which the infant interacts with its world (Figure 17.1). The first of these was to *assist*, in which the infant was helped to achieve its own goal. Delilah's supporting of her infant's head while it tried to crawl is an example. Similar behavior has been described for chimpanzees in the wild (Van-Lawick-Goodall, 1968) and in captivity (Yerkes & Tomalin, 1935). The most elaborate candidate for assisting in object-directed actions is that of the chimpanzee mother described earlier, who adjusted the orientiation of a nut her son was trying to crack, but other examples include that of a gorilla mother who simply plucked a food item her offspring was reaching for and gave it to him (Schaller, 1963).

In *reveal object properties*, the infant's attention is drawn to aspects of the environment it would not otherwise notice. This therefore includes Wood et al.'s *mark critical features*, and, as suggested earlier, Delilah's "dangling" actions appeared to be geared towards this function, which is otherwise not so apparent in nonhuman primates.

Create discovery environment is a very interesting category: as its label implies, the scaffolder sets up for the infant circumstances in which the infant might – according to how it responds – learn something it would otherwise not. An example would be a human parent building a tower of two blocks and handing the infant a third. This category, together with *reveal object properties*, accounted for most of the cases of *modify interaction* in Whiten and Milner's study on human infants. Almost paradoxically, it shares with Wood et al.'s *reducing degrees of freedom* a kind of channelling of the infant's activities towards interesting learning experiences: but it also *increases* degrees of freedom because it creates open-ended potentials beyond the infant's unaided experiences. In apes we appear to see only marginal candidates for this category. Cases of luring and expanding infants' locomotory horizons in other ways by hanging them on more advanced substrates do create relevant potentials, but, since the mother is essentially bringing infant and object toward each other rather than further influencing what form the interaction could take, these are perhaps more appropriately classed as only *support manipulation/locomotion*. The same appears to apply to chimpanzee mothers observed by Boesch (1991) who facilitate their infants' access to tools and nuts.

The category *demonstrate* is the same as for Wood et al. and has already been discussed. For Whiten and Milner, the additional category *teach* was defined as a demonstration followed by support of the infant's attempts to do what the parent did. When we observe this, it usually looks as though the parent is trying to get the infant to do something the parent can do. Thus perhaps the best approximation to this category (it did not appear to involve demonstration, as such) in apes is the observation by Fouts, Hirsch, & Fouts (1982), that the sign-using chimpanzee, Washoe, appeared to mold her infant to perform signs she had herself been taught earlier: for example, she was reported to stop signing "food" herself so as to take her infant's hand and mold it into the "food" configuration (Patterson & Cohn [1994] also report a gorilla performing molding, but with respect to a doll).

WHAT DOES IT ALL MEAN?

Are apes educators?

The preceding paragraphs show that, in many respects, apes have been observed to match several of the features of human scaffolding. However, even if we grant that particular categories of scaffolding occur, there is a more specific issue to address about the extent to which they are tailored to the infant's developmental needs. It seems intrinsic to the concepts of teaching and scaffolding that the particular forms they take will only be appropriate to their function if they are supportive at the specific level of competence the infant is at. This is why the term "scaffolding" is such a nice metaphor. While a building is being constructed it needs scaffolding to support it, and the form of the scaffolding must complement the shape of the building to support it well – indeed, it may be necessary to adjust the scaffolding to do this job as the building takes shape. Then, as the building can stand by itself, the scaffolding can be dismantled.

So with scaffolding for developing competencies in an infant. In fact, Delilah's behavior towards Daniel appeared to be tailored to his developing competencies at several levels. We can consider three. First, there was tailoring in that the form of each intervention was appropriate to the encouragement of infant behavior toward a level of competence *not yet manifest*, except as a poorly formed "attempt." Thus, Delilah could see only clumsy thrashing of limbs, yet her supporting and luring appeared to be designed as encouragement toward more advanced crawling. She held Daniel in a position for climbing after he had done little more than stand against the bars.

A second adjustment of scaffolding may be necessary to provide encouragement appropriate to *different levels* of developing competency. Having seen that her infant is trying to crawl, a human mother may be seen to steady his feet in initial attempts, and then later, when he can move without this support, attempt to lure him along. Putting this aspect of tailoring alongside the first, we can say that the parent is operating within what Vygotsky (1978) called the infant's "zone of proximal development" – within the immediate possibilities for modifiable developmental progress. Delilah, too, showed these different types of encouragement – supporting and luring – at appropriate stages in the development of Daniel's ability to crawl. Her encouragement of climbing was likewise developmentally tailored, starting with careful support against the lower part of the bars and progressing to leaving the infant hanging unsupported from the roof bars.

At a third level, encouragement may be tailored to differences which occur *during each performance* of the developing behavior. A human father may first present the appropriate building block to his infant and then steady the tower while she adds a block. Or, as the crawling infant approaches the "luring" toy, he may readjust the latter's position so as to maintain its function. Delilah clearly adjusted her luring position relative to Daniel in this moment-to-moment fashion.

Thus, taking into account the range of scaffolding functions observed together with these levels of tailoring to infant developmental progress, we are tempted to conclude: *apes teach* – at least in rudimentary fashion. However, this bald claim would be misleading, because the kinds of behavior reviewed in this chapter are either rare in the wild or even absent, only being observed in captive contexts like that of Delilah. Thus, in response to the question of whether behavior as elaborate as Delilah's is observed in wild mountain gorillas, K. Stewart (personal communication), after a study of early development, could only comment that "the mother may lift her infant off her body and place him on the ground next to her . . . whereas this may mean that the infant will most likely crawl onto her again (thus developing his climbing skill) the immediate goal as far as the mother was concerned seemed to be ridding herself of the infant so that she could feed or sleep." Others who have observed young gorillas extensively concur (Byrne, personal communication; Fletcher, personal communication). Nor do other detailed developmental studies of gorillas mention teaching-like behaviors (Fossey, 1979; Watts, 1985).

Similarly, Boesch's (1991) record of demonstration and assistance in wild chimpanzees is limited to just one case of each. Similar cases have not been reported in decades of research which have included developmental studies at Gombe and Mahale (Goodall, 1986; Nishida, 1990) and may be restricted just to certain mothers operating with a particularly difficult technology (nut-cracking). The literature on wild orangutans is similarly unforthcoming in its descriptions of scaffolding behaviors (e.g., Galdikas, 1982). In sum, it is difficult to sustain an argument that teaching, or even the wider category of scaffolding, is a general component in the repertoires of wild apes. This conclusion is consistent with a number of recent reviews (Caro & Hauser, 1992; Galef, 1992; King, 1994). King (1994) emphasizes that young nonhuman primates appear to be excellent information *acquirers*, which may usually suffice for their needs, without information *donation* from parents. Primates' powers of social learning and cultural transmission are reviewed from various perspectives in recent works by Tomasello and Call (1997), Russon (1997), Mc Grew (1992, in press) and Whiten (in press, a).

Is "scaffolding" in apes an occasional by-product of their advanced social cognition?

The conclusion that scaffolding has not emerged as a general feature of ape life is interesting because it would have supplied a ready explanation for the behavior of Delilah. If all gorillas performed the range of actions described earlier, then they might be subsumed under some concept of instinctive maternal responses elicited by particular developmental triggers on the part of the infant. This is the way in which the scaffolding actions of cats have been seen. As noted above, mother cats exhibit several of the scaffolding functions distinguished and they achieve all three levels described above with respect to tailoring their scaffolding to the level of competence and moment-to-moment achievements of their offspring, as did Caro's cheetahs. However, Ewer (1969) remarked that the cat mother "quite clearly did not comprehend the function of her own behavior ... the important change that takes place in the cat's behavior is a progressive inhibition of her predatory behavior, working from the terminal stages forwards. The cat first delays her eating until the food has been brought back to the young; later she kills but fails to eat and, finally, she captures but fails to kill." Whether this really shows that the cat mother does not have any cognitive control over her changed behavior is debatable, but the inhibition to which Ewer refers does appear to be a general feature of cat mothers, as good a candidate for "instinct" as any. That the luring behavior of rhesus and pigtail macaques (Maestripieiri, 1995b, 1996) appears to be a general pattern amongst reproductively experienced mothers is consistent with this kind of interpretation.

However, this appears not to be the case for Delilah's behavior, nor for several other of the relatively rare instances of quite elaborate scaffolding behavior described for apes in the body of this chapter; they appear far from automatic in ape mothers. This, however, does not mean we are dealing with incidents which can be dismissed as mere oddities. Yerkes and Tomilin (1935), for example, described in their captive chimpanzees actions so like Delilah's that it is worth recording I did not know of them when making my own observations. Yerkes and Tomilin noted that in the third month "the initial step commonly is a stretching exercise, in which the mother, usually lying on her back, grasps the infant's extremities in her own, lifts it into the air and slowly and gently stretches it by pulling its limbs." This is what is shown in Figure 17.6 (compare with Figure 7 and 8, from Yerkes, 1943). In addition:

> As soon as the infant has learned to stand with maternal support, its mother is apt to place it near or upon objects to which it can hold or on which it may climb. In our observational situation the two-inch mesh which constituted the side walls and roof of the cage admirably served this purpose. Several times we saw a mother carry her baby to the netting, place its hands against the wires, and then withdraw cautiously, leaving the baby standing beside or hanging from the netting... The mother may take the baby by one arm and half-lead, half-drag, it along, compelling it to follow as best it can on three extemities. Or again, crouching a few feet in front of the infant, she may vocally and posturally invite it to approach her ... if the legs are not used properly and effectively the mother may place a hand under the small body and, raising it from the floor, help the infant to walk standing on all fours.

These acts of mother chimpanzees are strikingly similar to those generated by the gorilla described here.

If thus not an obscure oddity, it might be objected that such behavior might

Figure 17.6 Mother holds infant by arms and manipulates it (compare Figures 17.7 and 17.8).

nevertheless be the actions of bored captive apes who are simply playing with their infants as live toys, as suggested for interest in infants shown by captive adult males by Mitchell (1989). It may be true that such apes have time to spare and are in some sense entertaining themselves by eliciting actions from their offspring. However, the important rejoinder to this interpretation must appeal to the tailoring of the actions to the infant's behavior as outlined earlier, particularly its aptness to behaviors not yet competently shown. This is what strikes an observer such as myself or Yerkes as so "teaching-like." It suggests that one way in which future studies can approach the issue more systematically is to assess in some detail the extent of the various aspects of tailoring, contrasting these with other aspects of exploration that would be more consistent with the boredom hypothesis.

If not a common pattern amongst apes, yet one that does appear repeatedly in some mothers with evidence of being tailored to infants developing needs, how, then, do we explain this kind of behavior? The explanation I shall explore is that these idiosyncratic scaffolding actions are the result of *a particular way in which these apes perceive certain*

Figure 17.7 "Maternal Tuition": Plate 14 from Yerkes, 1943. The original caption read: "Dita and her one-week-old infant Rosy (upper); from Yerkes and Tomilin, 1935. Wendy coaxing her four-month-old son Bob to walk toward her (middle); (from Yerkes and Tomilin, 1935). Fifi and her five-month-old daughter Beta; (from Yerkes and Yerkes, 1935).

Figure 17.8 "Josie exercising her son Mars, age three months" (Plate 15 from Yerkes, 1943).

actions of others. Specifically, they perceive certain actions in their infants as incompletely formed, and they attempt to help their infants complete them. Two ways in which they do this can perhaps be distinguished.

1. Perceiving attempts ("trying to")

In this case, the incompleteness in the infants' behavior is that it is "trying to" – trying to crawl, trying to climb, trying to reach things. Ape mothers sometimes appear to "extrapolate" from observable behavior to these goals, and then, in some cases, attempt to facilitate achievement of the goal by the infant, through various scaffolding acts. The latter are varied and idiosyncratic perhaps because the adult ape is, on the basis of its current store of knowledge and abilities, intelligently constructing acts judged possibly to help. In some cases these may be poor ideas – for example, the tugging/dragging of a relatively inert infant across the floor might help it "move" only in the simplest sense! – but such errors are themselves consistent with the operation of a relatively intelligent, if slightly misguided, cognitive process, rather than a wired-in system in which maladaptive efforts like dragging would be more surprising.

Of course, a tendency to perceive others' acts in this intentional fashion is exactly what Premack and Woodruff (1978) addressed in their first attempt at tackling the question "Does the chimpanzee have a theory of mind?" In their experiments, they tested whether a chimpanzee would perceive a human being trying to retrieve out-of-reach objects *as* trying (wanting or intending to get the objects), by seeing if she would pick out a picture representing the best way to achieve this. The peer commentary on that paper was largely unconvinced they had shown this using this particular technique, and, although several experiments testing for chimpanzee "theory of mind" have been conducted more recently, none have returned to the particular issue studied by Premack and Woodruff (Povinelli, 1996; Whiten 1997, for reviews). A tendency for apes to recognize "trying to achieve x" and attempt to help complete the infant's aim would make sense of most of the ape scaffolding described in this chapter, including Delilah's acts of luring, tugging, and supporting, and chimpanzees' facilitating and demonstrating behavior as described by Boesch (1991). In the case of removing nonfoods from the infant (Hiraiwa-Hasegawa, 1990), the mother appears to interpret the infant as aiming to eat the plant, but in this case she goes on to thwart the infant's goal rather than encourage it. Whether the same is true of wild gorilla mothers restricting infants locomotory exploration in situations about which they appear to be nervous (Fletcher, personal communication) is more difficult to judge.

An ability to read the uncompleted goals of the infant and help complete them is evident in all great ape species in the wild, even where this does not necessarily argue for encouragement of developing competencies. Orangutan mothers have been observed to use their own bodies to form "bridges" to allow their infants to cross gaps in the canopy (Rijksen, 1978). Schaller (1963) noted that "a five-month-old pulled with both hands at a one-foot section of broken lobelia stem rooted in the ground. It jerked and pulled but could not break it. The female watched. Finally she reached over with one hand, snapped the stem off, and laid it on the ground. The infant picked it up and gnawed it." Whether great apes are very special in this ability to assist their less competent offspring remains to be seen, but in acting in these ways they express a capacity which is likely to have been an important springboard for the elaboration of the teaching which evolved in hominids.

2. Perceiving ignorance

More controversially it could be argued that some maternal behavior suggests an appreciation that the infant is lacking knowledge. Delilah swinging Daniel over the edge of the ledge he had fallen off looked like an attempt to familiarize him with more about that particular location. The correcting and demonstrating actions of the chimpanzees described by Boesch could also be interpreted as responses to ignorance on the part of the infant, which the mother tries to correct by clarifying the correct way to do things. These interpretations, however, whilst of obvious relevance to the origins of teaching behavior, must remain speculative because the observations are so few.

A second reason for scepticism about interpretations resting on the attribution of ignorance is that attempts to experimentally test for such abilities in apes, whilst initially generating some positive results (Premack, 1988, Povinelli, Nelson, & Boysen, 1990), have with harder scrutiny and more refined investigation produced mostly negative conclusions (Heyes, 1993; Whiten, 1993; Povinelli, Rulf and Bierschwale, 1994; Povinelli and Eddy, 1996), although there are now also some strikingly positive ones (Whiten, in press b; Boysen, in preparation, Hauser, in preparation).

The possibility that at some level these apes are nevertheless recognizing an "incompleteness" – which they attempt to complete – is another matter. The subjects in these observations appear to be attempting to produce outcomes as they "should" be: for the chimpanzees, the hammer should be held thus, and the anvil placed so; for Delilah, the infant should not be falling off but should be swinging over the gap (as he well might in a foliage-rich environment). It may be that the ape is recognizing an "ignorant situation" – a pattern of behavior that is not right and needs correcting – rather than that she attributes a mental state of ignorance in the same explicit way a human adult might (cf. Perner, 1991; Whiten, 1996): but in any case, this kind of discrimination in the common great ape ancestor would have provided a powerful foundation for the evolution of human teaching.

In conclusion, then, I have described a variety of observations which show that many of the functions of human scaffolding – particularly those which occur in interactions with young children – have also been observed in great apes and, in several cases, in other primates and nonprimates. However, that many of the more elaborate cases of scaffolding are either rare or nonexistent in the wild argues against education being a significant part of the routine daily life of young apes, despite the fact that imitative ability, in principle, makes them highly "educable" (Whiten, in press a). What is perhaps more instructive in these actions, then, is what they tell us about the way apes perceive and interpret others, particularly in contexts where infants demonstrate only partially formed, developing competencies. A capacity for natural psychology, including a tendency to perceive what others are trying (unsuccesfully) to do and what information they need to act more compentently, may allow apes to generate teaching-like behaviors in relatively novel situations. Whilst not routine educators, apes may thus have powers of social cognition which in our common ancestor would have provided important foundations for the emergence of that scaffolding which became so commonplace in human evolution. This would mean that these powers originally evolved for other purposes which remain mysterious, but may have included deceptive interactions and imitation (Whiten, 1996).

ACKNOWLEDGMENTS

I am grateful to the following for comments and discussion relating to earlier versions of this work: J. Bruner, R. Byrne, T. Caro, A. Fletcher, B. King, A. Meltzoff, R. Mitchell,

S. Parker, R. Passingham, F. Plooij, A. Russon, and J. Tanner. For drawings from photographs taken in suboptimal lighting conditions, I am very grateful to R. J. Spijkerman.

REFERENCES

Ainsworth, M. D. S., Blehar, M. C., Waters, E., & Wall, S. (1978). *Patterns of attachment: A psychological study of the strange situation*. Hillsdale, NJ: Erlbaum.

Barnett, S. A. (1968). The "instinct to teach." *Nature*, **220**, 747–749.

Boesch, C. (1991). Teaching among wild chimpanzees. *Animal Behaviour*, **41**, 530–532.

Bruner, J. S. (1972). Nature and uses of immaturity. *American Psychologist*, **27**, 687–708.

Caro, T. M. (1980a). Predatory behaviour in domestic cat mothers. *Behaviour*, **74**, 128–147.

Caro, T. M. (1980b). Effects of the mother, object play and adult experience on predation in cats. *Behavioural and Neural Biology*, **29**, 29–51.

Caro, T. M. (1994). *Cheetahs of the Serengeti Plains: Group-living in an asocial species*. University of Chicago Press.

Caro, T. M. & Hauser, M. D. (1992). Is there teaching in nonhuman animals? *The Quaterly Review of Biology*, **67**, 151–171.

Ewer, R. F. (1969). The "instinct to teach." *Nature*, **222**, 570–607.

Fossey, D. (1979). Development of the mountain gorilla: the first thirty-six months. In D. A. Hamburg & E. R. McCown (eds.), *The great apes*. Menlo Park, CA: Benjamin/Cummings.

Fouts, R. S., Fouts, D. H., & Van Cantford, T. E. (1989). The infant Loulis learns signs from cross-fostered chimpanzees. In R. A. Gardner, B. T. Gardner, & T. E. van Cantford (eds.), *Teaching sign language to chimpanzees* (pp. 280–292). New York: State University of New York Press.

Fouts, R. S., Hirsch, A. D., & Fouts, D. H. (1982). Cultural transmission of human language in a chimpanzee mother–infant relationship. In H. E. Fitzgerald, J. A. Mullins, & P. Gage (eds.) *Child nurturance: Psychobiological perspectives*. New York: Plenum Press.

Galdikas, B. M. F. (1982). Orang-utan tool-use at Tanjung Puting Reserve, Central Indonesian Borneo. *Journal of Human Evolution*, **10**, 19–33.

Galef, B. G. (1992). The question of animal culture. *Human Nature*, **3**, 157–178.

Goodall, J. (1986). *"The chimpanzees of Gombe": Patterns of behaviour*. Cambridge, MA: Harvard University Press.

Harlow, H. F. (1959). Love in infant monkeys. *Scientific American*, **200**, 68–74.

Heyes, C. M. (1993). Anecdotes, training, trapping and triangulating: do animals attribute mental states? *Animal Behaviour*, **46**, 177–188.

Hiraiwa-Hasegawa, M. (1990). A note on the ontogeny of feeding. In T. Nishida, (ed.) *The chimpanzees of the Mahale Mountains: Sexual and life history strategies* (pp. 277–83). Tokyo: University of Tokyo Press.

King, B. J. (1991). Social information transfer in monkeys, apes and hominids. *Yearbook of Physical Anthropology*, **34**, 97–115.

King, B. J. (1994). *The information continuum: Evolution of information transfer in monkeys, apes and hominids*. Santa Fe, NM: School of American Research.

Lawick-Goodall, J. van (1968). The behaviour of free-living chimpanzees in the Gombe Stream Reserve. *Animal Behaviour Monographs*, **1**, 161–311.

Maestripieri, D. (1995a). Maternal encouragment in nonhuman primates and the question of animal teaching. *Human Nature*, **6**, 361–378.

Maestripieri, D. (1995b). First steps in the macaque world: do rhesus mothers encourage their infants' independent locomotion? *Animal Behaviour*, **49**, 1541–1549.

Maestripieri, D. (1996). Maternal encouragement of infant locomotion in pigtail macaques, *Macaca nemestrina*. *Animal Behaviour*, **51**, 603–610.

McGrew, W. C. (1992). *Chimpanzee material culture: Implications for human evolution*. Cambridge University Press.

McGrew, W. C. (in press). Culture in non-human primates? *Annual Review of Anthropology*.

Meder, A. (1989). Effects of handrearing on the behavioral development of infant and juvenile gorillas. *Developmental Psychology*, **22**, 357–376.

Mitchell, R. W. (1989). Functions and consequences of infant–adult male interaction in a captive colony of lowland gorillas (*Gorilla gorilla gorilla*). *Zoo Biology*, **8**, 125–137.

Nishida, T. (ed.) (1990). *The chimpanzees of the Mahale Mountains: Sexual and life history strategies* (pp. 277–283). Tokyo: University of Tokyo Press.

Parker, S. T. (1996). Apprenticeship in tool-mediated extractive foraging: the origins of imitation, teaching, and self-awareness in great apes. In A E. Russon, K. A. Bard, & S. T. Parker (eds.) *Reaching into thought: The minds of the great apes*. Cambridge University Press.

Parker, S. T. & Gibson, K. R. (1977). Object manipulation, tool use and sensorimotor intelligence as feeding adaptations in cebus monkeys and great apes. *Journal of Human Evolution*, **6**, 623–641.

Parker, S. T. & Gibson, K. R. (1979). A developmental model for the evolution of language and intelligence in early hominids. *Behavioral and Brain Sciences*, **2**, 367–408.

Parker, S. T. & Russon, A. E. (1996). On the wild side of culture and cognition in the great apes. In A. E. Russon, K. Bard, & S. T. Parker (eds.), *Reaching into thought: The minds of the great apes* (pp. 430–450). Cambridge University Press.

Patterson, F. & Cohn, R. (1994). Self-recognition and self-awareness in lowland gorillas. In S. T. Parker, R. W. Mitchell, & M. L. Boccia (eds.) *Self-awareness in animals and humans: Developmental perspectives* (pp. 373–390). Cambridge University Press.

Pearson, A. T. (1989). *The teacher: Theory and practice in teacher education*. New York: Routledge.

Perner, J. (1991). *Understanding the representational mind*. Cambridge, MA: MIT Press.

Povinelli, D. J. (1996). Chimpanzee theory of mind: the long road to strong inference. In P. Carruthers & P. K. Smith (eds.), *Theories of theories of mind*. Cambridge University Press.

Povinelli, D. J. & Eddy, T. J. (1996). What young chimpanzees know about seeing. *Monographs of the Society for Research in Child Development*, **61**, No. 2, Serial No. 247.

Povinelli, D. J., Nelson, K. E., & Boysen, S. T. (1990). Inferences about guessing and knowing by chimpanzees (*Pan troglodytes*). *Journal of Comparative Psychology*, **104**, 203–210.

Povinelli, D. J., Rulf, A. B., & Bierschwale, D. T. (1994). Absence of knowledge attribution and self-recognition in young chimpanzees (*Pan troglodytes*). *Journal of Comparative Psychology*, **108**, 74–80.

Premack, D. (1988). 'Does the chimpanzee have a theory of mind?' revisited. In R. W. Byrne & A. Whiten (eds.), *Machiavellian Intelligence* (pp. 160–178). Oxford University Press.

Premack, D. & Woodruff, G. (1978). Does the chimpanzee have a theory of mind? *Behavioral and Brain Sciences*, **1**, 515–526.

Rijksen, H. D. (1978). *A field study on Sumatran orangutans (Pongo pygmaeus abilii Lesson 1927): Ecology, behaviour and conservation*. Wageningen: H. Venman and Zonen.

Russon, A. E. (1997). Exploiting the expertise of others. In A. Whiten & R. W. Byrne (eds.), *Machiavellian Intelligence II* (pp. 174–206). Oxford University Press.

Schaller, G. B. (1963). The mountain gorilla: Ecology and behaviour. Chicago University Press.

Tomasello, M. & Call, J. (1997). *Primate cognition*. Oxford University Press.

Van de Rijt-Plooij, H. H. C. and F.X. Plooij (1987). Growing independence, conflict and learning in mother–infant relations in free-ranging chimpanzees. *Behaviour*, **101**, 1–86.

Vygotsky, L. S. (1978). Mind in society: the development of higher psychological processes. In M. Cole, J. Steiner, S. Scribner, & E. Sonberman (eds.) Cambridge, MA: Harvard University Press.

Watts, D. P. (1985). Observations on the ontogeny of feeding behaviour in mountain gorillas (Gorilla *gorilla beringei*). *American Journal of Primatology*, 8, 1–10.

Whitehead, J. M. (1986). Development of feeding selectivity in mantled howler monkeys, *Aloutta palliata*. In J. Else and P.C. Lee (eds.) *Primate ontogeny, cognition and social behaviour* (pp. 105–117. Cambridge University Press.

Whiten, A. (1975). Parental encouragement and teaching in the great apes and man. Unpublished MS (obtainable from author), University of Oxford.

Whiten, A. (1993). Evolving a theory of mind: The nature of non-verbal mentalism in other primates. In S. Baron-Cohen, H. Tager-Flusberg, & D. J. Cohen (eds.), *Understanding other minds* (pp. 367–396). Oxford University Press.

Whiten, A. (1996). Imitation, pretense and mindreading: secondary representation in comparative primatology and developmental psychology? In A E. Russon, K. A. Bard and S. T. Parker (eds.) *Reaching into thought: The minds of the great apes.* Cambridge University Press.

Whiten, A. (1997). The Machiavellian mindreader. In A. Whiten & R. W. Byrne (eds.) *Machiavellian intelligence II: Evaluations and extensions.* Cambridge University Press.

Whiten, A. (in press a). Primate culture and social learning. *Cognitive Science.*

Whiten, A. (in press b). Chimpanzee cognition and the question of mental re-representation. In D. Sperber (ed.) *Metarepresentations.* Oxford University Press.

Whiten, A. & Milner, P. (1984). The educational experiences of Nigerian infants. In H. V. Curran (ed.) *Nigerian children: Developmental perspectives.* London: Routledge and Kegan Paul.

Wood, D., Bruner, J. S. & Ross, G. (1976). The role of tutoring in problem-solving. *Journal of Child Psychology and Psychiatry*, **17**, 89–100.

Yerkes, R. M. (1943). *Chimpanzees.* New Haven, CT: Yale University Press.

Yerkes, R. M. & Tomalin, M. I. (1935). Mother–infant relations in chimpanzees. *Journal of Comparative Psychology*, **20**, 321–348.

18

The development of social roles in the play of an infant gorilla and its relationship to sensorimotor intellectual development

SUE T. PARKER

THE STUDY

This chapter reports on the development of social play during the first three years of life of an infant lowland gorilla living in a social group at the San Francisco Zoo (Parker, 1977a). It presents a hierarchical model for describing the temporal structure of play and its development in a single subject: at the highest level are the longest units, episodes of play. These are composed of shorter units, bouts of play. Bouts, in turn, are composed of games. Games are composed of the shortest, lowest-level units, schemes. Play bouts and episodes are punctuated by play regulators including enticements, terminations, and interventions. It also focuses on the relationship between social play and cognitive development.

Subjects

The subject of this study is an infant lowland gorilla (*Gorilla gorilla gorilla*) born and reared in a social group at the San Francisco Zoo: Mkumbwa was born May 1, 1975 to a wild-born mother, Jacqueline, and a wild-born father, Bwana. He is the younger brother of Koko, a gorilla who was taught to sign by Francine Patterson. At the time of the study, the social group was also composed of Mrs, a wild-born female, her male infant, Sunshine, also fathered by Bwana, and a nulliparous captive-born female, Pogo (referred to as Mkumbwa's "aunt"). Mkumbwa is currently the silverback male of the group which still includes Bwana, now a grandfather, and Pogo, as well as two younger females, Zura and Bawang. The group also includes Bawang's infant sons, Shango and Barney, fathered by Mkumbwa. (see Tanner, this volume) See Table 18.1 for a list of gorillas. See Figure 18.1a for a photograph of Jacqueline with Mkumbwa and Sunshine.

Setting

At the time of the study the gorilla group was exhibited in a cement grotto approximately 80 feet in diameter composed of a plateau area of about 50 feet in diameter furnished

Table 18.1 *Gorillas in the study group by age and family relationship*

Names	Relationship to infant	Ages during study (yrs)	Birthdates
Mkumbwa	self	0–3	May 1, 1975
Jacqueline	mother	19–22	1956
Bwana	father	17–20	1958
Sunshine	half brother	1.5–4.5	Jan 1974
Pogo	"aunt"	17–20	1958
Mrs	Sunshine's mother	adult	wild born

with large horizontally embedded tree trunks, some large rocks which formed small caves, and a shallow pool of water several feet in diameter. At its outer margin, the plateau at the top met a steep hill descending about 14 feet to a large cement moat area furnished with a large cement food box for the infants. One side of the hill had a cement stairway with a metal railing. The top inner side of the plateau had a door leading into the interior cages.

Observational methodology

Data were collected during a longitudinal study that began when the infant, Mkumbwa or "Kubie," was a few days old and continued until he was 3 years old, that is, from May 5, 1975 to May 2, 1978. Data were collected by the author using focal animal sampling with continuous recording of all behaviors while the infant was in view. Behavior was recorded at 1– or 2– week intervals during the first year, at 3–week intervals during the second year, and at approximately 6–week intervals during the third year. Behavior was recorded between the hours of 1000 and 1500 for an average of 95 minutes per session. A total of 83 hours of observational data were collected in 52 sessions. Data from two sessions were omitted from the analysis because they were taken at shorter intervals. Times were recorded at the beginning and end of sessions and sporadically as the demands of writing allowed. Behavioral descriptions were typed out and catalogued in a database program (RBase) by the author according to the following information: nature of activity, interactants, infant's motor patterns, interactant's related behavior (less complete), and location.

Analytic frameworks and methodology

This study uses a multilevel analysis of the organization of play behavior that relies upon behavioral samples sufficiently long (that is, an hour or more) to encompass the larger units of behavioral organization that I have called episodes. Likewise, the analysis of phases of play development attempted here relies upon longitudinal data on individual social development. Finally, analysis of developmental relationships among locomotoric, cognitive, and social aspects of play used here relies upon continuous comprehensive sampling of reasonably long chunks of behavior longitudinally in the same individual. Extended sampling periods, continuous comprehensive behavioral sampling in longitudinal studies of development of the same individual is a necessary, though insufficient, condition for determining the relationships among cognitive, social, and locomotoric variables. Comprehensive, long-term, longitudinal data on extended samples of gorilla social behavior with both parents and a sibling or half-sibling are sufficiently rare to make this study of some interest despite its focus on a single subject (De Waal, 1991). (See Tanner, this volume).

Diagnosing Play

Play behavior was distinguished from other kinds of behavior on the basis of the presence of certain key motor patterns including the relaxed open mouth (ROM) face, arm waving, wrist waving, bouncing, twisting, sliding, and rolling, as well as repetition and role reversal, and the absence of such other activities as feeding, resting, and aggression (Aldis, 1975; Fagen, 1981; Tomasello, Gust, & Frost, 1989). Play was "framed" or set off from other, nonplay behavior by onsets and offsets. Selected motor patterns indicative of play development were catalogued by observation date and concomitant age of the infant in lunar months (four 7–day weeks per month). Lists of the first occurrence of play behaviors (schemes) were compiled into a developmental ethogram of play behavior for the infant and each playmate. Play behavior was classified into four hierarchical levels: (1) *simple schemes* and *elaborated schemes*, (2) *games*, (3) *bouts*, and (4) *episodes*.

A *simple scheme* is a motor pattern, such as grasping or rolling over, which occurs in a variety of contexts; an *elaborated scheme* such as wrestling is a co-ordinated set of schemes which are performed as a unit either sequentially or simultaneously (Piaget, 1952).

A *game* is an intercoordinated set of elaborated schemes such as wrestling, sparring, and biting, which are repeatedly performed simultaneously or reciprocally by playmates in social play, or are repeatedly performed in solitary play. Elaborated schemes are similar to the "projects" of Simpson (1976) and Mitchell and Thompson (1991). Interactive games are "routines" (Mitchell & Thompson, 1991) or "scripts" (Mitchell, this volume). The concept of games has been widely used by ethologists (see e.g., Aldis, 1975; Fagen, 1981), if not always explicitly and systematically, to describe animal play. This usage dates back at least to the time of Groos (1976 [1898]) who used the term "plays" to describe what we would call games. I have systematically classified all forms of gorilla play into games.

The names of some games are common usage in discussions of animal play (for example, rough-and-tumble play or play fighting, keep away), others were named on the basis of their similarity with games human children play (e.g., hide-and-seek and showing off). Obviously, these game categories are merely convenient shorthand descriptions for complex structures; they should be viewed as tentative hypotheses about the organization of play which could be tested through some kind of grouping/ cluster analysis using a more comprehensive database. See Table 18.2 for a list and definition of gorilla games.

(a)

(b)

(c)

Figure 18.1 (a) Jacqueline with Mkumbwa at her breast and Sunshine on her side (b) Jacqueline holding Mkumbwa over her left arm as she regurigates into her right hand; Mrs standing with Sunshine clinging to her left shoulder (c) Bwana sitting with left hand touching Sunshine's rump

A *bout* of social play was defined as an onset, a game, and an offset, though in some cases the onset might be as simple as running away from a playmate while looking back, and an offset as simple as pausing. A bout might contain a single interchange or several interchanges. A bout might be followed by another behavior or by another bout of play behavior. If "G" stands for an elaborated play scheme or game, and "O" stands for an onset or offset (play regulator), and "N" stands for other nonplay behavior, then the structure of a play bout can be represented as "N O G ONO G ONO G ON," while the structure of a play episode can be represented as "N O G O G O G ON." According to this formulation, a play bout is the "O . . . O" unit, however many or few games and interchanges it may include (see Machlis [1977] and Berdoy [1993] for discussion of definitions of bouts).

An *episode* of social play was defined as several bouts of social play occurring in sequence. Episodes of social play in the two infant gorillas typically consisted of connected series of social play bouts (each bout consisted of a game initiated by enticements and stopped by an interruption). Episodes of social play were separated by prolonged bouts of solitary play or other nonplay behaviors. Interruptions which were followed by further enticements within a short time were classified as "bout" boundaries, while interruptions which were followed by other kinds of behavior were classified as episode boundaries.

In other words, an episode is a series of closely spaced bouts which are separated from

Table 18.2 *Definitions of gorilla games*

Name of game	Description of schemes and interactions in social games
Tickling/avoiding &/or submitting	an adult initiated game in which the adult tickles the infant who tries to avoid the tickling and/or rolls around submissively with a relaxed open mouth (ROM) face
Rough-and-tumble play (Play fighting)	sparring, grabbing, wrestling, rolling, biting, hitting, embracing, head locking with arm or leg, and kicking, characterized by rapid position changes and role reversals between playmates; play fighting in which the adult handicapped himself by using one hand (or rarely two) to spar with the infant;
Chasing/fleeing	running away while being chased and reciprocally chasing the run-away with changes in two roles, often preceded by acrobatics or tagging playmate
Keep Away	very similar to chasing and fleeing except that it was usually played on vertical cage bars and so involved climbing above playmate and bouncing and swinging and hanging by one or two hands or by feet moving out of reach as partner approached; adult playmates stayed on the ground and used one hand to pursue the infant
Hide-and-seek	very similar to chasing and fleeing except that it involved running into caves or the food box and hence out of sight of playmate, also it was more often preceded by an elaborate display of some sort including in some cases waving an object and then dropping it into the cave as a lure to the playmate (teasing game)
Sexy games	primarily touching the rump of playmate and self and mounting and pelvic thrusting with sibling, placing rump on hand of adult and bouncing rump on the hand of mother, and presenting to the father
Showing off	primarily chest beating, branch-waving and acrobatics apparently aimed at capturing the attention of a playmate, sometimes followed by imitative games, sometimes a prelude to hide-and-seek (difficult to distinguish from enticements)
Imitative play	simultaneous or overlapping performance of schemes, primarily chest beating and branch-waving but also acrobatics, e.g., sliding down railing, and object manipulation; occasionally this game involved manipulating and pulling the same object and hence became tug-of-war, other times it involved following the same circuit across various substrates and hence became follow the leader
Provoking	hitting, poking with hand or stick, pulling on piece of leather under father and running off as if to provoke chasing

Table 18.2 *Continued*

Name of game	Description of schemes and interactions in social games
Acrobatics	passive form while being dangled on mother's hand/foot; 1st active form involved crawling back and forth across mother's legs, then climbing on and off mother and bars, then sliding down mother, later primarily bouncing, climbing on and off, alternate dangling by hands and feet, jumping off bars, rocking, bouncing, twirling, somersaulting, sliding along flat hands on the floor, or sliding down railing; often solitary, but may have been directed at enticing a playmate; sometimes parallel social activity, i.e., imitation
Object play	in early infancy grasping and biting objects then removing them from mouth and examining them, later pulling and rotating, then tossing small objects while watching them fall, also scooping and sliding dirt on the floor, and carrying, lifting and dropping, climbing in and out of a tire, placing it against the wall, standing on it, and placing and piling burlap and branches on the floor often lying on them (nest-building games); usually solitary but sometimes social
Self manipulation play	in early infancy primary circular reactions such as fingering and pulling on own hands and face; in later infancy palpating and exploring nonvisible body parts especially own rump and face, usually solitary

each other only by play regulators. Episodes are separated from other episodes by a definite period of nonplay behavior. Berdoy (1993) makes a similar distinction between bites as bouts and meals as episodes. Analogically, a bout boundary is like the interval between fork lifts during a meal, while an episode boundary is like the interval between meals (Berdoy, 1993).

Bouts and episodes of social play were distinguished not only by the presence of games and the absence of other goal-directed activities, but by the presence of a class of onset and offset schemes I call *play regulators*. These are enticements and interruptions (or interventions) which act as onset and offset markers for play. They are named by analogy with such conversational regulators as postural shifts (Scheflen, 1964) and verbal pauses and shifts in pitch (Birdwhistell, 1972).

Enticements included play invitations and attempted redirections during an ongoing play bout (Mitchell & Thompson, 1991). *Interruptions* included returning to mother, cutting off attention, pausing, manipulating objects, changing partners, and other devices used by one or both playmates to stop play with a partner. *Interventions* were a special class of interruptions used by other animals to stop the play of the infants. In other words, the locus of regulation could be in an intervening adult, in the infant, or in his playmate. Since social play is a collaborative enterprise, interruptions by either playmate, or interventions by any adult, were sufficient to stop play, at least temporarily. See Table 18.3 for lists of play regulators.

Although the concept of a series of games punctuated by play regulators is helpful in analyzing the stream of gorilla play, in practice the structure of gorilla play is so highly variable and complex that the division between bouts and episodes was often difficult to distinguish. It is clear, however, that the number of interchanges within a bout, the number of bouts per play episode (hence the length of play episodes), and the kinds and frequencies of games and regulators employed varied from dyad to dyad (or triad) as well as changing developmentally.

Tracking developmental changes in the infant's patterns of play within and among various play dyads requires yet another, higher, level of analysis of the data. In an effort to explicate such changes, I developed the concept of *phases of development of play* (Harlow, 1969). Three global phases of play emerged from the data. These phases were distinguished primarily by the locus of control of Mkumbwa's play behavior, that is, control by the mother, the half-sibling, or the infant himself.

Sensorimotor intellectual development

Following the identification of phases of development of play, the phases were analyzed for their apparent cognitive correlates using concepts drawn from Piaget's stages of sensorimotor intelligence in various domains (Piaget, 1952, 1954, 1962; Parker, 1977b; Užgiris & Hunt, 1975). Particular emphasis was placed on Piaget's sensorimotor intelligence series. Interpretation of sensorimotor intelligence in the gorilla infant is based on an analysis of all the developmental data collected during this study. Ages represent the first occurrence of diagnostic behaviors. See Table 18.4 for a description of sensorimotor intelligence stages and domains (series) in human, chimpanzee, orangutan, and gorilla infants (see Chevalier-Skolnikoff's [1977, 1983] description of sensorimotor development in gorillas, chimpanzees, and orangutans; also see Miles on orangutans [1990], and Potì & Spinozzi on chimpanzees [1994]).

RESULTS AND DISCUSSION

Phases of play development in an infant gorilla

On the most global level, the development of play in this infant gorilla can be divided into three overlapping phases: (1) a mother-controlled phase when Mkumbwa was from about 2 to 12 months, (2) a playmate-controlled phase when he was from about 15 to 33 months, and (3) an infant-controlled phase when he was from about 25 to 36 months of age. See Table 18.5.

As previously indicated, *the mother-controlled phase* was characterized by almost total and exclusive maternal control over the times and amounts of social play Mkumbwa engaged in. Although Mkumbwa began to use play enticements during this phase, his mother rarely allowed his half-sibling to play with him. When she did allow the infants to play, she often intervened after a few bouts to stop them by distracting the older infant.

Table 18.3 *Play regulators employed by various gorillas*

Animal	Kind of Regulator Enticements	Interruptions	Interventions
Infant	slapping surface arm waving supine limb waving grabbing chestbeating object waving foot stamping acrobatics wrist waving tagging presenting hand clapping head waggling embracing looking over shoulder throwing object	going to mother manipulating object climbing on mother ignoring leaving pausing changing partners strategically positioning self	
Brother	branch waving jumps next to tagging presenting supine limb waving acrobatics slapping surface	leaving ignoring changing partners pausing	
Mother	touching w/ROM supine limb waving	ignoring chestbeating	wrestling w/brother grabbing infant pushing brother bringing to toys
Father	grabbing	ignoring	stops wrestling taking away toy growling
Aunt	supine limb waving taking object	ignoring changing partners	

The mother-controlled phase had several subphases. In the earliest playful interactions, the *passive subphase*, beginning when Mkumbwa was about 2 months old, his mother began to dangle him on her foot or hand while lying on her back. At this time he was quite passive, and spent most of his time nursing and sleeping. When he was slightly older she began to place him on his back on her belly which stimulated him to

Table 18.4 *Piaget's sensorimotor intelligence series stages in an infant gorilla, chimpanzees, orangutans, and human infants (in months of age)*

Patterns	Stages:	Humans	Gorilla	Chimps	Orangutans
Reflexive behaviors: sucking, grasping, righting	1st	birth	birth	birth	birth
First Voluntary adaptations & primary circular reactions (PCR): repeated schemes performed on self or to assimilate objects to self, e.g., thumb-sucking, watching own hands, manipulating own hands and feet	2nd	2.0	2.8	2.5	?
First voluntary prehensions & secondary circular reactions (SCR): repeated schemes performed on objects with apparent aim of recreating interesting contingent results, e.g., shaking objects	3rd	3.0 3.0	3.0 4.5	4.5 5.0	?
Coordination of secondary schemes and beginning differentiation of	4th				
means from ends: e.g., letting go of one object in order to		8.0	4.5		
grasp another, hand to hand		11.3			
transferring object, rotating objects to see other side		8.0		9.0	
pulling an string to reach attached object, dropping		8–9		14.0	
objects in order to watch them fall		8–9	16.0		
Derived secondary co-ordinations: e.g., serial application of familiar schemes, incipient definitions of objects by use	4th-5th	10	19.0		

Table 18.4 *(Continued)*

Patterns	Stages:	Humans	Gorilla	Chimps	Orangutans
Trial-and-error discovery of new means and tertiary circular reactions (TCR): e.g., (a) trial-and-error discovery of use of a stick to reach an object (b) use of object as a support (c) varied repetition of schemes on objects to discover contingently varied effects	5th	12.0 15–18 15–18	24.0 39.0		14.0
Representation of trial-and-error discovery of new means (insight): e.g., deferred imitation of novel actions	6th	18.0	39.0		

respond with a relaxed open mouth (ROM) face. At the same time, she began to distract the infant's half-brother from his increasingly frequent attempts to play with the infant. During this phase, which lasted for several months, the mother sparred vigorously with the older infant while pinning her infant down with her feet. The mother also used one arm to play a highly restrained form of keep away with the infant when he first clung to the bars of the cage. During this period he continued to spend most of his time nursing and cuddling, but he was much more active visually, indeed he spent a great deal of time watching his half-brother and mother playing and visually following other animals as they were moving about.

In the next subphase, *the self-absorbed subphase*, beginning when he was about 4 months old, the infant began climbing on and off and around his mother's body and engaging in playful self-directed actions (primary circular reactions typical of the 2nd stage of sensorimotor period development) such as repeatedly hitting himself and grabbing his own hands and feet often with a ROM face while he lay on his back in his mother's lap. These self-directed elaborated play schemes were often triggered by watching the play bouts between his mother and his half-brother which continued during this period. His observations of play wrestling and sparring between his mother and his half-brother seemed to be highly stimulating to him. He often tried to join in, but was unable to owing to his motoric immaturity and his mother's intervention to protect him from his half-brother's playful advances.

In the following, *exploratory subphase*, beginning when Mkumbwa was about 7 months old, his mother began tickling the infant and playfully biting his fingers and toes. She also began engaging in one- and two-arm play fighting and more vigorous keep

Table 18.5 *Global phases of social play development in an infant gorilla*

Phases	Sub-phases	Ages	Associated Developments	
		months	Sensorimotor stages	Life-history events
Mother-controlled phase	passive play	2.0	reflexive schemes (1st)	reflexive grasping, righting & sucking
	self-absorbed play	4.0	primary circular reactions (PCR) (2nd)	self-directed actions such as thumb-sucking & hand-grabbing
	exploratory play	7.0	secondary circular reactions (SCR) (3rd) (manual & gestural modalities only)	beginning prehensive & locomotoric control; leaving mother; security oriented behaviors; eating solid foods
Peer-controlled phase	protected play	9.0	secondary circular reactions (SCR) (3rd)	consolidating prehensive & locomotoric development; letting go of one object to pick up another; climbing
	imitative play	12.0	imitation of own familiar schemes; simple means/ends co-ordination (4th)	emergence of aggressive facial displays
	co-ordinated play	19.0	serial application of secondary schemes (derived SCR (transitional 4th-5th)	emergence of submissive & sexual displays; presenting
Self-controlled phase	communicative play	24.0	understanding of simple causality; imitation of novel schemes (5th)	emergence of purposive social communication
	strategic play	33.0	deferred imitation; mental anticipation of results (6th)	weaning; food preparation

away as he began to climb on the bars. By the time Mkumbwa was about 12 months old and had gained greater locomotoric skill, his mother was carrying him to a location such as the railing by the stairs or the food box in the moat where he could engage in acrobatics. On several occasions, she encouraged his play by pushing the infant into the food box where his half-brother often played. During this same period, she also began to engage in sexy play with him, holding his rump with her hand while he bounced up and down with a ROM face. The end of the mother-dominated play period coincided with and apparently occurred in response to the infant's growing locomotoric skills.

Analogously to the mother-controlled phase, *the playmate-controlled phase* was characterized by primary control of social play by his older half-sibling, Sunshine. Sunshine, in turn, was strongly influenced by Mkumbwa's mother to handicap himself by playing in a restrained manner appropriate to the infant's immaturity. Sunshine controlled play primarily by his responsiveness or lack of responsiveness to Mkumbwa's play invitations.

The playmate-controlled phase had at least three subphases. The first *transitional subphase*, characterized by protected play, developed gradually as the infant began to engage in rough-and-tumble play and chasing with his older half-brother under his mother's close supervision at about 9 months of age. The number of interchanges per play bout during this period were very few, the interruptions between bouts were long, the maternal interventions frequent, and therefore, the play episodes were short and sporadic. During this early period from about 9 to 15 months, Mkumbwa engaged in repeated chest beating in play enticements. He also remained on, or right next to, his mother as he played with his playmate, and turned to her to interrupt play bouts. He also interrupted play bouts by engaging in self-directed actions. He continued to spend a good deal of time nursing and cuddling with his mother.

The second, *imitative play, subphase* began at about 13 months of age when he started to move away from his mother and engage in hide and seek games in the caves and food box, to play sexy games of presenting and mounting. During this period he still touched his mother to regulate play bouts, and his play episodes were still relatively short. At about 12 months of age, he began to engage in imitative games of chest beating and branch waving with his half-brother. He interrupted his social play bouts with object play, and engaged in long bouts of solitary object play.

The third, *co-ordinated play, subphase* developed when Mkumbwa was 24 months old. At this time he began employing a richer repertoire of play enticements and interruptions including pausing and object play. His play entailed a greater number of interchanges per bout, a greater number of bouts per episode, and hence longer episodes of play. He also began to play teasing games which involved baiting his playmate by tossing objects into the caves. The gradual shift from mother-controlled to playmate-controlled play coincided with the infant's increasing locomotoric skills in running and acrobatics. It also coincided with the apparent achievement of the late 4th and early 5th phase in various sensorimotor intelligence series, that is, his understanding of object permanence, inside–outside relationships in space, simple causality, serial application of schemes, and the ability to imitate novel schemes as well as those from his own repertoire (Piaget, 1952, 1954). During this period, he also developed more complex solitary object manipulation games involving repeated elaborated object-throwing and nest-building schemes.

Finally, *the self-controlled phase* was characterized by Mkumbwa's increasing regulation of his own social play through more flexible and strategic use of a larger repertoire of play regulators. This phase was marked by two changes, (1) increasing interest in playing with his aunt and his father, and (2) an increasing understanding of the communicative significance of his own behaviors. The infant (self)-controlled phase began when Mkumbwa was between 24 and 33 months of age. By this time he spent most of his time playing with other animals or by himself. It had two subphases: a *communicative play subphase* and a *strategic play subphase*.

His understanding of the communicative potential of his actions during the *communicative subphase* was indicated by his watching and experimenting with other animal's responses to his behaviors. He watched responses to his displays, teasing playmates by waving and then dropping objects into the caves. This was apparently an expression of fifth-stage experimentation with means. Anticipation of outcomes of his actions during this phase is also suggested by strategically positioning himself between his father and his half brother. It was also indicated by Mkumbwa's efforts to provoke a reaction from his father by repeatedly poking and hitting him (see Adang, 1984 for description of similar behaviors in chimpanzees). This form of play seemed to be unique to the infant–father relationship.

Play interactions between Mkumbwa and his father were also characterized by repeated presenting by the infant and rump touching by the father, and by repeated showing off with play dominance displays by the infant. Because of the father's relatively low rate of interacting with the infant, much of the infant's behavior toward him involved showing off or enticements, which were sometimes difficult to distinguish in this context.

These father-specific features seemed to reflect ambivalent elements of aggression and fear in the infant's play with this very large figure. These elements could also explain the relatively late emergence of infant–father play. Otherwise the father played many of the games that the mother played (tickling and handicapped rough-and-tumble play), but at a much later period in the infant's development. Like his mother, Mkumbwa's father sometimes intervened in the play of the two infants, once by taking an object away from them. Like the mother, he also intervened on at least one occasion by joining in the play. The chief difference in this case was that Mkumbwa invited the intervention and seemed able to co-ordinate his play with two playmates.

His understanding of strategy during the *strategic subphase* was indicated by his strategic positioning and invitation to his father or mother to intervene during play with his half-brother. In other words, in the second subphase of the third phase, Mkumbwa controlled the triadic interaction whereas his mother controlled it in the first and second phases. This second subphase of the third phase in the development of social play coincided with the achievement of 6th-stage sensorimotor intelligence and especially with the capacity for deferred imitation of novel schemes.

It also coincided with two important life-history events, weaning and resumption of sexual cycling by his mother. In male infants this infant-controlled phase of development probably leads into apprenticeship in adult male dominance activities. Such apprenticeship in dominance is a common phenomenon among primates (Dolhinow & Bishop, 1970) Sadly, in this particular case, it also coincided with the death of the infant's mother, Jacqueline, who had been losing weight and behaving lethargically for some time before she was removed from the group on May 2, 1978, virtually on Mkumbwas third birthday. The close tie which was developing between Mkumbwa and his father at this time facilitated the father's subsequent adoption of the infant. A similar case of adoption of a 3-year-old infant Mountain gorilla is described by Fossey (1978), but not all gorilla fathers are as solicitous (Mitchell, 1989).

Mkumbwa's play with his "aunt," Pogo, began a little earlier than his play with his father, apparently because she repeatedly enticed him to play with elaborate displays of branch waving and supine limb waving, etc., which the father and mother rarely did. Unlike the aunt, the father seemed to let Mkumbwa take the initiative in their relationship. Perhaps this was because his earlier attempts to approach his son had met with resistance by the mother, or because the infant showed some fear. The aunt's play with both infants was also more vigorous and sibling-like than the more protective play of the infant's mother and father. Remarkably, Mkumbwa made only one play enticement to the other infant's mother, Mrs, during the 3-year period of observations (which she ignored). See Table 18.6 for a list of games played with each partner and Tables 18.7 and 18.8 for descriptions of onset of games with various play partners.

The kinds of games Mkumbwa played with his various playmates differed significantly, as did the frequencies with which he played them. These frequencies are measured in *game days*: number of days he was observed playing games with a particular playmate. He played with his half-brother, Sunshine, more frequently than with any other playmate. He played about 4 times more game days with his half-brother than with his father, Bwana. He played about 7 times more game days with his half-brother than with his mother; and about 12 times more game days with his half-brother than with his "aunt," Pogo. The lower frequency was partly a reflection of the relatively late development of play with his father and aunt and the early peaking of his play with his mother. See Table 18.9 for frequencies measured in game days.

The relative frequencies of the games Mkumbwa played with each of his playmates also differed. Rough-and-tumble play was the most frequent form of play with his half-brother, while showing off, imitation, and hide-and-seek were next most frequent, and keep away and sexy games were the least frequent, except for tug-of-war, which was seen only once. In the infant's play with his father, rough-and-tumble play was less frequent than showing off, while provoking his father was the next most frequent game after rough-and-tumble play. Because most of Mkumbwa's play with his mother was initiated by her, the frequencies of her games with him, rather than the converse, were examined. Overall, games with her occurred at the lowest frequencies except for those of Pogo, the "aunt," each being seen on only a few days. Given the small number of

Table 18.6 *Classification of games according to partners*

	Infant's games and schemes with and/or mediated by interactant	Interactant's games with and/or stimulated by the infant
Mother		
	passive acrobatics	baby dangling
	play submission	baby fingering
	play fighting	tickling attack
	keep away	1- & 2-handed play fighting
	mother as gym	1- & 2-handed keep away
	sexy games	body as jungle gym
		sexy games
Brother		
	play fighting	play fighting
	keep away	keep away
	showing off	showing off
	chasing & fleeing	chasing & fleeing
	hide-and-seek	hide-and-seek
	imitative play	imitative play
	sexy games	sexy games
	play submission	tickling attack
	teasing games	teasing games
	tug of war	tug-of-war
Father		
	provoking	ignoring
	sexy games	sexy games
	play submission	tickling attack
	play fighting	1- & 2-arm play fighting
	father as jungle gym	body as jungle gym
	showing off	watching
	fleeing	chasing
Aunt		
	play fighting	1- & 2-arm play fighting
	play submission	tickling attack
	keep away	keep away
	chasing & fleeing	chasing
	teasing games	teasing games

Table 18.7 *Mother's and father's play with infant by infant's age at first occurrence (in lunar months)*

Mother's game	Infant's age	Father's game	Infant's age
Dangling baby	0.7		
Distracts brother with arm wrestling	2.0		
Keep Away	2.8	Keep away	33.5
Mother serves as gym for baby	4.0	Father serves as gym for baby	28.0
Tickling	7.0	Tickling	32.0
1- & 2-arm play fighting	9.5	1- & 2-arm play	22.5
Sexy play	12.5	Sexy play	28.0
Gym school (placing infant by bars)	14.0		
		Chasing	25.0

Table 18.8 *Times of emergence of infant's schemes/games by partner (in lunar months of infant's age)*

Game	Mother	Brother	Aunt	Father
Keep away	2.8	8.0	24.5	33.5
Play fighting	9.5	9.0	20.0	22.5
Showing off				
play dominance		9.0	28.0	22.5
displaying				
w/acrobatics		13.0		22.5
w/objects		28.0		24.0
Chasing & fleeing games		9.5	20.0	
Imitation games				
displays		9.0	24.5	
w/objects	7.0	11.0		
acrobatic		12.0		
Sexy play	12.5	11.0		28.0
Hide-&-seek			15.0	
Teasing		20.0	39.0	35.0
Tug-of-War			37.0	

Table 18.9 *Frequencies of various games with different play partners in number of days observed (game days)*

Games	Half-brother	Father	Mother
Keep away	7	1	
Play fight	21	7	2
Chasing	4		
Showing off	11	8	
Imitation	10		
Hide-&-seek	9		
Sexy games	6	2	11
Teasing	4		
Provoking		3	
Baby dangling		3	
Body as gym			3
Tickling			2

observations, the relative frequencies are unlikely to be significant. Too few game days with the aunt were observed to warrant displaying.

The infant's solitary game days (days he played games with himself) were about half as frequent as all his other game days combined. The frequencies of various forms of solitary play also differed. Object play was the most frequent form of solitary play, followed by self-directed play, while displays and acrobatics tied for least frequent forms of solitary play. Mkumbwa also showed three categories of solitary play: self-directed play, object play, and acrobatic play which he displayed throughout the 3-year study. Self-directed play first occurred during the mother-controlled phase when the infant was motorically immature as well as protected from sibling play by his mother, that is, from 4 to 8 months of age. Later, after about 7 months of age, self-directed play had the character of self-exploration, for example, feeling invisible body parts such as the face and the rump.

Solitary object play included repetitive manipulation of small objects, that is, picking them up and putting them down, putting them in his mouth, pulling them out to examine, scooping and pushing around grains of sand and dirt, throwing and hitting small objects and watching them fall. Object-throwing games became a frequent occupation beginning when the infant was about 16 months old during the long transition from the 4th to the 5th stage and through the 5th stage in the sensorimotor intelligence series. Another elaborated solitary object scheme, which developed at about 28 months, involved repetitive bouts of dragging, laying out, placing on, pulling on, circling, stepping on and lying on burlap bags and/or leafy branches. Because these are actions that are involved in nest-building, I called them nest building games.

Overall, the ontogeny of play in Mkumbwa had the following course: (1) appearance of isolated schemes, for example, chest beating, slapping substrates, (2) repetition of those schemes across contexts, for example, chest beating to self, to various others, and slapping various surfaces; (3) application of those schemes as a means to an end, for example, as play regulators; (4) combination of those schemes with other schemes, for example, with branch waving, into elaborated play enticements, and show offs; (5) strategic use of games and regulators to organize play bouts and episodes. This course seems to be explained by his cognitive development: most self-directed schemes first appeared between 2 and 13 months; most object-directed schemes appeared between 4.5 and 12 months during the 2nd and 3rd stages of the sensorimotor period; most applications and simple co-ordinations of schemes appeared between 12 and 20 months during the 4th stage; finally the use of schemes in communication and organization of interactions occurred between 24 and 36 months during the 5th and 6th stages (refer to Table 18.4 for stage descriptions).

Imitation and showing-off games

While most of the games the gorilla infant played have been described in other species of mammals (e.g., Fagen, 1981), a few have not. In this section I focus on two less familiar games that the gorilla infant played, that is, imitative play, and showing off.

Imitative play is a peculiar category because it classifies play on the basis of its stimulus rather than the specific actions involved. The theoretical importance of imitation justifies this taxonomy. I use the term imitation in the broad sense to encompass all the forms of social imitation Piaget (1962) described in his 6 stages of the imitation series in the sensorimotor period: these include stage, 4 imitation of actions in one's own repertoire, as well as stage 5, trial-and-error imitation of novel schemes, and stage 6, deferred imitation of novel schemes (*contra* Galef, 1988). Imitation is notoriously difficult to detect because it usually requires comprehensive knowledge not only of the focal animal's behavior, but of his model's behavior as well. The occurrence of imitation is also contingent upon certain affectional and attentional relationships between imitator and model (Russon & Galdikas, 1995).

Fortunate observational conditions can turn these factors to advantage (Russon & Galdikas, 1993). In this study, the appropriate affectional relations obtained, but focal animal sampling by a single investigator severely limited the data on any model's behavior. The clearest cases in this study were those in which the infant's behavior immediately followed the model's behavior, and one case in which the infant's behavior inadequately mimicked the mother's oft repeated behavior of rubbing her stomach and regurgitating her food. This means that the amount of imitation is probably underestimated. The most commonly imitated actions were gestures and object manipulations. See Table 18.10 for a list of imitated actions both playful and otherwise.

Like imitative play, showing off is a peculiar category. In this case it is based on the focus on the receiver of the actions rather than the actions themselves. Showing off, like teasing and provoking, involves performing an action while watching the reaction of one or more observers, apparently with the aim of inciting a reaction. The reactions sought might be imitation or a switch to a new game such as chasing and fleeing. Showing off was sometimes difficult to distinguish from enticement. The most common motor

Table 18.10 *All instances of imitation by ages and stages of development in an infant gorilla (Mkumbwa) to three years of age*

Age in lunar months	Action imitated	Model imitated	Stage of imitation according to Piaget
7.0	dabbling in water with hand	mother	4th (imitating actions already in own repertoire)
9.0	scraping up grains	mother	same
9.5	licking slot in door	brother	same
9.5	leaf in mouth	brother	same
11.5	drinking water from pool	mother	same
15.4	sliding down railing	brother	same
21.0	going into food box	brother	same
24.5	chest beating	brother	same
28.0	stripping leaves off branch with teeth	mother	5th stage (trial & error matching own actions to novel actions of model)
29.5	scooping dirt	brother	same
29.5	slapping food box	brother	same
32.3	rubbing belly eating pretend regurgitate (saliva)	mother	6th stage (deferred imitation of novel schemes)
36.9	use of stick as a tool	brother	5th stage

patterns used in showing off were acrobatics, and chest beating, object waving, and throwing. Showing off was primarily aimed at the infant's half-brother and his father, but also at his aunt. It is interesting to note that the latter schemes, which play such a prominent role in gorilla play, are important components in the adult male silverback display (Schaller, 1963). See Table 18.11 for ages at which Mkumbwa first displayed various steps in the display.

Development of social roles

The presence of third parties is a common aspect of social play in primates which has important implications for the development of social roles (Kummer, 1967). Therefore, episodes of social play were analyzed in terms of both dyadic and triadic interactions, noting the role of the third party and the role of the infant relative to the third party, for example, whether the infant invited the intervention of the third party or not, or whether the three animals played together simultaneously or alternately. Developmental changes in tripartite relations were also noted.

Although the infant's play relationships with all his playmates involved some common elements, each dyad had unique interactional characteristics. On the grossest level, differences arose from disparities in size and strength and developmental levels between Mkumbwa and adults, and Mkumbwa and the older infant. On the next level they arose from the kin relations between the infant and his mother and father, who played

Table 18.11 *Stages of development of dominance display games and play enticements*

Sequence of performance of silverback display	Sequence and ages of emergence of components of sequence in infant gorilla
Hooting	?
"Symbolic feeding" with leaf between lips	7.0
Rising to bipedal stance	7.0
Rhrowing vegetation	10.5
Chest beating	10.0
Leg kicking	37.0
Running sideways	37.0
Slapping & tearing	16.0 (without tearing vegetation)
Dragging vegetation	18.0
Ground thumping	?

protective roles befitting their genetic investments. On another level, they reflected sex differences in reproductive and aggressive behaviors and roles.

Owing to these and other more subtle and individual variations, Mkumbwa gained experience with a variety of complementary communicative schemes and social roles: in playing with his half-brother, Sunshine, he learned the complementary roles of mounter and mountee, of chaser and fleer, attacker and attackee, hider and hidee, teaser and teasee, imitator and imitatee. Nine out of ten games he played with his half brother were reciprocal in nature. In playing with his father, Bwana, he learned the roles of show-off, provocateur, and sexual submissee. In playing with his half-brother in proximity to his mother and his father, he learned the roles of protectee and instigator of protection. Almost all the games he and his father, mother, and aunt played involved non-reciprocal roles.

In addition, Mkumbwa learned that his relationships with individuals changed over time. As he developed, the young gorilla learned to regulate interactions by playing the roles of enticer and enticee, and interrupter and interruptee. Both the infant and his older half-brother tried out various enticements that were unsuccessful in eliciting play, either because they were misunderstood or because they were insufficiently enticing in the particular context. During the first phase, most of the half-brother's early displays, for example, hand clapping and chest beating, to the infant went unanswered (e.g., when the infant was between 2.5 and 9 months of age). Likewise, many of Mkumbwa's early displays to his half-brother, for example, arm waving and slapping the shelf, were unsuccessful. Likewise his later attempt, for example, to use hand clapping to entice his half-brother's mother was unsuccessful. These unsuccessful attempts suggested trial-and-error learning about the effectiveness of potential enticements.

As he grew older, the infant also developed new means for interrupting play bouts: after an initial period of turning to his mother, or going next to her, he began to go into the food box or to turn and start manipulating an object, later still he positioned himself

Table 18.12 *Patterns of social play with brother at selected ages in an infant gorilla*

Age in lunar months	Number of interchanges per bout (typical no. & range)	Infant's enticements	Infant's interruptions	Kinds of games	Patterns of interaction
9.5	1	arm waving grabbing	grab mother; ignore brother; manipulate own body	playfight	few short bouts while on mother; sporadic
13.0	2 (1–3)	chest beating (repeated) manipulate object	onto mother;	playfight	on and/or next to mother repeated enticements for each interchange; several short bouts, some episodes
25.6	5 (2–10)	presenting chest beating wrist waving hand clapping embracing	go near mother; go into food box pause; manipulate object	playfight sexy games chasing imitation hide-&-seek	near mother several short & 1 medium episode; lots of enticements
33.5	7 (2–15)	same	pause; strategic position by father; changing partners	playfight chase hide-&-seek show off	away from mother for long time; near father; long play episodes; (with few enticements to brother; many to father)

strategically close to his father. At the same time that his play-regulating skills were developing, the length of his play bouts were increasing as measured by the number of interchanges between himself and his playmate between play interruptions. Likewise, the number of bouts per play episodes were increasing (see Table 18.12).

In a sense, then, playful interactions with each partner provided a microcosm for a particular role relationship. Playing with various partners instantiated Mkumbwa's developing roles as mother's son, brother's sibling, father's son, and "aunt's" playmate, at various ages. Being mother's son, for example, meant being protected from others, submitting to various restraints and punishments, being carried, cuddled, and nurtured, and very occasionally played with. Being a father's son, in contrast, meant being protected and restrained, but also being dominated, playing a submissive role, and having one's provocative behavior tolerated. Being a brother's sibling, on the other hand, meant almost nonstop playing of reciprocal alternating roles with only occasional submission.

In addition to these dyadic role relationships, some play interactions involved experiences in triadic relationships. Such triadic interactions are important in alliances and can involve considerable cognitive sophistication (Harcourt, 1988). These role-playing experiences simultaneously exercised Mkumbwa's emerging self-awareness and other awareness (Parker & Milbrath, 1994). These forms of awareness are indicated in watching other's reactions to his actions, strategically placing himself between his playmate and his father, and playing with the contingency of his mirror-image (see Table 18.13), as described in children (Lewis & Brooks-Gunn, 1979). See Table 18.14 for examples.

Finally, it is important to note that social roles are embedded in "scripts," that is, customary routines of interaction between two or more individuals (Nelson & Seidman, 1984; Nelson & Gruendel, 1986; Mitchell, this volume). The repetitive nature of these scripts provides a powerful frame for learning social roles (and language, in human children). In the case of youngsters, these routines are typically choreographed by the older interactant. Seen in this perspective, games are mini scripts that canalize social learning.

A note on variations in social play

Presumably, the phases of development of Mkumbwa's social play were constrained by species-typical ontogenetic patterns. Obviously, however, the content of the phases were shaped by the physical environment and the demographic composition of the group. Social interactions can vary considerably according to group composition, numbers of infants, their kin relationships, their ages and sexes (Hoff et al., 1981). In this setting (1) the playmate-controlled phase was conditioned by the presence of a playmate of a certain age and kin relation, (2) the mother-controlled phase may have been prolonged and intensified by his mother's status and the presence of an intrusive half-sibling, Sunshine, who was a year and a half older, and (3) Mkumbwa's interactions with his father were probably conditioned in part by the ages and temperaments of his mother and father and the relationship between them (Tilford & Nadler, 1978).

Table 18.13 *Developing indicators of self-awareness in an infant gorilla*

Behaviors	Ages of apearance
Directed dsplays	24.6
Strategic self-positioning relative to 3rd Party	25.0
Enticing 3rd Party to join play	33.0
Looking over shoulder at playmate while running away	25.0
Contingency play[a] with mirror-image	33.0

[a] Contingency play refers to behaviors apparently directed at testing the contingency between the actor's actions and the mirror-image's actions (Lewis & Brooks-Gunn, 1979; Parker, 1991).

Table 18.14 *Stages of sensorimotor intellectual development and categories of schemes/games by age of first appearance (in lunar months)*

Schemes and games	Age	Piagetian series	Stage
Solitary object play			
Small object manipulation	12.5	Sensorimotor	4th
Object-throwing games	16.0	Causality	4th or 5th
Nest-building games	28.0	Spatial	5th
Tool-use games	37.0	Causality	5th
Social games			
Imitation Games			
displays	9.0	Imitation	4th
w/objects	7.0	Imitation	4th
Hide-&-Seek	15.0	Object permanence	5th
		Spatial series	5th
Teasing	20.0	Sensorimotor series	?
	Causality series	?	
Showing off	24.0	Imitation series?	?
	60.0		

The influence of these factors is suggested, for example, by the quite different pattern of development of social play in Mkumbwa's son Shango. Shango had no sibling or other immature playmate until he was more than 4 years old (Tanner, 1993): Perhaps because his mother exerted less control over the infant, Shango's father, Mkumbwa, and his great "aunt," Pogo, played with him when he was much younger than Mkumbwa was when Bwana and Pogo played with him.

Virtually from the birth of his brother, Barney, Shango's mother, Bawang, allowed the then 4–year-old Shango and a nuliparous female, Zura, to carry the young infant under close supervision. She also played with Barney in a sitting position dangling him by both arms and tossing him onto her chest, a pattern quite different from that of Mkumbwa's mother Jacqueline (see Tanner, this volume).

SUMMARY AND CONCLUSIONS

Based on a multi-level framework, this chapter traces the development of social play in an infant gorilla, Mkumbwa, at the San Francisco Zoo from shortly after birth to 3 years of age. Three developmental phases were distinguished by contrasting the onset, duration, and nature of his play with his mother, half-brother, aunt and father: (1) a mother-controlled phase from birth to 8 months, (2) a playmate-controlled phase from 8 to 24 months, and (3) a self- controlled phase from 24 to 36 months. Phase transitions corresponded with shifts in the infant's locomotoric and sensorimotoric cognitive abilities, and to his mother's responses to him and his playmates. The mother-controlled phase encompassed the 2nd and 3rd sensorimotor stages; the playmate-controlled phase encompassed the 4th and 5th stages; and the self-controlled phase began with the 6th sensorimotor stage. Each phase was characterized by a different set of games which were played with different frequencies with different partners. The onset of particular games reflected the achievement of particular sensorimotor stages in specific series: object-throwing games, for example, reflected late 4th- and early 5th-stage causality, nest-building games reflected achievement of 5th-stage space, and hide-and-seek, 5th-stage object permanence and causality (see Table 18.14). Overall, the infant showed a trend toward increasing numbers of play schemes and games, and increasing duration of play bouts and episodes. By repeatedly engaging in games (mini scripts) with various playmates of varying ages and kin relations, the young gorilla developed a variety of social roles.

ACKNOWLEDGMENTS

I want to thank the San Francisco Zoo for allowing me to do this study. I especially thank Dr. Saul Kitchner, former Director of the San Francisco Zoo, and John Alcaraz, former keeper of the gorillas, and Mary Kerr, current keeper of the gorillas. I also thank the Hooper Foundation at University of California San Francisco for a small grant awarded to Suzanne Chevalier-Skolnikoff and myself in 1974–1975 during the early phase of this study, which began as part of a larger collaborative study of sensorimotor intelligence in great apes which ended in mid-1975. All the data used in this chapter were collected and analyzed solely by myself. I am grateful to the following colleagues for their helpful comments and suggestions: Martin Daly, Phyllis Dolhinow, Constance Milbrath, Robert Mitchell, Patrizia Potì, Anne Russon, and Joanne Tanner. Finally, I thank Sonoma State University for granting me funds for a semesters research leave in 1989.

REFERENCES

Adang, O. (1984). Teasing in young chimpanzees. *Behaviour*, **88**, 98–122.

Aldis, O. (1975). *Play fighting*. New York: Academic Press.

Berdoy, M. (1993). Defining bouts of behaviour: A three-process model. *Animal Behaviour*, **46**, 387–396.

Birdwhistell, R. (1972). *Kinesics in context*. New York: Ballentine Books.

Chevalier-Skolnikoff, S. (1977). A Piagetian model for comparing the socialization of monkey, ape, and human infants. In S. Chevalier-Skolnikoff & F. Poirier (eds.), *Primate biosocial development* (pp. 159–188). New York: Garland.

Chevalier-Skolnikoff, S. (1983). Sensorimotor development in orangutans and other primates. *Journal of Human Evolution*, **12**, 545–561.

De Waal, F. (1991). Complementary methods and convergent evidence in the study of primate social cognition. *Behaviour*, **118**, 297–320.

Dolhinow, P. C. & Bishop, N. (1970). The development of motor and social relationships among primates through play. In J. P. Hill (ed.), *Minnesota Symposia in Child Psychology* (Vol. 4, pp. 141–198). Minneapolis: University of Minnesota.

Fagen, R. (1981). *Animal play behavior*. New York: Oxford University Press.

Fossey, D. (1978). The first 36 months of life. In D. Hamburg & E. McCown (eds.), *The great apes* (pp. 139–185). Menlo Park, CA: Benjamin Cummings.

Galef, B. G., Jr. (1988). Imitation in animals: History, definition, and interpretation of data. In T. Zentall & J. Bennett C. Galef (eds.), *Social Learning*, (pp. 3–28). Hillsdale, NJ: Erlbaum.

Groos, K. (1976 [1898]). *The play of animals*. New York: Arno.

Harcourt, A. (1988). Alliances in contests and social intelligence. In R. Byrne & A. Whiten (eds.), *Machiavellian intelligence* (pp. 132–151). Oxford University Press.

Harlow, H. F. (1969). Age-mate or peer affectional system. In D. Lehrmann, R. Hinde, & E. Shaw (eds.), *Advances in the study of behavior* (Vol. 2, pp. 333–383). New York: Academic Press.

Hoff, M., Nadler, R., & Maple, T. (1981). Development of infant play in a captive group of lowland gorillas. *American Journal of Primatology*, **1**, 65–72.

Kummer, H. (1967). Tripartite relations in hamadryas baboons. In S. A. Altmann (ed.), *Social communication among primates* (pp. 63–71). University of Chicago Press.

Lewis, M. & Brooks-Gunn, J. (1979). *Social cognition and the acquisition of self*. New York: Plenum.

Machlis, L. (1977). An analysis of the temporal patterning of pecking in chicks. *Behaviour*, **60**, 1–70.

Miles, H. L. (1990). The cognitive foundations for reference in a signing orangutan. In S. T. Parker, R. W. Mitchell, & M. L. Boccia (eds.), *"Language" and intelligence in monkeys and apes* (pp. 511–539). Cambridge University Press.

Mitchell, R. W. (1989). Functions and consequences of infant–adult male interactions in a captive group of lowland gorillas. *Zoo Biology*, **8**, 125–137.

Mitchell, R. W. & Thompson, N. (1991). Projects, routines, and enticements in dog-human play. In P. P. G. Bateson & P. Klopfer (eds.), *Perspectives in ethology* (vol. 9, pp. 189–261). New York: Plenum.

Nelson, K. & Gruendel, J. (1986). Children's scripts. In K. Nelson (ed.),
Event Knowledge (pp. 21–45). Hillsdale, NJ: Erlbaum.

Nelson, K. & Seidman, S. (1984). Playing with scripts. In I. Bretherton (ed.), *Symbolic play* (pp. 45–72). New York: Academic Press.

Parker, S. T. (1977a). Comparative behavioral development in human, gorilla, and macaque infants, . Seattle, WA: Meeting of the American Association of Physical Anthropology.

Parker, S. T. (1977b). Paiget's sensorimotor series in an infant macaque: A model for comparing unstereotyped behavior and intelligence in human and nonhuman primates. In S. Chevalier-Skolnikoff & F. Poirier (eds.), *Primate biosocial development* (pp. 43–112). New York: Garland.

Parker, S. T. & Milbrath, C. (1994). Contributions of imitation and role-playing games to the

construction of self in primates. In S. T. Parker, R. W. Mitchell & M. L. Boccia (eds.), *Self-awareness in animals and humans* (pp. 108–128). Cambridge University Press.
Piaget, J. (1952). *The origins of intelligence in children*. New York: International Universities Press.
Piaget, J. (1954). *The construction of reality in the child*. New York: Basic Books.
Piaget, J. (1962). *Play, dreams, and imitation*. New York: W. W. Norton.
Potì, P. & Spinozzi, G. (1994). Early sensorimotor development in chimpanzees (*Pan troglodytes*). *Journal of Comparative Psychology*, **108**(1), 93–103.
Russon, A. & Galdikas, B. (1993). Imitation in free-ranging rehabilititant orangutans (*Pongo pygmaeus*). *Journal of Comparative Psychology*, **107**(2), 147–161.
Russon, A. & Galdikas, B. (1995). Constraints on great ape imitation: Model and action selectivity in rehabilitant orangutans (*Pongo pygmaeus*) imitation. *Journal of Comparative Psychology*, **109**, 5–17.
Schaller, G. (1963). *The mountain gorilla: Biology and behavior*. University of Chicago Press.
Scheflen, A. (1964). The significance of posture in communication systems. *Psychiatry*, **27**, 316–331.
Simpson, M. J. A. (1976). The study of animal play. In P. P. G. Bateson & R. A. Hinde (eds.), *Growing points in ethology* (pp. 385–400). Cambridge University Press.
Tanner, J. (1993). *Development of play in an infant gorilla, Shango*. Unpublished MS.
Tilford, B. L. & Nadler, R. D. (1978). Male parental behavior in a captive group of lowland gorillas (*Gorilla gorilla gorilla*). *Folia primatologica*, **29**, 218–228.
Tomasello, M., Gust, D., & Frost, G. T. (1989). A longitudinal investigation of gestural communication in young chimpanzees. *Primates*, **30**(1), 35–50.
Užgiris, I. & Hunt, M. (1975). *Assessment in infancy: Ordinal scales of psychological development*. Urbana: University of Illinois Press.

PART V

Epilogue

19

The mentalities of gorillas and orangutans in phylogenetic perspective

SUE T. PARKER AND ROBERT W. MITCHELL

This chapter is based on the premise that it is impossible to understand human evolution unless we study gorillas and orangutans as well as our closest living relatives, chimpanzees and bonobos. It is designed to redress the chimpocentric bias of primatologists (Beck, 1982) by reviewing recent data on the cognitive abilities of gorillas and orangutans, especially data presented in this volume. It is also designed to highlight aspects of gorilla and orangutan behavior and abilities heretofore neglected, misrepresented, or unknown. Finally, it is designed to review the similarities and differences among the four species of great apes using the cladistic method to show the evolutionary implications of these comparative data.

THE CLADISTIC APPROACH TO PHYLOGENETIC RECONSTRUCTION

The cladistic method for reconstructing the evolution of behavior involves comparing the taxonomic distribution of behaviors among all the species closely related to our target species. (In order to avoid circularity, it is important to use a cladogram previously constructed from molecular data.) Through this systematic process of comparison, it is possible to determine which characteristics are (1) shared with members of other taxonomic groups (shared characters), (2) shared only by a group of sister species (shared–derived characters), and (3) unique to a given species (derived characters). From this it is possible to identify the putative common ancestors in whom particular shared derived characteristics must have arisen.

These patterns can only be determined through comparisons among the ingroup species, and between those species and more distantly related outgroup species Brooks & Mclennan, 1991; Harvey & Pagel, 1991). In other words, we can only know ourselves through systematic comparisons of our (human) characteristics with those of our close relatives (the great apes) as contrasted with those of our more distant relatives (lesser apes and monkeys).

Molecular phylogenies indicate that all the great apes shared a common ancestor approximately 12 million years ago, the African apes approximately 6 million years ago,

Pan and *Homo*, about 5 million years ago, and the two *Pan* species, chimpanzees and bonobos, about 1.5 million years ago (see Begun, this volume). The group as a whole, and each of the sister species, joined at a branching point constitute an adaptive array.

Our evolutionary analysis is based on the cladistic approach to reconstructing the evolution of behavior. The characteristics we compare include life-history features, locomotion, diet, brain size, sexual behavior, temperament, and cognition. Ingroup and outgroup comparisons allow us to classify various character states as shared among all the great apes, shared–derived among all great apes, shared–derived among African apes, or derived in one species of great ape. This, in turn, allows us to identify the characteristics of the putative common ancestor (Brooks & McLennan, 1991).

ANATOMICAL PATTERNS

Structurally, all apes display the short, inflexible backs, broad, shallow chests, long arms and fingers, and flexible elbows and wrists that facilitate the form of suspensory locomotion known as brachiation (Clark, 1959; Fleagle, 1988). These characteristics are shared characteristics among great apes, and shared–derived characteristics of all apes, in other words, characters that first arose in the common ancestor of great apes.

Consequent to their movement on the ground, the great apes evolved two forms of terrestrial quadrupedalism using the long-armed and short-legged anatomy of brachiators. Orangutans display a form known as fist-walking, and the African apes, a form known as knuckle-walking. Knuckle-walking, which involves specialized ligaments that stabilize the knuckles, is a shared–derived characteristic of African apes, whereas, the sloppier, less-specialized fist-walking, is a derived characteristic of orangutans (Tuttle, 1986). These are not forms of locomotion found among the lesser apes.

An analysis of these data shows that brachiation arose in the common ancestor of all apes, and that a specialized form of terrestrial quadrupedalism secondary to brachiation evolved in the common ancestor of all the great apes. Subsequently, derivatives of specialized terrestrial quadrupedalism, fist-walking and knuckle-walking, evolved in the ancestor of orangutans and in the common ancestor of the African apes, respectively.

Great apes are also characterized by brain sizes which exceed those of any other nonhuman primates. Their brains are 2 to 3 times larger than those of lesser apes, and three times smaller than those of modern humans, ranging from about 350 to 500 cc, as compared to about 180 cc in gibbons, and about 1250 in humans. Great ape brains also share morphological features with human brains. The large brain size of great apes correlates with their prolonged period of immaturity and their low reproductive rate, key life-history features. They are shared–derived characters that evolved in the common ancestor of great apes.

LIFE HISTORY PATTERNS

All the great apes display similar patterns of development and life history, gorillas having the most rapid development, and orangutans the least. Great apes display gestation periods of approximately 8 months, as compared to 7 months in lesser apes and 9 months in humans. Gorillas and bonobos wean at about 3.5 years, chimpanzees at about 5 years, and orangutans as late as 6 or 7 years. Female great apes enter puberty at about 8 years. Gorilla, bonobo and chimpanzee females have their first birth at about 10 years, orangutan females as late as 14 years. Great apes enjoy life spans of between 40 and 50 years (Harvey, Martin, & Clutton-Brock, 1987; Parker, chapter 2, this volume). In highly sexually dimorphic species, orangutans and gorillas, males take several years longer to mature than females. Even in slightly dimorphic chimpanzees, male maturation takes longer than female maturation. Among barely dimorphic bonobos, male and female sexual maturation occurs at roughly the same ages. The slight variations in life histories place gorillas and bonobos at the more rapid end of the maturation scale, orangutans at the slower end, and chimpanzees intermediate. The common features must have been present in the common ancestor of the great apes because their variations do not correspond to order of phyletic branching.

DIETS AND FORAGING PATTERNS

Gorillas, the largest great apes, are primarily herbivores. The mountain gorillas are the most extreme in this regard, while the eastern and western lowland gorillas are more catholic in their tastes, eating diets of fruits and herbaceous plants that overlap with those of sympatric chimpanzees. Both mountain and lowland gorillas eat ants, but apparently no vertebrate prey.

Bonobos are primarily frugivores, but also herbivores. They also eat a variety of plant parts, insects, and some small vertebrate prey. The number of plant species they eat is apparently smaller than the number eaten by common chimpanzees. This pattern relates to their smaller species range and the greater homogeneity and stability of their habitat. Bonobo diets are intermediate between those of gorillas and chimpanzees.

Chimpanzees are the most omnivorous of the great apes. They eat the largest number of plant parts of the greatest number of plant species. They also eat a large number of insect species, especially social insects, and a large number of vertebrate species. In most locations, their diets are highly seasonal. Seasonal differences in the dispersion and density of foods has favored the well-known fission–fusion social organization of this species. Consistent with their wide range and various habitats, chimpanzee populations differ widely in the proportions of various foods they consume (McGrew, 1992).

Orangutans are primarily frugivorous, specializing particularly on unripe and toxic plants. They also eat a variety of plant parts, and consume social insects. The energy demands imposed by their large body size, combined with the widely scattered distribution of foods, enforces a semi-solitary life on this species. It is impossible to tell the ancestral pattern from these comparative data alone. Difference in dietary niches do not correlate with phyletic distance, though some differences may reflect competitive exclusion between sympatric species.

SOCIAL ORGANIZATION AND MATING PATTERNS

Wild gorillas live in relatively stable bisexual groups usually composed of 1 or 2 adult males, a few adult females, and their young. Both sexes disperse at adulthood, but females are often kidnapped and brought into groups by males. Males are intolerant of other adult males (except close relatives) when females are present. Immature males often live temporarily in all-male groups. Male competition sometimes results in mortality. Females are attracted to males and largely uninterested in one another (Fossey, 1983).

Wild chimpanzees live in large groups of 50 to more than 100 individuals who share a defended home range. Seasonal variation in the density and dispersion of their food sources has favored a flexible fission–fusion pattern of groupings. The size of these groupings varies dramatically from very large to very small depending on food densities (Chapman, White, & Wrangham, 1994). The mother–infant subgroup is the most constant and stable. Other common subgroups include brothers and male friends, and mating pairs. Males are philopatric while females disperse at sexual maturity and even after bearing young. Females remain essentially solitary within the unit group, each female having her own home range within the larger home range of the group males. Related males patrol and defend territorial boundaries. Males form political alliances and are dominant over females. Male competition is substantial and results in a higher death rate in males as compared to females (Goodall, 1986).

Wild bonobos also live in large unit groups of 60 to 150 individuals, which fission and fuse allowing them to respond to seasonal differences in food dispersion and density. Although their average group sizes are similar to those of chimpanzees, they do not display the very small and very large groupings typical of chimpanzees (Chapman et al., 1994). Mother–offspring subgroups which include adult sons are the most common and stable subgroups. Unlike chimpanzees, they are generally found in bisexual groups and have overlapping home ranges that are not defended against other groups. Like chimpanzees, bonobo males are philopatric and females dispersing (Pusey & Pakcer, 1987); however, only nulliparous female bonobos disperse. After they bear offspring, bonobo females remain in the new group. Bonobo females also differ from chimpanzee females in their gregariousness. Emigrating females select a resident female to whom they bond through sexual behaviors. Finally, bonobos differ from chimpanzees in the greater social equality of males and females (Kano, 1986).

Wild orangutans show an exploded version of the chimpanzee social pattern. Large adult males defend large territories that overlap the smaller territories of several adult females and their young. Subadult and smaller adult males range covertly in these defended territories, feeding and trying to gain copulations. Unlike chimpanzee males, orangutan males do not band together to defend territories. Adult females apparently defend smaller feeding territories against competing female and young. Young orangutans remain with their mothers until they are almost 7 years old. Young females form territories close to their mothers. Orangutans are most gregarious during the juvenile phase (Galdikas, 1995). Again, it is impossible to tell the ancestral pattern from these data. It is clear, however, that the varieties of social organization do not correspond to phyletic branching.

TEMPERAMENTS

Although in general chimpanzees might be described as gregarious, hyper-excitable, loud, aggressive, and insecure, individuals differ greatly. Some chimpanzees are quite calm, others are quite jittery (Hebb, 1949; Nissen, 1956; Maple, 1980). The great noisiness of chimpanzees may derive from the temporal instability of their group structure, which predisposes them to communicate vocally over long distances (Maple, 1980). Male chimpanzees prefer the company of other males except during mating, whereas females are less gregarious than males and good at using friendly gestures to cover nasty intentions. Kin-based male solidarity is important in the patrolling and defense of territorial boundaries (Goodall, 1986).

Like chimpanzees, bonobos are gregarious and excitable, but less aggressive than chimpanzees. Their greater mellowness derives from their extensive use of sexuality as a means for tension reduction and bonding between individuals of both sexes and all ages. In contrast to chimpanzees, bonobo males are more attracted to females than to other males and rely on their mothers for support. Their use of sex for bonding between unrelated females distinguishes them from other apes (De Waal & Lanting, 1997).

Gorillas are calm but socially responsive. They are shy and introverted, emotionally unexpressive, moody (Maple, 1980; Mitchell, this volume). They are secretive and self-conscious (Patterson & Cohn, 1994), apparently uncomfortable being watched (Yerkes, 1929). Their low vocalizations are appropriate for communication in relatively stable, spatially contiguous social groupings (Maple, 1980).

Orangutans are quiet, emotionally unexpressive, and secretive as befits their largely solitary life and the sneaker strategy of subadult males. Typically, captive orangutans are inactive and quiet (Yerkes & Yerkes, 1929; Maple, 1980). The phelgmatic or apathetic disinterest shown by some zoo orangutans may be a result of hyper-parathyroidism rather than being species typical (see Röhrer-Ertl, 1988).

In problem solving, orangutans and gorillas show great persistence and patience, a pattern which is less common among chimpanzees (Yerkes & Yerkes, 1929; Nissen, 1956; Reynolds, 19670). Orangutans are very curious about mechanical objects. Many chimpanzees are easily frustrated in problem-solving situations, readily throwing temper tantrums (Köhler, 1927; Nissen, 1956; Reynolds, 1967).

Temperament correlates better with social organization than with phyletic distance. Social organization depends upon attractions and antipathies within and among various age and sex categories within a species as well as activity levels and proclivities (Kummer, 1971). Temperaments apparently play an important role in generating species-typical forms of social interaction.

COGNITIVE ABILITIES

One of the aims of this volume is to re-examine the traditional view that gorillas are less intelligent than other great apes. This judgment has been expressed in various ways. First, gorillas have been characterized as poor tool users (Natale, 1989). This view, first expressed in Yerkes' (1927; 1929) report on Congo, was later supported by Antinucci's (1990) report on Roma. Both gorillas were solitary individuals. McGrew (1992) acknowledged that captive gorillas use tools, but he suggested that tool use is a poorer index of intelligence than mirror self-recognition. Second, gorillas have been reported to be unable to recognize themselves in mirrors (Gallup, 1977; Ledbetter & Basen, 1982). McGrew, consistent with his criterion, suggested that gorillas' failure at MSR was indicative of lower intelligence than that of other great apes. Reports of MSR, imitation, and tool use in Koko, the signing gorilla, have been largely discounted by primatologists Third, the symbolic abilities of the signing gorillas, Koko and Michael, have been discounted by those critical of Patterson's approach. Contrary to the general view, evidence in this volume suggests that gorillas are similar to other great apes in their intellectual abilities.

Tool use

Few if any publications have focused on the tool-using abilities of gorillas. Two chapters on tool use in this book help fill this gap. Both suggest that gorillas readily use tools and, when given the opportunity, use a variety of tools for a variety of functions (see Boysen, Kuhlmeier, Halliday, & Halliday and Parker, Kerr, Markowitz, & Gould, this volume). The remaining question is whether they are capable of the kinds of complex tool use recently reported among wild chimpanzees (Sugiyama, 1997). The hierarchical skills revealed in the analysis of plant preparation in wild gorillas supports the notion that they are capable of such hierarchically organized forms of tool use (Byrne, this volume).

Tool use in captive orangutans has been well documented (e.g., Lethmate, 1982), but the discovery of tool use in populations of wild orangutans reported in this volume (see Fox, Sitompul, & Van Schaik) was unexpected. Even more unexpected is its frequency and facility, and the apparent use of tool sets (i.e., using more than one tool in a given task). This discovery implies that the common ancestor of all the great apes had this capacity and used it in the wild (Parker & Gibson, 1977). See Table 19.1 for a summary of tool use in great apes, and Table 19.2, for advanced tool use.

Following Goodall's (1964) report, wild chimpanzees have been widely observed to use a variety of regionally specialized forms of tool use especially in the context of feeding (McGrew, 1992). Both chimpanzees and orangutans use tools specifically in the context of extractive foraging on a variety of embedded and encased foods (Parker & Gibson, 1977). It is notable that the occurrence of tool use in wild great apes does not accord perfectly with the order of phyletic branching.

Imitation

The ability to imitate has been viewed in contradictory ways. Imitation of new behaviors has been viewed alternatively as a low-level rote activity (mimicry) (Terrace, Petitito, Saunders, & Bever, 1979) or as a cognitively complex activity requiring at least

Table 19.1 *Kinds of simple tool use among wild, rehabilitant, and captive great apes*

Kinds of tool use	Chimpanzees	Bonobo	Gorillas	Orangutans
Throwing missiles	W, C	C	W, C	W C
Dropping missiles	W, C	C	W	W C
Sponging up fluids	W, C	C	C	C
Probing with stick	W, C		C	W, R, C
Grooming other with object	C			R
Grooming self with object	W, C	C	C	W, R, C
Jabbing, searing or stabbing	C			W, R
Hitting (clubbing)	W, C		C	W, R, C
Using object as barrier	W		C	W, C
Using ladder/stool	C		C	R, C
Raking in objects	C		C	W, R, C
Making bridge or pole-vault	C	C	C	R
Levering	W, C		C	R, C
Digging with stick	W, C		C	
Using container	C	C	C	R, C
Using stick to hold food	W		C	
Using stick to prop up sore limb			C	
Using objects as trays or plates	W, C			
Stimulating self with objects	N, C			W, R, C
Hammering open objects	W, C		?	C
Cutting or sawing open	C			C
Wrapping or enclosing in envelope, & crushing = +	W +			W + , R
Brushing or scraping away with objects	W	C	C(?)	W
Pushing object away with object				W
Siphoning or sucking with tube	C		C	R, C
Using leaves as shelter from elements	W (rare)	W		W
Stirring substances	W			

Sources: Alp (1997); Beck (1980); Ingmanson (1996); Maple (1982); McGrew (1992); Parker, this volume; Russon, this volume.

Table 19.2 *Advanced forms of tool use in great apes*

Kinds of tool use	Chimpanzees	Bonobo	Gorillas	Orangutans
Using anvil to stabilize nut	W			
Using wedge to stabilize anvil	W			
Joining sticks together to form tool	C			C, R
Solving tube trap task	C	C		
Stacking boxes to reach goal	C		C	C
Using tool set	W			
Hanging cloth to suspend self				C
Weaving rope to suspend self				C
Wrapping rope to suspend hammock				R
Baiting prey with food	C			
Making stone flake to open	C	C		C

Sources: Beck (1980); Matsuzawa & Yamakoshi (1996)

fifth-stage sensorimotor stage abilities (Piaget, 1952, 1962). Recently, some primatologists have argued that great apes can imitate facial and other gestures, but not tool use (Call, this volume). Others have argued that apes are less skilled at imitation of facial and other gestures than at imitation of tool use (Byrne & Russon, in press).

Several experimental studies have focused specifically on gestural imitation in chimpanzees (Custance & Whiten, in press; Piaget, 1952, 1962; Russon & Galdikas, 1993, 1995; Whiten & Custance, 1996; Whiten et al., 1996) and orangutans (Miles, Mitchell, & Harper, 1996). There have been few if any studies of imitation in gorillas or bonobos. An observational study of the sign-using gorilla, Koko, (Chevalier-Skolnikoff, 1977) revealed some gestural imitation (also see Parker, chapter 18, this volume). Data on sign-language acquisition in gorillas (Patterson, 1980; Patterson & Linden, 1981) provides evidence for gestural imitation similar to that in a signing chimpanzee and orangutan (Fouts, 1973; Miles, 1990; Mitchell, 1993) and children (Guillaume, 1971).

The controversy surrounding imitation of tool use in great apes is reviewed in the chapters by Russon and Call. Call notes that in the rake task (in which the human model turns the rake and pulls a desired item toward himself with the rake) chimpanzees and orangutans typically fail to turn the rake over before pulling it toward themselves, whereas human children tend to imitate both actions. Russon argues that imitation is revealed in the fact that the apes pull the rake toward themselves. She notes that imprecise forms of imitation are typical of young human children as well as great apes. Russon also notes that in nonexperimental situations, orangutans tend to replicate only parts of an entire event, though often in the same sequence as presented by the model.

Call argues that great apes have greater difficulty exactly imitating tool use than gestures because the first task requires a third-person perspective (the model acts toward an object or toward another, which the ape observes), whereas the latter requires a second-person perspective (the model acts toward the observer). In contrast, Russon suggests that great apes can imitate a model's actions on objects as well as a model's gestures, but they fail to achieve the fidelity of response of older human children because they are able to hold fewer components in mind. She also notes that imitation is only one of several processes by which great apes and humans develop specific applications of tool use. The fact that orangutans and chimpanzees imitate novel actions suggests the presence of this ability in their common ancestor.

Self-awareness and mirror self-recognition

Gallup (1977) and others (Ledbetter & Basen, 1982) have claimed that gorillas lack the capacity to recognize their own images in mirrors. Ironically, evidence of gorillas using mirrors to look at parts of themselves otherwise not visible has been available for almost three decades (Riopelle, 1970; Riopelle, Nos, & Jonch, 1971; Patterson, 1978), if not even earlier. Indeed, in their review of all research on self-recognition in great apes since 1970, Swartz, Sarauw, & Evans (this volume) show that differences in the frequencies with which the four species display self-recognition are not statistically significant. This accords with the observation that self-recognition is individually variable in all great ape species including chimpanzees (50% or more of those tested from each great ape species failed to self-recognize by passing the mark test).

Given that MSR seems to depend upon kinesthetic–visual matching of actions of the self with those of the mirror-image, we should expect species with the greatest propensity to imitate facial and gestural movements to be the more likely to display MSR (Mitchell, 1993; Parker, 1991). The occurrence of MSR in all the great apes suggests that the capacity evolved in their common ancestor.

Language and iconicity

As in other cognitive domains, the great apes are very similar in their linguistic abilities (Miles, this volume). All four great apes can learn signs and symbolic communication at a level approximating that of 2– to 2.5–year-old human children. Bonvillian and Patterson (this volume) describe similarities in early sign-learning in gorillas and humans. They find similar initial vocabularies in the two species, but note that Koko and Michael use more iconic signs than human children do. Tanner and Byrne (this volume) describe the spontaneous development and use of iconic signs by a male gorilla, Mkumbwa (Kubie), in his frustrated efforts to elicit mating from a female in his group in the San Francisco Zoo. These iconic signs seem to have arisen through a process of conventionalization of intention movements similar to that reported in captive bonobos (Savage-Rumbaugh, Wilkerson, & Bakeman, 1977). The repeated use of stylized gestures by Kubie and other captive gorillas, as well as by chimpanzees (Goodall, 1986), suggests an innate foundation upon which sign-language abilities develop. This foundation is also revealed in the spontaneous creation of self-imitative (iconic) and deictic (pointing) signs in the rehabilitant orangutan observed by Shapiro (this volume). Again, the presence of symbolic abilities in all the great apes suggest its presence in their common ancestor.

Deception

There are many reports of deception and concealment in wild chimpanzees (Bryne & Whiten, 1990). In contrast, there are few reports in wild gorillas and orangutans, and none in bonobos. In captivity, however, all four species show extensive and comparable skill in deceit. Mitchell (this volume) surveyed the literature on deception and found significant overlap in deceit even to the specific behaviors apes use to deceive or conceal. Various models have been proposed for analyzing cognitive levels involved in deception in animals (Mitchell, 1986; Whiten, 1991). Mitchell (this volume) elaborates on previous models, suggesting that deceptions and concealments are strategic violations of event schemas, or scripts. Such violations require an understanding of behavioral sequences and the consequences of behavior.

Species and population differences in rates of deception might be expected to follow upon differences in kinds and degrees of resource competition. If so, chimpanzees should show higher rates than other great apes (Whiten & Byrne, 1988). If on the other hand the success of deception depends upon the deceiver's level of arousal (and hence "leakage" of clues – Ekman & Friesen, 1969), their relatively low degree of emotional arousal and expressiveness might make gorillas and orangutans more successful deceivers than chimpanzees and bonobos. In any case, the capacity for deception must have been present in the common ancestor of the great apes.

PLAY, SOCIALIZATION, AND ENCULTURATION

The lion's share of attention to social learning in great apes has gone to chimpanzees, who have been reported to learn tool use through imitation of movements demonstrated by their mothers (Boesch, 1991, 1993; Caro & Hauser, 1992). Several chapters in this volume focus on less well-known cases of socialization in gorillas and orangutans.

Whiten (this volume) and Parker (chapter 18, this volume) both describe mother gorillas coaching their infants to crawl toward them, a behavior described in chimpanzees by Plooij (1984). Parker also describes strategic placement of gorilla infants by their mothers near substrates that provide practice in locomotor skills. The intentional teaching function of these placements is suggested by their timing in accord with the locomotor readiness of the infants. Although wild gorillas must learn elaborate hierarchical motor patterns for plant preparation which may depend upon program-level imitation (Byrne, this volume), no instances of demonstration teaching comparable to that in chimpanzees has been reported in this species. The closest approximation is the report by Patterson and Cohn (1994) of Koko's molding the hands of her doll.

Fox et al. (this volume) describe observational learning of tool use by wild infant

Table 19.3 *Great ape characteristics*

	Chimpanzees	Bonobos	Gorillas	Orangutans
Life histories	first female reproduction at about 10 yrs	first female reproduction at about 10 years	first female reproduction at about 10 years	first female reproduction at about 14 years
Diets	omnivorous & highly seasonal	frugivorous & herbivorous	herbivorous	frugivorous
Mating systems	promiscuous mating & consortships	promiscuous mating	harem polygyny	harem polygyny & sneak copulations
Social organization	highly variable fission–fusion groups; male philopatry & male bonding	limited fission–fusion groups; male philopatry & male–female & female–female bonding	harems with 1 or 2 males & several females & young; both sexes disperse; male–female bonding	semi-solitary male territory overlaps that of several females; adult females remain close to mother's territory
Temperaments	excitable, gregarious, aggressive	excitable, gregarious, sexual	gregarious, calm, shy, and covert	solitary, calm, & shy
Cognitions	tool & symbol use, MSR, & imitation	tool & symbol use, MSR, & imitation	tool & symbol use, MSR, & imitation	tool & symbol use, MSR, & imitation

orangutans. Galdikas (1995) has suggested that the prolonged association lasting as long as 7 years between mother and infant orangutans serves an important socialization function in a species that is semi-solitary in some habitats. Further observations may reveal more about the teaching role of wild mother orangutans, and possible cultural transmission of tool-use techniques.

Cultural transmission of tool-using techniques in wild chimpanzees apparently occurs through females who disperse to new groups and demonstrate it, intentionally or not, to their infants and other individuals in the new group (Goodall, 1973; Boesch, 1993; Matsuzawa & Yamakoshi, 1996). The short distance dispersal of female orangutans combined with their less gregarious social life reduces the scope of this kind of transmission in this species.

CONCLUSIONS: PRECHIMPANZEE ORIGINS OF COGNITIVE ABILITIES IN GREAT APES

This brief survey reviews evidence that the great apes share many anatomical, life-history, and intellectual characteristics that are not present in lesser apes or Old World monkeys. The taxonomic distribution of these characters indicates that they must have evolved in the common ancestor of the great apes (including humans). Further research may confirm subtle differences in cognitive ability, particularly in advanced or complex tool use seen among wild chimpanzees. If so, we may infer that these more advanced abilities evolved only in this species, or alternatively that it was lost in other species.

In contrast, great apes differ markedly in body size, diet, social-sexual behavior, and temperament. The distribution patterns suggest that these latter characteristics have evolved more rapidly than the former characteristics. Various lines of evidence suggest that mating behavior and social organization are secondary adaptations related to the defensibility of resources. Resource deferability, in turn, is related to the temporal and spatial distribution of foods, which, in turn, depend upon the diet . Temperament may have arisen as a secondary adaptation for social organization (Kummer, 1971). See Table 19.3 for a summary of these characteristics.

A cladistic analysis of comparative data on the great apes yields two conclusions: first, given the common intellectual abilities of the great apes, the first major step in cognitive evolution must have occurred not in the common ancestor of chimpanzees (and bonobos) and humans, but in the common ancestor of all the living great apes. Second, it is only possible to discern this pattern by comparing data on all the living great apes and their outgroups. In other words, a chimpocentric (Beck, 1982) approach to evolutionary reconstruction is inadequate.

REFERENCES

Alp, R. (1997) "Stepping Sticks" and "seat sticks": New types of tools used by wild chimpanzees (*Pan troglodytes*) in Sierra Leone *American Journal of Physical Anthropology*. **44**, 45–52.

Antinucci, F. (1990). The comparative study of cognitive ontogeny in four primate species. In S. T. Parker & K. R. Gibson (eds.), *"Language" and intelligence in monkeys and apes* (pp. 157–171). New York: Cambridge University Press.

Beck, B. (1980). *Animal tool behavior*. New York: Garland.

Beck, B. (1982). Chimpocentrism: Bias in cognitive ethology. *Journal of Human Evolution*, **11**, 3–17.

Boesch, C. (1991). Teaching among wild chimpanzees. *Animal Behaviour*, **41**, 530–532.

Boesch, C. (1993). Aspects of transmission of tool use in wild chimpanzees. In K. R. Gibson & T. Ingold (eds.), *Tools, language and cognition in human evolution* (pp. 171–183). Cambridge University Press.

Brooks, D. & McLennan, D. (1991). *Phylogeny, ecology, and behavior*. University of Chicago Press.

Byrne, R. W. & Russon, A. (in press). Learning by imitation: A hierarchical approach. *Behavioral and Brain Sciences*.

Bryne, R. & Whiten, A. (1990). Tactical deception in primates: the 1990 database. *Primate Report*, **27**, 1–101.

Caro, T. M. & Hauser, M. D. (1992). Is there teaching in nonhuman animals? *The Quarterly Review of Biology*, **67**(2), 151–174.

Chapman, C. A., White, F. J., & Wrangham, R. W. (1994). Party size in chimpanzees and bonobos: A reevaluation of theory based on two similarly forested sites. In R. W. Wrangham, W. C. McGrew, F. de Waal, & P. G. Heltne (eds.), *Chimpanzee cultures* (pp. 41–58). Cambridge, MA: Harvard University Press.

Chevalier-Skolnikoff, S. (1977). A Piagetian model for comparing the socialization of monkey, ape, and human infants. In S. Chevalier-Skolnikoff & F. Poirier (eds.), *Primate biosocial development* (pp. 159–188). New York: Garland.

Clark, W. W. L. G. (1959). *The antecedents of man*. New York: Harper and Row.

Custance, D. & Whiten, A. (1995). Can young chimpanzees *(Pan troglodytes)* imitate arbitrary actions? Hayes and Hayes (1952) revisited. *Behaviour*, **132**, 839–858.

De Waal, F. & Lanting, F. (1997). *Bonobos: The forgotton ape*. Berkeley: University of California Press.

Ekman, P. & Friesen, W. (1969). The repertoire of nonverbal behavior – categories, origins, usage, and coding. *Semiotica*, **1**, 49–98.

Fleagle, J. G. (1988). *Primate adaptation and evolution*. New York: Academic Press.

Fossey, D. (1983). *Gorillas in the mist*. New York: Houghton Mifflin.

Fouts, R. (1973). Acquisition and testing of gestural signs in four young chimpanzees. *Science*, **180**, 978–80.

Galdikas, B. (1995). *Reflections of Eden: My years with the orangutans of Borneo*. New York: Little Brown.

Gallup, G. G., Jr. (1977). Self-recognition in primates. *American Psychologist*, **32**, 329–338.

Goodall, J. (1964). Tool using and aimed throwing in a community of free-living chimpanzees. *Nature*, **201**, 1264–12646.

Goodall, J. (1973). Cultural elements in a chimpanzee community. In J. E. W. Menzel (ed.), *Precultural primate behavior* (pp. 144–184). Basle: Karger.

Goodall, J. (1986). *Chimpanzees of the Gombe*. Cambridge: Harvard University Press.

Guillaume, P. (1971). *Imitation in children* (Elaine Halperin, trans.). (2nd edn). University of Chicago Press.

Harvey, P. H. & Pagel, M. D. (1991). *The comparative method in evolutionary biology*. Oxford University Press.

Harvey, P., Martin, R. D., & Clutton-Brock, T. (1987). Life histories in comparative perspective. In B. Smuts, D. Cheney, R. Seyfarth, R. Wrangham, & T. Struhsaker (eds.), *Primate societies* (pp. 181–196). University of Chicago Press.

Hebb, D. (1946). Emotion in man and animal: An analysis of the intuitive processes of recognition. *Psychological Review*, **53**, 88–106.

Ingmanson, E. (1996). Tool-using behavior in wild *Pan paniscus*: Social and ecological considerations. In A. E. Russon, K. Bard & S. T. Parker (eds.), *Reaching into thought: The minds of the great apes* (pp. 190–210). Cambridge University Press.

Jones, C. & Sabater-Pi, J. (1971). *Comparative ecology of Gorilla gorilla (Savage and Wyman) and Pan troglodytes (Blumenbach) in Rio Muni, West Africa*. Basle: Karger.

Kano, T. (1986). *The last ape: Pygmy chimpanzee behavior and ecology* (Evelyn Ono Vineberg, trans.). Stanford University Press.

Köhler, W. (1927). The mentality of apes. New York: Vinage.

Kummer, H. (1971). *Primate societies*. New York: Aldine.

Ledbetter, D. H. & Basen, J. A. (1982). Failure to demonstrate self-recognition in gorillas. *American Journal of Primatology*, **2**, 307–310.

Lethmate, J. (1982). Tool-using skills of orangutans. *Journal of Human Evolution*, **11**(49–64).

Maple, T. (1980). *Orang-utan behavior*. New York: Van Nostrand Reinhold.

Maple, T., & Hoff, M. (1982). *Gorilla behavior*. New York: Van Nostrand Reinhold.

Matsuzawa, T. & Yamakoshi, G. (1996). Comparison of chimpanzee material culture between Bossou and Nimba, West African. In A. E. Russon, K. Bard, & S. T. Parker (eds.), *Reaching into thought: The minds of the great apes*, (pp. 211–232). Cambridge University Press.

McGrew, W. (1992). *Chimpanzee material culture*. New York: Cambridge University Press.

Miles, H. L. (1990). The cognitive foundations for reference in a signing orangutan. In S. T. Parker & K. R. Gibson (eds.), *"Language" and intelligence in monkeys and apes* (pp. 511–539). New York: Cambridge University Press.

Miles, H. L., Mitchell, R., & Harper, S. (1996). Simon says: The development of imitation in an enculturated orangutan. In A. Russon, K. Bard, & S. T. Parker (eds.), *Reaching into thought: The minds of the great apes* (pp. 278–299). Cambridge University Press.

Mitchell, R. W. (1986). A framework for discussing deception. In R. W. Mitchell & N. Thompson (eds.), *Deception: Perspectives on human and nonhuman deceit* (pp. 3–40). Albany, NY: State University of New York Press.

Mitchell, R. W. (1993). Mental models of mirror-self-recognition: Two theories. *New Ideas in Psychology*, **11**, 295–325.

Natale, F. (1989). Causality II: The stick problem. In F. Antinucci (ed.), *Cognitive structure and development in nonhuman primates* (pp. 121–135). Hillsdale, NJ: Erlbaum.

Nissen, H. W. (1956). Individuality in the behavior of chimpanzees. *American Anthropologist*, **58**, 407–413.

Parker, S. T. (1991). A developmental approach to the origins of self-awareness in great apes and human infants. *Human Evolution*, **6**, 435–449.

Parker, S. T. & Gibson, K. R. (1977). Object manipulation, tool use, and sensorimotor intelligence as feeding adaptations in cebus monkeys and great apes. *Journal of Human Evolution*, **6**, 623–641.

Patterson, F. (1978). Conversations with a gorilla. *National Geographic*, **154**, 438–465.

Patterson, F. (1980). Innovative use of language by gorilla: A case study. In K. Nelson (ed.), *Children's Language* (vol. 2, pp. 497–561). New York: Gardner.

Patterson, F. & Cohn, R. (1994). Self-recognition and self-awareness in lowland gorillas. In S. T. Parker, R. W. Mitchell, & M. L. Boccia (eds.), *Self-awareness in animals and humans* (pp. 273–290). New York: Cambridge University Press.

Patterson, F. & Linden, E. (1981). *The education of Koko*. New York: Holt, Rinehart & Winston.

Piaget, J. (1952). *The origins of intelligence in children*. New York: International Universities Press.

Piaget, J. (1962). *Play, dreams, and imitation*. New York: W. W. Norton.

Plooij, F. X. (1984). *The behavioral development of free-living chimpanzee babies and infants*. Norwood, NJ: Ablex.

Pusey, A. E. & Packer, C. (1987). Dispersal and philopatry. In B. Smuts, D. Cheney, R. Seyfarth, R. Wrangham, & T. Struhsaker (eds.), *Primate societies* (pp. 250–265). University of Chicago Press.

Reynolds, V. (1967). *The apes*. New York: Harper.

Riopelle, A. J. (1970). *Growing up with Snowflake. National Geographic*, **138**, 491–503. Riopelle, A. J., Nos, R., & Jonch, A. (1971). Situational determinants of dominance in captive young gorillas, *International Congress of Primatology*, **3**, 86–91.

Röhrer-Ertl, O. (1988). Research history, nomenclature, and taxonomy of the orangutan. In J. H. Schwartz (ed.), *Orangutan biology* (pp. 7–18). New York: Oxford University Press.

Russon, A. E. & Galdikas, B. (1993). Imitation in free-ranging rehabilititant orangutans (*Pongo pygmaeus*). *Journal of Comparative Psychology*, **107**(2), 147–161.

Russon, A. E. & Galdikas, B. (1995). Constraints on great ape imitation: Model and action selectivity in rehabilitant orangutans' (*Pongo pygmaeus*) imitation. *Journal of Comparative Psychology*, **109**, 5–17.

Savage-Rumbaugh, E. S., Wilkerson, B., & Bakeman, R. (1977). Spontaneous gestural communication among conspecifics in pygmy chimpanzees. In G. Bourne (ed.), *Progress in ape research* (pp. 287–309). New York: Academic Press.

Sugiyama, Y. (1997). Social tradition and the use of tool-composites by wild chimpanzees. *Evolutionary Anthropology*, **6**, 23–27.

Terrace, H., Petitito, L., Saunders, R. J., & Bever, T. (1979). Can an ape create a sentence? *Science*, **206**, 891–902.

Tomasello, M., Kruger, A. C., & Ratner, H. H. (1993). Cultural learning. *Behavioral and Brain Sciences*, **16**, 495–552.

Tuttle, R. (1986). *Apes of the world: Their socialization, behavior, mentality, and ecology*. Park Ridge, NJ: Hayes.

Whiten, A. (ed.). (1991). *Natural theories of mind: Evolution, development and simulation of everyday mindreading*. Cambridge, MA: Basil Blackwell.

Whiten, A. & Byrne, R. (eds.), (1988). *Machiavellian intelligance*. London: Oxford University Press.

Whiten, A. & Custance, D. (1996). Studies of imitation in chimpanzees and children. In B. G. G. & C. M. Heyes (eds.), *Social learning in animals: The roots of culture* (pp. 291–318). New York: Academic Press.

Whiten, A., Custance, D., Gomez, J.-C., Teixidor, P., & Bard, K. A. (1996). Imitative learning of artificial fruit processing in children (*Homo sapiens*) and chimpanzees (*Pan troglodytes*). *Journal of Comparative Psychology*, **110**(1), 3–14.

Wrangham, R. (1986). Evolution of social structure. In B. Smuts, D. Cheney, R. Seyfarth, R. Wrangham, & T. T. Struhsaker (eds.), *Primate societies*, (pp. 382–305). University of Chicago Press.

Yerkes, R. M. (1927). The mind of a gorilla: Part I and Part II. *Genetic Psychology Monographs*, **2**, 1–193; 337–551.

Yerkes, R. M. (1929). The mind of a gorilla: Part III. Memory. *Comparative Psychology Monographs*, 8(24), 1–92.

Yerkes, R. M., & Yerkes, A. (1929). *The great apes*. New Haven, CT: Yale University Press.

Index of authors

Index of subjects